EESES

Electric Energy Systems and Engineering Series

Editors: J. G. Kassakian · D. H. Naunin

Guy Séguier · Francis Labrique

Power Electronic Converters

DC-AC Conversion

With 259 Figures

Springer-Verlag
Berlin Heidelberg NewYork
London Paris Tokyo
Hong Kong Barcelona
Budapest

Professor Dr. Guy Séguier
Université des Sciences et Techniques de Lille
59655 Villeneuve d'Ascq Cedex
France

Professor Dr. Francis Labrique
Université Catholique de Louvain
1348 Louvain-la-Neuve
Belgique

Translated from the French by the authors
with the help of N. Quayle

This is the fourth volume in a series. The first two were published in English by McGraw-Hill
the first—on AC/DC conversion, in 1986,
the second—on AC/DC conversion, in 1987,
the third—on DC/DC conversion—was published by Springer-Verlag in 1993.
These three volumes are often referred to in the present work.

Title of the original French edition:
Les convertisseurs de l'électronique de puissance
Volume 4: La conversion continu – alternatif

ISBN 978-3-642-50324-5 ISBN 978-3-642-50322-1 (eBook)
DOI 10.1007/978-3-642-50322-1

Typesetting: Macmillan (India) Ltd., Bangalore, India
61/3020 - 5 4 3 2 1 0 - Printed on acid-free paper

Introduction to the Electric Energy Systems and Engineering Series

Concerns for the continued supply and efficient use of energy have recently become important forces shaping our lives. Because of the influence which energy issues have on the economy, international relations, national security, and individual well-being, it is necessary that there exists a reliable, available and accurate source of information on energy in the broadest sense. Since a major form of energy is electrical, this new book series titled *Electric Energy Systems and Engineering* has been launched to provide such an information base in this important area.

The series coverage will include the following areas and their interaction and coordination: generation, transmission, distribution, conversion, storage, utilization, economics.

Although the series is to include introductory and background volumes, special emphasis will be placed on: new technologies, new adaptations of old technologies, materials and components, measurement techniques, control – including the application of microprocessors in control systems, analysis and planning methodologies, simulation, relationship to, and interaction with, other disciplines.

The aim of this series is to provide a comprehensive source of information for the developer, planner, or user of electrical energy. It will also serve as a visible and accessible forum for the publication of selected research results and monographs of timely interest. The series is expected to contain introductory level material of a tutorial nature, as well as advanced texts and references for graduate students, engineers and scientists.

The editors hope that this series will fill a gap and find interested readers.

John G. Kassakian · Dietrich H. Naunin

Preface

This is the final volume in a four-volume series concerning POWER ELECTRONIC CONVERTERS.

The first volume studies AC/DC conversion,
the second studies AC/AC conversion,
and the third DC/DC conversion.
This final volume deals with DC/AC conversion, i.e. with inverters.

At the output of an inverter fed by a DC voltage supply, this voltage is alternatively found with one polarity and then with the other; in other words, an AC voltage made up of square pulses is obtained. Filtering must be carried out if, as is normally the case, a virtually sinusoidal voltage is required: this problem of *filtering* underlies the entire study of inverters.

In some applications, the load itself provides the filtering. In others, a filter is installed between the inverter and the load; however, as it will be shown in Chap. 2, in cases where the filtered voltage is at industrial network frequency and comprises only a single square-wave pulse per half-cycle, the filter becomes bulky and costly, and the results obtained are poor.

Filtering problems explain the considerable development of inverters during the last years:

- Firstly there is increasing use of pulse width modulation: each half-cycle is cut up into several pulses of suitable widths; this greatly simplifies filtering. The use of a chopping frequency which is much greater than the frequency of the fundamental components of the inverter output voltage and current has only been made possible by progress in the field of semiconductor devices.
- Secondly, attempts are made to widen the field of application of resonant inverters, operating at frequencies much higher than those of the industrial network. Raising the frequency enables the size and cost of passive elements (transformers, inductors, capacitors) to be reduced in applications where these are required.

On account of the development of inverters, the first chapter is relatively long. It begins with a rapid outline of the various inverter types and of their applications, before attempting to justify the choice of the following chapters and the differing emphases laid on various points in their contents.

Chapter 2 concerns voltage-source inverters with only one square-wave pulse per half-cycle. Filtering problems are presented at this point.

Chapter 3 deals with voltage-source inverters operating in pulse width modulation.

Chapter 4 concerns the drive and snubber circuits of transistors and GTO thyristors in voltage-source inverters.

Chapter 5 presents current-source inverters with a single square-wave pulse per half-cycle or operating in pulse width modulation.

Chapter 6 deals with resonant inverters used either directly or as intermediate stages in energy conversion cascade.

In order to avoid an unnecessary increase in the length of some chapters, the following have been included in the appendices at the end of the volume:

- the detailed study of the input filter in voltage-source inverters,
- the study of the "sinusoidal" inverters,
- the study of the thyristor turn-off circuits in voltage-source inverters.

Two further appendices have been added which do not deal with inverters but involve techniques used by the latter. One appendix concerns voltage-fed rectifiers operating in PWM and the other the AC current-fed chopper operating as a four-quadrant reversible rectifier.

The titles of the various chapters and appendices clearly show that the problems which arise in the study of inverters are very similar to those of choppers. Frequent reference will thus be made to the preceding volume to avoid needless repetition.

Such consideration have led us to have recourse to the same authors as in the volume on choppers. Robert Bausiere was responsible for Chap. 4, Francis Labrique for Chap. 5 as well as for Appendices C and E.

This collection aims to provide a useful working tool for power electrotechnicians and Vol. 4 has tried to remain faithful to this aim. This has been more difficult than for the preceding volumes insofar as there was a great deal to say and we did not wish to go beyond the length of the preceding three.

Our intention to provide a detailed presentation of the basic principle and initial applications has led us to be more succinct at the end of certain chapters. For similar reasons we have not been able to show all the diagrams which the user may require but have indicated how these could be drawn, with the aid of some examples[1].

Lille, July 1993 G. Séguier

[1] We are very grateful to Mrs. Reine Del Vitto for typing the manuscript and to Mr. Paul Leroy for drawing the figures of this book.

Notes

The publications indicated in the *Bibliography* date, as in the first three volumes of this series, from 1970 onwards. This bibliography thus remains within reasonable limits.

Books are indicated at the end of Chapter 1.

Articles have been limited to those published in the *Proceedings* and *Transactions of Electrical and Electronic Engineers* (USA), the *Proceedings of the Institution of Electrical Engineers* (GB) and the *Revue Générale de l'Electricité* (France). Furthermore we have given preference to articles containing bibliographies themselves.

References to the following have been grouped together:

- voltage-source inverters, at the end of Chap. 3,
- current-source inverters, at the end of Chap. 5,
- resonant inverters, at the end of Chap. 6.

A number of articles on PWM rectifiers and active filters have been noted after the appendices.

Contents

Chapter 4
Commutations in Voltage-Source Inverters 191

Chapter 5
Current-Source Inverters 248

Chapter 6
Resonant Inverters . . . 325

Nomenclature

Primes are used to distinguish quantities relating to the output from the corresponding quantities relating to the input.

c	number of chops per half-cycle (PWM)
C	capacitor
D	diode
e	instantaneous value of an electromotive force
E	rms value of an electromotive force
f	frequency
G	conductance
h	existence function of a switch
h	rank of current harmonics
i	instantaneous value of a current
I	mean value of a DC current
I	constant DC current
I	rms value of an AC current
j	positive integer
j	phase current (delta-connection)
J	rms value of j
k	ratio
k	positive integer
K	switch
l or L	inductance
m	transformation ratio
m	frequency–modulation ratio (PWM)
M	modulation wave or carrier (PWM)
n	number of turns in a winding
n	rank of voltage harmonics
N	leakage inductance
p	pulse number per half-cycle (PWM)
p	step number per half-cycle
P	power
Q	GTO thyristor
Q	reactive power
r	amplitude–modulation ratio (PWM)

r	equation root
r or R	resistance
t	time
t_c	commutation time
t_f	fall time
t_p	reverse-bias time
t_q	turn-off time
T	cycle
T	thyristor, transistor
TC	controlled turn-on/off switch
u	instantaneous voltage
U	mean value of a DC voltage
U	constant DC voltage
U	rms value of an AC voltage
v	instantaneous voltage
V	rms value of an AC voltage
W	energy
X	reactance
Y	admittance
Z	impedance
α	angle
α	ratio
α	damping coefficient
β	angle
β	angular frequency
δ	angle
Δi	current ripple; overcurrent
Δu	voltage ripple, overvoltage
ζ	damping coefficient
θ	angle
τ	harmonic rate
φ	phase shift
ϕ	flux
ψ	phase shift
ω	angular frequency

Subscripts

A	auxiliary
A, B, C	relating to the three phases of a 3-phase circuit
c	of commutation
C	of capacitor C
d	of rectifier output
D	of diode D

f, F	of filter
K	of switch K
max or M	maximum
min or m	minimum
mean	average
nom or n	rated
N	of neutral point
o	at initial instant
o	at full output
o	of resonance
O	of mid-point
OFF	off-state
ON	on-state
Q	of semiconductor device Q
rms	root mean square
r or R	of resistance r or R
R	reverse
S	of supply
T	of device T or TC
w	wanted
μ	magnetizing

Chapter 1

Introduction and Presentation

An inverter is a static converter providing DC–AC conversion. For example, if there is a DC voltage at the input and if, by use of semiconductor switches, each terminal of the load is periodically connected either to one or the other of the input terminals, an AC output voltage is obtained. The frequency of this voltage is given by the frequency of the connection changes.

The inverter is said to be autonomous if the turn-on, maintaining and then turn-off of these connections between the input and output depends only on the switch control. This volume deals essentially with autonomous inverters.

- Like all static converters, an autonomous inverter is largely dependent on the characteristics of the generator and the load between which it is connected. On these characteristics depend notably the nature of the commutations to be carried out and, as a result, the semiconductor switches required.

  We will use the notions developed at the beginning of Vol. 3 of this series. As in the case of the choppers, the DC or AC generators and loads will be characterized as voltage sources or as current sources from the point of view of the commutations. We will also use the study made of power semiconductor devices in commutation. The "switches" will be characterized and the basic diagrams laid down in the same way.

By considering the structure of converters in this way, two types of inverter can be identified:

- *voltage-source inverters* (VSI),
- *current-source inverters* (CSI);

the former are fed by a DC-voltage supply and the latter by a DC-current supply.

The nature of the AC load must match with that of the DC supply: voltage-source inverters supply current receptors and current-source inverters supply voltage receptors.

Inverters supplying a resonant circuit operating close to its resonant frequency require special treatment. Known as *resonant inverters*, these are voltage or current-source inverters with a specific control mode and field of application.

Finally it should be noted that the output current or voltage of a voltage or current-source inverter can be made up of not only one but several pulses per half-cycle. When this occurs, the inverter is referred as a *pulse width modulated inverter*.

- We will confine our study to *symmetrical structures*, i.e. those delivering an AC signal with two half-cycles – positive and negative – which are identical to within a sign. Such structures require at least two controlled semiconductor devices for a single-phase output and at least six for a three-phase output.
  Asymmetrical structures of the type presented for Switch Mode Power Supplies (see Vol. 3, Chap. 6) enable the number of switches required to be halved. But their use is almost entirely limited to low-power applications.

- The first part of this chapter presents the principle of voltage and current-source autonomous inverters and of resonant inverters.
  The second part deals with the semiconductor switches used in each case, the consequences of imperfections in the supply and the load, and problems linked to filtering and reversibility.
  In the third part we will show the inverter to be used, as well as the conditions of use, in each major field of inverter applications. This will indicate the points to be developed in the following chapters, dealing with a detailed study of each major class of inverter.

1.1 Types of Inverter: Principles

We present the principle of the various inverter types for a single-phase output, by using the four "switch" (K_1, K_2, K'_1 and K'_2) configuration shown in Fig. 1.1.

The supply placed on the DC side is assumed to be perfect, as are the switches (zero forward voltage drop, zero leakage current, instantaneous commutation).

i and u are used to denote the input current and voltage,
i' and u', the output current and voltage,
i_{K_1} and v_{K_1}, the current and voltage across K_1,
i_{K_2} and v_{K_2}, the current and voltage across K_2,
T, the operating cycle of the inverter and its output values.

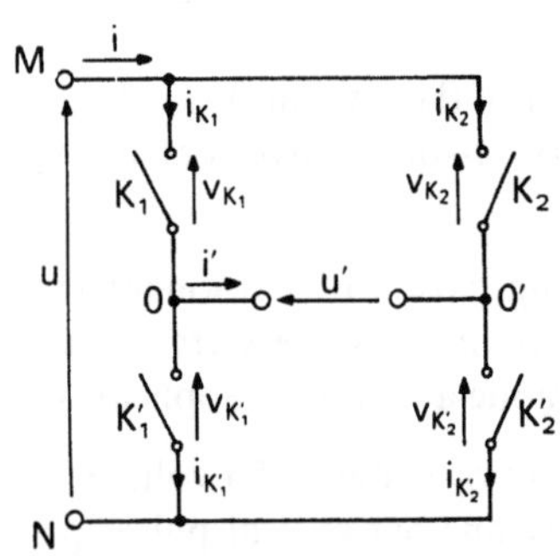

Fig. 1.1

The inverter is assumed to be in "full-wave" control[1], i.e.:

– for $0 < t < T/2$, K_1 and K'_2 are on and K_2 and K'_1 off:

$$u' = u \qquad i = i'$$

$$i_{K_1} = i', \qquad v_{K_1} = 0.$$

– for $T/2 < t < T$, K_2 and K'_1 are on, K_1 and K'_2 are off:

$$u' = -u, \qquad i = -i'$$

$$i_{K_1} = 0, \qquad v_{K_1} = u.$$

If the variations in i_{K_1} and v_{K_1} are followed, the currents in the other switches and the voltages across them can be directly deduced:

$$i_{K'_2} = i_{K_1} \qquad v_{K'_2} = v_{K_1}$$

$$i_{K_2} = i_{K'_1} \quad \text{with } i_{K'_1}(t + T/2) = i_{K_1}(t)$$

$$v_{K_2} = v_{K'_1} \quad \text{with } v_{K'_1}(t + T/2) = v_{K_1}(t).$$

1.1.1 Voltage-Source Inverters

- A voltage-source inverter is an inverter *fed by a DC-voltage supply*, i.e. by a supply with negligible internal impedance; its voltage u is not affected by the variations in current i flowing through it. The DC supply *imposes the voltage* at the input – and thus at the output – of the inverter.

  If u is constant and equal to U, whatever the value of i,

  for $0 < t < T/2$, $u' = +U$
  for $T/2 < t < T$, $u' = -U$.

– The output *current* i' and thus the input current i depend on the load placed on the AC side. This load can be of any type, provided that it is not another voltage source (capacitor or AC emf) connected directly across the output terminals.

– *Switch* K_1 must be made up of a fully controlled (turn-on and turn-off) switch TC_1. It must connect terminal 0 to terminal M at instant $t = 0$ and separate these two terminals at instant $t = T/2$, whatever the AC load may be. For i_{K_1} to flow via K_1, if the latter is momentarily negative, an antiparallel-connected diode D_1 has to be added to TC_1.

[1] With the full-bridge configuration, the shifted control can be used (see Chap. 2, Sect. 2.3). We present the case of the full-wave control so that the results obtained can be applied to single-phase inverters with only two switches.

The basic diagram of the autonomous voltage inverter can thus be proposed, as in Fig. 1.2.

– To show the influence of the load on the operation required of the semiconductor devices, we will consider the *load to be perfect*, i.e. that it takes a sinusoidal current. The cases of the inductive and capacitance loads are considered.

● Figure 1.3 provides the waveforms of u', i', i, v_{K_1} and i_{K_1} when the *load is inductive* (current i' lags voltage u' fundamental component).

For switch K_1,

- from $t = 0$ to $t = t_1$, diode D_1 enables i', which is then negative, to flow;
- from $t = t_1$ to $t = T/2$, TC_1 enables i' to flow;
- at instant $t = T/2$, turning-off TC_1 leads to i' being transferred from TC_1 to D'_1.

Three remarks can be made about this operational mode:

– For $t = t_1$, TC_1 begins conducting when the current is zero; no overcurrent is produced in this device and no overvoltage across it.
– Controlled devices TC_1 and TC'_2 turn-on drive signals must be applied throughout period $[0, T/2]$ (and that of TC_2 and TC'_1 throughout period $[T/2, T]$), since the value of t_1 depends on the load. Although TC_1 is controlled at turn-on from $t = 0$, it remains non-conducting until $t = t_1$, when it starts to conduct.
– As was shown in the case of choppers, a short period of time must be left between the turn-on of TC'_1 (or TC'_2) and the turn-off of TC_1 (or TC_2) and vice-versa, in order to avoid short-circuiting the DC voltage supply. On account of the diodes, there is no risk of the AC current load circuit being opened.

● Figure 1.4 shows the waveforms of u', i', i, v_{K_1} and i_{K_1} when inverter feeds a *capacitive load* (current i' leads voltage u' fundamental component).

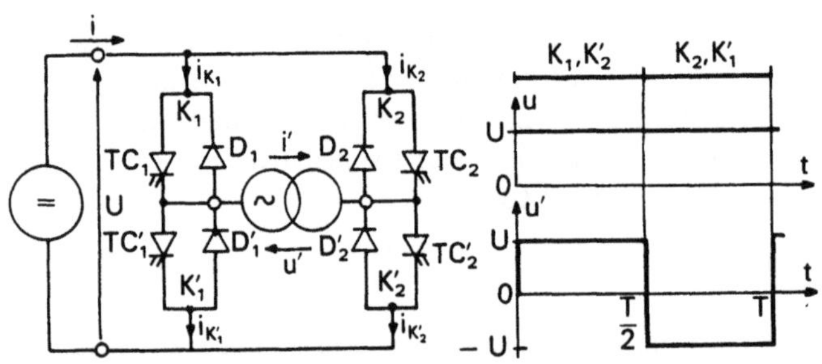

Fig. 1.2

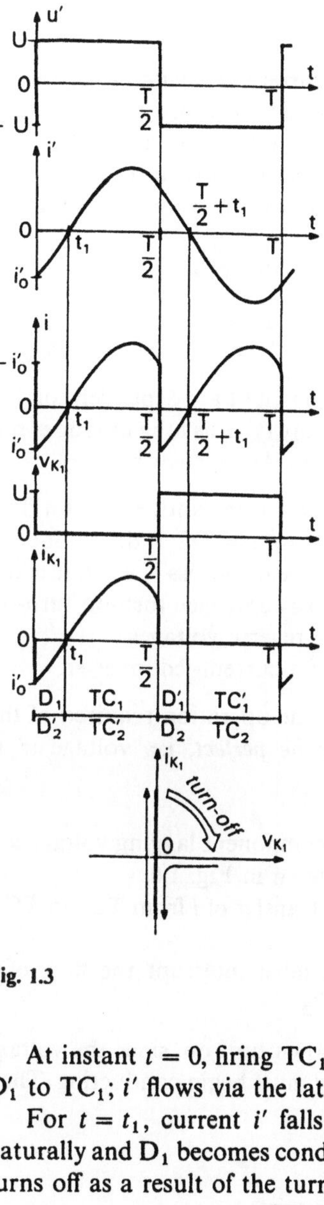

Fig. 1.3

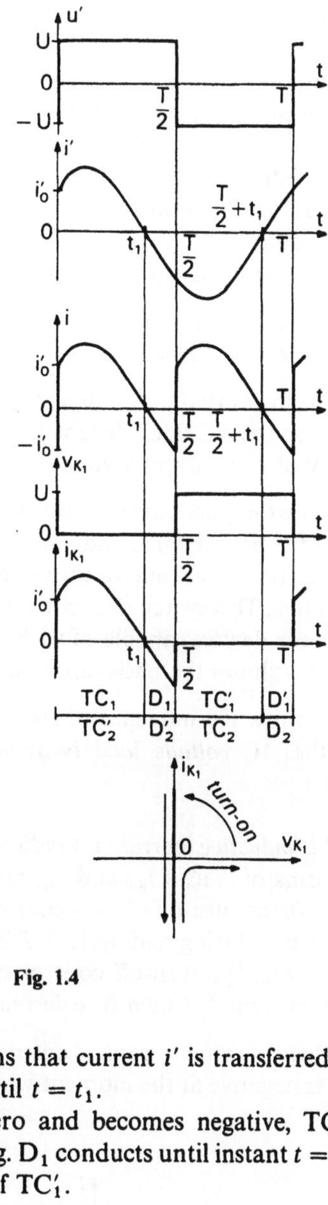

Fig. 1.4

At instant $t = 0$, firing TC_1 means that current i' is transferred from diode D'_1 to TC_1; i' flows via the latter until $t = t_1$.

For $t = t_1$, current i' falls to zero and becomes negative, TC_1 turns off naturally and D_1 becomes conducting. D_1 conducts until instant $t = T/2$ when it turns off as a result of the turn-on of TC'_1.

As in the case of a flow across an inductive load, there is one commutation without overcurrent nor overvoltage.

The other commutation is ensured by the turn-on of a controlled device. The ability for turn-off is not used in this case.

- If the load can be either inductive or capacitive, the semiconductor switches must be controllable at both turn-on and turn-off.

1.1.2 Current-Source Inverters

- A current-source inverter (also known as a current-*commutator*) is *fed by a DC-current supply*, i.e. by a supply whose internal inductance is so great that current i through it cannot be affected by variations of voltage u across it – more especially by the discontinuities in u, corresponding to the commutations. The DC supply *imposes the current waveform* at the converter input and thus at its output.

  If i is constant and equal to I, whatever the value of u,

  for $0 < t < T/2$, $i' = I$
  for $T/2 < t < T$, $i' = -I$.

- *Voltage* u' at the output and thus voltage u at the input are dependent on the load placed on the AC side. This load can be of any type, as long as it does not consist only of one current source (inductance or AC current source).

- *Switch* K_1 must impose current I through the load at instant $t = 0$, let it flow throughout the period $0, T/2$ and cut it off at instant $t = T/2$. It does not need to be current-reversible, but voltage u, which appears across it when it is off, can be negative. This switch thus corresponds to a controlled turn-on/turn-off semiconductor device, capable of withstanding reverse voltages.

  Figure 1.5 shows the basic configuration of a current-fed inverter.

- In order to show the influence of the load on the operation required of the switches, the *AC voltage load is assumed to be perfect*, i.e. voltage u' is sinusoidal.

- *If the load is inductive* (current i' fundamental component lagging voltage u'), the waveforms of i', u', u, i_{K_1} and v_{K_1} are as shown in Fig. 1.6.

  For $t = 0$, the turn-off of TC_2 controls the transfer of i from TC_2 to TC_1. TC_1 conducts I during half-cycle $0, T/2$.

  For $t = T/2$, the turn-off control of TC_1 must interrupt the flow of I through it; current I is then transferred to TC_2.

Semiconductor switches cannot ensure their own turn-on, since the voltage across them is negative at the moment when they must begin conducting. Their

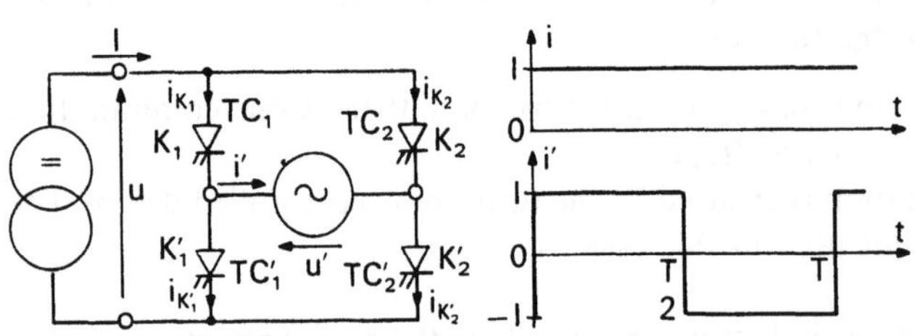

Fig. 1.5

role is simply to allow the flow of current I which was previously flowing via the other semiconductor device connected to the same terminal of the current supply.

The K_1 turn-on control must thus come slightly before the turn-off control of K_2 and vice versa. This avoids the current supply circuit being opened. The same applies to K_1' and K_2'.

- *If the load is capacitive* (current i' fundamental component leading voltage u'), the waveforms in Fig. 1.7 are to be obtained for i', u', u, i_{K_1} and v_{K_1}.

  For $t = 0$, turning TC_1 on leads to a negative voltage being applied across TC_2 and thus to the latter's being turned off.

  TC_1 conducts throughout period $0, T/2$. When TC_2 is turned on, TC_1 is turned off.

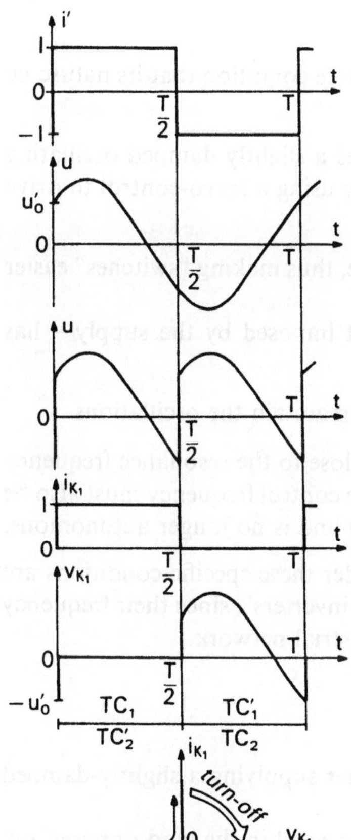

Fig. 1.6

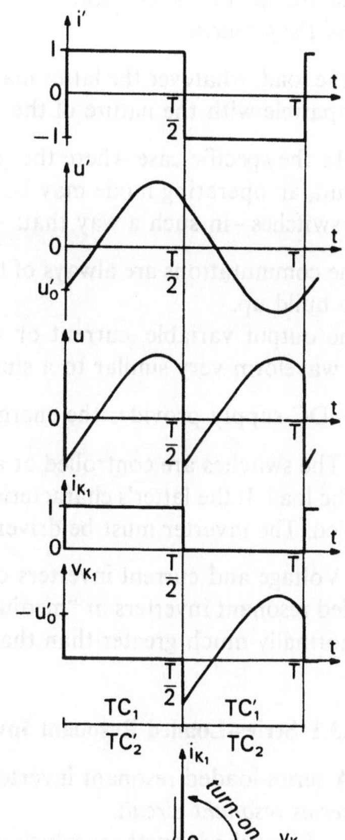

Fig. 1.7

During this operating mode, the blocking capacity of the controlled devices is no longer used. These may be ordinary thyristors. As in rectifier configurations, voltage u' ensures the transfer of current I' from TC_1 to TC_2 or vice-versa, or from TC_1' to TC_2' or vice-versa, whenever a switch is turned on.

• *Remark.* As in the case of the voltage-source inverter, it can be seen that control ability of controlled switches, during full-wave operation, can be reduced if it is known for certain that the load will always remain inductive or capacitive.

1.1.3 Resonant Load Inverters

The voltage or current-source inverters which we have presented behave like real voltage or current sources with regard to AC load. They impose:

- the voltage or the current,
- *and the frequency*

on the load, whatever the latter may be, on the sole condition that its nature be compatible with the nature of the DC supply.

In the specific case where the load comprises a slightly damped oscillating circuit, an operating mode may be obtained – by using a servo-control to drive the switches – in such a way that:

- the commutations are always of the same type, thus making "switches" easier to build up.
- the output variable, current or voltage – not imposed by the supply – has a waveform very similar to a sinewave.

The DC supply provides the energy needed to maintain the oscillations.

The switches are controlled at a frequency close to the resonance frequency of the load. If the latter's characteristics vary, the control frequency must also be varied. The inverter must be driven by the load and is no longer autonomous.

Voltage and current inverters operating under these specific conditions are called resonant inverters or "medium frequency inverters", since their frequency is normally much greater than that of the industrial network.

1.1.3.1 Series-Loaded Resonant Inverter

• A series-loaded resonant inverter is an inverter supplying a slightly-damped *series resonant circuit.*

Since the inductance which is serially connected in the load opposes any discontinuity in current i', the inverter feeds a current receptor and must thus be powered by a voltage supply. It is thus a *voltage-source inverter* (Fig. 1.8).

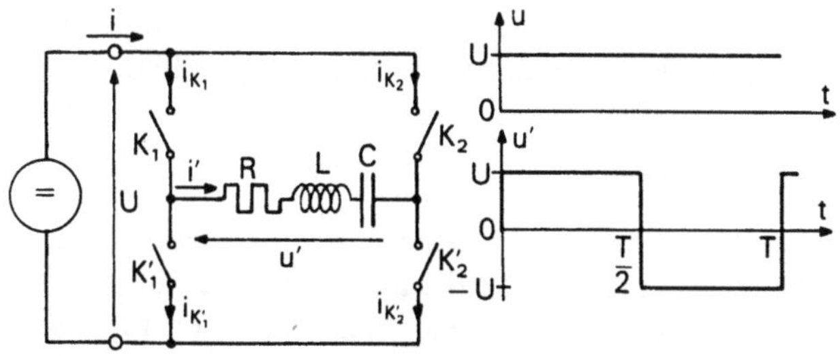

Fig. 1.8

The study of the RLC circuit supplied by square-waves of voltage u' will be developed in Chap. 6. For the present, we confine our analysis to some results of this study.

The RLC circuit can be characterized by

- its angular resonance frequency when undamped ($R = 0$):

$$\omega_0 = 1/\sqrt{LC}$$

- its damping coefficient:

$$\alpha = R/2L$$

- or the value of the latter related to ω_0:

$$\zeta = \frac{\alpha}{\omega_0} = \frac{R}{2}\sqrt{\frac{C}{L}}.$$

- The apparent nature of the load depends on the relative value of angular frequency ω, equal to $2\pi/T$, of the voltage u' fundamental.

- If ω is *greater* than $\omega_0\sqrt{1-\zeta^2}$, the load is *inductive* and value i'_0 of current i' for $t = 0$ is negative; the waveforms of Fig. 1.3 are obtained.

  Each of the four switches must be made up of a controlled turn-off/spontaneous turn-on semiconductor device, with a diode in antiparallel.

- If ω is *less* than $\omega_0\sqrt{1-\zeta^2}$, the load is *capacitive*, i'_0 has a positive value and the waveforms shown in Fig. 1.4 are obtained.

  Each switch is made up of a semiconductor controlled only at turn-on – e.g. an ordinary thyristor – with a diode in antiparallel.

- This second solution long remained the chosen solution as it allowed the use of semiconductor devices controlled only at firing (operating in natural commutation). However, in view of the developments of semiconductor device technology, the first is adopted more and more frequently.

- Whatever the solution chosen, if ζ is low and ω close to $\omega_0\sqrt{1-\zeta^2}$, the current wave is very similar to a sinewave.

By varying the difference between the two angular frequencies, it is possible to control the amplitude of the current supplied to the load.

1.1.3.2 Parallel-Loaded Resonant Inverter

- A parallel-loaded resonant inverter is a resonant inverter connected to a *slightly-damped parallel resonant circuit.* As the load comprises a capacitor connected directly between the output terminals, it resists sudden variations in voltage u'; the inverter must thus be fed by a DC current supply: it is a *current-source inverter* (Fig. 1.9).

If the consumption of active power by the load is accounted for by using a conductance G, the *duality* enables immediate use to be made of the results obtained for the series RLC circuit which is supplied by a voltage-source inverter.

The only substitutions required are as follows:

I for U, u for i, i' for u', u' for i'
i_{K_1} for v_{K_1}, v_{K_1} for i_{K_1},
G for R, C for L, L for C.

The angular resonance frequency, the damping coefficient and the ratio between these two have the following values:

$$\omega_0 = \frac{1}{\sqrt{LC}} \qquad \alpha = \frac{G}{2C} \qquad \zeta = \frac{G}{2}\sqrt{\frac{L}{C}}$$

Depending on whether the current i' fundamental angular frequency ω is lower or greater than $\omega_0\sqrt{1-\zeta^2}$, the load behaves inductively or capacitively, and the waveforms are those shown in Figs. 1.6 or 1.7.

- For natural commutation operation, i.e. with ordinary thyristors, the initial value u'_0 of voltage u' must be negative (see Fig. 1.7); angular frequency ω must thus be greater than $\omega_0\sqrt{1-\zeta^2}$. Operation usually takes place just above

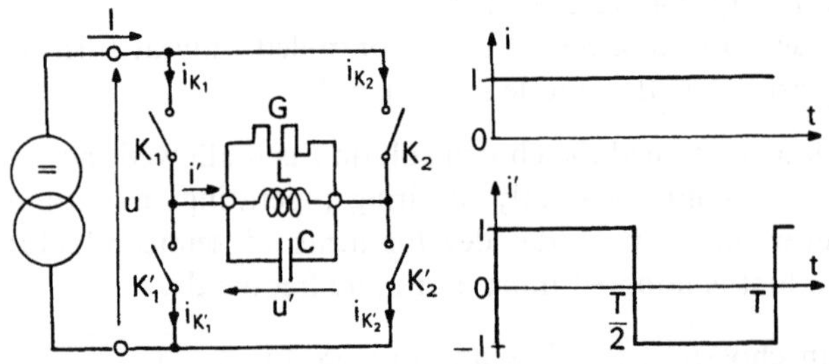

Fig. 1.9

this frequency limit, which enables a waveform close to a sinusoid to be obtained for u'. The magnitude of u' can be varied by acting on the difference between the two frequencies.

If operation was required at ω lower than $\omega_0\sqrt{1-\zeta^2}$, controlled turn-off devices, able to withstand negative voltages, would have to be used (see Fig. 1.6).

1.2 Types of Semiconductor Devices. Supply and Load Imperfections. Filters

In order to prepare the third part of this chapter, in which, the type of inverter best suited to each main field of application will be shown, the following need to be indicated:

- the types of semiconductor devices which may be used,
- the effects of supply and load imperfections and the way in which these may be corrected.

1.2.1 Types of Semiconductor Devices

1.2.1.1 Summary Table

The preceding simplified presentation of various types of inverter showed that the inverter operation can be characterized by:

- the nature of the DC supply,
- the nature of the AC load,
- the switch control.

- Concerning the DC supply, it has been assumed to be a perfect voltage or current source. The output thus consists of voltage square-waves or current square-waves.

  The nature of the DC supply imposes to the load to be a current source or a voltage source, since the supply and the load must be of different natures.

Leaving aside the type of control (autonomous or load-resonant), only two types of inverter need to be considered:

the *voltage*-source inverter,
the *current*-source inverter.

However, for each inverter, two cases can be distinguished, as the type of operation required of the switches varies depending on the load. It is thus necessary to take into account

the case of the *inductive load,*
the case of the *capacitive load.*

- The theoretical operating modes of both inverters for both types of load are shown, in tabular form, in Figs. 1.10 and 1.11.

- Figure 1.10 deals with the voltage-source inverter. At the top is a diagram of the inverter connected between the DC-voltage supply and the AC-current load. Below are shown the waveforms of u', i', i, of v_{K_1}, i_{K_1} and the switching loci of K_1,
  - for an inductive load, on the left (Fig. 1.10a)
  - for a capacitive load, on the right (Fig. 1.10b).

  The lower part of the figure indicates the operation of switch K_1 when the inverter must fed either type of load (Fig. 1.10c).

- For the current-source inverter, the waveforms of i', u', u, of i_{K_1}, v_{K_1} and the operation required of K_1 have been plotted, when feeding
  - an inductive load (Fig. 1.11a),
  - a capacitive load (Fig. 1.11b),
  - any type of load (Fig. 1.11c).

- It can be seen that there is a complete *duality* between the voltage-source inverter feeding an inductive load and the current-source inverter feeding a capacitive load, and between the voltage-source inverter feeding a capacitive load and the current-source inverter feeding an inductive load.

  This duality is not only applicable to the input and output variables:

  $(U \rightarrow I,\ i \rightarrow u,\ u' \rightarrow i',\ i' \rightarrow u')$
  (leading → lagging, lagging → leading)

  but also to switches:

  $(v_{K_1} \rightarrow i_{K'_1},\ i_{K_1} \rightarrow v_{K'_1})$
  (turn-on → turn-off, turn-off → turn-on).

1.2.1.2 Notes on the Dual Thyristor

In taking duality to its logical conclusion, H. Foch introduced the notion of the *dual thyristor* and showed that this was the optimal "switch" for the implementation of some structures.

An ordinary thyristor can withstand a negative voltage; it remains in the off state when the voltage across it becomes positive. The gate then enables it to be fired. After firing, the thyristor forward current evolution depends on the circuit into which the thyristor is embedded. When this current falls to zero, the thyristor turns off naturally. Figure 1.12a shows the locus of the operating point in the voltage–current plane.

The dual thyristor lets the negative current flow; it remains on when the current becomes positive. It can then be blocked by means of the control.

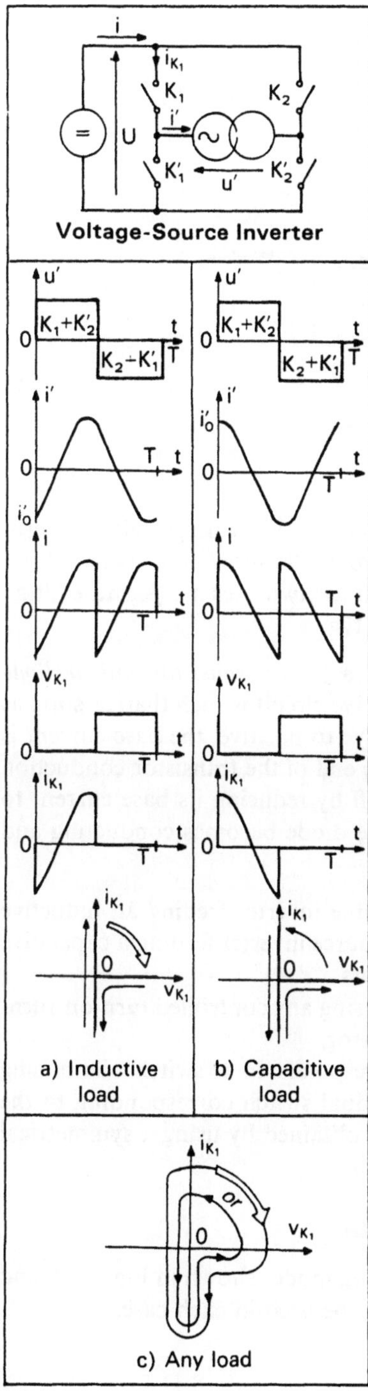

Fig. 1.10

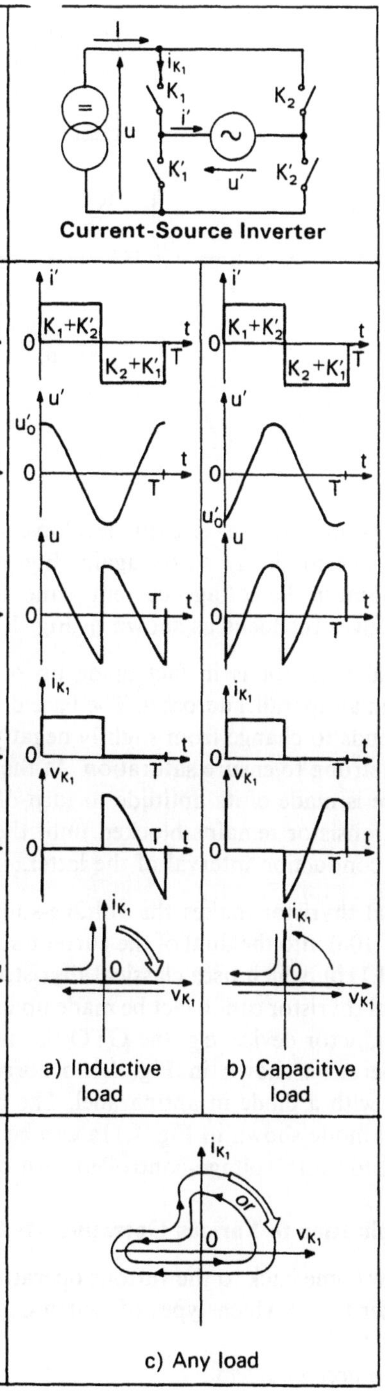

Fig. 1.11

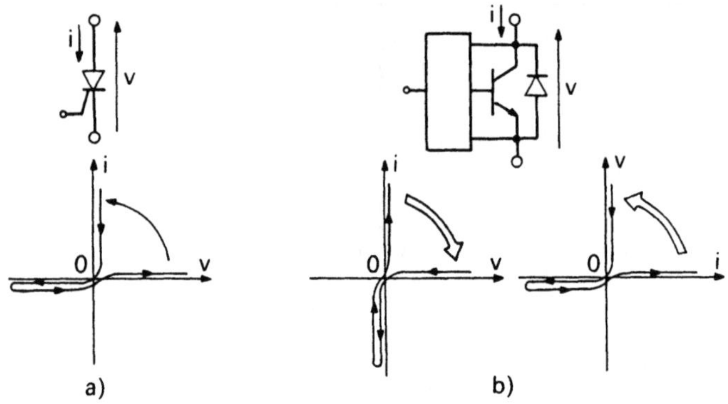

Fig. 1.12

A positive voltage across it then develops. When this voltage falls to zero, conduction automatically starts again. Figure 1.12b shows the locus of the operating point in the voltage–current plane. If the two axes are permuted, this once more gives the locus as shown in Fig. 1.12a.

The dual thyristor is in fact made up of a *power transistor with a diode* connected in antiparallel across it. The base drive circuit is such that, as soon as voltage v tends to change from slightly negative to positive, the base current is sufficiently strong to ensure saturation. At the end of the transistor conduction interval, use is made of its aptitude to turn-off by reducing its base current to zero. The transistor remains blocked until the diode becomes conducting and during the conduction interval of the latter.

The dual thyristor makes the voltage-source inverter feeding an inductive load (Fig. 1.10a) into the dual of the current-source inverter feeding a capacitive load (Fig. 1.11b), which uses classical thyristors.

The dual thyristor can in fact be made up using any controlled turn-on/turn-off semiconductor device, e.g. the GTO thyristor.

The operation shown in Fig. 1.10b corresponds to a switch comprising a thyristor with a diode in antiparallel. The dual switch corresponding to the operational mode shown in Fig. 1.11a can be obtained by using a symmetrical GTO thyristor with voltage-controlled turn-on.

1.2.1.3 Application to Various Operating Modes

We can now come back to the various operating modes shown in Figs. 1.10 and 1.11 in order to see which types of switch can be used in each case.

- *Voltage-source inverters*

Controlled semiconductor devices must have diodes connected in anti-parallel across; they do not have to withstand reverse voltage.

- *If the load is always inductive*, transistors with "long" control signals are used for low and medium power equipment; operation as a dual thyristor is particularly well-adapted to this operational mode.

  Ordinary thyristors with auxiliary turn-off circuits or GTO thyristors can be used at high power levels. Care must be taken to supply the gate until the moment when the forward current builds up.

- *If the load is always capacitive*, classical thyristors are suitable, since there is natural commutation operation. Asymmetrical thyristors can be used.

  Only when the turn-off time t_q of the thyristors leads to problems of switching frequency rise can the replacement of them by controlled turn-off semiconductor devices (power transistors or GTO thyristors) be considered. The latter's aptitude to turn-off is not used.

- Controlled turn-on/turn-off devices must be used in cases where *the load can be of any nature.*

  The following can be used: power transistor for low- and medium-power applications, thyristor with auxiliary turn-off circuit or GTO thyristor for high-power applications.

- *Current-source inverters*

There is an unidirectional current in each switch: there is no need to add a diode in antiparallel across the controlled semiconductor device. However the latter must be capable of withstanding a reverse voltage.

- *If the load is inductive*, switch K_1, for example, cannot turn on naturally, since voltage v_{K_1} is negative for $t = 0$. K_1 turn-on is caused by switching K_2 off and by the fact that current I, equal to $i_{K_1} + i_{K_2}$, fed by the current supply must remain constant. The K_1 turn-on control must overlap the K_2 turn-off control to allow I to be transferred from K_2 to K_1. Controlled turn-on/turn-off semiconductor devices must therefore be used, *as in the case where the load can be of any nature.*

  The following can be used: symmetrical thyristors with auxiliary turn-off circuits, symmetrical GTO thyristors, power transistors. In the latter case, a diode must be series connected with each transistor in order to withstand the reverse voltage.

- *If the load is capacitive*, the natural commutation operating mode which occurs is particularly suited to the use of the symmetrical thyristor.

  Only the limitation of frequency, linked to turn-off time t_q, can lead to use of controlled turn-on/turn-off devices, whose turn-off possibilities remain unused.

1.2.1.4 Remarks

- In the four operating modes shown in Figs. 1.10 and 1.11, use is only made of either the turn-on aptitude *or* the turn-off aptitude of the controlled semiconductor devices.

This results from *the control frequency of the switches being equal to the output variables frequency.* During each half-cycle, the output variable which is not imposed by the supply falls to zero. For the case of the voltage-source inverter, i' and thus i_{K_1} fall to zero between $t = 0$ and $t = T/2$. For the case of the current-source inverter, u' and thus v_{K_1} fall to zero between $t = T/2$ and $t = T$.

This was not the case with choppers whose output frequency was zero in steady state, and is not the case for pulse width modulated inverters in which the output frequency is lower than the switching frequency. For a same operating mode, the aptitude of switches to turn on or off is thus used.

- In plotting the waveforms, it has been assumed that the AC-current load or the AC-voltage load is perfect, i.e. that i' or u' are sinusoidal. If this is not the case, no change ensues in the commutations to be carried out or in the semiconductor devices to be used.

  For example, in the case of the voltage-source inverter, if the AC-current load is imperfect, this gives once again $i'(T/2) = -i'_0$ and the current falls to zero during each half-cycle of u'. The commutations only depend on the positive or negative sign of i'_0.

  The same applies to the current-source inverter; the commutations depend only on the sign of u'_0.

1.2.2 DC-Supply Imperfection. Input Filter

In the volume on choppers, we presented the effects of supply and load imperfections (Vol. 3, Chap. 4). Similarly, in the case of inverters, the effects giving the most problems arise from *the inductance in the path where the current must be suddenly reversed*:

- DC supply inductance in a voltage-source inverter,
- AC load inductance in a current-source inverter.

1.2.2.1 DC-Voltage Supply

- *Voltage spikes*

In order to take into account the non zero value of inductance L_S of an imperfect voltage supply u, this supply is replaced by a perfect voltage source U_S serially connected with this inductance L_S (Fig. 1.13a).

$$u = U_S - L_S \, \mathrm{d}i/\mathrm{d}t.$$

- *When feeding an inductive load,* during each commutation, current i must go from $|i'_0|$ to $-|i'_0|$ (see Fig. 1.3); this rapid decrease in i leads to a voltage peak Δu.

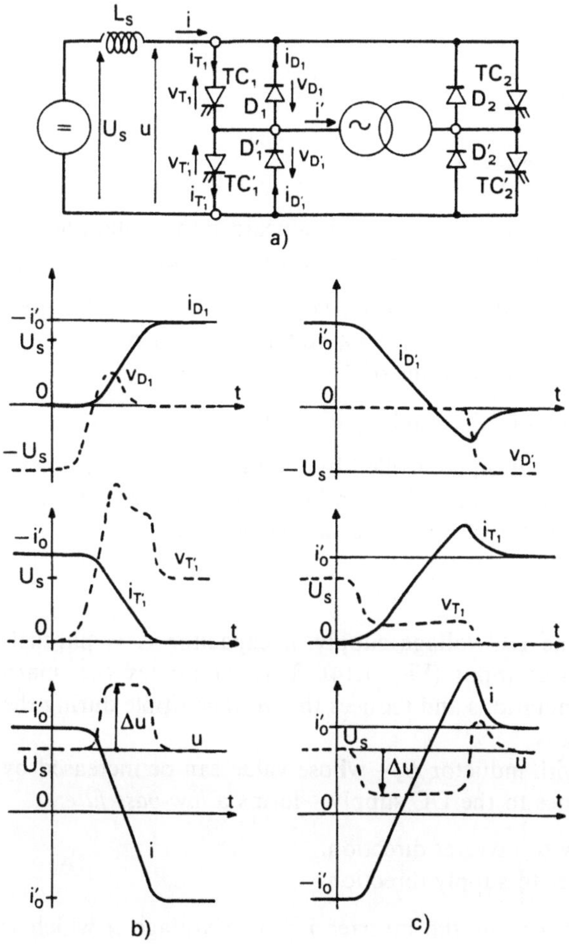

Fig. 1.13

Figure 1.13b indicates[1] the waveforms

of v_{D_1} and i_{D_1} (or of $v_{D'_2}$ and $i_{D'_2}$),
of $v_{T'_1}$ and $i_{T'_1}$ (or of v_{T_2} and i_{T_2}),
of i, equal to $i_{T'_1} - i_{D_1}$, since $i_{T'_1}$ equals i_{T_2},
of u, equal to $U_S - L_S\,\mathrm{d}i/\mathrm{d}t$.

during the TC'_1–D_1 (or TC_2–D'_2) commutation, which occurs when, for $t = 0$, the turn-off of TC'_1 (and TC_2) is controlled.

---

[1] Since this commutation from a controlled semiconductor device to a diode and the reverse commutation are the most common, we have used them throughout Chap. 2 of Vol. 3 to characterize the behaviour of semiconductor devices.

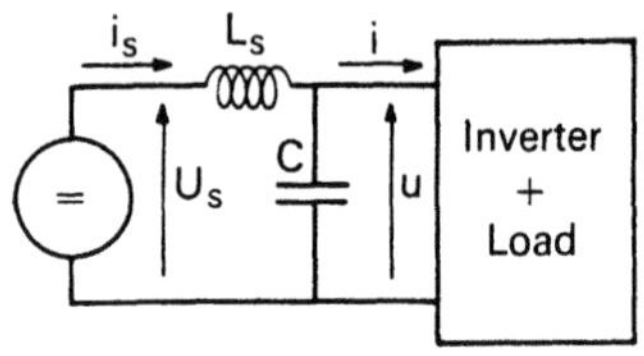

Fig. 1.14

Since v_{T_1} equals $u + v_{D_1}$, the peak voltage Δu appears in the voltage across TC_1'; moreover, it can be increased by the forward voltage peak of diode D_1.

- *When feeding a capacitive load*, during each commutation, current i goes from $-|i_0'|$ to $|i_0'|$ (see Fig. 1.4); this leads to a voltage through Δu.

  Figure 1.13c indicates the waveforms of

$$v_{D_1'} \text{ and } i_{D_1'},\ v_{T_1} \text{ and } i_{T_1}, u \text{ and } i$$

during the D_1'–TC_1 commutation controlled by the turn-on of TC_1.

The reverse current peak in diode D_1' can be found in the waveform of i_{T_1} and, multiplied by 2, in that of i; it causes a voltage peak.

- *The input filter*

In order to "correct" the DC-voltage supply, a capacitor C is parallel-connected across the inverter input (Fig. 1.14). This eliminates the sharp variations in u during commutations and reduces this voltage ripple during the period between commutations.

Capacitor C, together with inductor L_S – whose value can be increased by adding an external inductance to the DC supply – forms a *low-pass filter*.

- for voltages, in the supply to inverter direction,
- for currents, in the inverter to supply direction.

This enables to be obtained at the inverter input, a voltage u which is practically constant, despite variations in i and even if voltage U_S is affected by a ripple (e.g. in the case of voltage supplied by a rectifier).

It also enables current i_s taken from the supply to be kept practically constant, while current i at the inverter input shows an important ripple.

The calculation required for the input filter is very similar to that required in the case of choppers. The cost of the filter decreases as the inverter operating frequency increases.

1.2.2.2 DC-Current Supply

- *Input current fluctuations*

Imperfection in a DC-current supply results from the finite value of its internal inductance. This supply can be represented as a DC-voltage supply (with a value U_S equal to the mean value of u) serially connected with an inductance L_S (Fig. 1.15a). This once again gives the diagram in Fig. 1.13a except

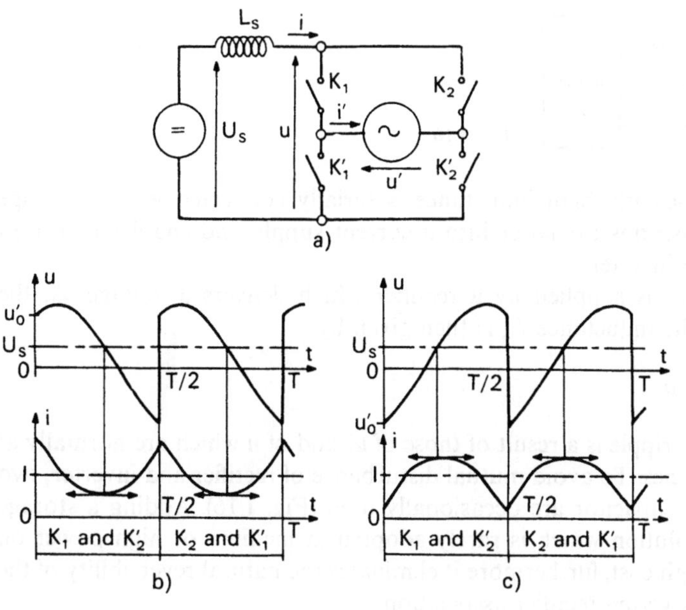

Fig. 1.15

that L_S is now much greater and, mainly, that the inverter output is connected to a voltage load.

The ripple in current i, on either side of its mean value I, can be determined from

$$L_S \frac{di}{dt} = U_S - u.$$

Figure 1.15b shows the waveform of i in the case of an inductive load; Fig. 1.15c in that of a capacitive load.

This problem is similar to that of a rectifier feeding an emf via an inductance (see Vol. 1, Chap. 7, Sect. 6).

Current i ripple can be a problem for the inverter since it increases the current stresses imposed on semiconductor devices.

Furthermore, when mean current I, taken from the DC supply, falls below a certain limit I_{lim}, the discontinuous conduction phenomenon, already encountered for rectifiers and choppers, appears; current i is periodically zero.

- *Input DC current smoothing inductance*

When the inductance of a current source is increased by adding a smoothing inductance in series, this enables the inverter input current i ripple to be reduced, as well as reducing the drawbacks linked to this ripple.

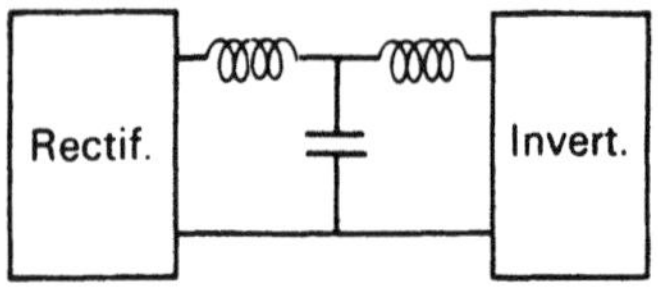

Fig. 1.16

When a sufficiently high inductance is serially connected with a voltage source, this transforms the latter into a current supply and enables it to feed a current-source inverter.

If the inverter is supplied by a rectifier which delivers a voltage u_d', the voltage across the inductance L_S is then given by

$$L_S \frac{di}{dt} = u_d' - u.$$

The current i ripple is a result of those of u_d' and of u which are normally at different frequencies. To avoid mutual disturbance of rectifier and inverter, two inductors and a capacitor are occasionally used (Fig. 1.16). Adding a storage capacitor is a solution which is rarely adopted at medium or high power on account of its high cost; furthermore it eliminates the natural reversibility of the current-source inverter–rectifier association.

1.2.3 AC-Load Imperfection. Output Filter

1.2.3.1 AC-Current Load

If the load placed at the voltage-source inverter output were be a perfect sinusoidal AC-current receptor, current i' would be sinusoidal although voltage u' is not. The load should show infinite impedance for all the terms of the series expansion of voltage u', except for the fundamental.

This is the case – more or less – for a series resonant circuit supplied at a frequency very close to its resonance frequency (see Sect. 1.1.3.1).

However, in many applications, the load imperfection poses few problems. Concerning the operation of the inverter, the essential point is that the AC load should not be a voltage receptor, i.e., it should not oppose the sharp variations of voltage u' applied to it.

- *Current smoothing; output voltage filtering*

– If the load is of the R, L, e type where e denotes an AC emf with the same angular frequency ω as voltage u', current i' is such that

$$Ri' + L\frac{di'}{dt} = u' - e.$$

If inductance L is not great enough to reduce current i' variations on both sides of its fundamental component, the current can be "smoothed" by increasing L. But this increase leads to a similar one in the apparent power supplied to

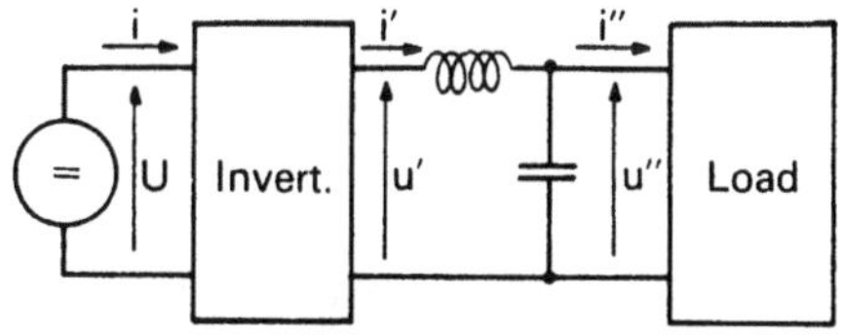

Fig. 1.17

the load and thus, for a given active power, to an increase in the various voltages.

- For other applications, the aim might be to obtain a quasi-sinusoidal voltage from the square-wave voltage u' supplied by the inverter. The load should then be supplied via a low-pass filter (Fig. 1.17).

  It should be noted that the low-pass filter for voltages in the inverter-to-load direction is also a low-pass filter for currents in the load-to-inverter direction. Current i' is closer to a sinewave than is current i''.

- Smoothing current i' or filtering voltage u' becomes easier as the output frequency of the inverter increases.

- *Commutations*

The imperfection of the AC-current load has practically no effect on the commutation process. As remarked above (Sect. 1.2.2.1), in a voltage-source inverter, the current transfers occur between switches serially connected under voltage u. It is in the DC supply that fast current variations occur. If the load inductance has a value which is low but sufficient to make the variations of i' negligible during commutations, the latter are unaffected by an increase in the value of this inductance.

1.2.3.2 AC-Voltage Load

In order to be considered as a perfect sinusoidal AC-voltage receptor, the load connected to the current-source inverter output should show zero impedance for all the terms of current i' Fourier series expansion, except for its fundamental. This is approximately the case for a parallel-resonant circuit supplied by a frequency close to its resonance frequency.

However such "perfection" of the AC-voltage load is normally far from being achieved. This is of little importance for operation if the load does not behave as a current receptor, i.e. if it does not oppose sudden variations in the current which flows through it.

- *Voltage spikes*

In most cases, the AC load is of R, L, e type with a non-negligible value of inductance L. As the principal imperfection of the load, this inductance has two effects:

- *it makes the commutations last longer,*
- *it causes voltage spikes.*

In order to show the effects of L, a current source inverter is shown (Fig. 1.18a) connected to a load made up of a perfect voltage source in series with an inductance L. The effect of R is ignored. However, it is assumed that L is sufficiently small and the commutations sufficiently short to enable e to be assumed constant during the commutations and equal to value u'_0 of voltage u' when the latter begins.

The commutations in a current-source inverter are different from those in a voltage-source inverter: the current must be reversed in the AC load and no

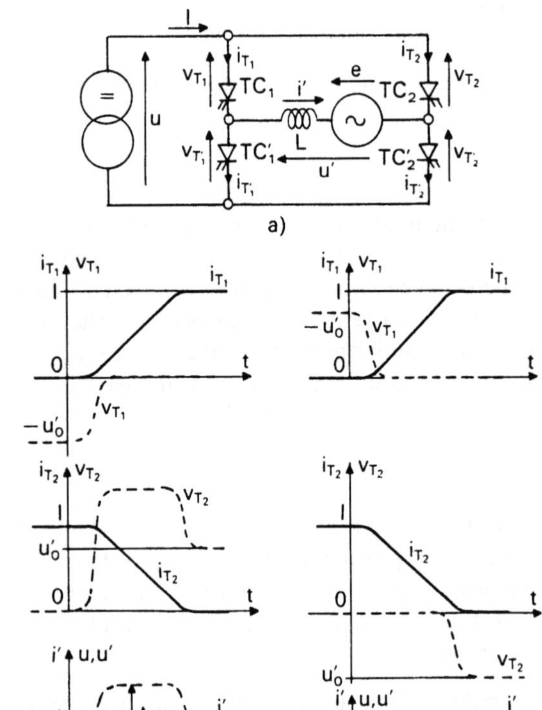

Fig. 1.18

longer in the DC supply. Because of the constant nature of input current I, the transfers take place between semiconductor devices connected to the same input terminal:

between TC_1 and TC_2,
between TC'_1 and TC'_2.

– *When the load is inductive* (see Fig. 1.6), the commutations are forced.

For $t = 0$, when u' is positive, the turn-off of TC_2 and TC'_1 is controlled. As soon as TC_1 and TC'_2 become conducting, current i' can be reversed, passing from $-I$ to I. This results in a voltage peak u, equal to $L\,di/dt$, across the load. This peak appears across the semiconductor devices which are being turned off.

Figure 1.18b shows the waveforms of i_{T_1} and v_{T_1}, of i_{T_2} and v_{T_2}, of u', u and i.

– *When the load is capacitive* (see Fig. 1.7), the commutations are natural.

For $t = 0$, when u' is negative, the turn-on of TC_1 and TC'_2 is controlled. As long as the load is short-circuited by the simultaneous conduction of the four semiconductor devices, voltage u' is zero, and the increase in current i' is given by $L\,di'/dt = e = u_0$.

Figure 1.18c shows the waveforms of i_{T_1} and v_{T_1}, of i_{T_2} and v_{T_2}, of u', u and i during this commutation.

● *Correcting the AC load*

– In order to reduce the effects of load inductance, a capacitor must be connected across the load. If this capacitor is large enough, the variations of voltage u' during the commutations can be ignored.

– In general, there is no attempt to make the current waveform supplied to the load, or the voltage waveform across it, approximately sinusoidal. To achieve this, a filter would need to be added. This addition is normally too costly if the load shows considerable variations and if the filter has to play an effective role in spite of this.

1.2.4 Remarks on Reversibility

1.2.4.1 The Two Types of Reversibility

● *Instantaneous reversibility*

The symmetrical structure inverters, which are being studied here, have an AC output voltage u' and an AC output current i'. The load is thus a "source" (according to the definition of sources given at the beginning of Vol. 3), which is voltage- and current-reversible.

Apart from the exceptional case where the load is a pure resistance, voltage u' and current i' do not always have the same polarity: instantaneous power p, equal to $u'i'$, is now positive, now negative. Since the losses in the switches are ignored, the input instantaneous power ui is equal to p. This assumes that the DC supply shows at least one reversibility. It has been observed that:

- the DC voltage supply feeding a voltage-source inverter must be current-reversible
- the DC current supply feeding a current-source inverter must be voltage-reversible.

- *Functional reversibility*

Functional reversibility is said to occur when the energy transfer direction (i.e. power P equal to $(u'i')_{\text{mean}}$ or $(ui)_{\text{mean}}$) can be reversed.

Using the accepted conventional signs, if P is positive, the energy transfer takes place from the DC side towards the AC side and the converter operates as an inverter in the accepted sense. If P is negative, the transfer takes place in the opposite direction and the converter is operating as a rectifier: the DC source recovers energy from the AC source.

If the output voltage or the output current is assumed to be sinusoidal, and using φ to denote the phase shift of current i' or of its fundamental component in relation to voltage u' or to its fundamental component[1],

P is positive, for $-\pi/2 < \varphi < \pi/2$
P is negative, for $\varphi < -\pi/2$ or $\varphi > \pi/2$.

The problem of functional reversibility only arises if a generator ($P < 0$) can be placed on the AC side instead of a receptor ($P > 0$).

- *Where does the problem of reversibility arise?*

The study of "switches" (see Sect. 1.2.1) has shown that inverters can operate for any value of φ, providing that controlled turn-off/turn-on semiconductor devices are used. However, if the load is inductive ($0 < \varphi < \pi$), only the turn-off aptitude is used and, if the load is capacitive ($0 > \varphi > -\pi$), only the turn-on aptitude.

The problem of reversibility arises from the *DC supply*. We will illustrate this on the two most common types of DC supply:

- the battery supply,
- the rectified industrial AC network.

Any remarks concerning these types of supply can easily be adapted to others.

[1] Throughout this book, φ will be taken as positive when the current lags the voltage; this enables φ to be positive for the majority of standard loads.

1.2.4.2 Battery Supply

- A battery is a current-reversible DC voltage supply. It is thus *particularly well-adapted to supplying a voltage-source inverter* and can ensure the functional reversibility of the latter.

  For operation in the inverter mode, the mean value of i is positive and the battery discharges. During recovery, the mean value of i is negative and the battery charges up.

  A capacitor with a very low inductance is usually connected in parallel across the battery as the latter opposes very fast variations in the current.

  An input filter can be used (see Fig. 1.14) to reduce the current ripple supplied or received by the battery.

- If a current-source inverter is to be supplied, a large inductance has to be placed between the battery and the inverter. The full-bridge inverter with four controlled semiconductor devices (Fig. 1.19) does not allow current i to be reversed; it requires voltage u to be reversed.

  Inductance L_S can bring about the *necessary instantaneous reversiblity*, with the voltage drop $L_S\,(di/dt)$ absorbing the difference between voltages U_s and u. But $U_S i$ is always positive: using the configuration in Fig. 1.19, functional reversibility cannot be achieved.

- If a current-source inverter has to be supplied by a battery and if the *functional reversibility* must be guaranteed, the current-reversibility of the battery must be used,
  - either by placing a current reverser between the battery and the inverter (Fig. 1.20a):
    - for $P > 0$, T_A and T'_A are on ($i_S = i > 0$)
    - for $P < 0$, T_B and T'_B are on ($i_S = -i < 0$).
  - or by doubling the number of controlled semiconductor devices forming the inverter (Fig. 1.20b):
    - for $P > 0$, TC_1, TC'_1, TC_2, TC'_2 are used ($i_S = i > 0$),
    - for $P < 0$, TC_3, TC'_3, TC_4, TC'_4 are used ($i_S = i < 0$).

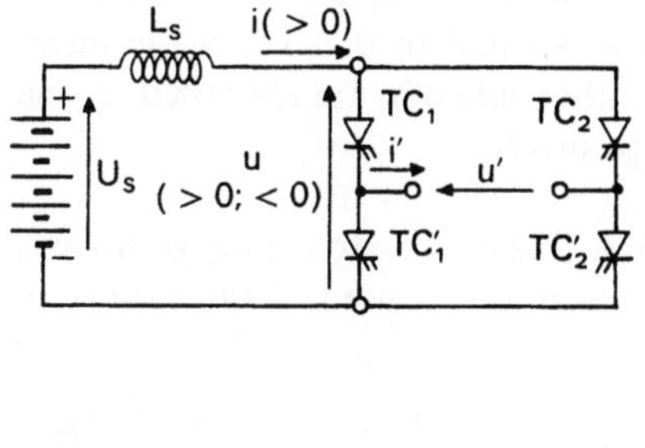

Fig. 1.19

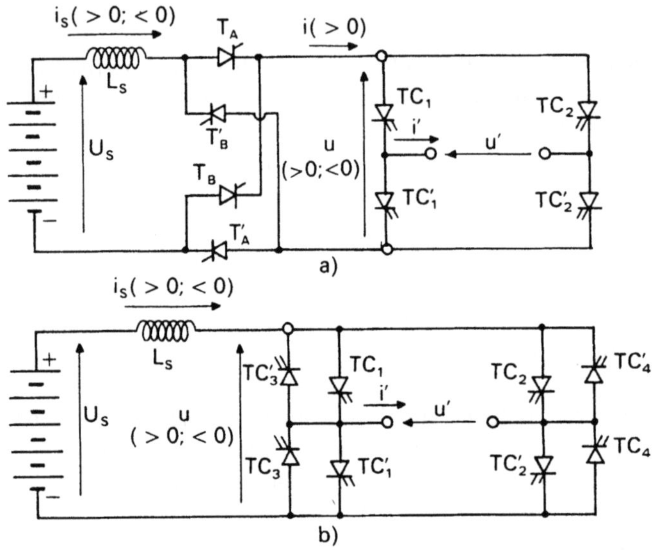

Fig. 1.20

1.2.4.3 Rectifier Supply

In order to achieve good characteristics, a rectifier must operate in the continuous mode. This means that a rectified current smoothing inductance must be connected in series at the rectifier output. The rectifier–inductance assembly behaves as a voltage-reversible DC-current source. It is thus *well-adapted to supplying a current-source inverter*, since it ensures the functional reversibility in that case.

For $P > 0$, the firing angle α of the rectifier thyristors is less than $\pi/2$, the mean value U'_d of the rectified voltage is positive. For $P < 0$, angle α is greater than $\pi/2$ and U'_d is negative.

- When a voltage-source inverter is fed by the industrial network through a rectifier serially connected with an inductance L_S, a capacitor C must be parallel-connected across the inverter input (Fig. 1.21). This capacitor ensures the *instantaneous reversibility*, by absorbing the difference between the unidirectional current i_d and current i which can be positive or negative. The mean value of i_d and i being the same, the former must be positive.

  The configuration in Fig. 1.21 does not allow for functional reversibility. Voltage u must be positive. Its mean value is equal to that U'_d of the mean rectified voltage. The variations of u'_d on either side of u are absorbed by the inductance: like u, U'_d must therefore be positive.

- Two rectifiers connected in antiparallel (Fig. 1.22a) – one to supply positive i_d current and the other negative i_d current – must be used, if the supply of

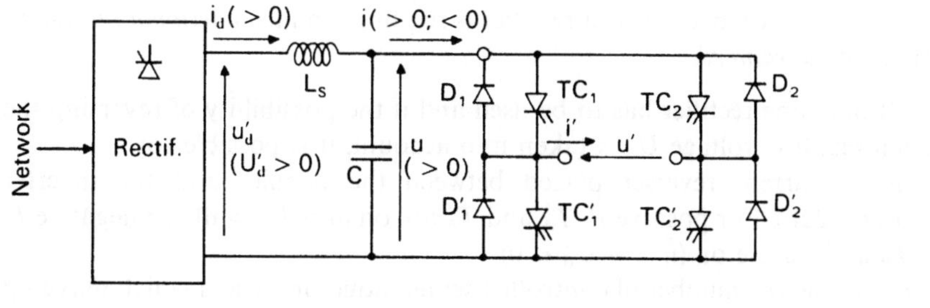

Fig. 1.21

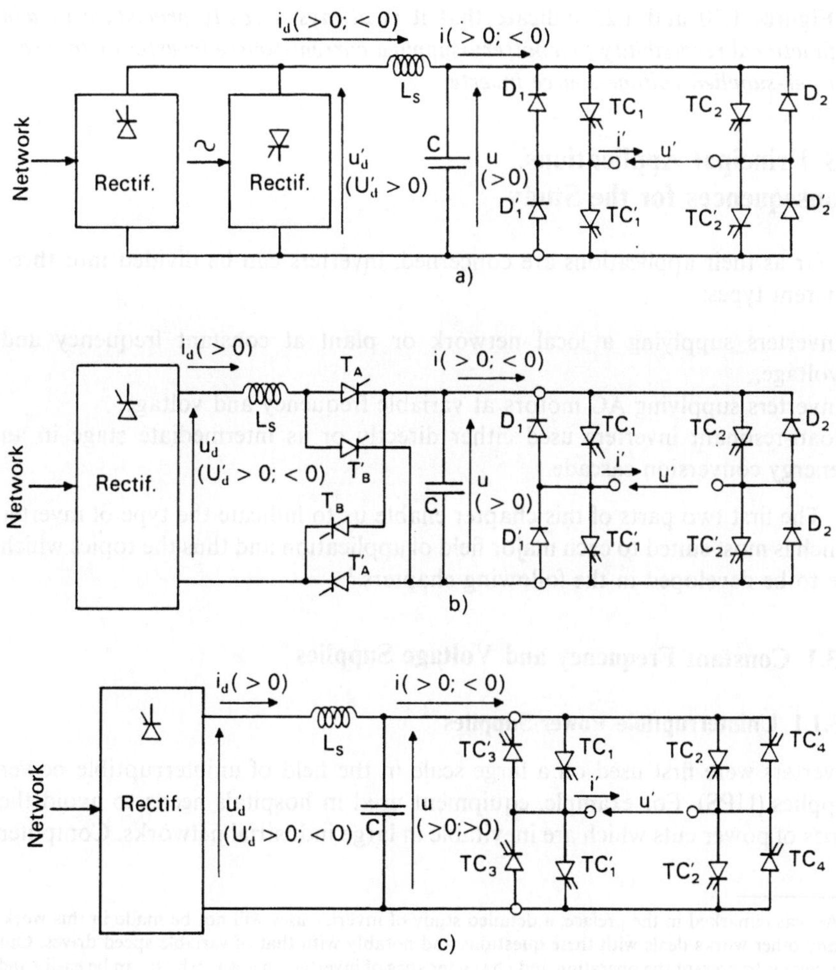

Fig. 1.22

a voltage-source inverter by a rectifier is required and its *functional reversibility to be guaranieed.*

If only one rectifier has to be used and if the possibiliity of reversing the mean rectified voltage U_d' is taken into account, it is possible to:

- use a current reverser placed between the rectifier and the inverter (Fig. 1.22b). For positive P, T_A and T_A' are on ($u \simeq U_d' > 0$); for negative P, T_B and T_B' are on ($u \simeq -U_d' < 0$)
- or double the number of controlled semiconductor devices which make up the inverter (Fig. 1.22c) and use half of these as diodes (with permanently supplied control electrode). For positive P, TC_3, TC_3', TC_4 and TC_4' are used as diodes ($u \simeq U_d' > 0$); for negative P, TC_1, TC_1', TC_2 and TC_2' are used as diodes ($u \simeq U_d' < 0$).

- Figures 1.20 and 1.22 indicate that it is *always a costly procedure to add functional reversibility to a battery-supplied current-source inverter or to a rectifier-supplied voltage-source inverter.*

1.3 Principal Applications. Consequences for the Study

As far as their applications are concerned, inverters can be divided into three different types:

- inverters supplying a local network or plant at constant frequency and voltage,
- inverters supplying AC motors at variable frequency and voltage,
- load-resonant inverters used either directly or as intermediate stage in an energy conversion cascade.

The first two parts of this chapter enable us to indicate the type of inverter which is most suited to each major field of application and thus the topics which are to be developed in the following chapters.[1]

1.3.1 Constant Frequency and Voltage Supplies

1.3.1.1 Uninterruptible Power Supplies

Inverters were first used on a large scale in the field of uninterruptible power supplies (UPS). For example, equipment used in hospitals needs to avoid the types of power cuts which are inevitable in large industrial networks. Computer

[1] As was remarked in the preface, a detailed study of inverter uses will not be made in this work. Many other works deals with these questions and notably with that of variable speed drives. Our aim here is to present the operation and characteristics of inverters in a way which can be easily and readily used in a study of standard applications.

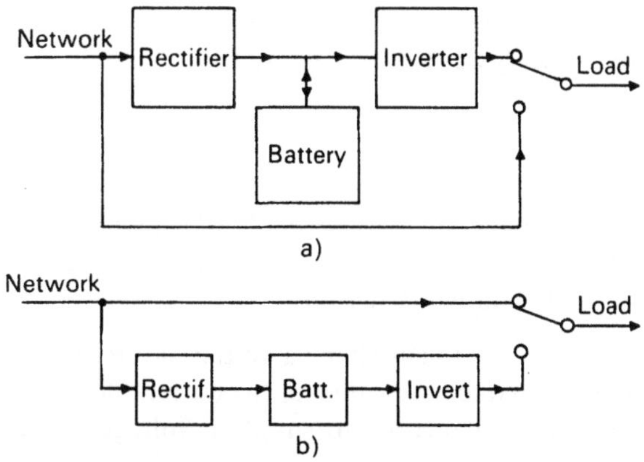

Fig. 1.23

energy supplies are often unable to withstand voltage blackouts in their supply networks, even if these last only a few cycles.

Such types of equipment are provided with uninterruptible power supplies intended to take over from the network in cases where the latter fails. This type of supply must provide a voltage of the same frequency and value as that of the network; its harmonic rate must also be of the same order of value.

Such equipment can also be permanently supplied by an inverter (Fig. 1.23a); the latter can be supplied either by the network via a rectifier, or by a battery connected as a storage element between the rectifier and the inverter. A direct network-load by-pass enables the supply to be taken over by the network if the inverter fails. Such a solution is very costly.

Normally, the network is used to supply the equipment directly (Fig. 1.23b); if the latter fails, the load is rapidly switched to an inverter powered by a battery, charged by an auxiliary rectifier.

For obvious reasons of reliability, some redundant elements are necessary and the actual configurations are more complicated than those of Fig. 1.23. As soon as they reach a certain power level, these uninterruptible power supplies must have a three-phase output; in that case they must supply balanced voltages, as does the network, even if the load is unbalanced.

1.3.1.2 Other Constant Frequency and Voltage Supplies

Inverters with an output voltage of constant frequency and magnitude are to be found in railway vehicles or aircrafts where they supply the internal lighting circuits, air-conditioning circuits . . . , from a battery.

They are also to be found in portable equipments which are AC-supplied from a battery or which can be supplied either by the network or by internal battery.

Note that, at low power, inverters have single-phase output and that at very low power, the type of symmetrical assembly being studied here will not necessarily be used.

For uninterruptible power supply systems, as well as other inverters included in this first group, *functional reversibility need not be ensured.*

1.3.1.3 Choosing the Inverter

For these inverters, intended to supply an AC load normally of *highly variable impedance* from a battery or a rectifier, *a voltage-source inverter* should be chosen. Such a choice is made for the following reasons:

1° The battery is particularly well adapted to supplying a voltage-source inverter. If necessary, an *L–C* filter can reduce the ripple of the current delivered by the battery. Such a filter enables a rectifier to supply a voltage-source inverter.

2° The operation of a voltage-source inverter is highly affected by the imperfections in the DC supply and hardly at all by those of the AC load. The reverse applies to the current-source inverter. For the uses being considered here, the DC supply can be selected but not the AC load.

3° Neither of the inverters directly gives a sinusoidal output voltage and an output filter must be used. In the case of the voltage-source inverter, the voltage to be filtered is imposed by the DC supply and the exact waveform to be filtered is known. With the current-source inverter, the voltage to be filtered depends on the load.

Chapter 2 will deal with single-phase or three-phase output *voltage-source inverters* with *one square-wave pulse per half-cycle*, and with input-current and output-voltage filtering (the input filter is studied in detail in *Appendix A*).

1.3.1.4 Output Filter. PWM Inverters

In the study of the output voltage filtering at the end of Chap. 2, it will be shown that this filtering is awkward and costly, on account of the load variations. The rank of the voltage first harmonic at the filter input should therefore be raised.

To achieve this, inverters of the "sinusoidal output voltage" type can be used: each half-cycle of the output voltage is made up of a succession of steps of various amplitudes which approximate the form of the required sinusoid.

Increasingly frequent use is made of pulse width modulated (PWM) inverters; in this case, each half-cycle of the output voltage is made up of several square-pulses of equal amplitude but different widths. This leads to the controlled semiconductor devices being turned on and off several times per cycle.

Chapter 3 will study *pulse width modulated voltage-source inverters* and in *Appendix B* "sinusoidal" inverters will be briefly described.

• *Remarks*

- Commutations found in the voltage-source inverters closely resemble those noted in the case of choppers in Vol. 3. However, in the majority of these inverters, two assemblies, each formed of a controlled device and a diode connected in antiparallel, are to be found in series between the DC voltage supply terminals. This series connection offers new possibilities in the field of snubber circuits for transistors and GTO thyristors and in the field of auxiliary turn-off circuits for conventional thyristors.

  Chapter 4 deals with the *control and snubber circuits of both transistors and GTO thyristors* in voltage-source inverters.

  Since the study of turn-off circuits for conventional thyristors was developed in Vol. 3, we will simply add some complementary informations concerning voltage-source inverters in *Appendix C*.

- In their great majority, rectifiers use diodes and thyristors which operate in natural commutation. However, in certain cases where performances on the network side are to be improved, use is often made of PWM rectifiers operating in forced commutation. This was mentioned at the end of Vol. 1 and will be further studied in *Appendix D*.

1.3.2 Speed Variation of AC-Current Motors

1.3.2.1 General Characteristics of Inverters

The DC-current motor remains frequently used in variable speed drive systems. Nevertheless the use of AC-current motors in this field is becoming more and more frequent.

The procedure which enables the AC synchronous motor speed to be varied consists of varying the frequency of its supply voltage, i.e. supplying it by an inverter. The same procedure is also the most widely used for induction motors.

- The rms value of the supply voltages must be varied together with the frequency. Usually a "constant flux" control is chosen, which enables the same torque to be developed at every speed; the value of voltages then increases more or less proportionally to their frequency. The inverter must thus produce *variable frequency and voltage.*
- In most cases *three-phase* motors are used; as the three phases are balanced, a three-phase inverter whose three phases are not independent can be used.
- The equipment is nearly always required to allow fast regenerative braking, i.e. to ensure *functional reversibility*. It should be noted that the inverter is usually supplied by the industrial network via a rectifier.
- The inverter output frequency is variable, starting from a very low value, which corresponds to the starting of the motor, up to the rated frequency

corresponding to full-speed operation. It is virtually *impossible to filter* the inverter output voltages as the value of the filter elements would need to be varied as the frequency varies.

1.3.2.2 Induction Motor Drive

The interest of the squirrel-cage induction motor lies in the mechanical robustness of its rotor. It inevitably absorbs reactive power and thus behaves as an inductive load: the inverter supplying it – whether it be a voltage-source or a current-source inverter – operates in forced commutation.

As a first approximation (see Vol. 2, Chap. 4, Sect. 2.3.1.) each phase can be assimilated to an emf of the same angular frequency as the supply voltage fundamental components, connected in series with a resistance and an inductance.

- *Voltage-Source Inverter Drive*

When the motor is fed by a voltage-source inverter, its inductance (cf. Sect. 2.3.1.) plays no role in the commutations and, produces no voltage peak. If the DC supply is virtually not inductive and if fast semiconductor devices are used, it becomes possible to operate at a relatively high frequency.

If the rated frequency of the motor is close to that of the industrial network, pulse width modualtion can be used. This enables the motor losses due to harmonics to be reduced and, more particularly, the machine torque pulsations to be lessened.

The main disadvantage of the voltage-source inverter drive becomes apparent when the latter is supplied by a rectifier; achieving functional reversibility is a costly task (cf. Sect. 2.4.3.).

In certain applications and especially in electric traction, the AC–DC conversion is sometimes carried out by replacing the rectifier by a four-quadrant chopper. This has the structure of a voltage-source inverter, operates in forced commutation and delivers power on the DC side via a capacitive filter. This implies that the AC network acts as a "current-source" and has an adequate inductance; this inductance can be obtained by acting on the flux leakages of the transformer which supplies the chopper. This system will be studied in *Appendix E.* It allows operation in all four quadrants.

The induction motor is supplied by a voltage-source inverter

- for small machines with a high rated frequency, especially if the inverter is supplied by a battery,
- for medium- and high-power machines when the inverter is supplied by a DC-voltage network; PWM is normally used in such cases,
- for medium- and high-power machines powered by a rectifier when a high switching frequency is required, e.g. for high-speed machines.

• *Current-Source Inverter Drive*

Using a current-source inverter supplied by a rectifier has the great advantage of making functional reversibility easy to achieve (see Sect. 1.2.4.3.).

The disadvantage of the current-source inverter is that the inductances of the machine are involved in commutations (see Sect. 1.2.3.2), thus increasing their duration and the associated losses.

Chapter 5 deals with autonomous *current-source inverters.* As these inverters are essentially intended to supply AC current motors, the study will concentrate on the *three-phase* inverter.

The current-source inverter is used for medium- and high-power drives supplied by a rectifier and not requiring an excessively high frequency of commutations.

When the torque oscillations become too great, the machine used has two stator windings shifted by $\pi/6$ and supplied by three-phase systems of currents shifted by $\pi/6$. The frequency of oscillations changes from 6 to 12 times the motor supply frequency and these oscillations are greatly reduced.

The same result can be obtained by using an inverter and a motor with a number of phases higher than 3.

In the case of motors supplied at low frequency or for low-speed operation of medium rated frequency motors, Pulse Width Modulation can be used.

• *Remark.* Most variable speed drives with induction motors are said to be "*self-controlled*". Frequency and amplitude of the voltages applied to the machine are determined from variabtes which characterize the state of the latter: rotation speed and absorbed current.

1.3.2.3 Synchronous Motor Drive

The rotor of the synchronous motor contains the field winding (or bears permanent magnets); it is therefore more expensive than that of the induction motor. However, in the stator windings, emf's can be found which are proportional to the speed and increase with the field current. Such a motor can operate without consuming reactive power at the fundamental frequency and may even generate such power if over-excited (emf fundamental component greater than that of the supply voltage). It then behaves like a capacitive load and can be supplied by an inverter operating in natural commutation.

• *Current-Source Inverter Drive*

High- and very-high power equipment supplied by a rectifier uses a current-source inverter; the synchronous motor is over-excited. This drive has two main advantages: functional reversibility is inherent in the configuration and the inverter operates in natural commutation.

The firing of the inverter thyristors must be controlled by the position of the rotor, so that the torque due to stator–rotor interaction should always have the required sign. This drive is referred as a "*self-controlled*" synchronous motor or, since the inverter operates in natural commutation, as a "*self-commutated*" synchronous motor or, since the inverter-machine assembly is the equivalent of a DC-current motor, as a "*commutatorless DC motor*".

In fact the current-source inverter which supplies an over-excited synchronous motor should not form part of a study of the autonomous inverters examined in the present work. It is a rectifier acting as a non-autonomous load-commutated inverter (cf. Vol. 1, Chap. 7). It is the machine which ensures that commutations are carried out, thanks to its emf.

When energy flows from the mains towards the driven load, the rectifier operates effectively as a rectifier and the inverter as a non-autonomous inverter. When the energy transfer direction is reversed, the inverter acts as a rectifier and the rectifier as a non-autonomous inverter. In both cases, the natural commutations of the rectifier are ensured by the mains and those of the inverter by the machine.

At low speed, the machine emf's are insufficient to ensure the commutations. Various procedures enable the current delivered to the inverter to be turned off between each operating phase of the latter.

In order to reduce the torque pulsations, the same procedures as those indicated for the induction motor can be used: use of two three-phase stator windings, shifted by $\pi/6$ and supplied by two series-connected inverters which generate current square-waves shifted by $\pi/6$; increase in the number of motor and inverter phases.

- *Voltage-Source Inverter Drive*

Some low-power equipments with synchronous motor are fed by a voltage-source inverter. In such cases, the motor is often made with a permanent magnet rotor (*brushless DC motor*) and the inverter can operate in forced commutation at all speeds.

In most cases the motor is still self-controlled; it is not usually self-commutated.

This type of drive is used for equipment where a single inverter supplies several motors which have to run at the same speed. This explains the use of synchronous motors or synchronised asynchronous motors.

- *Remark.* A synchronous machine has a high inductance at the supply fundamental frequency (direct-axis and quadrature-axis synchronous reactances). However, especially in the case of a current-source inverter fed motor, the inductances involved in the commutation processes have a low value; since the rotor is then equipped with damper windings, they correspond approximately to the subtransient inductances. The commutation difficulty is similar to that of an induction motor.

1.3.3 Load-Resonant Inverter Applications

It has been noted (Sect. 1.1.3) that load-resonant inverters are voltage-source inverters (series-loaded resonant inverters) or current-source inverters (parallel-loaded resonant inverters), operating at a relatively high frequency and supplying a slightly damped oscillating load close to its resonance frequency.

1.3.3.1 "Medium Frequency" Inverter Applications

The traditional use of load-resonant inverters corresponds to the "medium frequency" supply of inductive loads (e.g. induction heating) or capacitive loads (e.g. ozone producing). To the load, there need only to be added a reactive element (capacitor in the first case and inductor in the second) in order to obtain an oscillating circuit. The inverter supplies it with the energy required to start and then maintain oscillations.

When operating at high power, the inverter uses thyristors. The operating frequency is chosen in relation to the oscillating circuit frequency in such a way that the latter behaves like a capacitive load: the thyristors then operate in natural commutation.

When operating at low or medium power with transistor controlled as dual thyristors, the operating frequency can be placed on the other side of the resonance frequency, so that only the turn-off rather than turn-on are controlled.

1.3.3.2 Intermediate Converter Applications

In converters supplied by the industrial network, a transformer is normally needed to ensure galvanic insulation between the input and the output and to fit the voltages. If supplied at the frequency of the industrial network, this transformer forms the heaviest and often one of the most expensive part of the equipment.

For a given conversion, this conventional configuration is thus often replaced by a more complicated configuration with an intermediate stage operating at a much higher frequency than the input and output frequency. The transformer which works at this intermediate stage frequency is much lighter.

It has been seen (Vol. 3, Chap. 6) that switch-mode power supplies generate a DC voltage from the industrial network by means of a rectifier–chopper–transformer–rectifier cascade. *Resonant supplies* are often used instead of switch-mode power supplies as soon as the power in question rises above a minimal value: the resonant supplies use the rectifier-medium frequency inverter–transformer–rectifier cascade. Seen via a diode rectifier and a transformer, the DC-supplied load appears virtually as a resistance. To obtain an oscillating circuit at the inverter output, an inductor and a capacitor must be added.

Chapter 6 deals with *load-resonant inverters.* The results of the study of oscillating circuits supplied by voltage or current square-waves are directly applied to medium-frequency inverters in single- or three-phase configuration and used directly or in an energy conversion cascade.

Bibliography

All general works on power electronics contain one or more chapters on inverters. A number of these are indicated in the bibliography at the end of Chap. 1 of first volume of this series.

There are few works entirely devoted to *inverters*.

The book by B.D. Bedford and R.G. Hoft [1] is still frequently referred to, despite its age. The work by F. Brichant [2] is also of interest.

Resonant inverters and high-frequency link converters are more specifically dealt with in Y. Cheron [3].

However, many works can be found on *adjustable-speed AC drives* [4–12] with asynchronous or synchronous motor, the latter having a wound rotor or permanent magnets. All these books begin with lengthy remainders of the operation of inverters.

In the bibliography at the end of Chap. 3, there are several general articles on adjustable-speed AC drives. The majority of these contain a large number of references.

Inverters

1. Bedford BD, Hoft RC (1964) *Principles of Inverter Circuits*, Wiley, New York.
2. Brichant F (1982) *Les Onduleurs Autonomes*, Dunod, Paris.
3. Cheron Y (1989) *La Commutation Douce dans la Conversion Statique de l'Energie Electrique* Technique et Documentation, Paris.

Adjustable – Speed AC Drives

4. Murphy JMD (1973) *Thyristor Control of AC Motors*, Pergamon, New York.
5. Chauprade R, Milsant F (1980) *Commande Electronique des Moteurs à Courant Alternatif*, Eyrolles, Paris.
6. Bose BK (1981) *Adjustable Speed AC Drives Systems*, Wiley, New York.
7. Kenjo T, Nagamori S (1985) *Permanent Magnet and Brushless DC Motors*, Clarendon Press, Oxford.
8. Dewan SB, Slemon GR, Straughen A (1985) *Power Semiconductor Drives*, Wiley, New York.
9. Leonhard W (1985) *Control of Electrical Drives*, Springer, Berlin.
10. Bose BK (1986) *Power Electronics and AC Drives*, Prentice-Hall, Englewood Cliffs, NJ.
11. Murphy JMD, Turnbull FG (1988) *Power Electronic Control of AC Motors*, Pergamon, New York.
12. Dubey GK (1989) *Power Semiconductor Controlled Drives*, Prentice-Hall, Englewood Cliffs, NJ.

Chapter 2

Voltage-Source Inverters with One Square Pulse per Half-Cycle

- In order to obtain an AC voltage from a DC voltage and if only two "switches" are used, a mid-point is necessary, on either the AC output or DC input side.

  This corresponds to:

  - *the single-phase inverter with centre-tapped primary output transformer* (often called push–pull inverter),
  - *the single–phase inverter with a capacitive divider at the input* (often called half-bridge inverter).

    If the relative width of the square-wave pulses forming the output voltage half-cycles has to be varied, four "switches" are necessary. This is

  - *the single-phase bridge* (or full bridge) *inverter.*

    After studying these three types of single-phase inverters, we will examine the three-phase inverter with six "switches", known as

  - *the three-phase bridge inverter.*

    Its behaviour when supplying an unbalanced three-phase load will show the cases in which it must be replaced by three separate single-phase inverters.

    This chapter will end with a study of

  - *the input current filtering*
  - *the output voltage filtering.*

- As in the case of all the converters studied in previous volumes, we will distinguish the various difficulties:

  - For the basic study of each of the four voltage-source inverters, it will be assumed that the DC voltage supply is perfect, i.e. supplying a constant DC voltage, and that the AC load is perfect, i.e. absorbing a sinusoidal current.
  - For the input current filtering study, it will be assumed that the output current is sinusoidal.
  - For the output voltage filtering study, it will once again be assumed that the input voltage is constant.

2.1 Single-Phase Inverter with Centre-Tapped Transformer

Figure 2.1 provides the configuration of the single-phase voltage-source inverter with two "switches" and a centre-tapped primary transformer; it shows the notations used. The DC voltage supply has a terminal connected to the mid-point O. Its other terminal is connected either to A by the conduction of TC_1 or D_1, or to B by the conduction of TC'_1 or D'_1.

2.1.1 Principle

If the transformer is assumed to be perfect, i.e. if the resistances and leakage inductances of its windings are ignored, the voltages are in relation to the number of turns:

$$v_1 = v'_1, \qquad u' = v_1 \frac{2n_2}{n_1}. \tag{2.1}$$

If the magnetizing m.m.f. is ignored, the primary and secondary currents are linked by the compensation in m.m.f.:

$$\frac{n_1}{2}(i_{K_1} - i_{K'_1}) = n_2 i'. \tag{2.2}$$

2.1.1.1 Expressions. Waveforms

- For $0 < t < T/2$, K_1 is on, K'_1 is off:

$$v_1 = +U, \qquad u' = +\frac{2n_2}{n_1}U$$

$$v'_1 = +U$$

$$v_{K_1} = 0, \qquad v_{K'_1} = v_1 + v'_1 = +2U$$

$$i_{K_1} = i = \frac{2n_2}{n_1}i', \qquad i_{K'_1} = 0.$$

i_{K_1} goes via TC_1 if positive and via D_1 if negative.

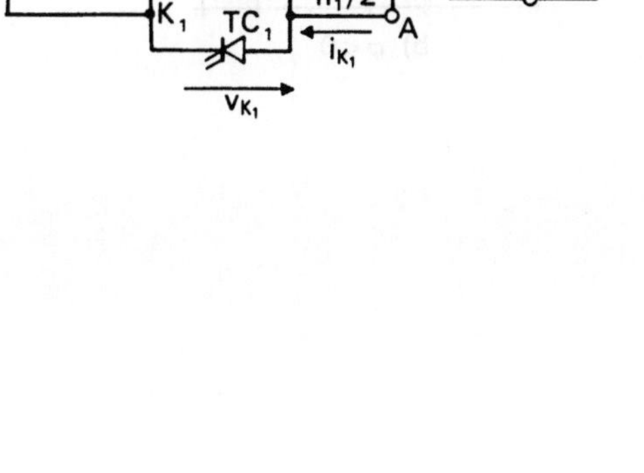

Fig. 2.1

– For $T/2 < t < T$, K'_1 is on and K_1 off:

$$v'_1 = -U, \qquad u' = -\frac{2n_2}{n_1}U$$

$$v_1 = -U$$

$$v_{K_1} = -v_1 - v'_1 = +2U, \qquad v_{K'_1} = 0$$

$$i_{K_1} = 0, \qquad i_{K'_1} = i = -\frac{2n_2}{n_1}i'.$$

$i_{K'_1}$ goes via TC'_1 if positive and via D'_1 if negative.

The phase shift φ between the output current, which is assumed to be sinusoidal, and the fundamental component of the output voltage is taken as positive in the case of a lagging current. A leading current thus corresponds to a negative value of φ.

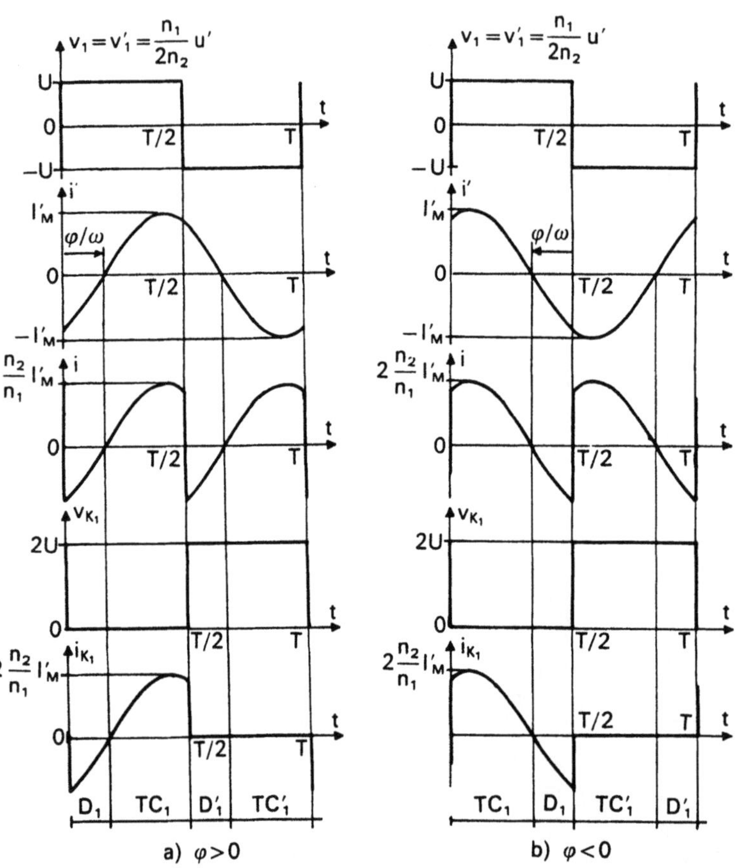

Fig. 2.2

Figure 2.2 shows the waveforms of v_1 and u', of i', of i, of v_{K_1} and i_{K_1} as well as the conduction diagram of the semiconductor devices.

(a) for the case of a lagging current ($\varphi > 0$),
(b) for the case of a leading current ($\varphi < 0$).

This gives exactly the same waveforms as those found in Fig. 1.10.

2.1.1.2 Commutations Required

- In the case of an inductive load, the commutations are forced. For $t = T/2$, when TC_1 is on, its conduction must be interrupted. This results in a voltage equal to $+2U$ across it; the current $(2n_2/n_1)i'$ is transferred from TC_1 to D'_1.
- In the case of a capacitive load, the commutations are natural. For $t = 0$, when D'_1 is on, TC_1 is fired. This results in a voltage equal to $-2U$ across diode D'_1 which turns off. Current $(2n_2/n_1)i'$ is transferred from D'_1 to TC_1.
- In both cases, during each half-cycle, there is a free natural commutation between the semiconductor devices of a same switch – from D_1 to TC_1 or vice-versa, or from D'_1 to TC'_1 or vice-versa. This occurs when i' reverts its polarity.

2.1.1.3 Doubling the Voltage

The presence of a centre-tapped transformer means that the reverse voltage across the diodes and the forward voltage across the controlled devices are equal to *twice* the DC supply voltage:

$$-v_{\mathrm{D\,max}} = v_{\mathrm{T\,max}} = 2U. \tag{2.3}$$

2.1.1.4 The Magnetizing Current

Flux ϕ in the magnetic circuit of the transformer is linked to voltage v_1 across the $n_1/2$ turns by

$$\frac{n_1}{2}\frac{\mathrm{d}\phi}{\mathrm{d}t} = v_1 .$$

For $0 < t < T/2$, $v_1 = +U$ gives $\phi = \dfrac{U}{n_1/2}\left(t - \dfrac{T}{4}\right)$.

For $T/2 < t < T$, $v_1 = -U$ gives $\phi = \dfrac{U}{n_1/2}\left(-t + \dfrac{3T}{4}\right)$.

The flux has a triangular waveform represented at the top of Fig. 2.3. If the material of the magnetic circuit should have a constant permeability, the

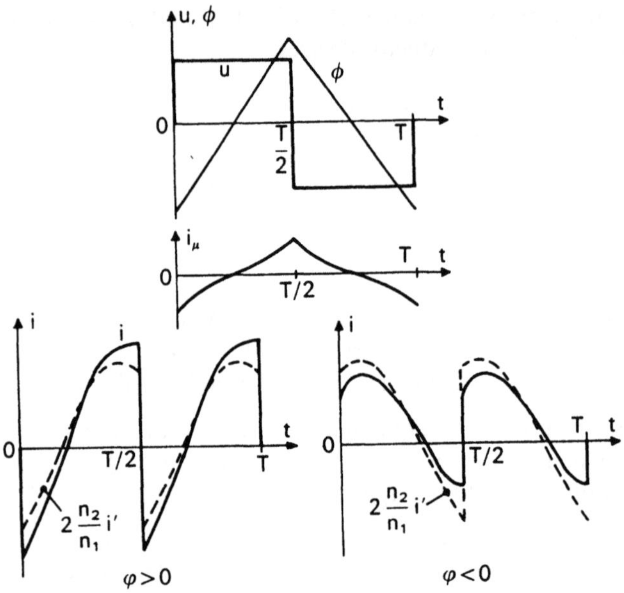

Fig. 2.3

magnetizing current i_μ would also be triangular. On account of the saturation, it has the waveform effectively shown.

$$\text{For}\quad 0 < t < \frac{T}{2},\quad i = i_{K_1} = \frac{2n_2}{n_1} i' + i_\mu$$

$$\text{For}\quad \frac{T}{2} < t < T,\quad i = i_{K_1'} = -\frac{2n_2}{n_1} i' - i_\mu\,.$$

In the case of an inductive load, i_μ increases the rms value of i, of i_{K_1} and the value of the current to be commuted. In the case of a capacitive load, i_μ reduces these values.

2.1.2 Characteristics

In characterizing the input and output values as well as those which determine the converter components, the semiconductor devices and the transformer are assumed to be perfect and the magnetizing current of the latter is ignored.

2.1.2.1 Output Values

- *Output voltage* u', equal to $+(2n_2/n_1)U$ during one half-cycle and to $-(2n_2/n_1)U$ during the other, has a rms value of

$$U' = \frac{2n_2}{n_1} U. \tag{2.4}$$

The rms value of its fundamental component is

$$U'_1 = \frac{2\sqrt{2}}{\pi} 2 \frac{n_2}{n_1} U. \tag{2.5}$$

Its series expansion

$$u' = \frac{8}{\pi} \frac{n_2}{n_1} U\left(\sin \omega t + \frac{1}{3} \sin 3\omega t + \frac{1}{5} \sin 5\omega t + \frac{1}{7} \sin 7\omega t + \ldots\right)$$

indicates the presence of all the uneven harmonics. The value of each one is inversely proportional to its frequency:

$$\frac{U'_{2k+1}}{U'_1} = \frac{1}{2k+1}. \tag{2.6}$$

The harmonic ratio, given by either of the equations

$$\tau_{u'} = \frac{1}{U'_1}\left(\sum_{k=1}^{\infty} U'^2_{2k+1}\right)^{1/2} = \frac{1}{U'_1}\sqrt{U'^2 - U'^2_1}$$

is equal to

$$\tau_{u'} = 0.483. \tag{2.7}$$

- If the output *current* with a peak value I'_M or a rms value I', is sinusoidal

$$i' = I'_M \sin(\omega t - \varphi) = I' \sqrt{2} \sin(\omega t - \varphi)$$

the series expansion of the output instantaneous power $u'i'$ comprises terms with angular frequency $2\omega, 4\omega, 6\omega, \ldots$ in addition to the constant term $U'_1 I' \cos \varphi$. The mean value of input current i corresponds to the latter and its harmonics to the former.

2.1.2.2 Characteristics Related to the Converter

- *The transformer*, whose secondary voltage and current are u' and i' respectively, has a voltage equal to $(n_1/2n_2)u'$ across each of its primary half-windings. The currents in the latter have the following rms value:

$$\frac{1}{\sqrt{2}} \frac{2n_2 I'}{n_1}.$$

The power rating for the primary is $\sqrt{2}$ times the power rating of the secondary.

- For example, the current in the *controlled semiconductor* device TC_1, has the following expressions when the load is inductive:

$$i_{T_1} = \frac{2n_2}{n_1} I' \sqrt{2} \sin(\omega t - \varphi) \quad \text{for } \varphi < \omega t < \pi$$

$$i_{T_1} = 0 \quad \text{for } \pi < \omega t < 2\pi + \varphi.$$

From these equations, the following can be deduced:

– its maximum value:

$$i_{\mathrm{T\,max}} = \frac{2n_2}{n_1} I' \sqrt{2} \qquad \text{if } 0 < \varphi < \frac{\pi}{2}$$

$$i_{\mathrm{T\,max}} = \frac{2n_2}{n_1} I' \sqrt{2} \sin \varphi \quad \text{if } \frac{\pi}{2} < \varphi < \pi \tag{2.8}$$

– its rms value:

$$I_{\mathrm{T\,rms}} = \frac{2n_2}{n_1} I' \sqrt{2} \sqrt{\frac{\pi - \varphi}{4\pi} + \frac{\sin 2\varphi}{8\pi}} \tag{2.9}$$

– its mean value:

$$I_{\mathrm{T\,mean}} = \frac{2n_2}{n_1} I' \sqrt{2} \frac{1 + \cos \varphi}{2\pi} \tag{2.10}$$

and the current to be commuted:

$$i_\pi = \frac{2n_2}{n_1} I' \sqrt{2} \sin \varphi. \tag{2.11}$$

- The current in *diode* D_1, for example, has the following expressions, when the load is inductive:

$$i_{D_1} = -\frac{2n_2}{n_1} I' \sqrt{2} \sin(\omega t - \varphi) \quad \text{for } 0 < \omega t < \varphi$$

$$i_{D_1} = 0 \qquad \text{for } \varphi < \omega t < 2\pi.$$

Hence,

– its maximum value:

$$i_{\mathrm{D\,max}} = \frac{2n_2}{n_1} I' \sqrt{2} \sin \varphi \quad \text{if } 0 < \varphi < \frac{\pi}{2}$$

$$i_{\mathrm{D\,max}} = \frac{2n_2}{n_1} I' \sqrt{2} \qquad \text{if } \frac{\pi}{2} < \varphi < \pi \tag{2.12}$$

– its rms value:

$$I_{\mathrm{D\,rms}} = \frac{2n_2}{n_1} I' \sqrt{2} \sqrt{\frac{\varphi}{4\pi} - \frac{\sin 2\varphi}{8\pi}} \tag{2.13}$$

– its mean value:

$$I_{\mathrm{D\,mean}} = \frac{2n_2}{n_1} I' \sqrt{2} \frac{1 - \cos \varphi}{2\pi}. \tag{2.14}$$

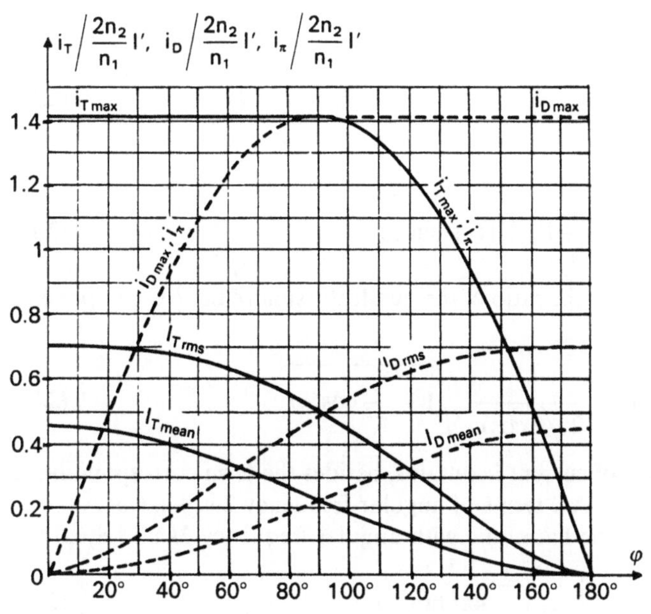

Fig. 2.4

- In Fig. 2.4, the variations of $i_{T\max}$, $I_{T\mathrm{rms}}$, $I_{T\mathrm{mean}}$, i_π, of $i_{D\max}$, $I_{D\mathrm{rms}}$, $I_{D\mathrm{mean}}$ have been plotted as functions of φ. All these values are referred to $(2n_2/n_1)I'$.

  These curves can be used if the supply of a capacitive load is analyzed. The only difference is that i_π no longer indicates the current which has to be turned off but that which has to be turned on.

For φ greater than $\pi/2$, the various values of the current in the diodes are greater than the corresponding values in the controlled devices; this is due to the fact that the energy recovery periods ($i < 0$) are then longer than the energy supply periods ($i > 0$).

2.1.2.3 Input Values

Input current i has a cycle equal to the half that of the output variables.

When the load is inductive, it can be expressed as

$$\text{for} \quad 0 < \omega t < \pi, \quad i = \frac{2n_2}{n_1} I' \sqrt{2} \sin(\omega t - \varphi).$$

– Its maximum instantaneous value is equal to $i_{T\max}$ and its minimal instantaneous value to $-i_{D\max}$.

- its mean value is

$$I = \frac{2n_2}{n_1} I' \sqrt{2} \frac{2}{\pi} \cos\varphi \tag{2.15}$$

• and its rms value

$$I_{rms} = \frac{2n_2}{n_1} I' . \tag{2.16}$$

This gives its ripple

$$\Delta i = i_{max} - i_{min} = \frac{2n_2}{n_1} I' \sqrt{2}(1 + \sin\varphi) \tag{2.17}$$

and its harmonic ratio – the ratio of the quadratic sum of the AC components to the mean value

$$\tau_i = \frac{1}{I}\sqrt{I_{rms}^2 - I^2} = \frac{\pi}{2\sqrt{2}\cos\varphi}\sqrt{1 - \frac{8}{\pi^2}\cos^2\varphi} . \tag{2.18}$$

– The series expansion of current i contains, besides the term corresponding to its average value I, the AC terms of angular frequency $2\omega, 4\omega, 6\omega, \ldots$

The rms value of the harmonic with angular frequency $2k\omega$ is equal to

$$I_{2k} = \frac{2n_2}{n_1} I' \frac{4|\cos\varphi|}{\pi(4k^2 - 1)} \sqrt{1 + 4k^2 \tan^2\varphi} . \tag{2.19}$$

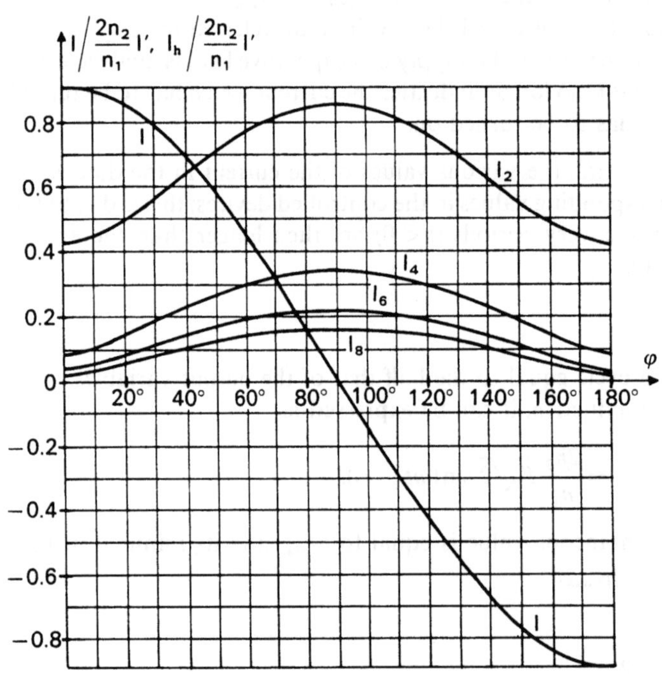

Fig. 2.5

It is minimum for φ equal to 0 or π, and maximum for φ equal to $\pi/2$:

$$I_{2k\,\min} = \frac{2n_2}{n_1} I' \frac{4}{\pi(4k^2 - 1)}, \qquad I_{2k\,\max} = \frac{2n_2}{n_1} I' \frac{8k}{\pi(4k^2 - 1)}.$$

Since all losses in the converter have been ignored, the various terms of the series expansion of i correspond to those of the series expansion of the instantaneous power at the output:

$$i = \frac{1}{U} u' i'.$$

This shows, in particular, that I is equal to $(U'_1 I' \cos\varphi)/U$.

- Figure 2.5 shows the variations – as a function of φ – of the mean value I of the input current and of the rms values I_h of its first harmonic components. All the values are referred to $(2n_2/n_1)I'$.

  For φ equal to $\pi/2$, the mean value is reversed, while all harmonics reach their maximum.

  These characteristics can be used when supplying a capacitive load, since neither I nor I_h are affected by the sign of φ.

2.2 Single-Phase Half-Bridge Inverter

The single-phase half-bridge inverter uses two "switches" which are bidirectional in current and a centre-tapped voltage supply.

The mid-point can usually be obtained by using two capacitors C_1 and C'_1 with the same capacitance C. If the latter is sufficient, two voltages which are virtually constant and equal to $U/2$ are obtained.

Figure 2.6 shows the configuration and the notations used.

2.2.1 Principle

As the sum of the voltages $u_{C_1} + u_{C'_1}$ across the two capacitors is equal to U and as this voltage is assumed to be constant,

$$u_{C_1} + u_{C'_1} = U$$

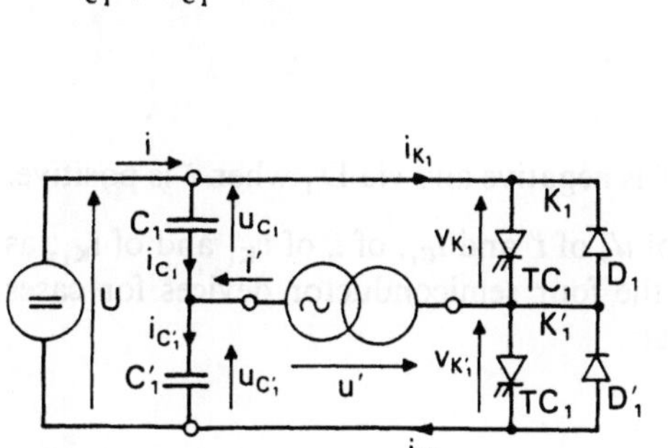

Fig. 2.6

gives

$$C\frac{du_{C_1}}{dt} = -C\frac{du_{C'_1}}{dt}$$

$$i_{C_1} = -i_{C'_1}.$$

The charging (or discharging) current of C_1 is equal to the discharging (or charging) current of C'_1.

Since $i' = i_{C'_1} - i_{C_1}$

$$i' = 2i_{C'_1} = -2i_{C_1}. \tag{2.20}$$

Current i', which reaches the midpoint of the capacitive divider has *twice* the value of the current in one of the capacitors.

2.2.1.1 Expressions. Waveforms

- For $0 < t < T/2$, K_1 is on and K'_1 off:

$$u' = u_{C_1} \simeq +\frac{U}{2}$$

$$v_{K'_1} = +U$$

$$i = i_{C'_1} = \frac{i'}{2}$$

$$i_{K_1} = i' = 2i.$$

Current i_{K_1} goes via TC_1 when i' is positive and via D_1 when i' is negative.

- For $T/2 < t < T$, K_1 is off and K'_1 on:

$$u' = -u_{C'_1} \simeq -\frac{U}{2}$$

$$v_{K_1} = +U$$

$$i = i_{C_1} = -\frac{i'}{2}$$

$$i_{K'_1} = -i' = 2i.$$

Current $i_{K'_1}$ goes via TC'_1 when i' is negative and via D'_1 when i' is positive.

Figure 2.7 shows the waveforms of u', of i' and i_{C_1}, of i, of v_{K_1} and of i_{K_1}, as well as the conduction diagrams of the four semiconductor devices for cases where the output current is sinusoidal:

(a) for a lagging current ($\varphi > 0$),
(b) for a leading current ($\varphi < 0$).

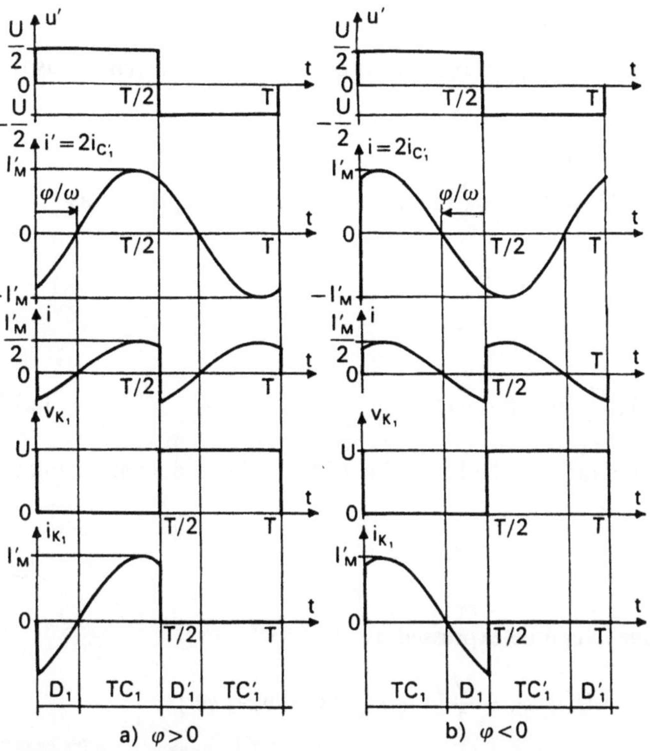

Fig. 2.7

The same waveforms as in Fig. 1.10 or 2.2 are to be found; only some amplitudes are modified.

2.2.1.2 Commutations Required

The commutations required are the same as those for the push–pull inverter:

- forced commutations TC'_1–D_1 and TC_1–D'_1 when supplying an inductive load
- natural commutations D'_1–TC_1 and D_1–TC'_1 when supplying a capacitive load.

2.2.1.3 Doubling the Current

Owing to the capacitive divider, the current in the on-switch is *twice* the current in the DC supply.

$$i_{K_1} \quad \text{or} \quad i_{K'_1} = 2i. \tag{2.21}$$

2.2.1.4 Rating the Input Capacitances

- The current in capacitor C'_1, for example, is equal to $i'/2$. If current i' has the form

$$i' = I'\sqrt{2}\sin(\omega t - \varphi)$$

since $C\,du_{C_1}/dt = i'/2$, voltage $u_{C'_1}$ in steady state can be expressed as

$$u_{C'_1} = \frac{U}{2} - \frac{I'\sqrt{2}}{2C\omega}\cos(\omega t - \varphi)$$

or

$$u_{C'_1} = \frac{U}{2}[1 - k\sqrt{2}\cos(\omega t - \varphi)] \tag{2.22}$$

by denoting as k the ratio of the rms value of the AC component of voltage $u_{C'_1}$ to $U/2$:

$$k = \frac{I'}{C\omega U}. \tag{2.23}$$

- The output voltage u' can be expressed as

$$0 < t < \frac{T}{2}, \quad u' = U - u_{C'_1} = \frac{U}{2}[1 + k\sqrt{2}\cos(\omega t - \varphi)]$$

$$\frac{T}{2} < t < T, \quad u' = -u_{C'_1} = \frac{U}{2}[-1 + k\sqrt{2}\cos(\omega t - \varphi)].$$

The ripple of $u_{C'_1}$ affects the waveform of the output voltage. Figure 2.8 shows the modified shape of voltage u':

(a) for an inductive load,
(b) for a capacitive load.

- This modification of the waveform of u' affects the different characteristics of this voltage, particularly the value of its Fourier series fundamental.
  The rms value of the fundamental, equal to

$$U'_1 = \frac{\sqrt{2}}{\pi}U$$

for k equal to zero, becomes

$$U'_1 = \frac{\sqrt{2}}{\pi}U\sqrt{1 + \frac{\pi}{2\sqrt{2}}k\sin\varphi + \frac{\pi^2}{8}k^2}. \tag{2.24}$$

Figure 2.9 shows, for some values of k, the variations of $U'_1/(\sqrt{2}U/\pi)$ as a function of φ.

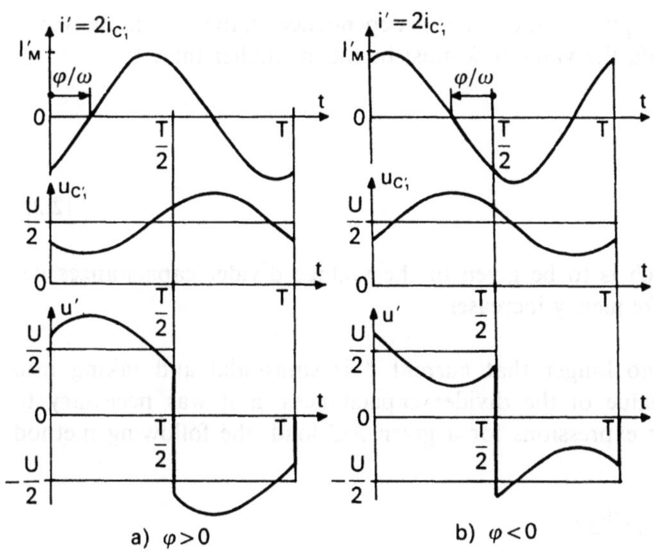
i′ = 2i_C′1
I′_M
φ/ω
t
0
T/2
T
u_C′1
U/2
u′
−U/2
a) φ>0
b) φ<0

Fig. 2.8

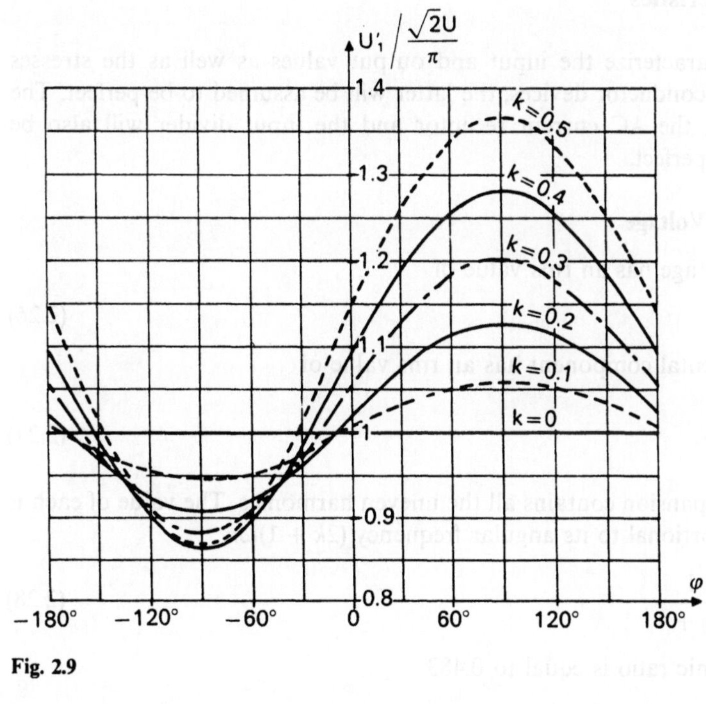
U′_1 / (√2U/π)
1.4
1.3
1.2
1.1
1
0.9
0.8
k=0.5
k=0.4
k=0.3
k=0.2
k=0.1
k=0
φ
−180°
−120°
−60°
0
60°
120°
180°

Fig. 2.9

- In order that the output voltage has little dependence on the current delivered or on its phase shift, the value of k must be much smaller than 1:

$$\frac{I'}{C\omega U} \ll 1$$

$$C \gg \frac{I'}{\omega U} \tag{2.25}$$

The value which is to be given to the voltage divider capacitances decreases when the frequency increases.

Remark. Assuming no longer that current i' is sinusoidal and taking into account the finite value of the divider capacitances, if it was necessary to calculate the current expressions for a given AC load, the following method would be used:

$$\text{since } i' = 2i_{C_1'} = 2C\frac{du_{C_1'}}{dt}$$

the voltage made up of rectangular square-wave pulses ($\pm U/2$) may be considered to feed a circuit formed by the load itself in series with a capacitor of capacitance $2C$.

2.2.2 Characteristics

In order to characterize the input and output values as well as the stresses applied to semiconductor devices, the latter will be assumed to be perfect. The voltage supply, the AC current receptor and the input divider will also be assumed to be perfect.

2.2.2.1 Output Voltage

The output voltage has an rms value of

$$U' = U/2. \tag{2.26}$$

Its fundamental component has an rms value of

$$U'_1 = \frac{\sqrt{2}}{\pi}U. \tag{2.27}$$

Its series expansion contains all the uneven harmonics. The value of each is inversely proportional to its angular frequency $(2k+1)\omega$:

$$\frac{U'_{2k+1}}{U'_1} = \frac{1}{2k+1}. \tag{2.28}$$

The harmonic ratio is equal to 0.483.

2.2.2.2 Semiconductor Devices

All the equations [(2.8)–(2.14)] laid down for the inverter with centre-tapped transformer can be used, provided that $2n_2/n_1$ is replaced by 1.

This same operation enables the characteristics of Fig. 2.4 to be used.

The maximum forward voltage across TC_1 and TC'_1 and the maximum reverse voltage across D_1 and D'_1 are now equal to the DC supply voltage:

$$-v_{D\,max} = v_{T\,max} = U. \tag{2.29}$$

2.2.2.3 Input Current

The input current is equal to $i'/2$ during one half-cycle and to $-i'/2$ during the other.

All the equations laid down in paragraph 2.1.2.3 can be used provided that $(2n_2/n_1)I'$ is replaced by $I'/2$.

The same operation enables the graphs in Fig. 2.5 to be used.

2.3 Single-Phase Full-Bridge Inverter

This is the full-bridge structure that was used in the previous chapter to present the various inverters. The configuration of the voltage-source inverter is drawn again on Fig. 2.10.

In the previous chapter, it was assumed that switches K_1 and K'_2 were on during the first half-cycle of voltage u', making the latter equal to $+U$; switches K_2 and K'_1 were on during the other half-cycle, making u' equal to $-U$.

This control mode uses only a part of the possibilities of the full-bridge voltage-source inverter, since the latter enables not only the voltage frequency but also its value to be varied.

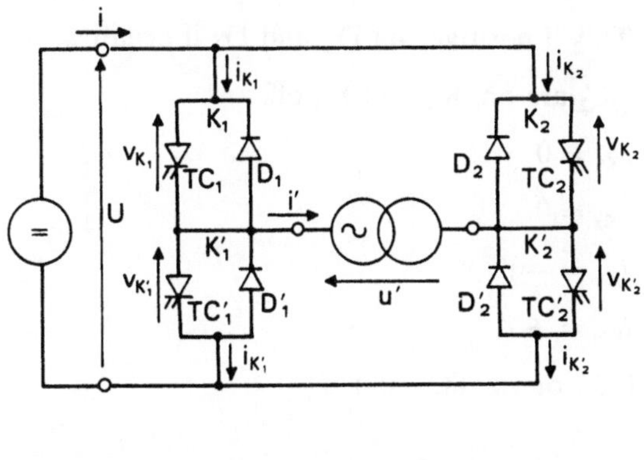

Fig. 2.10

2.3.1 Principle

The control signals of K_1 and K'_1 must be complementary in order to avoid short-circuiting voltage U supply and opening current i' circuit. Similarly, for the other half-bridge, the controls of K_2 and K'_2 must be complementary. However the controls of both half-bridges need not be simultaneous; there may be a phase-shift between the two.

If ω is once again used to denote the angular frequency of voltage u' fundamental component, angle β can be used to characterize the *phase shift* between the control signals:

K_1 is on for $0 < \omega t < \pi$; K'_2 for $\beta < \omega t < \pi + \beta$.
K'_1 is on for $\pi < \omega t < 2\pi$; K_2 for $\pi + \beta < \omega t < 2\pi + \beta$.

2.3.1.1 Expressions. Waveforms

- For $0 < \omega t < \beta$, K_1 and K_2 are on, K'_1 and K'_2 off; the current i' source is short-circuited:

$$u' = 0, \quad v_{K_1} = 0, \quad v_{K'_1} = +U$$
$$v_{K_2} = 0, \quad v_{K'_2} = +U$$
$$i = 0 \quad i_{K_1} = +i', \quad i_{K'_1} = 0$$
$$i_{K_2} = -i', \quad i_{K'_2} = 0.$$

Current i' goes via TC_1 and D_2 if positive, via TC_2 and D_1 if negative.

- For $\beta < \omega t < \pi$, K_1 and K'_2 are on, K'_1 and K_2 off; the current source is connected to the voltage source:

$$u' = +U, \quad v_{K_1} = 0, \quad v_{K'_1} = +U$$
$$v_{K_2} = +U, \quad v_{K'_2} = 0$$
$$i = +i' \quad i_{K_1} = +i', \quad i_{K'_1} = 0$$
$$i_{K_2} = 0, \quad i_{K'_2} = +i'.$$

Current i' goes via TC_1 and TC'_2 if positive, via D'_2 and D_1 if negative.

- For $\pi < \omega t < \pi + \beta$, K'_1 and K'_2 are on, K_1 and K_2 off:

$$u' = 0, \quad v_{K_1} = +U, \quad v_{K'_1} = 0$$
$$v_{K_2} = +U, \quad v_{K'_2} = 0$$
$$i = 0 \quad i_{K_1} = 0, \quad i_{K'_1} = -i'$$
$$i_{K_2} = 0, \quad i_{K'_2} = +i'.$$

Current i' flows via TC'_2 and D'_1 or via TC'_1 and D'_2.

- For $\pi + \beta < \omega t < 2\pi$, K'_1 and K_2 are on, K_1 and K'_2 off:

$$u' = -U, \quad v_{K_1} = +U, \quad v_{K'_1} = 0$$

$$v_{K_2} = 0, \quad v_{K'_2} = +U$$

$$i = -i' \quad i_{K_1} = 0, \quad i_{K'_1} = -i'$$

$$i_{K_2} = -i', \quad i_{K'_2} = 0.$$

Current i' goes via D'_1 and D_2 and D_2 or via TC_2 and TC'_1.

- Figure 2.11 shows the waveforms of u' and i', of i, of v_{K_1} and i_{K_1}, of v_{K_2} and i_{K_2} when the load is inductive ($0 < \varphi < 180°$):

  (a) for $\beta = 120°$, $\varphi = 30°$,
  (b) for $\beta = 120°$, $\varphi = 90°$.

  In both cases, the control periods for switch turn-on are shown at the top of the figure and the semiconductor devices conduction periods at the bottom.
  Furthermore, F and N are used to indicate the nature – forced or natural – of the commutations.

In Fig. 2.12, waveforms of the same voltages and currents have been plotted for a capacitive load ($0 < -\varphi < 180°$):

  (a) for $\beta = 120°$, $\varphi = -90°$,
  (b) for $\beta = 120°$, $\varphi = -150°$.

- Whether the load is inductive or capacitve, the waveforms of $v_{K'_1}$, $i_{K'_1}$, $v_{K'_2}$ and $i_{K'_2}$ need not to be plotted, since the following always occurs:

$$v_{K'_1}(\omega t + \pi) = v_{K_1}(\omega t), \quad i_{K'_1}(\omega t + \pi) = i_{K_1}(\omega t)$$

$$v_{K'_2}(\omega t + \pi) = v_{K_2}(\omega t), \quad i_{K'_2}(\omega t + \pi) = i_{K_2}(\omega t).$$

2.3.1.2 The Three Types of Period. The Four Zones

- Concerning the energy exchange between the DC source and the AC load, three types of period can be observed.

  - the *supply* periods during which the DC source delivers power to the load. Current i is then positive. Both controlled semiconductor devices conduct simultaneously

    TC_1 and TC'_2 or TC_2 and TC'_1

  - the *recovery* periods during which the AC load sends energy back towards the DC source. Current i is negative; two diodes are simultaneously on

    D'_2 and D_1 or D'_1 and D_2

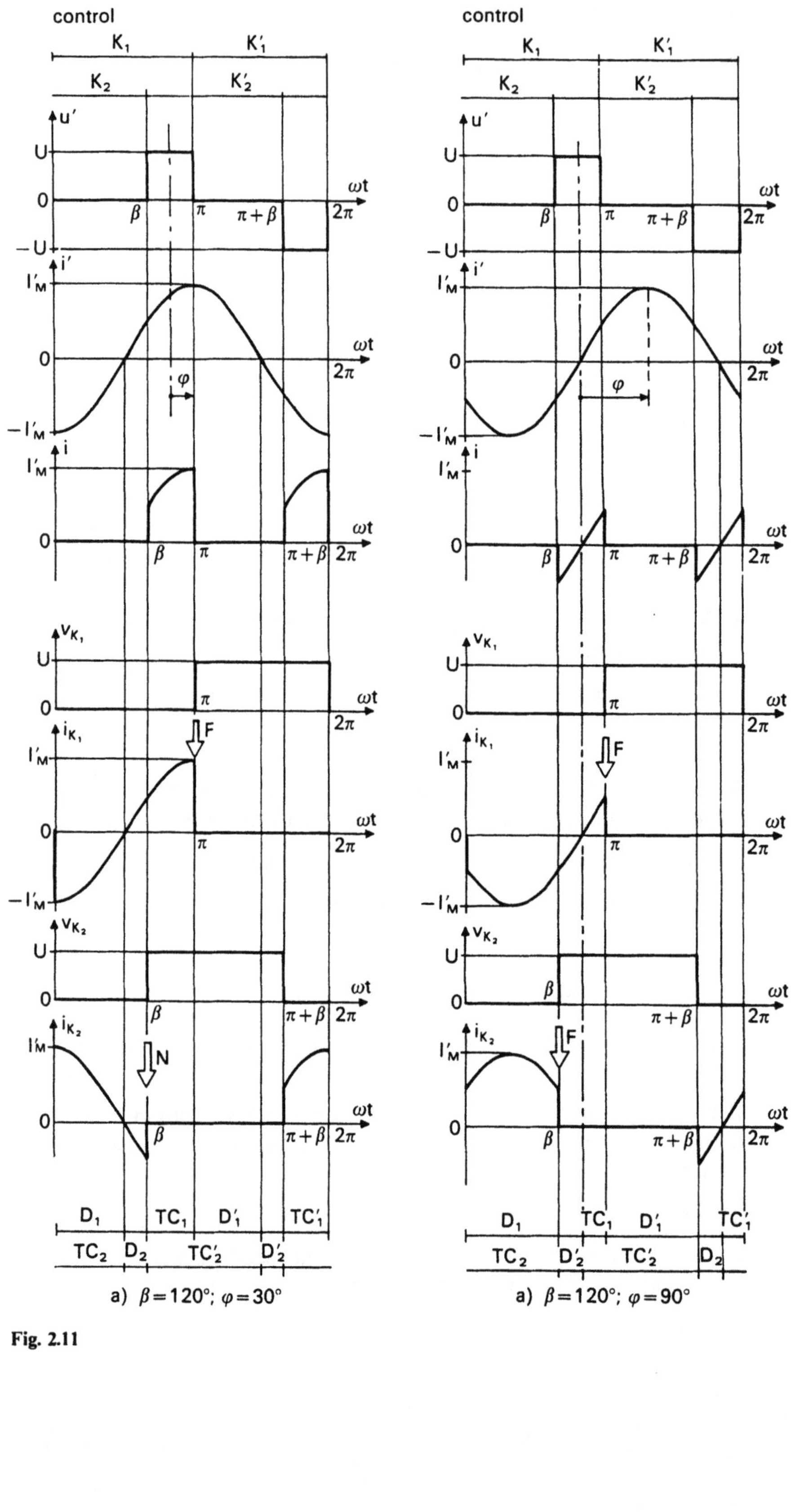
control
K₁
K'₁
K₂
K'₂
u'
U
0
−U
ωt
β
π
π+β
2π
i'
I'M
−I'M
φ
i
vK₁
iK₁
F
vK₂
iK₂
N
D₁
TC₁
D'₁
TC'₁
TC₂
D₂
TC'₂
D'₂
a) β=120°; φ=30°
a) β=120°; φ=90°

Fig. 2.11

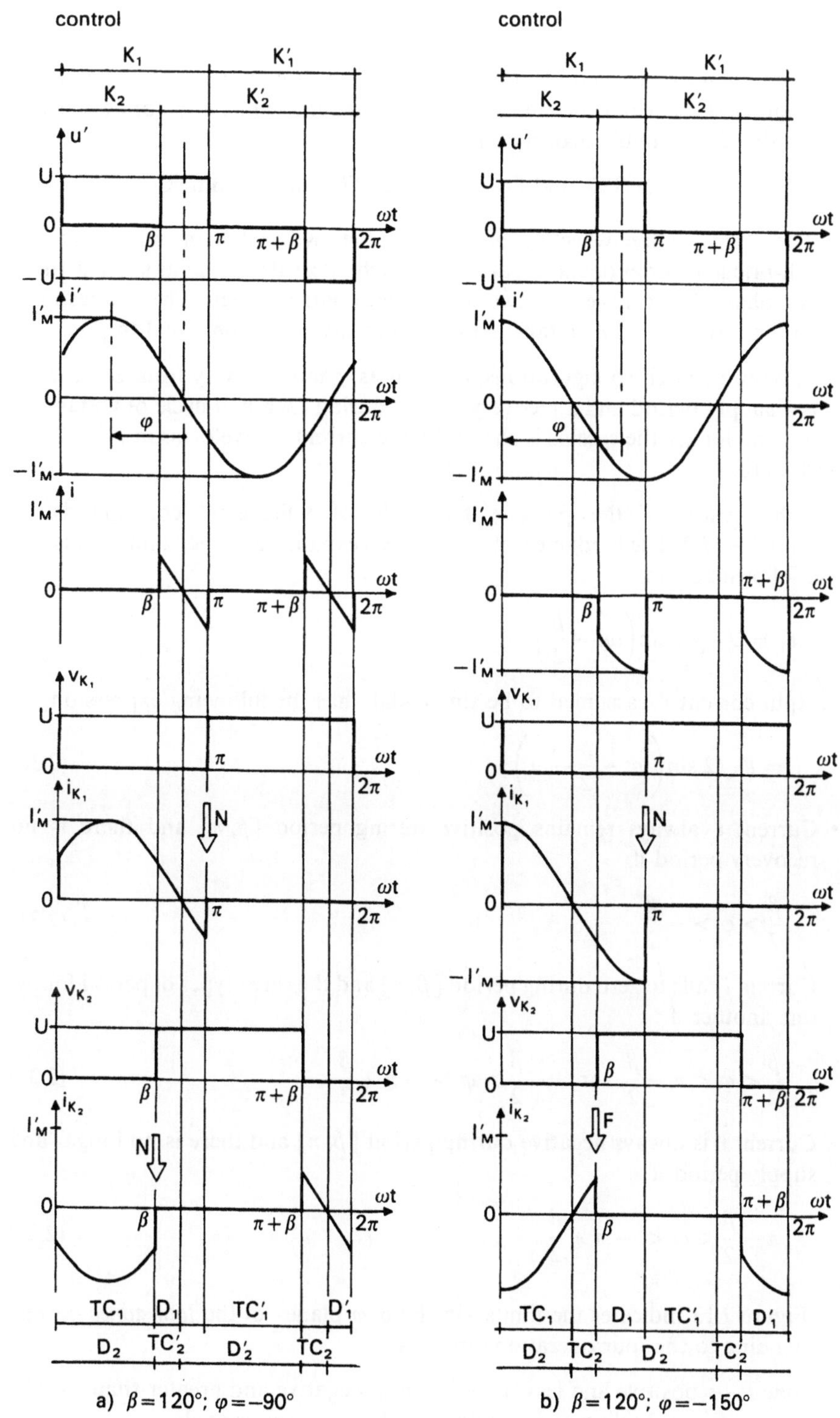

control
K1
K'1
K2
K'2
u'
U
0
−U
ωt
β
π
π+β
2π
i'
I'M
−I'M
φ
i
vK1
iK1
N
vK2
iK2
F
TC1
D1
TC'1
D'1
D2
TC'2
D'2
TC2
a) β=120°; φ=−90°
b) β=120°; φ=−150°

Fig. 2.12

- the *free-wheeling* periods: the AC load is short-circuited, the DC voltage source is open and current i is zero. A controlled semiconductor device and a diode are simultaneously on

  TC_1 and D_2 or TC_2 and D_1 or D'_1 and TC'_2 or D'_2 and TC'_1.

- There are no free-wheeling periods for the two-switch inverters and the full-bridge inverter operating at zero β. Such intervals appear as soon as there is a phase shift between the two half-bridge control signals. They correspond to the zero output u' voltage periods; they have duration equal to β/ω.

For two-switch configurations, as soon as φ moved away from zero, there was a supply period and a recovery period during each half-cycle of voltage u'. This is no longer the case as is shown by the current i waveforms in Figs. 2.11a and 2.11b.

The centre of the positive half-cycle of voltage u' corresponds to $\omega t = \pi/2 + \beta/2$. The fundamental of the series expansion of the voltage thus has the following form:

$$u'_1 = U'_1\sqrt{2}\sin\left(\omega t - \frac{\beta}{2}\right).$$

Output current i' – assumed to be sinusoidal – has the following expression:

$$i' = I'\sqrt{2}\sin\left(\omega t - \frac{\beta}{2} - \varphi\right). \tag{2.30}$$

- Current i' always remains positive during period $[\beta, \pi]$ and there is no recovery period if

$$\frac{\beta}{2} > \varphi > -\frac{\beta}{2}. \tag{2.31}$$

- Current i' falls to zero during period $[\beta, \pi]$ and the three types of period follow one another if

$$\frac{\beta}{2} < \varphi < \pi - \frac{\beta}{2} \quad \text{or} \quad -\frac{\beta}{2} > \varphi > -\pi + \frac{\beta}{2}. \tag{2.32}$$

- Current i' is always negative during period $[\beta, \pi]$ and there is no longer any supply period if

$$\pi - \frac{\beta}{2} < \varphi < -\pi + \frac{\beta}{2}. \tag{2.33}$$

- Figure 2.13 indicates the limits – in the β, φ plane – of the *four zones* corresponding to the four operational modes:

  zone 1: (φ positive and less than $\beta/2$ or φ negative and greater than $-\beta/2$) Free-wheeling and supply periods occur successively.

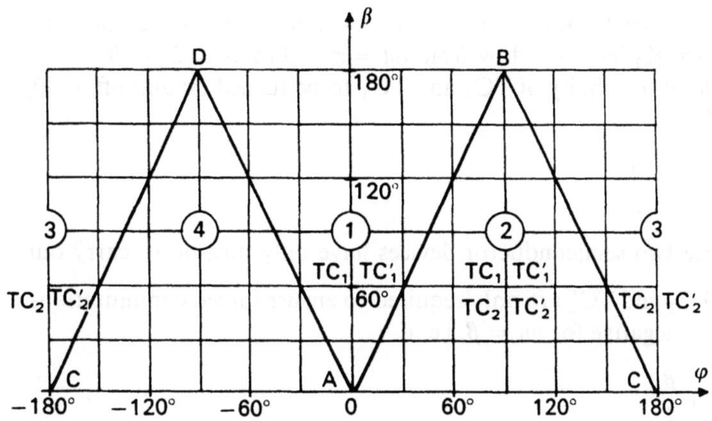

Fig. 2.13

zone 2: (φ positive taken between $\beta/2$ and $\pi - \beta/2$)
Free-wheeling, recovery and supply periods follow one another

zone 3: (φ positive taken between $\pi - \beta/2$ and π or φ negative taken between $-\pi + \beta/2$ and $-\pi$).
Free-wheeling and recovery periods occur successively.

zone 4: (φ negative taken between $-\beta/2$ and $-\pi + \beta/2$)
Free-wheeling, supply and recovery periods follow one another.

Segment AB corresponds to zero i' for $\omega t = \beta$, segment BC to zero i' for $\omega t = \pi$. Segment AD corresponds to zero i' for $\omega t = \pi$, segment DC to zero i' for $\omega t = \beta$.

Each drawing on Figs. 2.11 and 2.12 shows the operation in one of these zones:

zone 1 for Fig. 2.11a,
zone 2 for Fig. 2.11b,
zone 3 for Fig. 2.12b,
zone 4 for Fig. 2.12a.

2.3.1.3 Commutations Required

The types of commutation required no longer depend solely on the nature of the AC load (positive or negative φ) but also on angle β.

Switch K_1 lets current i flow from $\omega t = 0$ to $\omega t = \pi$; this current reverses its polarity during this period. If i' is negative for $\omega t = 0$, D_1 conducts first of all and TC_1 then begins conducting when i' becomes positive; for $\omega t = \pi$, TC_1 must turn off current i' and perform a forced commutation. If i' is positive for $\omega t = 0$, TC_1 takes the current which then goes via D_1 and is naturally transferred to TC'_1 when the latter is turned on at instant $t = \pi/\omega$.

The commutations to be carried out by TC_2 depend on the sign of $-i'$ for $\omega t = \pi + \beta$, since K_2 lets $-i'$ flow from $\omega t = \pi + \beta$ to $\omega t = 2\pi + \beta$.

Use is made of the ability of TC_1 and TC'_1 to be turned on and off if $i'(0)$ is negative, i.e. if

$$-\frac{\beta}{2} < \varphi < \pi - \frac{\beta}{2}.$$

Otherwise, these two semiconductor devices have only turn-on to carry out.

Similarly, TC_2 and TC'_2 are only required to ensure forced commutations in cases where i' is negative for $\omega t = \beta$, i.e. if

$$\frac{\beta}{2} < \varphi < \pi + \frac{\beta}{2}.$$

The limits of the zones in Fig. 2.13 also indicate the type of commutations to be carried out, depending on φ and β.

The following devices must operate in controlled turn-off:

TC_1 and TC'_1 for zone 1,
TC_1 and TC'_1, TC_2 and TC'_2 for zone 2
TC_2 and TC'_2 for zone 3,
no device for zone 4.

In Fig. 2.13, the semiconductor devices which must carry out the forced commutations are marked underneath each of the zone numbers.

Remarks

(a) During operation in zone 1, an instantaneous current reversibility of voltage U supply is not necessary, as current i is always positive or zero. Similarly, for operation in zone 3, current i' is always negative or zero.

(b) If, instead of delaying the turn-on period of K'_2 by β/ω compared with that of K_1, the period was advanced, the waveforms of the output (u', i') and input (U, i) values would not be modified. The half-bridge K_1, K'_1 would simply play the role previously held by the other half-bridge K_2, K'_2 and vice-versa.

In the labelling of the switches semiconductor devices, currents flowing through the latter and voltages across them, the subscripts 1 and 2 need only be permuted.

(c) When the load is almost capacitive, all commutations are natural only when in zone 4.

For a capacitive load, with a phase shift which can have any value between 0 and $\pi/2$, in order to ensure commutations in the part which corresponds to zone 1, two of the switches must have controlled turn-on and turn-off.

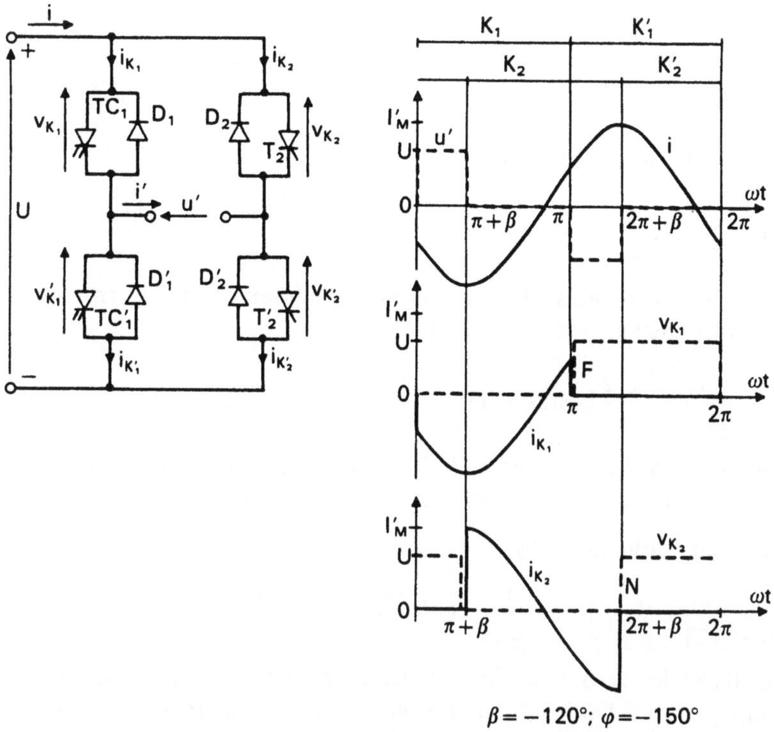

Fig. 2.14

Moreover, if the energy transfer direction has to be reversed ($-\pi < \varphi < -\pi/2$), in order to make operation possible in zone 3 the other two switches must also be controlled at turn-off. It is, however, possible to use the ability of the first two to turn off, if the direction of phase shift β is reversed.

Figure 2.14 represents an asymmetrical full-bridge inverter which ensures all operations on a capacitive load, providing that a negative β is used for operation in zone 3. The various waveforms have been plotted for a negative value of β; comparison with Fig. 2.12b shows that the roles of the two half-bridges have been permuted.

2.3.2 Characteristics

2.3.2.1 Output Voltage

If the sources and switches are assumed to be perfect, output u' voltage *depends only on β.*

Voltage u' is thus equal to $+U$ for $\beta < \omega t < \pi$, to $-U$ for $\pi + \beta < \omega t < 2\pi$, and zero during the remainder of its cycle.

Its rms value is

$$U' = U\sqrt{1 - \frac{\beta}{\pi}}. \tag{2.34}$$

Its fundamental has a rms value of

$$U'_1 = U\frac{2\sqrt{2}}{\pi}\cos\frac{\beta}{2}. \tag{2.35}$$

Its Fourier series contains all the harmonics of uneven order. The rms value of the term with angular frequency $(2k + 1)\omega$ is

$$U'_{2k+1} = \frac{U}{2k+1}\frac{2\sqrt{2}}{\pi}\cos(2k + 1)\frac{\beta}{2}. \tag{2.36}$$

The curves in Fig. 2.15 show how U', U'_1 and the first harmonic components vary when β goes from zero to π. All these values are referred to U.

The harmonic ratio, given by

$$\tau_{u'} = \frac{1}{U'_1}\left(\sum_{k=1}^{\infty} U'^2_{2k+1}\right)^{1/2} = \frac{1}{U'_1}\sqrt{U'^2 - U'^2_1}$$

is equal to 0.483 for zero β; at first, it decreases when β increases, falls to a minimum equal to 0.290 for β equal to 46°4, and then increases once more, moving towards infinity as β moves towards π.

In order to see the overall weight of the harmonics, their quadratic sum should be computed:

$$\left(\sum_{k=1}^{\infty} U'^2_{2k+1}\right)^{1/2} = \sqrt{U'^2 - U'^2_1}.$$

The variation of this sum is shown as a function of β in Fig. 2.15. Equal to $0.435U$ for zero β, it decreases at first, increases and then decreases once more, falling to zero when β reaches 180°. Its first minimum value is obtained for $\beta = 51°8$ and is equal to $0.237U$.

2.3.2.2 Semiconductor Devices

As with two-switch voltage-source inverters, each of the four switches which make up the bridge lets current i' flow during a half-cycle. But, in this case, the limits of this half-cycle depends on φ and on β; the same remark applies to the values characterizing the currents in the semiconductor devices.

We shall firstly establish the equations and then plot characteristics for switch K_1, made up of TC_1 and D_1, in the case of operation in zones 1 and 2:

$$0 < \varphi + \frac{\beta}{2} < \pi.$$

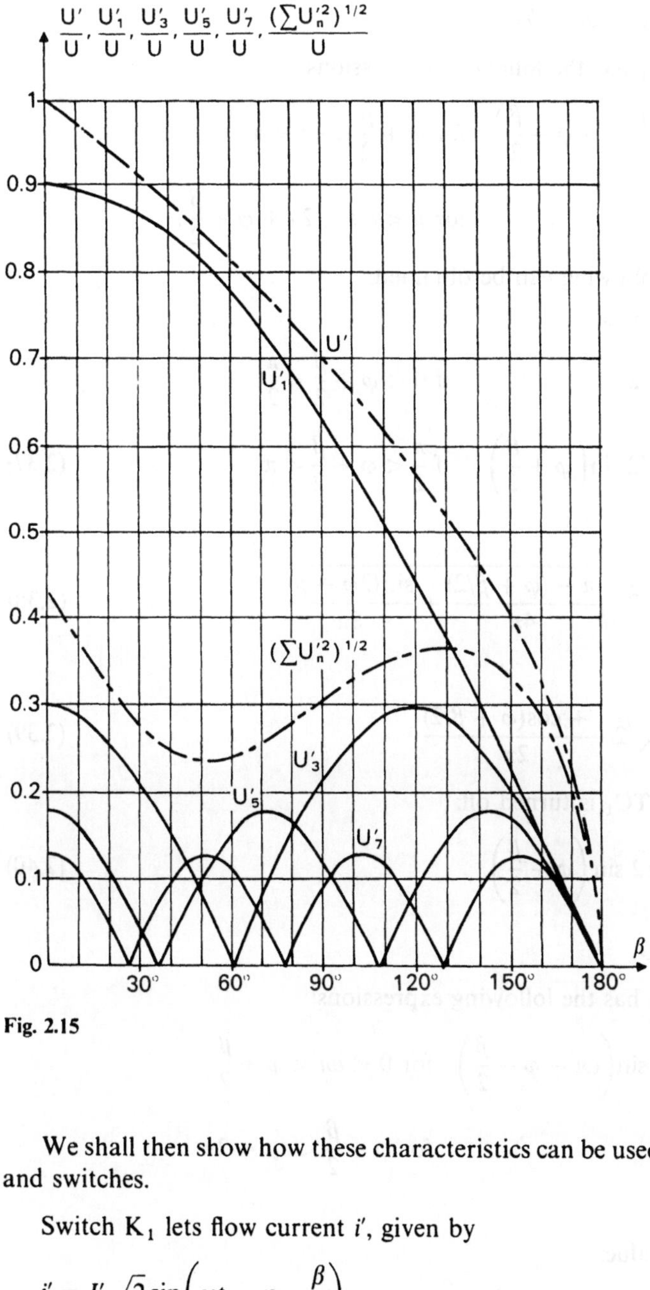

Fig. 2.15

We shall then show how these characteristics can be used for the other zones and switches.

Switch K_1 lets flow current i', given by

$$i' = I'\sqrt{2}\sin\left(\omega t - \varphi - \frac{\beta}{2}\right)$$

during the period $[0, \pi/\omega]$.

For $0 < \varphi + \beta/2 < \pi$, current i_{K_1} is negative at the beginning of this period and positive at the end: diode D_1 conducts first.

- *Current in controlled device TC_1.*

Current i_{T_1} in TC_1 has the following expressions:

$$i_{T_1} = I'\sqrt{2}\sin\left(\omega t - \varphi - \frac{\beta}{2}\right) \quad \text{for } \varphi + \frac{\beta}{2} < \omega t < \pi$$

$$i_{T_1} = 0 \quad \text{for } \pi < \omega t < 2\pi + \varphi + \frac{\beta}{2}.$$

From these, the following can be obtained:

– its maximum value:

$$i_{T_1\,\max} = I'\sqrt{2} \quad \text{if } 0 < \varphi + \frac{\beta}{2} < \frac{\pi}{2}$$

$$i_{T_1\,\max} = I'\sqrt{2}\sin\left(\varphi + \frac{\beta}{2}\right) \quad \text{if } \frac{\pi}{2} < \varphi + \frac{\beta}{2} < \pi \tag{2.37}$$

– its rms value:

$$I_{T_1\,\mathrm{rms}} = I'\sqrt{2}\sqrt{\frac{\pi - (\varphi + \beta/2)}{4\pi} + \frac{\sin(2\varphi + \beta)}{8\pi}} \tag{2.38}$$

– its mean value:

$$I_{T_1\,\mathrm{mean}} = I'\sqrt{2}\,\frac{1 + \cos(\varphi + \beta/2)}{2\pi} \tag{2.39}$$

– its value when TC_1 is turned off:

$$i_{T_1}(\pi) = I'\sqrt{2}\sin\left(\varphi + \frac{\beta}{2}\right). \tag{2.40}$$

- *Current in diode D_1*

Current i_{D_1} in D_1 has the following expressions:

$$i_{D_1} = -I'\sqrt{2}\sin\left(\omega t - \varphi - \frac{\beta}{2}\right) \quad \text{for } 0 < \omega t < \varphi + \frac{\beta}{2}$$

$$i_{D_1} = 0 \quad \text{for } \varphi + \frac{\beta}{2} < \omega t < 2\pi.$$

Giving

– its maximum value:

$$i_{D_1\,\max} = I'\sqrt{2}\sin\left(\varphi + \frac{\beta}{2}\right) \quad \text{if } 0 < \varphi + \frac{\beta}{2} < \frac{\pi}{2}$$

$$i_{D_1\,\max} = I'\sqrt{2} \quad \text{if } \frac{\pi}{2} < \varphi + \frac{\beta}{2} < \pi \tag{2.41}$$

– its rms value:

$$I_{D_1 rms} = I'\sqrt{2}\sqrt{\frac{(\varphi - \beta/2)}{4\pi} - \frac{\sin(2\varphi + \beta)}{8\pi}} \quad (2.42)$$

– its mean value:

$$I_{D_1 mean} = I'\sqrt{2}\,\frac{1 - \cos(\varphi + \beta/2)}{2\pi}. \quad (2.43)$$

Providing that $\varphi + \beta/2$ is taken as a variable, the same curve enables the development of each value characterizing currents i_{T_1} and i_{D_1} to be followed. It would in fact be possible to use the characteristics of Fig. 2.4, if $(2n_2/n_1)I'$ were replaced by I' and φ by $\varphi + \beta/2$.

In Fig. 2.6, however, we have once more plotted these curves in order to be able to continue them in directions corresponding to operation in zones 3 and 4. The right-hand side of Fig. 2.16 shows the variations of $i_{T_1 max}$, $I_{T_1 rms}$, $I_{T_1 mean}$, of $i_{D_1 max}$, $I_{D_1 rms}$, $I_{D_1 mean}$ of $i_{T_1}(\pi)$ when $\varphi + \beta/2$ moves from zero to π. All the current values are referred to I'.

- *Current values i_{T_1} and i_{D_1} in the other zones*

If the phase shift of current i' is decreased (or increased) by π in relation to the voltage u' fundamental, the sign of its instantaneous value is reversed:

$$I'\sqrt{2}\sin\left(\omega t - \varphi - \frac{\delta}{2} - \pi\right) = -I'\sqrt{2}\sin\left(\omega t - \varphi - \frac{\delta}{2}\right).$$

At each instant in period $(0 < \omega t < \pi)$ where switch K_1 lets current i' flow, the current flowing through D_1 is equal to that which was flowing through TC_1 and vice-versa.

The characteristics plotted for $\varphi + \beta/2$ going from π to 0 can be used for $\varphi + \beta/2$ going from 0 to $-\pi$, as long as the following substitutions are made:

$i_{T_1 max}$ by $i_{D_1 max}$, $I_{T_1 rms}$ by $I_{D_1 rms}$, $I_{T_1 mean}$ by $I_{D_1 mean}$,

$i_{D_1 max}$ by $i_{T_1 max}$, $I_{D_1 rms}$ by $I_{T_1 rms}$, $I_{D_1 mean}$ by $I_{T_1 mean}$,

$i_{T_1}(\pi)$ by $i_{D_1}(\pi)$.

This provides the characteristics on the left-hand side in Fig. 2.16. Current i_{T_1} is now at zero at the end of period $(0, \pi)$ and $i(\pi)$ indicates the current transferred from D_1 to TC'_1 when the latter is turned on.

- *Current in TC_2 and D_2*

Switch K_2 lets $-i$ flow for $\pi + \beta < \omega t < 2\pi + \beta$.
The current wave in this switch, crossed by

$$-I'\sqrt{2}\sin\left(\omega t - \varphi - \frac{\beta}{2}\right) \quad \text{for } \pi + \beta < \omega t < 2\pi + \beta$$

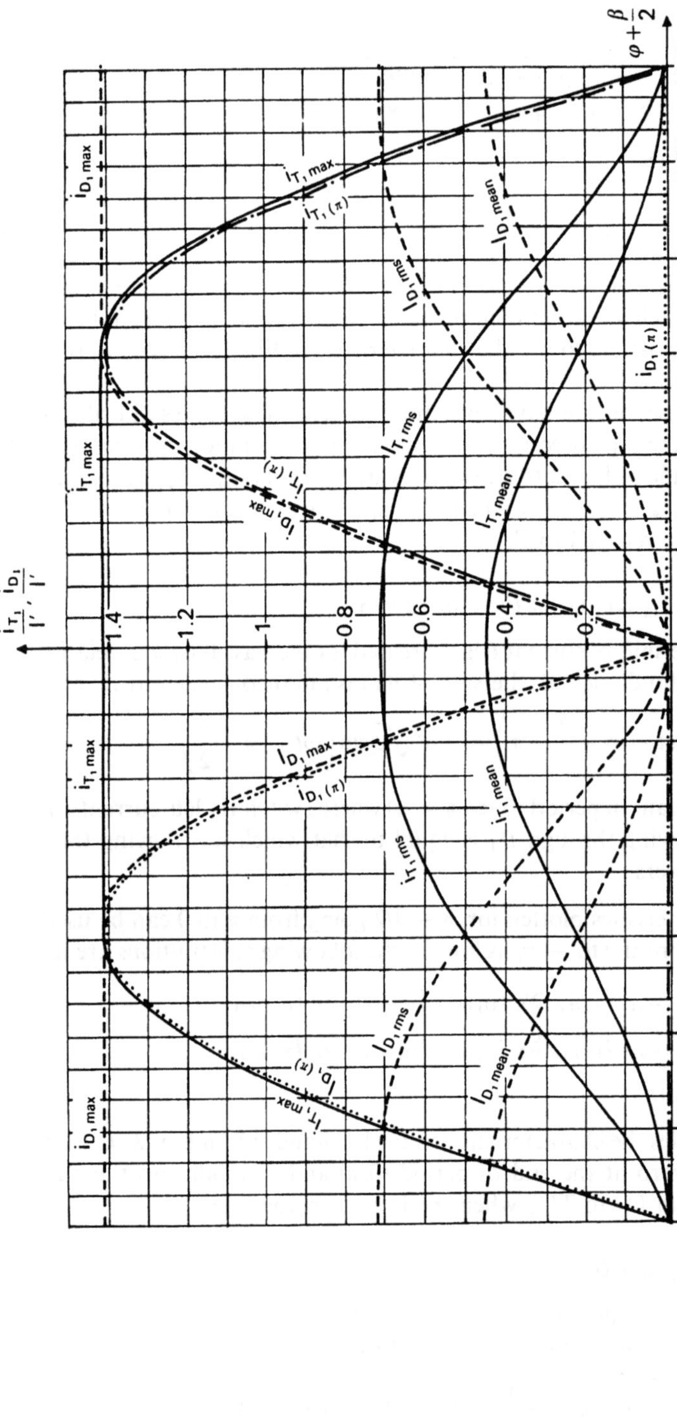
$\frac{i_{T_1}}{I'}, \frac{i_{D_1}}{I'}$
1.4
1.2
1
0.8
0.6
0.4
0.2
0
$\varphi + \frac{\beta}{2}$
−180°
−150°
−120°
−90°
−60°
−30°
0
30°
60°
90°
120°
150°
180°
$i_{D, max}$
$i_{T, max}$
$i_{T, max}$
$i_{D, max}$
$I_{T, max}$
$I_{T, (\pi)}$
$I_{D, rms}$
$I_{D, mean}$
$I_{T, rms}$
$I_{T, mean}$
$I_{D, max}$
$I_{T, (\pi)}$
$i_{D, (\pi)}$
$I_{D, max}$
$I_{D, (\pi)}$
$I_{T, mean}$
$I_{T, rms}$
$I_{D, rms}$
$I_{D, mean}$
$I_{T, max}$
$I_{D, (\pi)}$
Fig. 2.16

is the same as if it was crossed by

$$I'\sqrt{2}\sin\left(\omega t - \varphi - \frac{\beta}{2}\right) \quad \text{for } \beta < \omega t < \pi + \beta$$

or by

$$I'\sqrt{2}\sin\left(\omega t - \varphi + \frac{\beta}{2}\right) \quad \text{for } 0 < \omega t < \pi.$$

For ratings of semiconductor devices TC_2 and D_2, Eqs. (2.37)–(2.43) – worked out for TC_1 and D_1 – can be used providing that $\varphi + \beta/2$ is replaced by $\varphi - \beta/2$.

Similarly, the curves in Fig. 2.16 giving the values related to i_{T_1} and i_{D_1} can give the values related to i_{T_2} and i_{D_2} if $\varphi + \beta/2$ is replaced by $\varphi - \beta/2$.

- *Voltages*

The voltage across all the switches when they are off is equal to $+U$. Therefore, for the four diodes and for the four controlled devices, this yields

$$-v_{D\,max} = v_{T\,max} = U. \tag{2.44}$$

2.3.2.3 Input Current

The cycle of input current i is equal to half that of output voltage u'. During one of its cycles, it can be expressed as

$$i = 0 \qquad \text{for } 0 < \omega t < \beta$$

$$i = i' = I'\sqrt{2}\sin\left(\omega t - \varphi - \frac{\beta}{2}\right) \quad \text{for } \beta < \omega t < \pi.$$

- In order to determine current i *ripple*, it can be observed that
  - the maximum value of i equals $I'\sqrt{2}$ if

    $$\frac{\beta}{2} - \frac{\pi}{2} < \varphi < \frac{\pi}{2} - \frac{\beta}{2}$$

    in that case, the minimum value of i corresponds to $\omega t = \beta$, if φ is positive, and to $\omega t = \pi$, if φ is negative.
  - the minimum value of i is equal to $-I'\sqrt{2}$ if

    $$\varphi > \frac{\pi}{2} + \frac{\beta}{2} \quad \text{or} \quad \varphi < -\frac{\pi}{2} - \frac{\beta}{2}$$

    in that case, the maximum value of i corresponds to $\omega t = \pi$, if φ is positive and to $\omega t = \beta$, if φ is negative.
  - if i_{max} is less than $I'\sqrt{2}$ and i_{min} greater than $-I'\sqrt{2}$, the ripple is given by

    $$\Delta i = |i(\beta) - i(\pi)| = 2I'\sqrt{2}\cos\frac{\beta}{2}\sin|\varphi|.$$

For zero β and φ equal to $\pm\pi/2$, the ripple is maximum and is then equal to $2I'\sqrt{2}$.

- Current i has the following *mean value*:

$$I = \frac{1}{\pi}\int_{\beta}^{\pi} I'\sqrt{2}\sin\left(\omega t - \varphi - \frac{\beta}{2}\right) d\omega t$$

$$I = \frac{2}{\pi} I'\sqrt{2}\cos\varphi\cos\frac{\beta}{2}. \tag{2.45}$$

The same result could have been obtained by writing

$I = (U'_1 I' \cos\varphi)/U.$

The mean value of i does not depend on the phase-shift direction (φ lagging or leading) but on the sign of $\cos\varphi$. It is reversed when the energy transfer direction is reversed with $|\varphi|$ becoming greater than $\pi/2$.

Figure 2.17 shows the variations – as a function of β – of the I/I' ratio for $\varphi = 0$, $\varphi = \pm\pi/6$, $\varphi = \pm\pi/3$, $\varphi = \pm\pi/2$, $\varphi = \pm 2\pi/3$, $\varphi = \pm 5\pi/6$ and $\varphi = \pi$.

- Current i has the following *rms value*:

$$I_{\text{rms}} = I'\sqrt{\frac{\pi - \beta}{\pi} + \frac{\sin\beta\cos 2\varphi}{\pi}}. \tag{2.46}$$

This value decreases from I' to zero when β goes from zero to π. It decreases when φ moves away from either zero or π; it is independent of the sign of $\cos\varphi$ since $\cos 2\varphi$ equals $2\cos^2\varphi - 1$.

Figure 2.18 shows the variations of I_{rms}/I' as a function of β for four values of $|\cos\varphi|$.

- The series expansion of current i contains – in addition to the average term I – all the *harmonics* with angular frequencies which are even multiples of that of the output current.

  The rms value of the $2k\omega$ angular frequency harmonic is given by

$$I_{2k} = I'\frac{2}{\pi}\left\{\left|\frac{\cos(2k-1)\frac{\beta}{2}}{2k-1}\right|^2 + \left|\frac{\cos(2k+1)\frac{\beta}{2}}{2k+1}\right|^2 - 2\frac{\cos(2k-1)\frac{\beta}{2}\cos(2k+1)\frac{\beta}{2}\cos 2\varphi}{(2k-1)(2k+1)}\right\}^{1/2}. \tag{2.47}$$

Since angle φ appears once more through $\cos 2\varphi$, the value of I_{2k}/I' depends only on β and on $|\cos \varphi|$.

At the top of Fig. 2.19 can be seen the variations of the quadratic sum of the harmonics, as a function of β for several values of $|\cos \varphi|$:

$$\left(\sum I_h^2\right)^{1/2} = \left(\sum_{k=1}^{\infty} I_{2k}^2\right)^{1/2} = \sqrt{I_{\text{rms}}^2 - I^2}.$$

Below, there can be found the curves showing the variations in rms values I_2, I_4, I_6 of the first harmonics. All these values are referred to the rms value I' of the output current.

All the harmonics reach their maximum value for $\beta = 0$ and $\varphi = \pm \pi/2$.

2.3.3 Comparison of the Three Single-Phase Voltage-Source Inverters

2.3.3.1 Comparison of the Required Components and Characteristics

- The two types of two-switch voltage-source inverters – push–pull and half-bridge inverter – require only two controlled semiconductor devices and two diodes, whereas the full-bridge inverter requires four controlled devices and four diodes. This advantage of the first two types of inverter is, however, reduced by the doubling of the voltage or the current:

- for a given voltage U of the DC supply,

  for the push–pull inverter, $v_{\text{T max}} = -v_{\text{D max}} = 2U$,
  for the other two, $v_{\text{T max}} = -v_{\text{D max}} = U$.

- for a given current i of the DC supply, the current in switch K_1, when it is on, has the value

  $i_{K_1} = |i|$ for the push–pull inverter,
  $i_{K_1} = 2|i|$ for the half-bridge inverter,
  $i_{K_1} = |i|$ (or $|i'|$) for the full-bridge inverter.

The two-switch inverters require half as many semiconductor devices but the voltage or current rating of the latter is doubled.

- The basic principle of the push–pull inverter requires it to have a transformer and that of the half-bridge inverter to have a capacitive divider, whereas the full-bridge inverter requires neither. These *disadvantages* of the two-switch inverters are less important that it would initially appear.

- Virtually all inverters require a transformer to ensure galvanic isolation and to obtain the wanted output voltage amplitude. In the case of the push–pull inverter, the transformer is inherent in the operating principle. However the fact that it has a centre-tapped primary and that each half-primary is only used during one half-cycle means that the primary power rating is multiplied by $\sqrt{2}$.

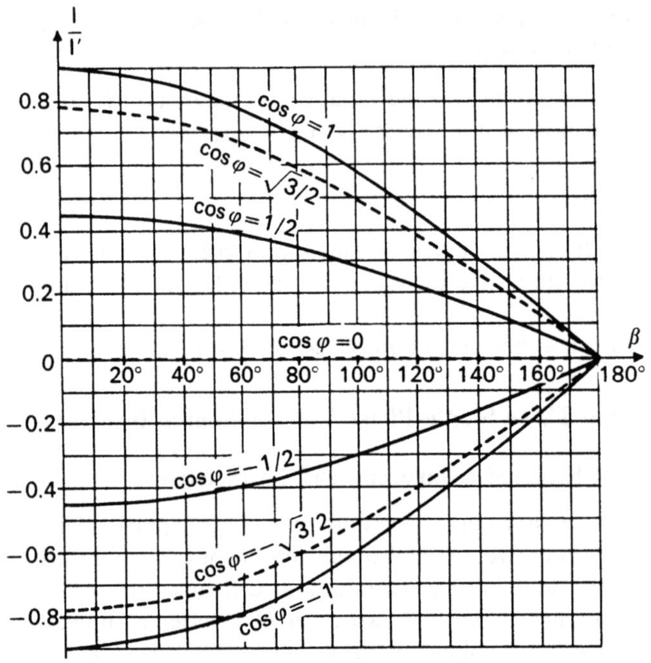

Fig. 2.17

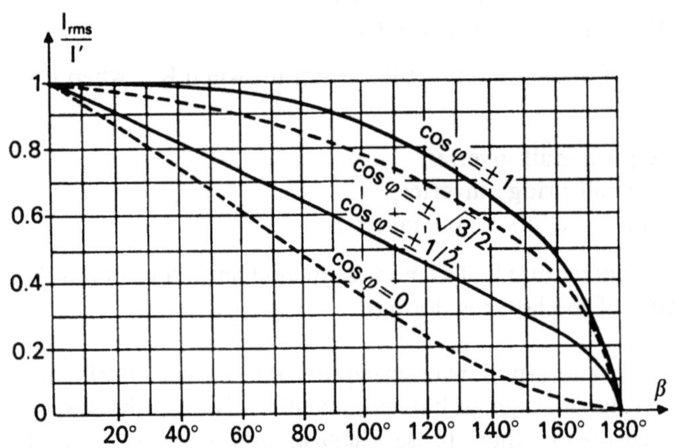

Fig. 2.18

– Virtually all inverters require an input filter comprising an inductor and a capacitor. The capacitors of the half-bridge inverter can also act as a filtering capacitor (see Sect. 2.5.3). Instead of one capacitor of capacitance C submitted to voltage U, two capacitors – each with a capacitance $2C$ and submitted to a voltage $U/2$ – are used.

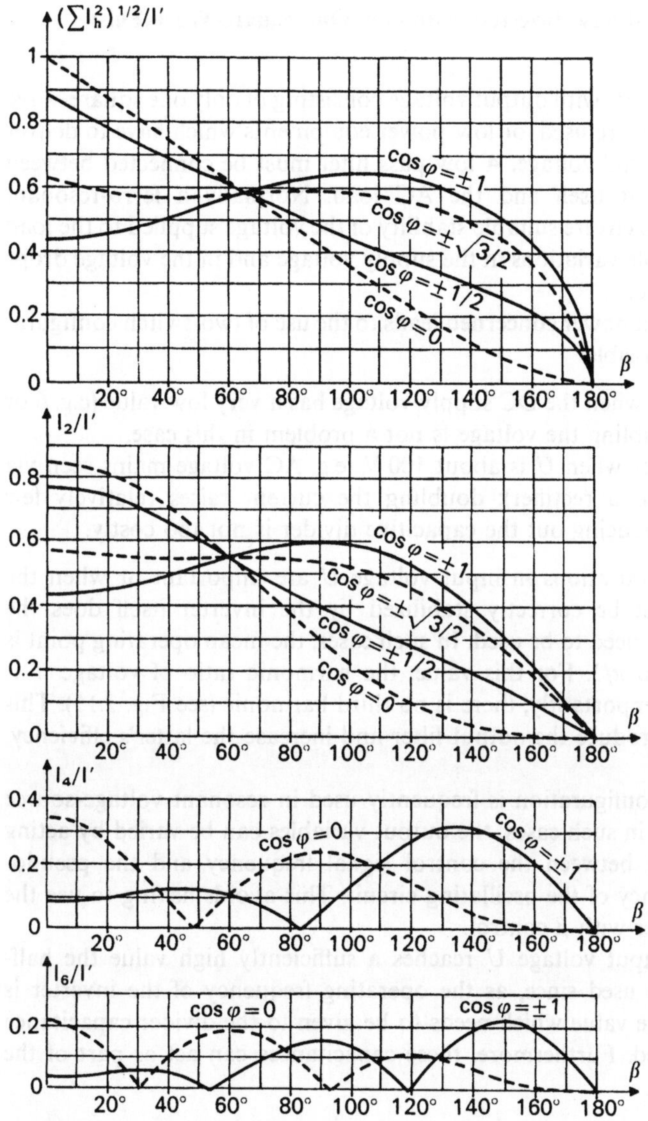

Fig. 2.19

- The main difference is that the full-bridge inverter – more expensive overall – enables *the output voltage to be varied* and thus, more precisely, to be regulated.

The full-bridge inverter requires the conduction of two series connected switches; the corresponding voltage drop and losses lead to a highly reduced efficiency when the voltage supply U has a relatively low value.

2.3.3.2 Use of Single-Phase Inverters with only One Square-Wave Pulse per Half-Cycle

- Single-phase inverters with output voltage consisting of only one square-wave pulse per half-cycle are used for low-power equipments which have to deliver a relatively stable AC voltage. A low-pass filter must be connected between the inverter output itself and the AC load. Normally a ferro-resonant stabiliser is used to ensure suitable stability of the voltage supplied to the load despite unavoidable variations in the supply voltage and in the voltage drops inside the converter.

  The low level of power concerned leads to the use of two-switch configurations wherever possible:

- push-pull inverter when the DC supply voltage has a very low value (e.g. 6 or 12 V battery); doubling the voltage is not a problem in this case.
- half-bridge inverter when U is about 100 V (e.g. AC voltage mains seen via a transformer and a rectifier); doubling the current raises relatively few problems and producing out the capacitive divider is not too costly.

Only when the variations in input voltage U are important or when the output voltage must be correctly stabilized by the inverter itself does the full-bridge converter need to be used. In such cases, the mean operating point is placed at β equal to $\pi/3$. For this value, the harmonic ratio of voltage u' is reduced and, most importantly, there is no third harmonic (see Fig. 2.15). This makes it easier to produce the output filter and increase the latter's efficiency.

- The two-switch configuration is frequently used in resonant voltage-source inverters. Indeed, in such cases, the output variables can be varied by acting on the difference between the control signal frequency and the pseudo-resonance frequency of the oscillating circuit. This avoids having to use the full-bridge inverter with β control.

  As soon as input voltage U reaches a sufficiently high value the half-bridge inverter is used since, as the operating frequency of the inverter is relatively high, the value which needs to be given to the divider capacitance [(2.25)] is reduced. Furthermore, these capacitances can act as part of the oscillating circuit.

2.3.3.3 Use in the Pulse Width Modulation Inverters

Increasing use is now made of pulse width modulation. This enables to obtain a rough estimate of a sinusoidal voltage by forming each of the output voltage half-cycles of a succession of suitable-sized square-wave pulses. The advantage of the full-bridge inverter, which enables the output voltage value to be varied, disapears since the other types can now also guarantee this variation.

However, the two-switch inverters only provide two voltage levels ($u' = +U$ or $-U$), whereas the full-bridge inverter gives three voltage levels ($u' = +U$ or

$-U$ or zero). The full-bridge inverter obviously make it easier to approximate the required sinusoid. This allows a more effective use of components and an easier filtering. As soon as the power being used is important and the importance attached to performances (especially in reducing the overall weight of the equipment) become paramount, the full-bridge inverter will be used.

2.4 Three-Phase Full-Bridge Inverter

- A three-phase voltage-source inverter can be obtained by grouping together three single-phase half-bridges and using a common capacitive divider (Fig. 2.20).

  To ensure that the AC output currents i'_A, i'_B and i'_C remain continuous, control signals of switches K_1 and K'_1, K_2 and K'_2, K_3 and K'_3 must be complementary in pairs.

In order that the output voltages v'_A, v'_B and v'_C should be identical to within a third of cycle T of their fundamental, each half-bridge must be controlled with a delay of $T/3$ compared to the previous one. Turn-on signals are thus applied to

K_1 for $\omega t = 0$, K'_1 for $\omega t = \pi$

K_2 for $\omega t = 2\pi/3$, K'_2 for $\omega t = \pi + 2\pi/3$

K_3 for $\omega t = 4\pi/3$, K'_3 for $\omega t = \pi + 4\pi/3$

- If currents i'_A, i'_B and i'_C are sinusoidal and form a balanced three-phase system

$$\begin{aligned} i'_A &= I'\sqrt{2}\sin(\omega t - \varphi) \\ i'_B &= I'\sqrt{2}\sin(\omega t - \varphi - 2\pi/3) \\ i'_C &= I'\sqrt{2}\sin(\omega t - \varphi - 4\pi/3) \end{aligned} \tag{2.48}$$

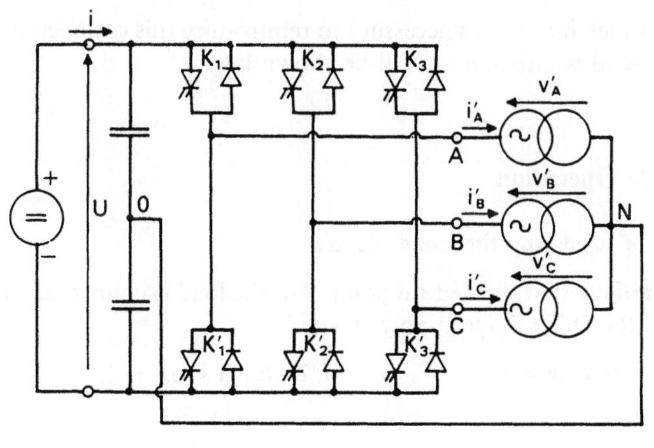

Fig. 2.20

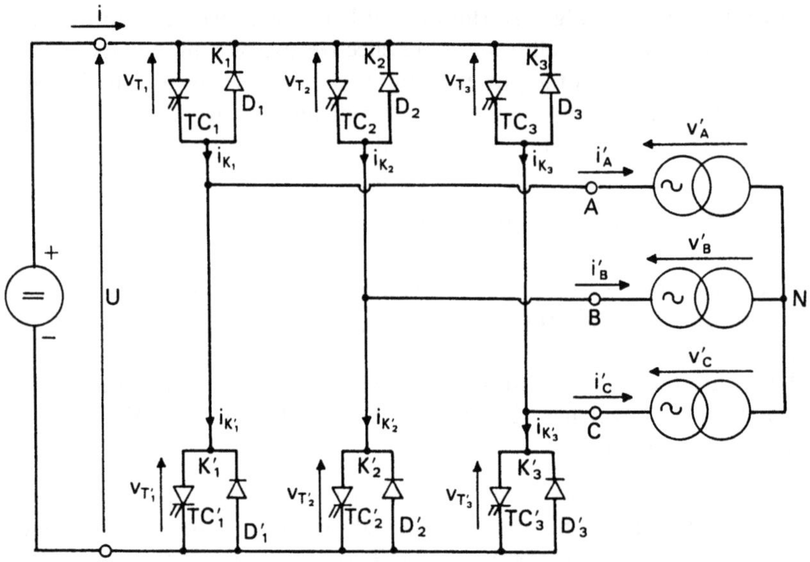

Fig. 2.21

their instantaneous sum is zero. No current flows through the connection between the neutral point N and the mid-point O of the capacitive divider: this connection can be eliminated and thus the divider, without modifying the output voltages and currents. This gives the standard configuration of the six-switch voltage-source inverter shown in Fig. 2.21, in which the notations used are indicated.

It will be shown that, in cases where the currents are balanced but not sinusoidal, it is not only possible but a positive advantage to suppress connection O–N.

The conditions in which it becomes necessary to reintroduce this connection when the three-phase load is unbalanced will be shown later.

2.4.1 Balanced Mode Operation

2.4.1.1 Consequences of Insulating the Load Neutral

- Let it be assumed initially that the neutral point N of the load is connected to the mid-point O of the DC voltage supply. Then

$$v'_A = +U/2 \quad \text{for } 0 < \omega t < \pi, \qquad v'_A = -U/2 \quad \text{for } \pi < \omega t < 2\pi.$$

As

$$v'_B(\omega t) = v'_A(\omega t - 2\pi/3), \qquad v'_C(\omega t) = v'_A(\omega t - 4\pi/3)$$

the Fourier series of the three output voltage can be written as

$$v'_A = V'_1\sqrt{2}\sin\omega t + V'_3\sqrt{2}\sin 3\omega t + V'_5\sqrt{2}\sin 5\omega t + \ldots$$

$$v'_B = V'_1\sqrt{2}\sin\left(\omega t - \frac{2\pi}{3}\right) + V'_3\sqrt{2}\sin 3\omega t + V'_5\sqrt{2}\sin\left(5\omega t + \frac{2\pi}{3}\right) + \ldots$$

$$v'_C = V'_1\sqrt{2}\sin\left(\omega t - \frac{4\pi}{3}\right) + V'_3\sqrt{2}\sin 3\omega t + V'_5\sqrt{2}\sin\left(5\omega t + \frac{4\pi}{3}\right) + \ldots$$

with $V'_{2k+1} = V'_1/(2k+1)$.

The real three-phase system is formed by superimposing the balanced systems corresponding to the fundamental and the different harmonics:

- positive sequence systems for the fundamental and harmonics 7, 13, 19, . . .
- negative sequence systems for harmonics 5, 11, 17, . . .
- zero sequence systems for harmonic 3, 9, 15, . . .

If the load is balanced, each system in the voltages yields a similar system in current i'_A, i'_B and i'_C. Deriving one from the other depends on the load impedance for a particular frequency.

The currents in the three phases are thus formed by superimposing positive sequence, negative sequence, and zero sequence balanced systems.

The positive sequence and negative sequence systems are zero-sum systems. The zero sequence terms of the same frequency are in phase in all three phases, their sum which is the triple of each current flows via connection N–O.

- If this connection is suppressed, then harmonics 3, 9, . . . in the phase currents are, as are *the uneven harmonics with a rank multiple of* 3 in voltage v'_A, v'_B, v'_C applied to the load phases. Such a suppression – especially that of harmonic 3 – greatly facilitates the filtering of AC voltages.

The sum of voltages v'_A, v'_B, v'_C must be zero. To obtain the instantaneous values of these voltages, voltages $v_A - v_O$, $v_B - v_O$, $v_C - v_O$ between the three output terminals of the inverter and the (fictitious) mid-point of the DC voltage supply are considered, and then the instantaneous zero-sequence component of this voltage system[1] is removed.

For the first phase, this gives

$$v'_A = v_A - v_O - \frac{1}{3}[(v_A - v_O) + (v_B - v_O) + (v_C - v_O)]$$

as well as for other phases.

---

[1] The notion of an instantaneous zero-sequence component has already been used to describe the equations of m.m.f. in the transformers feeding certain rectifiers (Vol. 1; Chap. 2, Sect. 2.3.3).

This yields

$$v'_A = \frac{2}{3}(v_A - v_O) - \frac{1}{3}(v_B - v_O) - \frac{1}{3}(v_C - v_O)$$

$$v'_B = -\frac{1}{3}(v_A - v_O) + \frac{2}{3}(v_B - v_O) - \frac{1}{3}(v_C - v_O) \tag{2.49}$$

$$v'_C = -\frac{2}{3}(v_A - v_O) - \frac{1}{3}(v_B - v_O) + \frac{1}{3}(v_C - v_O).$$

The zero sequence component which is suppressed in v'_A, v'_B, v'_C corresponds to the potential difference between the points N and O which are now disconnected.[1]

$$v_N - v_O = (v_A - v_O) - (v_A - v_N) = (v_A - v_O) - v'_A$$

$$= \frac{1}{3}[(v_A - v_O) + (v_B - v_O) + (v_C - v_O)].$$

2.4.1.2 Expressions. Waveforms

According to the turn-on and turn-off of the six switches, six periods can be distinguished during cycle T of the output variables. However, a study of the first two periods will be sufficient, if the identity, to within $T/3$ or $2T/3$, of the voltages and currents in the three phases is taken into account.

- For $0 < \omega t < \pi/3$, K_1, K'_2 and K_3 are on:

$$v_A - v_O = \frac{U}{2}, \quad v_B - v_O = -\frac{U}{2}, \quad v_C - v_O = \frac{U}{2}$$

$$v'_A = \frac{U}{3}, \quad v'_B = -\frac{2}{3}U, \quad v'_C = \frac{U}{3}$$

$$v_{K_1} = v_{K'_2} = v_{K_3} = 0, \quad v_{K'_1} = v_{K_2} = v_{K'_3} = U$$

$$i_{K_1} = i'_A, \quad i_{K'_2} = -i'_B, \quad i_{K_3} = i'_C$$

$$i_{K'_1} = i_{K_2} = i_{K'_3} = 0, \quad i = i'_A + i'_C = -i'_B.$$

- For $\pi/3 < \omega t < 2\pi/3$, K_1, K'_2 and K'_3 are on:

$$v_A - v_O = \frac{U}{2}, \quad v_B - v_O = -\frac{U}{2}, \quad v_C - v_O = -\frac{U}{2}$$

$$v'_A = \frac{2}{3}U, \quad v'_B = -\frac{U}{3}, \quad v'_C = -\frac{U}{3}$$

$$v_{K_1} = v_{K'_2} = v_{K'_3} = 0, \quad v_{K'_1} = v_{K_2} = v_{K_3} = U$$

$$i_{K_1} = i'_A, \quad i_{K'_2} = -i'_B, \quad i_{K'_3} = -i'_C$$

$$i_{K'_1} = i_{K_2} = i_{K_3} = 0, \quad i = i'_A = -i'_B - i'_C, \ldots$$

---

[1] Assuming that the currents are sinusoidal – as when presenting the suppression of the N–O connection – means that load impedance is assumed to be infinite for all harmonics:

During the following third of a cycle, in the previous expressions, the switches subscripts must be increased by one. In the equations, A must be replaced by B, B by C and C by A in the output current and voltage subscripts.

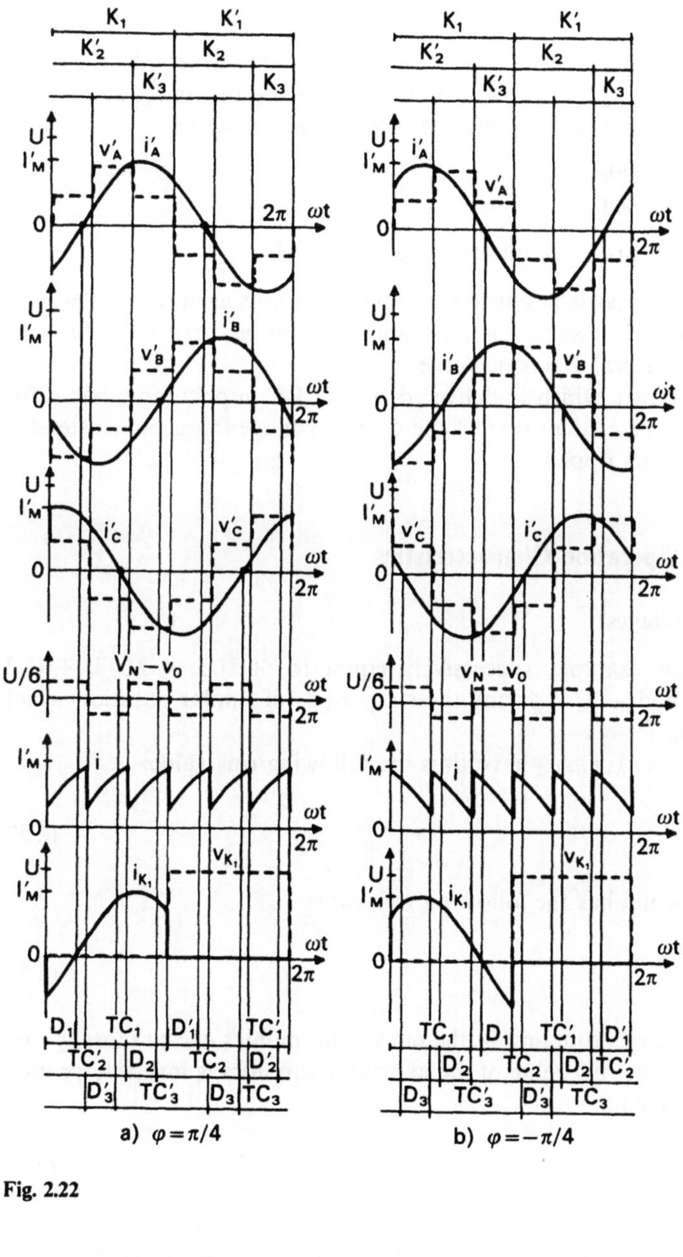

Fig. 2.22

Thus, for $2\pi/3 < \omega t < \pi$, K_2, K'_3 and K_1 are on:

$$v'_B = \frac{U}{3}, \quad v'_C = -\frac{2}{3}U, \quad v'_A = \frac{U}{3}$$

$$v_{K_2} = v_{K'_3} = v_{K_1} = 0, \quad v_{K'_2} = v_{K_3} = v_{K'_2} = U$$

$$i_{K_2} = i'_B, \quad i_{K'_3} = -i'_C, \quad i_{K_1} = i'_A$$

$$i_{K'_2} = i_{K_3} = i_{K'_1} = 0, \quad i = i'_B + i'_A = -i'_C.$$

Figure 2.22 shows, from top to bottom, the switches on periods, the waveforms of v'_A and i'_A, of v'_B and i'_B, of v'_C and i'_C, of $v_N - v_O$, of i, of v_{K_1} and i_{K_1} and the conduction diagrams of the different semiconductor devices

(a) for a lagging shift,
(b) for a leading shift.

The output currents are assumed to be sinusoidal.

The currents in the switches have the same waveforms as in the single-phase two-switch inverters. The same commutations have to be carried out: forced if φ is positive and natural if φ is negative.

However, current i, which is delivered by the DC supply, now has a frequency six times greater than that of the output voltage frequency, instead of twice. This reduces its ripple.

2.4.2 Balanced Operation Characteristics

2.4.2.1 Output Voltages

Voltage v'_A is, for example, successively equal to $+U/3$, $+2U/3$, $+U/3$, $-U/3$, $-2U/3$ and $-U/3$ during the six periods of similar duration which make up its cycle.

Output voltages v'_A, v'_B, v'_C have thus the following rms value:

$$V' = \frac{\sqrt{2}}{3}U. \tag{2.50}$$

Their fundamental has the following rms value:

$$V'_1 = \frac{\sqrt{2}}{\pi}U. \tag{2.51}$$

Their Fourier series contains all the uneven harmonics *except those of rank 3 or multiple of 3*. The rms value of the existing harmonics is inversely proportional to their frequency:

$$\frac{V'_{2k+1}}{V'_1} = \frac{1}{2k+1}. \tag{2.52}$$

The harmonic ratio is equal to 0.311. The difference between this value and that (0.483) found for the single-phase two-switch inverters arises from the suppression of harmonics of rank 3 or multiple of 3.

2.4.2.2 Semiconductor Devices

In order to compute the values relating to the currents, the output currents are assumed to be sinusoidal and thus to have the following expression form:

$$i'_A = I'\sqrt{2}\sin(\omega t - \varphi).$$

All values relating to the currents in the semiconductor devices are the same as for the half-bridge single-phase inverter.

- The currents in the *controlled devices* have
  - a maximum value of

$$i_{T\,max} = I'\sqrt{2} \quad \text{if } 0 < \varphi < \frac{\pi}{2}$$

$$i_{T\,max} = I'\sqrt{2}\sin\varphi \quad \text{if } \frac{\pi}{2} < \varphi < \pi \tag{2.53}$$

  - an rms value of

$$I_{T\,rms} = I'\sqrt{2}\sqrt{\frac{\pi - \varphi}{4\pi} + \frac{\sin 2\varphi}{8\pi}} \tag{2.54}$$

  - a mean value of

$$I_{T\,mean} = I'\sqrt{2}\,\frac{1 + \cos\varphi}{2\pi} \tag{2.55}$$

and the current to be switched is equal to

$$i_\pi = I\sqrt{2}\sin\varphi. \tag{2.56}$$

- The currents in the *diodes* have
  - a maximum value of

$$i_{D\,max} = I'\sqrt{2}\sin\varphi \quad \text{if } 0 < \varphi < \frac{\pi}{2}$$

$$i_{D\,max} = I'\sqrt{2} \quad \text{if } \frac{\pi}{2} < \varphi < \pi \tag{2.57}$$

  - an rms value of

$$I_{D\,rms} = I'\sqrt{2}\sqrt{\frac{\varphi}{4\pi} - \frac{\sin 2\varphi}{8\pi}} \tag{2.58}$$

– a mean value of

$$I_{\mathrm{D\,mean}} = I'\sqrt{2}\frac{1-\cos\varphi}{2\pi}. \tag{2.59}$$

The characteristics shown in Fig. 2.4 can be used once again, providing that the different values are referred to I' rather than to $(2n_2/n_1)I'$. Once more these values are independent of the sign of φ.

- The maximum voltages across the semiconductor devices are the same as in the case of the half-bridge single-phase inverter:

$$-v_{\mathrm{D\,max}} = v_{\mathrm{T\,max}} = U. \tag{2.60}$$

2.4.2.3 Input Current

Input current i has a cycle equivalent to a sixth of that of the output variables.
When feeding an inductive load, it has the following expression:

$$\text{for } 0 < \omega t < \frac{\pi}{3}, \quad i = -i'_{\mathrm{B}} = -I'\sqrt{2}\sin\left(\omega t - \frac{2\pi}{3} - \varphi\right)$$
$$= I'\sqrt{2}\sin\left(\omega t + \frac{\pi}{3} - \varphi\right).$$

- It has the following maximum instantaneous value:

$$\begin{aligned} i_{\max} &= I'\sqrt{2} && \text{if } 0 < \varphi < \frac{\pi}{6} \\ &= I'\sqrt{2}\sin\left(\frac{2\pi}{3} - \varphi\right) && \text{if } \frac{\pi}{6} < \varphi < \pi \end{aligned} \tag{2.61}$$

and minimum instantaneous value:

$$\begin{aligned} i_{\min} &= I'\sqrt{2}\sin\left(\frac{\pi}{3} - \varphi\right) && \text{if } 0 < \varphi < \frac{5\pi}{6} \\ &= -I'\sqrt{2} && \text{if } \frac{5\pi}{6} < \varphi < \pi. \end{aligned} \tag{2.62}$$

The ripple Δi of the input current – equal to $i_{\max} - i_{\min}$ – is minimal for φ equal to 0 or π. Its maximum value equals $I'\sqrt{2}$ and corresponds to $\varphi = \pi/2$, while, in the case of single-phase inverters, the maximum value of Δi equals $2I'\sqrt{2}$.

It should be noted that for $0 < \varphi < \pi/3$ (or for $2\pi/3 < \varphi < \pi$) current i has always the same polarity, and that the current reversibility of the DC voltage supply is not necessary.

- The input current has a mean value of

$$I = \frac{3}{\pi} I' \sqrt{2} \cos\varphi \tag{2.63}$$

and an rms value of

$$I_{\text{rms}} = I' \sqrt{2} \sqrt{\frac{1}{2} + \frac{3\sqrt{3}}{4\pi} \cos 2\varphi}. \tag{2.64}$$

This gives the quadratic sum of the harmonics:

$$\sqrt{I_{\text{rms}}^2 - I^2} = I' \sqrt{2} \sqrt{0.044 - 0.0424 \cos 2\varphi}.$$

- In addition to the average term I, the series expansion of current i includes the AC terms of 6ω, 12ω, 18ω, . . . angular frequency.

  The rms value of the $6k\omega$ angular frequency term is equal to

$$I_{6k} = I' \frac{6|\cos\varphi|}{\pi(36k^2 - 1)} \sqrt{1 + 36k^2 \tan^2 \varphi}. \tag{2.65}$$

It reaches its minimum for φ equal to 0 or π and its maximum for φ equal to $\pi/2$:

$$I_{6k\,\text{min}} = I' \frac{6}{\pi(36k^2 - 1)}, \qquad I_{6k\,\text{max}} = I' \frac{36k}{\pi(36k^2 - 1)}.$$

Figure 2.23 shows the variations, as a function of φ, of the mean value I of the input current and the rms value I_6, I_{12}, I_{18} of the first harmonics of this current. All these values are referred to I'.

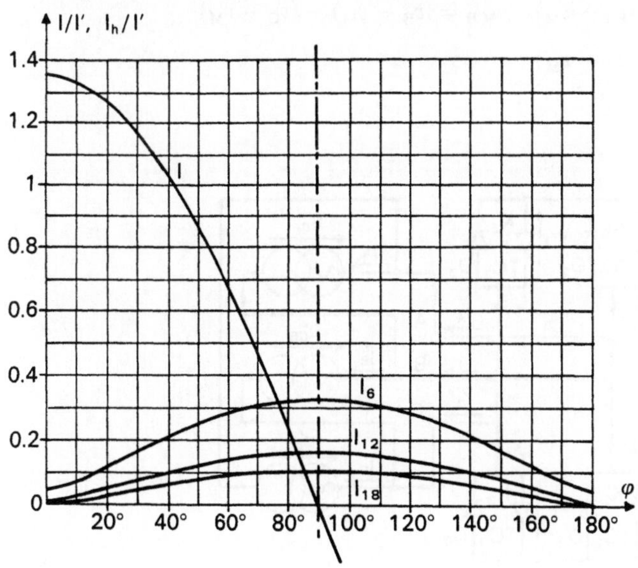

Fig. 2.23

The average value of the current is reversed for φ equal to $\pi/2$ since $I(\pi/2+\varphi)$ is equal to $-I(\pi/2-\varphi)$.

All these curves can be used for negative φ, since the expressions of I and I_{6k} indicate that these values do not depend on the sign of φ.

A comparison with Fig. 2.5 shows the important reduction of the harmonics when going from a single-phase two-switch inverter to a three-phase six-switch inverter.

2.4.3 Delta-Connected AC Load

Instead of being star-connected as in Fig. 2.21, the AC load may be delta-connected. In this case, the diagram becomes that shown in Fig. 2.24; on the latter, in addition to the notations previously used, phase currents j'_A, j'_B, j'_C and phase voltages u'_{BA}, u'_{CB}, u'_{AC} have been added.

2.4.3.1 Voltages

Turning on the switches imposes the voltage directly across the AC load.

If the negative terminal M of the DC voltage supply is taken as reference point,

when K_1 is on, $v_A - v_M = U$; when K'_1 is on, $v_A - v_M = 0$;
when K_2 is on, $v_B - v_M = U$; when K'_2 is on, $v_B - v_M = 0$;
when K_3 is on, $v_C - v_M = U$; when K'_3 is on, $v_C - v_M = 0$.

Since

$$u'_{BA} = (v_A - v_M) - (v_B - v_M), \quad u'_{CB} = (v_B - v_M) - (v_C - v_M)$$

$$u'_{AC} = (v_C - v_M) - (v_A - v_M)$$

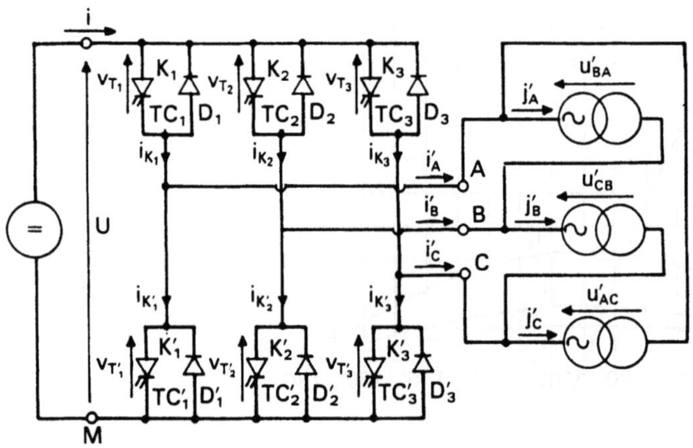

Fig. 2.24

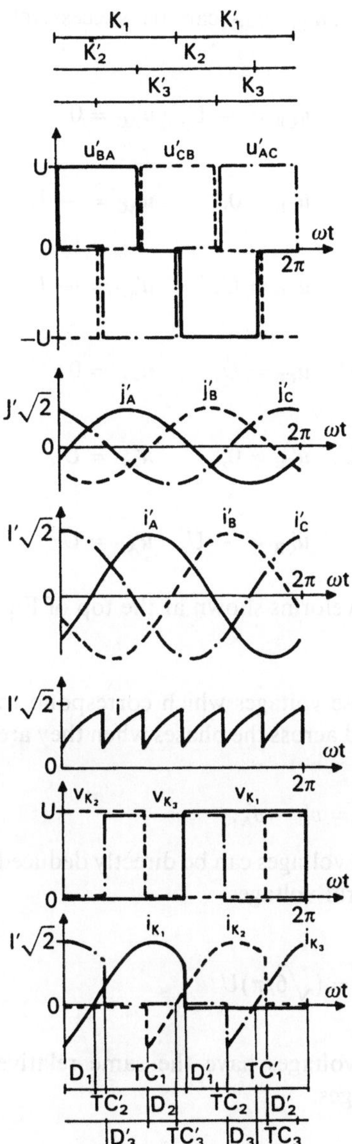

Fig. 2.25

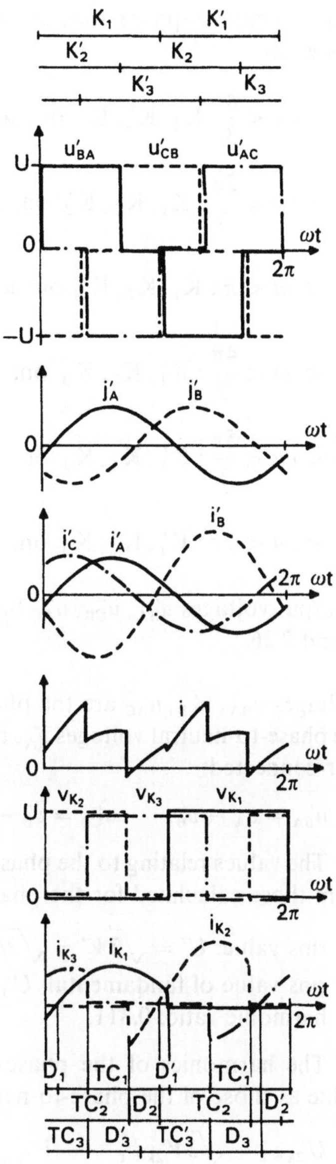

Fig. 2.26

the three phase-to-phase output voltages $u'_{BA}, u'_{CB}, u'_{AC}$ can be successively expressed as

$$0 < \omega t < \frac{\pi}{3}: \mathrm{K}_1, \mathrm{K}'_2, \mathrm{K}_3 \text{ on}; \; u'_{BA} = U \qquad u'_{CB} = -U, \quad u'_{AC} = 0$$

$$\frac{\pi}{3} < \omega t < \frac{2\pi}{3}: \mathrm{K}_1, \mathrm{K}'_2, \mathrm{K}'_3 \text{ on}; \; u'_{BA} = U, \qquad u'_{CB} = 0, \qquad u'_{AC} = -U$$

$$\frac{2\pi}{3} < \omega t < \pi: \mathrm{K}_1, \mathrm{K}_2, \mathrm{K}'_3 \text{ on}; \; u'_{BA} = 0, \qquad u'_{CB} = U, \qquad u'_{AC} = -U$$

$$\pi < \omega t < \frac{4\pi}{3}: \mathrm{K}'_1, \mathrm{K}_2, \mathrm{K}'_3 \text{ on}; \; u'_{BA} = -U, \quad u'_{CB} = U, \qquad u'_{AC} = 0$$

$$\frac{4\pi}{3} < \omega t < \frac{5\pi}{3}: \mathrm{K}'_1, \mathrm{K}_2, \mathrm{K}_3 \text{ on}; \; u'_{BA} = -U, \quad u'_{CB} = 0, \qquad u'_{AC} = U$$

$$\frac{5\pi}{3} < \omega t < 2\pi: \mathrm{K}'_1, \mathrm{K}'_2, \mathrm{K}_3 \text{ on}; \; u'_{BA} = 0, \qquad u'_{CB} = -U, \quad u'_{AC} = U$$

Output voltages $u'_{BA}, u'_{CB}, u'_{AC}$ have the waveforms shown at the top of Fig. 2.25 and 2.26.

- Voltages $u'_{BA}, u'_{CB}, u'_{AC}$ are the phase-to-phase voltages which correspond to the phase-to-neutral voltages v'_A, v'_B, v'_C found across the phases when they are star-connected:

  $$u'_{BA} = v'_A - v'_B, \qquad u'_{CB} = v'_B - v'_C, \quad u'_{AC} = v'_C - v'_A.$$

  The values relating to the phase-to-phase voltages can be directly deduced from those calculated for the phase-to-neutral voltages:

  rms value: $U' = \sqrt{3}\,V' = \sqrt{2/3}\,U$
  rms value of fundamental: $U'_1 = \sqrt{3}\,V'_1 = (\sqrt{6}/\pi)U$
  harmonic ratio: 0.311.

  The harmonics of the phase-to-phase voltages have the same relative value as those of the phase-to-neutral voltages:

  $$\frac{U'_{2k+1}}{U'_1} = \frac{\sqrt{3}\,V'_{2k+1}}{\sqrt{3}\,V'_1} = \frac{1}{2k+1}.$$

2.4.3.2 Currents

If currents j'_A, j'_B, j'_C, are assumed to form a balanced system, they can be characterized by their rms value J' and their phase shift φ with respect to the fundamental of the corresponding voltages.

The fundamental of u'_{BA} has the form $U'_1\sqrt{2}\sin(\omega t + \pi/6)$ and the phase currents can thus be written:

$$j'_A = J'\sqrt{2}\sin\left(\omega t + \frac{\pi}{6} - \varphi\right)$$

$$j'_B = J'\sqrt{2}\sin\left(\omega t + \frac{\pi}{6} - \varphi - \frac{2\pi}{3}\right)$$

$$j'_C = J'\sqrt{2}\sin\left(\omega t + \frac{\pi}{6} - \varphi - \frac{4\pi}{3}\right).$$

This gives the currents at the inverter output terminals:

$$i'_A = j'_A - j'_C = I'\sqrt{2}\sin(\omega t - \varphi)$$

$$i'_B = j'_B - j'_A = I'\sqrt{2}\sin\left(\omega t - \varphi - \frac{2\pi}{3}\right)$$

$$i'_C = j'_C - j'_B = I\sqrt{2}\sin\left(\omega t - \varphi - \frac{4\pi}{3}\right)$$

with $I' = \sqrt{3}J'$.

Since, for a given value I', the currents at the inverter terminals are the same as those delivered when the load was star-connected, the currents in the switches, the semiconductor devices and input current i will also remain unchanged.

In order to show the operation on a balanced delta-connected load, Fig. 2.25 represents the on-periods of the switches as well as the waveforms

of u'_{BA}, u'_{CB}, u'_{AC}
of j'_A, j'_B, j'_C
of i'_A, i'_B, i'_C
of i
of v_{K_1}, v_{K_2}, v_{K_3}
of i_{K_1}, i_{K_2}, i_{K_3}

together with the conduction diagrams of the different semiconductor devices. The values of I' and φ are those used in drawing Fig. 2.22a. It can be seen that, in these circumstances, there have been no modifications as far as the inverter is concerned.

This is a simple application of the star-to-delta conversion. Replacing the star-connected load by the equivalent delta-connected one has no effect on the operation of the circuit to which this load is connected.

2.4.4 Operation with an Unbalanced Load

The delta-connection of the three phases of the load enables them to be supplied with balanced voltages even if these phases absorb unbalanced currents.

In cases where, on the AC side, a neutral conductor is needed to feed certain equipments between phase and neutral and where a transformer must also be used, by using a suitable connection of the transformer windings, balanced output voltages can still be obtained even if the load is unbalanced.

Only in the case where a neutral conductor is required and where there is no transformer does a capacitive input divider need to be reintroduced.

2.4.4.1 Delta-Connected Load

If the three phases are delta-connected (see Fig. 2.24) turning on and off the switches imposes voltages between the output terminals A, B, C and thus across the load.

The unbalance of the latter leads to an unbalance in the load phase currents j'_A, j'_B, j'_C and the line currents i'_A, i'_B, i'_C. The currents flowing through the various switches are different and, in certain cases, the types of commutation which they have to ensure may differ. The input current i now has a cycle equivalent to twice the output voltage cycle instead of six times.

Figure 2.26 shows, by way of example, an operation mode where two phases absorb sinusoidal currents j'_A and j'_B, of the same amplitude and shifted between themselves by $2\pi/3$; the third phase absorbs a zero current. The output voltages and the voltages across the switches are not affected by any current unbalance.

2.4.4.2 Use of an Output Transformer

When a transformer is needed and if the secondary windings are star-connected in order to have a neutral conductor, the primary windings must be delta-connected (Fig. 2.27).

- The switches impose the phase-to-phase primary voltages $u'_{BA}, u'_{CB}, u'_{AC}$ and thus the phase-to-neutral secondary voltages v''_A, v''_B, v''_C and, as a result, the phase-to-phase secondary voltages.
- The unbalance of the load leads to a non zero sum of the secondary currents i''_A, i''_B, i''_C and to the presence of a current i''_N in the neutral.

  The delta-connection of the primary allows the compensation of m.m.f. on each transformer leg:

$$j'_A = \frac{n_2}{n_1} i''_A, \quad j'_B = \frac{n_2}{n_1} i''_B, \quad j'_C = \frac{n_2}{n_1} i''_C$$

thus: $j'_A + j'_B + j'_C = \frac{n_2}{n_1} i''_N$.

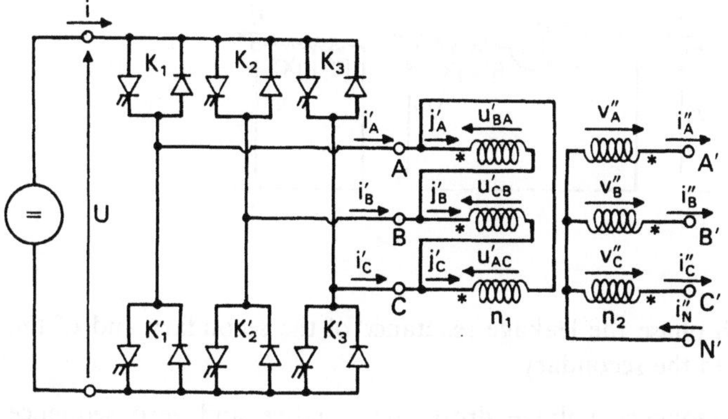

Fig. 2.27

The currents supplied by the inverter are thus

$$i'_A = j'_A - j'_C = \frac{n_2}{n_1}(i''_A - i''_C)$$

$$i'_B = j'_B - j'_A = \frac{n_2}{n_1}(i''_B - i''_A)$$

$$i'_C = j'_C - j'_B = \frac{n_2}{n_1}(i''_C - i''_B).$$

Their sum is zero whatever the value of i''_N.

The corresponding magnetizing currents must be added to these primary currents.

- For the fundamental frequency terms, the voltage drops and the *voltage unbalance* due to internal impedance of the transformer has still to be estimated.

  The unbalanced system of secondary currents at fundamental frequency has to be split into three balanced systems of components $\underline{I}''_d$, $\underline{I}''_i$, $\underline{I}''_0$ respectively; the first one is of normal positive sequence, the second is of negative sequence and the third of zero sequence.

  The positive and negative sequence components of the secondary currents are balanced out by the positive and negative components in primary windings. They bring corresponding components in the line currents.

  The secondary zero sequence component is balanced out by a current flowing in the triangle formed by the primary windings.

  This leads to the single-phase equivalent circuits of the transformer for the three systems (Fig. 2.28): $\underline{Z}_\mu$ represents the magnetizing impedance.

  Since the secondary m.m.f. is balanced out by the primary one in the three systems, impedance $R'' + jX''$ is the same in all three configurations:

  R'' is for each phase the resistance of the secondary and of the primary referred to the secondary,

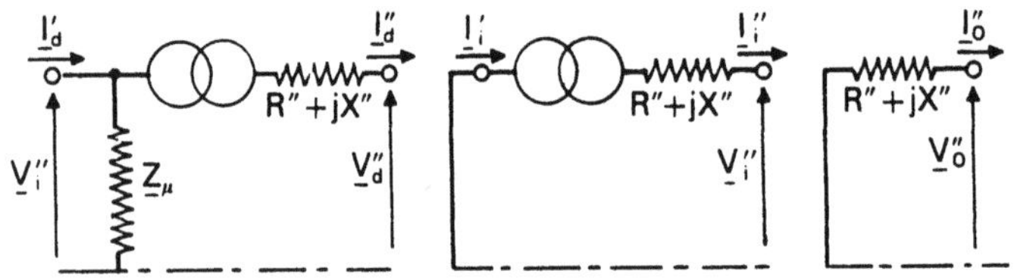

Fig. 2.28

X'' is for each phase the leakage reactance of the secondary and of the primary referred to the secondary.

The positive sequence voltage drop, the negative and zero sequence secondary voltages are thus given by

$$\Delta \underline{V}''_d = (R'' + jX'')\underline{I}''_d, \qquad \underline{V}''_i = (R'' + jX'')\underline{I}''_i, \quad \underline{V}''_0 = (R'' + jX'')\underline{I}''_0.$$

If $R'' + jX''$ is such that the voltage drop due to I''_d is small, then the unbalance of the secondary voltages due to I''_i and I''_0 will also be of reduced value[1].

2.4.4.3 Reintroducing a Capacitive Input Divider

If the inverter has to feed the load through a four-wire connection and if there is no transformer, the centre point of the capacitive divider forms the neutral (Fig. 2.29).

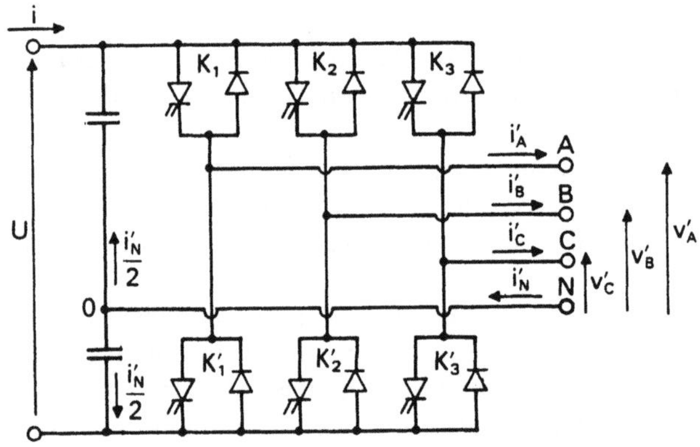

Fig. 2.29

[1] Things would be very different if the primary windings were star-connected with an insulated neutral. The zero sequence component of the secondary currents could no longer be balanced. The impedance for the zero sequence system would be much larger than for the two others (especially in the case of a free flux transformer). At a given I''_0, voltage V''_0 and thus the unbalance of the phase secondary voltages would be greatly increased.

- Connecting neutral N to the centre point O of the divider modifies the waveform and the value of the phase-to-neutral voltages. If the capacitors are large enough for the voltages across them to be assumed constant and equal to $U/2$:

  when K_1 is on, $v'_A = U/2$; when K'_1 is on, $v'_A = -U/2$;
  when K_2 is on, $v'_B = U/2$; when K'_2 is on, $v'_B = -U/2$;
  when K_3 is on, $v'_C = U/2$; when K'_3 is on, $v'_C = -U/2$.

  The phase-to-neutral voltages now have the following rms value:

$$V' = \frac{U}{2}$$

  with their fundamental equal to

$$V'_1 = \frac{\sqrt{2}}{\pi} U.$$

  Their harmonic ratio is equal to 0.483.

  Such an increase in the harmonic content is due to the presence of harmonics of rank 3 and multiple of 3.

  The phase-to-phase voltages remain unchanged.

- If the unbalance of the output currents i'_A, i'_B, i'_C (or the presence in these currents of harmonics which form zero sequence systems) makes the sum $i'_A + i'_B + i'_C$ different from zero, a current flows in the neutral conductor.

$$i'_N = i'_A + i'_B + i'_C.$$

  Current i'_N corresponds to a current $i'_N/2$ through each capacitor of the capacitive divider.Current i delivered by the DC voltage supply thus becomes

$$i = i_{K_1} + i_{K_2} + i_{K_3} - \frac{1}{2} i'_N = i_{K'_1} + i_{K'_2} + i_{K'_3} + \frac{1}{2} i'_N.$$

- The study of the value to be given to the capacitor which was carried out in the case of the half-bridge inverter (Sect. 2.2.1.4) can be used provided that i' is replaced by i'_N.

  If the current in the neutral conductor has an rms value of I'_N and if its angular frequency is ω then, in order to make the voltage ripple across the capacitors negligible, the latter must be given a value C such that

$$C \gg \frac{I'_N}{\omega U}$$

- As an illustration of these remarks, we have shown, in Fig. 2.30, the waveforms of voltages v'_A, v'_B, v'_C, of currents i'_A, i'_B, i'_C, of current i'_N and of current i for the case where i'_A and i'_B are sinusoidal currents of the same amplitude, shifted by $2\pi/3$ and where i'_C is zero.

  K_1 and K'_1, K_2 and K'_2 operate as single-phase half-bridges with output currents i'_A and i'_B. The input current has a fundamental angular frequency equal to 2ω.

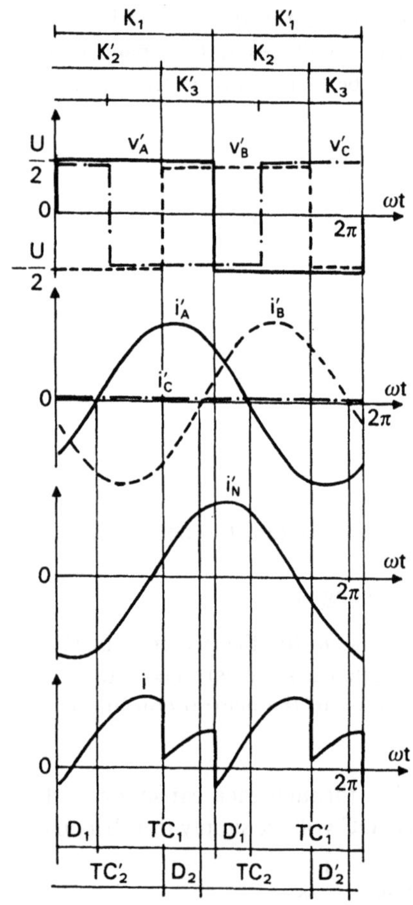

Fig. 2.30

2.5 Input Filter

The study of the input filter of voltage-source inverters is very similar to that carried out in the case of choppers (see Vol. 3, Chap. 4, Sect. 2).

If the DC voltage supply is inductive, the sharp variations of current i which it delivers to the inverter lead to voltage spikes at the latter's input and, as a result, across the semiconductor devices of which it is made up. A capacitor C must be connected across the inverter input in order to correct the DC voltage supply.

Normally a series inductor is added to reduce the ripple of the current taken at the supply.

The series inductor L_S and the parallel capacitor C make up the input filter. The latter is analyzed using the equivalent circuit shown in Fig. 2.31. Voltage source U_S is assumed to be perfect, since the inductance of the real supply has

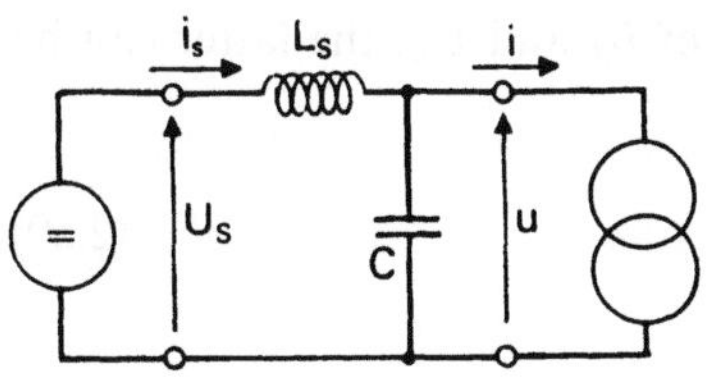

Fig. 2.31

been combined with the additional inductance in L_S. The inverter and AC load are assumed to be perfect: current i of the current source is thus as defined in the previous paragraphs.

For the present, we will confine ourselves to a rapid analysis in the frequency domain. A more detailed study of the filtering and particularly of the ripple $\Delta\, i_S$ in the input current and the ripple Δu in the output voltage will be provided in Appendix A.

2.5.1 Input Current and Output Voltage Harmonics

- Current i_S at the filter input and voltage u at its output are related to voltage U_S of the voltage source and to current i of the current source by

$$L_S \frac{di_S}{dt} + u = U_S$$

$$i_S - C\frac{du}{dt} = i. \tag{2.66}$$

Current i is the sum of a mean value term I and of harmonics with rms values I_h and angular frequencies $h\omega$.

Constant voltage U_S provides the mean value U of u; the mean value I_S of i_S is equal to that of i:

$$U = U_S, \qquad I_S = I. \tag{2.67}$$

The current i harmonics bring the same frequency terms in i_S and in u. The rms values I_{Sh} and U_h of terms with angular frequency $h\omega$ are related to I_h by

$$I_{Sh} = \frac{I_h}{|1 - L_S C h^2 \omega^2|} \qquad U_h = \frac{L_S h\omega I_h}{|1 - L_S C h^2 \omega^2|}$$

The resonance frequency ω_f of the filter is equal to

$$\omega_f = \frac{1}{\sqrt{L_S C}} \tag{2.68}$$

If this is transferred into the expressions of I_{Sh} and U_h, the latter can be written as

$$I_{Sh} = I_h \frac{1}{\left|1 - \left(\frac{h\omega}{\omega_f}\right)^2\right|} \tag{2.69}$$

$$U_h = \sqrt{\frac{L_S}{C}} I_h \frac{\frac{h\omega}{\omega_f}}{\left|1 - \left(\frac{h\omega}{\omega_f}\right)^2\right|}. \tag{2.70}$$

Figure 2.32 shows the variations as a function of $h\omega/\omega_f$, of I_{Sh}/I_h and $U_h/(I_h\sqrt{L_S/C})$. For the higher values of $h\omega/\omega_f$, they show, respectively, little difference from $1/(h\omega/\omega_f)^2$ and from $1/(h\omega/\omega_f)$.

- In the case of the first harmonic of current i, the ratio $h_1\omega/\omega_f$ must be well above unity, in order to obtain a small value for *current* i_S *ripple*. This indicates the value to be given to the product $L_S C$.

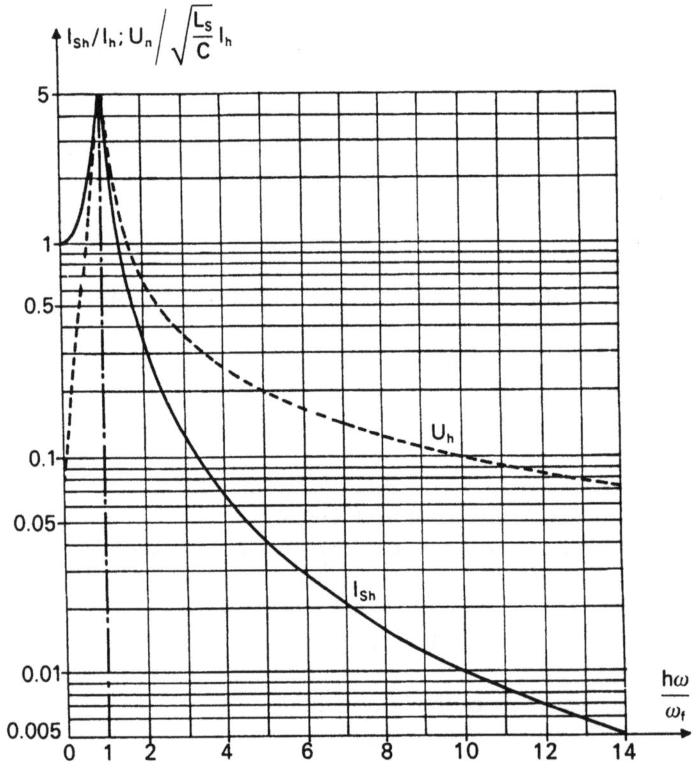

Fig. 2.32

For the first harmonic,

$$\sqrt{L_S C}\, h_1 \omega \gg 1 \tag{2.71}$$

If this condition is satisfied, the I_{Sh}/I_h ratio will be even lower for the harmonics of higher rank. As an initial approximation, only the first one may be considered. The peak-to-peak ripple Δi_S of current i_S will not be very different from the peak-to-peak value of its first harmonic.

The *voltage u ripple* indicates how the product $L_S C$ should be distributed between L_S and C. If condition (2.71) is satisfied, Eq. (2.70) becomes

$$U_h \simeq \frac{I_h}{Ch\omega}.$$

This means that virtually the whole of current I_h flows through the capacitor. For the voltage ripple across the latter to be small compared to its mean value U_S, the following must be true for the first harmonic:

$$\frac{I_{h_1}}{Ch_1\omega} \ll U_S \quad \text{or} \quad Ch_1\omega \gg \frac{I_{h_1}}{U_S}. \tag{2.72}$$

If this condition is satisfied for the first harmonic, it will be all the more so for terms of higher angular frequency. The peak-to-peak ripple Δu of voltage u may be supposed to be equal to the peak-to-peak value of its first harmonic.

2.5.2 Application to the Different Inverters

If current i' is assumed to be sinusoidal at the inverter output, current i at the latter's input (i.e. at the filter output) depends on the rms value I' of i' as well as on its phase shift φ with respect to the fundamental component of u'.

At a given I', the mean value I increases the nearer $\cos\varphi$ is to 1 and the value of the first harmonic increases the closer $\cos\varphi$ is to zero. The maximum ripples $\Delta i_{S\,max}$ and Δu_{max} must thus be evaluated for the rated value of I' and the smallest value of $\cos\varphi$. Ripple Δi_S of current i_S will be referred to its mean value I for $\cos\varphi$ equal to 1; ripple Δu will be referred to the mean value U_S of voltage u.

2.5.2.1 Single-Phase Inverter with Centre-Tapped Transformer

The first harmonic of current i has an angular frequency 2ω. The mean value I of this current and the rms value I_2 of its 2nd harmonic are given by Eqs. (2.15) and (2.19):

$$I = \frac{2n_2}{n_1} I' \frac{2\sqrt{2}}{\pi} \cos\varphi$$

$$I_2 = \frac{2n_2}{n_1} I' \frac{4}{3\pi} |\cos\varphi| \sqrt{1 + 4\tan^2\varphi}.$$

I, equal to I_S, is maximum for $\cos\varphi$ equal to 1 and has then the following value:

$$I_{S\max} = \frac{2n_2}{n_1} I' \frac{2\sqrt{2}}{\pi}.$$

I_2 is at its maximum for $\cos\varphi$ equal to 0 and has then the following value.

$$I_{2\max} = \frac{2n_2}{n_1} I' \frac{8}{3\pi}.$$

If $\Delta i_{S\max}$ and $\Delta u_{\max}$ are computed for $\cos\varphi = 0$ and if Eqs. (2.69) and (2.70) are taken into account, the relative values of the ripples can be writtèn as

$$\frac{\Delta i_{S\max}}{I_{S\max}} = \frac{2\sqrt{2}\, I_{S_2\max}}{I_{S\max}} = \frac{8}{3} \frac{1}{4L_S C\omega^2 - 1} \tag{2.73}$$

$$\frac{\Delta u_{\max}}{U_S} = \frac{2\sqrt{2}\, U_{2\max}}{U_S} = \frac{2n_2}{n_1} I' \frac{1}{U_S} \frac{32\sqrt{2}}{3\pi} \frac{L_S\omega}{4L_S C\omega^2 - 1}. \tag{2.74}$$

For example, if both ripples must have a relative value of 10%, the following is necessary:

$$\sqrt{L_S C}\, 2\omega = 5.26, \quad L_S = 0.555 \frac{1}{\omega} \frac{U_S}{\frac{2n_2}{n_1} I'}, \quad C = 12.45 \frac{1}{\omega} \frac{\frac{2n_2}{n_1} I'}{U_S}.$$

2.5.2.2 Single-Phase Half-Bridge Inverter

The expression of the half-bridge inverter input current i is the same as that of the push–pull inverter input current, provided only that $2n_2 I'/n_1$ is replaced by $I'/2$.

If this is carried out, all the previous equations can be used.

2.5.2.3 Single-Phase-Full-Bridge Inverter

In the case of the single-phase full-bridge inverter, the first harmonic has once again an angular frequency of 2ω. But current i at the inverter input depends not only on I' and φ in this case, but also on β.

- If β may vary from zero to π, the highest value of the 2nd harmonic is obtained – for a given value of I' – for β and $\cos\varphi$ equal to zero. The equations laid down for the push–pull inverter can once again be used, providing that $2n_2 I'/n_1$ is now replaced by I'.
- If the range of variations of β is restricted, then curves I/I' in Fig. 2.17 and I_2/I' in Fig. 2.19 show that this restriction must be considered in computing the ripples.

For example, if β is kept equal to $\pi/3$ in order to suppress the third harmonic of the output voltage, Eq. (2.45) and (2.47) give

$$I = \frac{2\sqrt{2}}{\pi}\frac{\sqrt{3}}{2}\cos\varphi\, I', \qquad I_2 = \frac{2}{\pi}\frac{\sqrt{3}}{2} I' \quad \text{whatever the value of } \varphi.$$

By considering only the first harmonic, this gives

$$\frac{\Delta i_S}{I_{S\,max}} = \frac{2\sqrt{2}\,I_{S_2}}{I_{S\,max}} = \frac{2}{4L_S C\omega^2 - 1}$$

$$\frac{\Delta u}{U_S} = \frac{2\sqrt{2}\,U}{U_S} = \frac{I'}{U_S}\frac{4\sqrt{6}}{\pi}\frac{L_S\omega}{4L_S C\omega^2 - 1}.$$

A comparison with (2.73) and (2.74) shows that, for given values of the ripples, the filter elements can be slightly reduced.

2.5.2.4 Three-Phase Full-Bridge Inverter

– If the three-phase load is *balanced*, the first harmonic of current i at the inverter input is that of angular frequency 6ω. Equations (2.63) and (2.64) give

$$I = I'\frac{3\sqrt{2}}{\pi}\cos\varphi$$

$$I_6 = I'\frac{6|\cos\varphi|}{35\pi}\sqrt{1 + 36\tan^2\varphi}.$$

The mean value I of current i or of current i_S is at its maximum for $\cos\varphi$ equal to 1. The rms value I_6 of the first harmonic of current i is at its maximum for zero $\cos\varphi$.

$$I_{S\,max} = I'\frac{3\sqrt{2}}{\pi}, \qquad I_{6\,max} = I'\frac{36}{35\pi}.$$

This gives the approximate relative values of the maximum values of current i_S and voltage u ripples:

$$\frac{\Delta i_{S\,max}}{I_{S\,max}} = \frac{2\sqrt{2}\,I_{6\,max}}{I_{S\,max}} = \frac{24}{35}\frac{1}{36L_S C\omega^2 - 1}$$

$$\frac{\Delta u_{max}}{U_S} = \frac{2\sqrt{2}\,U_{6\,max}}{U_S} = \frac{I'}{U_S}\frac{432\sqrt{2}}{35\pi}\frac{L_S\omega}{36L_S C\omega^2 - 1}$$

In order to limit these two maximum ripples to a relative value of 10% as in the example given for the single-phase inverter, the following is necessary:

$$\sqrt{L_S C}\,6\omega = 2.80, \quad L_S = 0.123\frac{1}{\omega}\frac{U_S}{I'}, \quad C = 1.77\frac{1}{\omega}\frac{I'}{U_S}$$

A comparison with the values found in the single-phase inverter shows that multiplying the first harmonic frequency by 3 enables the per-unit value of the inductance to be divided by 4.5 and the per-unit value of the capacitance by 7.

- If the load is *unbalanced*, the ripples of i_S and u depend mainly on the 2nd harmonic of current i as soon as this component may not be neglected. The filter must then be calculated using the value of the 2nd harmonic.

- *Remark.* Even if the load is considered to be balanced, as a result of unavoidable imperfections in the load, in the converter and in its control, current i shows component of angular frequency ω and 2ω. Such *abnormal harmonic* frequencies must not be located too close the filter resonance frequency, or they will be greatly amplified by the filter.

  In the example given for the three-phase inverter, $6\omega/\omega_f = 2.80$ is not suitable since $2\omega/\omega_f$ is too close to 1. If operation is only required on a balanced load, a value about 4.5 could be used for $6\omega/\omega_f$, giving $\omega/\omega_f = 0.75$ and $2\omega/\omega_f = 1.5$. For operation on an unbalanced load, in order to obtain a sufficiently high value of $2\omega/\omega_f$ without ω/ω_f being too close to 1, ω/ω_f must be chosen above 1 – e.g. $\omega/\omega_f = 1.5$ – giving $2\omega/\omega_f = 3$ and $6\omega/\omega_f = 9$.

  This remark concerning abnormal harmonics shows that a filter which is more efficient but more expensive than one which would sufficiently reduce the normal first harmonic is often chosen.

Similarly, in the case of single-phase inverters, $2\omega/\omega_f$ must be sufficiently high for ω/ω_f to be considerably greater than 1. However, in this case, the calculation made using the 2ω term usually gives a value of ω_f which satisfies this condition.

2.5.3 Remark Concerning the Use of the Half-Bridge Inverter Divider as an Input Filter Element

The two capacitors of the capacitive divider of the half-bridge single-phase inverter can act as an input filter capacitance.

Figure 2.33 shows the circuit equivalent to an inverter fed by a constant voltage source U_S through an inductance L_S during the first half-cycle of u' (switch K_1 on), and then during the second half-cycle (K_2 on).

The following still applies:

$$L_S \frac{di_S}{dt} + u = U_S.$$

When K_1 is on:

$$i_S = C_1 \frac{du_{C_1}}{dt} + i' = C'_1 \frac{du_{C'_1}}{dt} \quad \text{with } C_1 = C'_1 = C$$

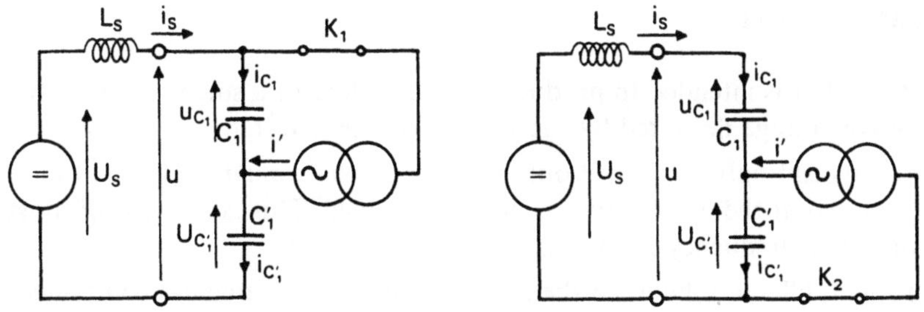

Fig. 2.33

thus:

$$i_S = \frac{1}{2}\left[C\left(\frac{du_{C_1}}{dt} + \frac{du_{C'_1}}{dt}\right) + i'\right] = \frac{1}{2}C\frac{du}{dt} + \frac{i'}{2}.$$

When K_2 is on:

$$i_S = C_1\frac{du_{C_1}}{dt} = C'_1\frac{du_{C'_1}}{dt} - i' = \frac{1}{2}C\frac{du}{dt} - \frac{i'}{2}.$$

This therefore gives, throughout the cycle,

$$L_S\frac{di_S}{dt} + u = U_S$$

$$i_S - \frac{1}{2}C\frac{du}{dt} = i \tag{2.75}$$

using i to denote the current, equal at times to $i'/2$ and at others to $-i'/2$, which is taken by the half-bridge inverter assumed to be perfect.

Equations (2.75) are identical to those of the system (2.66) providing only that C is replaced by $C/2$, i.e. that the capacitance of the capacitor equivalent to two series-connected capacitors C_1 and C'_1 is used.

Equations (2.68), (2.69), and (2.70) become

$$\omega_f = \frac{1}{\sqrt{L_S\frac{C}{2}}} \qquad I_{Sh} = I_h\frac{1}{\left|1 - \left(\frac{h\omega}{\omega_f}\right)^2\right|} \qquad U_h = \sqrt{\frac{2L_S}{C}}\,I_h\frac{\frac{h\omega}{\omega_f}}{\left|1 - \left(\frac{h\omega}{\omega_f}\right)^2\right|}$$

and enable an approximate evaluation to be made of the voltage u and current i_S ripples from 2nd harmonic of current i.

The value which must be given to C in order to reduce the voltage u ripple on either side of U_S is close to that needed to reduce the ripple of u_{C_1} and $u_{C'_1}$ on either side of $U_S/2$ to a comparable relative value (see Sect. 2.2.1.4).

2.6 Output Filter

The output filter is intended to produce a voltage close to a sine wave from the square-wave voltage delivered by the voltage-source inverter.

The study can only be carried out in the *frequential domain*, if the nature of the load placed at the filter output is not well defined. This is the case of most constant output frequency inverters.

The small difference between the frequency of the fundamental component and that of the first harmonic when the latter is of a rank equal to 3, makes it *very difficult to obtain effective filtering with a simple filter*, even for a fixed frequency output. We shall show this by a study of the L–C filter.

For the case where the output is at a variable frequency, the only option is to "smooth" the current wave supplied to the load.

2.6.1 General Equations

The DC voltage source feeding the inverter is assumed to be perfect, as are the semiconductor devices which make up the latter. Under these assumptions, voltage u' at the filter input is as was determined during the study of the various inverters.

The filter comprises an inductor L and a capacitor C. The load is represented by an admittance $\underline{Y}$ (Fig. 2.34). The modulus and phase angle of the latter depend on the rank n of the harmonic under consideration, without there being necessarily any simple relation between the admittances which correspond to the various terms of the series expansion of voltage u'.

Using complex numbers, for $n\omega$ angular frequency component, the filter input and output values are linked by

$$\underline{U}''_n = \underline{U}'_n \frac{\dfrac{1}{\underline{Y}_n + \mathrm{j}Cn\omega}}{\mathrm{j}Ln\omega + \dfrac{1}{\underline{Y}_n + \mathrm{j}Cn\omega}}$$

$$\underline{I}''_n = \underline{I}'_n \frac{\underline{Y}_n}{\underline{Y}_n + \mathrm{j}Cn\omega}$$

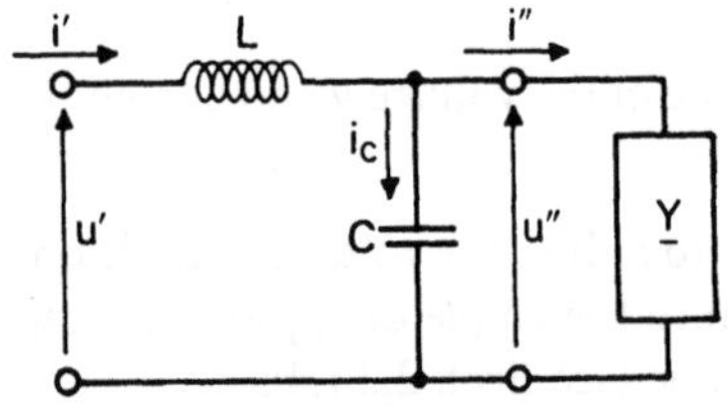

Fig. 2.34

with $\underline{Y}_n$ being used to denote the load admittance for $n\omega$ angular frequency component

$$\underline{Y}_n = G_n - \mathrm{j}B_n = Y_n(\cos\varphi_n - \mathrm{j}\sin\varphi_n) \tag{2.76}$$

with φ_n positive if the nth harmonic of current i'' is lagging the nth harmonic of voltage u''.

The filter can be characterized by L and C or by its angular resonance frequency ω_f and characteristic conductance G_f:

$$\omega_f = \frac{1}{\sqrt{LC}} \qquad G_f = \sqrt{\frac{C}{L}} \tag{2.77}$$

According to our needs we will use one or another of the following systems of equations.

$$\underline{U}''_n = \underline{U}'_n \frac{1}{1 - \left(\dfrac{n\omega}{\omega_f}\right)^2 + \mathrm{j}Ln\omega\,\underline{Y}_n} \tag{2.78a}$$

$$\underline{I}''_n = \underline{I}'_n \frac{1}{1 + \dfrac{\mathrm{j}Cn\omega}{\underline{Y}_n}} \tag{2.78b}$$

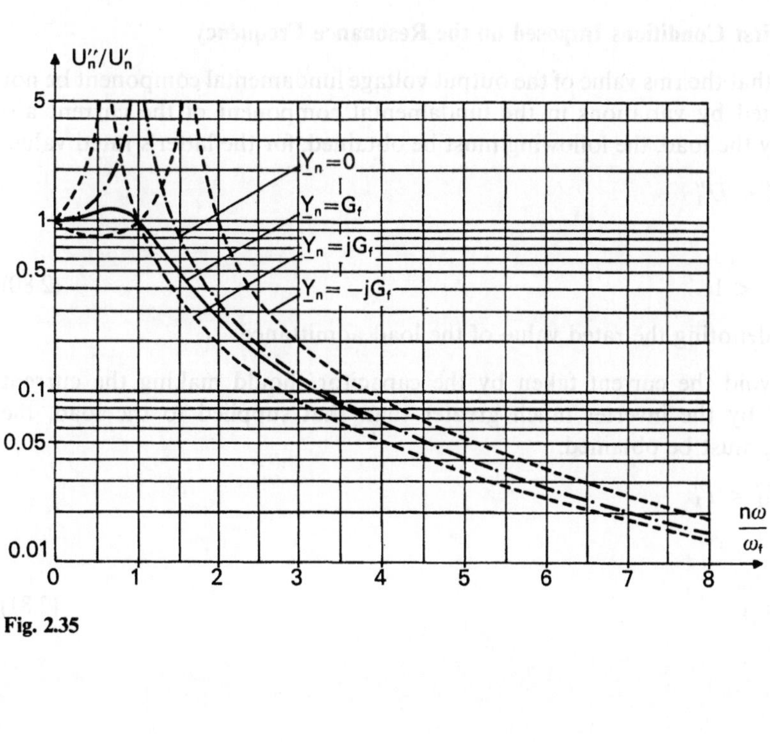

Fig. 2.35

$$\underline{U}''_n = \underline{U}'_n \frac{1}{1 - \left(\frac{n\omega}{\omega_f}\right)^2 + j\left(\frac{n\omega}{\omega_f}\right)\frac{\underline{Y}_n}{G_f}} \tag{2.79a}$$

$$\underline{I}''_n = \underline{I}'_n \frac{1}{1 + j\left(\frac{n\omega}{\omega_f}\right)\frac{G_f}{\underline{Y}_n}} \tag{2.79b}$$

To begin to show the influence of the load, we have plotted (Fig. 2.35) the variations of ratio U''_n/U'_n as a function of $n\omega/\omega_f$,

for a no load operation ($Y_n = 0$)
then for $\underline{Y}_n$ successively equal to G_f, to jG_f and to $-jG_f$.

It can be seen that, for angular frequencies close to ω_f, the effect of the filter is closely dependent on the load. It is only for angular frequencies much greater than ω_f that the influence of the load is reduced and that the curves plotted for a loaded filter come close to the no load curve.

2.6.2 Effects of the Filter on the Fundamental

Before studying the effects of the filter on ratios U''_1/U'_1 and I'_1/I''_1, we shall determine the zone in which ratio ω/ω_f of the fundamental component frequency to the resonance frequency must be located.

2.6.2.1 First Conditions Imposed on the Resonance Frequency

In order that the rms value of the output voltage fundamental component be not too affected by variations in the fundamental component of the current absorbed by the load, the following must be obtained, for the latter's rated value:

$$L\omega I''_1 < U''_1$$

or

$$L\omega Y_1 < 1 \tag{2.80}$$

with Y_1 denoting the rated value of the load admittance.

To avoid the current taken by the capacitor should making the current delivered by the inverter much greater than that supplied to the load, the following must be obtained:

$$C\omega U''_1 < I''_1$$

or

$$\frac{C\omega}{Y_1} < 1. \tag{2.81}$$

By combining conditions (2.80) and (2.81), we have

$$LC\omega^2 < 1$$

or

$$\omega < \omega_f. \tag{2.82}$$

The two effects of the filter must thus be examined for a fundamental component frequency considerably lower than the resonance frequency of the filter.

2.6.2.2 Variation in the Output Voltage

At no load, the output voltage U''_{10} is linked to the input voltage by

$$\frac{U''_{10}}{U'_1} = \frac{1}{1 - (\omega/\omega_f)^2}. \tag{2.83}$$

For I''_1 (or Y_1) other than zero, the voltage ratio becomes

$$\frac{U''_1}{U'_1} = \frac{1}{\left\{\left[1 - \left(\frac{\omega}{\omega_f}\right)^2 + L\omega Y_1 \sin\varphi_1\right]^2 + (L\omega Y_1 \cos\varphi_1)^2\right\}^{1/2}}. \tag{2.84}$$

It depends on the current amplitude and phase shift φ_1.

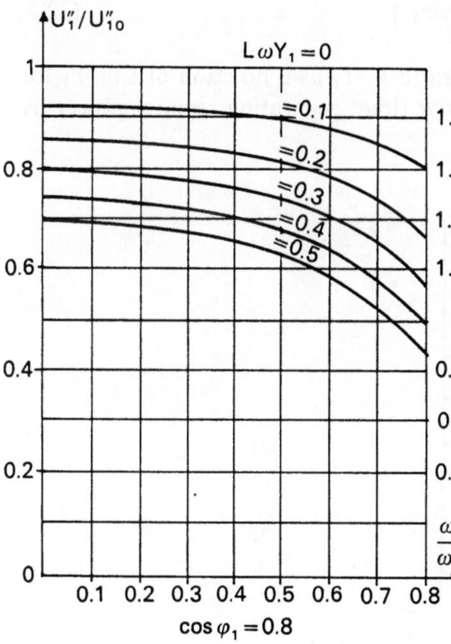

Fig. 2.36

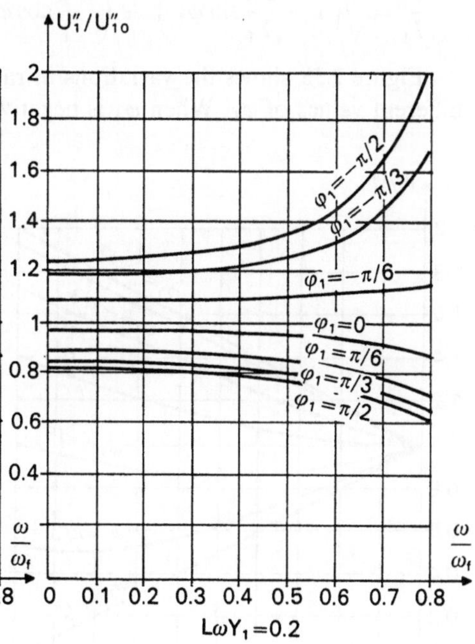

Fig. 2.37

In order to show the influence of the value and phase shift of the current delivered, we have plotted the variations of the ratio U''_1/U''_{10} at constant U'_1, as a function of ω/ω_r.

- firstly (Fig. 2.36) with a constant lagging shift φ_1, corresponding to $\cos\varphi_1$ equal to 0.80, for different values of $L\omega Y_1$,
- and then (Fig. 2.37) at constant $L\omega Y_1$ equal to 0.20, for different values of φ_1.

When ω/ω_r is zero (i.e. when there is no capacitance), the voltage variation caused by the inductance obviously depends on the value and the phase shift of the current delivered by the inverter. When ω/ω_r increases, the variation of U'_1/U''_{10} is firstly quite slow, and then greatly increases when ω approaches ω_r.

2.6.2.3 Increase in the Input Current

Because of the current taken by the capacitor, fundamental component I'_1 of the current delivered by the inverter, is different from that – I''_1 – corresponding to the current supplied to the load.

Equations (2.78b) and (2.76) give

$$\frac{\underline{I}'_1}{\underline{I}''_1} = 1 + \frac{\mathrm{j}C\omega}{\underline{Y}_1} = 1 + \frac{\mathrm{j}C\omega}{Y(\cos\varphi_1 - \mathrm{j}\sin\varphi_1)}$$

from which the following can be deduced:

$$\frac{I'_1}{I''_1} = \sqrt{\left(1 - \frac{C\omega}{Y_1}\sin\varphi_1\right)^2 + \left(\frac{C\omega}{Y_1}\cos\varphi_1\right)^2}. \tag{2.85}$$

Figure 2.38 shows the variations in ratio I'_1/I''_1 as a function of $C\omega/Y_1$ for different values of φ_1. When φ_1 is negative (load generating reactive power by

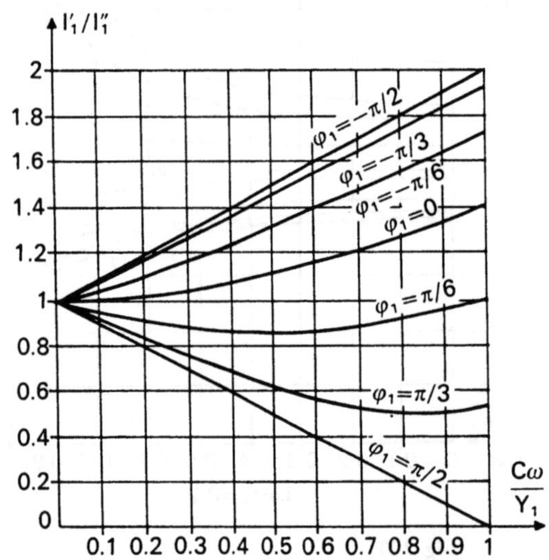

Fig. 2.38

itself), the increase in $C\omega/Y_1$ increases the ratio I'_1/I''_1. When φ_1 is positive (inductive load) the increase in C first of all decreases I'_1/I''_1; this ratio is at its minimum for $C\omega Y_1$ equal to $\sin\varphi_1$ when the reactive power generated by the capacitor compensates the reactive power absorbed by the load; after that, ratio I'_1/I''_1 increases.

2.6.3 Effects of the Filter on the Harmonics

In general, the load connected at the filter output cannot be easily characterized, as far as the harmonics are concerned. However, it can be noted that most AC loads are inductive and that their admittance thus decreases as the harmonics rank increases.

2.6.3.1 Output Voltage Harmonic Reduction

The aim of the filter is to reduce the voltage harmonic components between input and output. The first harmonic, of rank n_1 determines the minimal value to be given to LC for any filtering to be adequate.

If $n\omega/\omega_f$ is considerably greater than 1 – even in the case of the first harmonic component – and especially if Y_n is small compared to G_f, Eq. (2.79a) can be simplified and written as

$$\frac{U''_n}{U'_n} \simeq \frac{1}{(n\omega/\omega_f)^2 - 1}. \tag{2.86}$$

This equation corresponds to the part where $n\omega/\omega_f$ is greater than 1 on the curve plotted on Fig. 2.35 for zero Y_n.

The maximum ratio U''_n/U'_n which can be allowed for the first harmonic and frequency of the latter impose the minimum value to be given to the product LC. Indeed

$$\frac{1}{(n_1\omega/\omega_f)^2 - 1} < \left(\frac{U''_{n_1}}{U'_{n_1}}\right)_{\max}$$

gives

$$LC > \frac{1}{n_1^2\omega^2}\left[1 + \frac{1}{(U''_{n_1}/U'_{n_1})_{\max}}\right]. \tag{2.87}$$

The value to be given to LC increases as the angular frequency $n_1\omega$ of the first harmonic component and ratio $(U''_{n_1}/U'_{n_1})_{\max}$ decrease.

For the harmonics of higher rank, Eq. (2.84) indicates that the reduction will be greater.

2.6.3.2 Remarks on the Input Current Harmonic Reduction

The relation between the filter output voltage u'' harmonics and the output current i'' harmonics obviously depends on he characteristics of the AC load.

- Usually the load is less or more inductive and the relative value of the harmonic components of current i'' is even smaller than that of the harmonics of voltage u''.

  As a result, this brings about a reduction in current i' harmonic components at the filter input. If Y_n has the form $-jY_n$, Eq. (2.78b) in fact gives

  $$\frac{I'_n}{I''_n} = 1 - \frac{C n\omega}{Y_n}.$$

- If the load is not linear or contains non-linear elements, it acts as a harmonic current generator. A part of these currents is shunted by capacitor C and the rest transmitted to the inverter through inductance L. The higher the frequency of these harmonic currents, the more the part which is transmitted to the inverter is reduced.

  This second remark indicate the well-known fact that a low-pass filter for the voltage in the input to output direction of the filter is also a low-pass filter for the current in the output to input direction.

If the filtering of the voltage supplied by a voltage-source inverter is effective, the current at the output of this inverter has a nearly sinusoidal form. This justifies the hypothesis of the sinusoidal waveform which has adopted for current i' in the study of the different inverters presented in the first four parts of this chapter.

2.6.4 Choosing L and C

The study of the effects of the filter on the fundamental and harmonic components allows the determination not only of the value to be given to ω_f or the product LC, but also the value to be given to each element – L and C.

2.6.4.1 Method

The system of axes used is shown in Fig. 2.39. (Y_1 is used to denote the rated admittance of the load placed at the filter output.) The vertical axis corresponds to U''_1/U''_{10} eual to 1, and the horizontal axis to I'_1/I''_1 equal to 1.

In this system, the following loci are plotted:

- curve $\mathscr{C}_1$ which corresponds to the maximum acceptable variation for U''_1 between no load and full load operation,
- straight line $\mathscr{C}_2$ which corresponds to the maximum acceptable value of ratio I'_1/I''_1 under full load operation,
- curve $\mathscr{C}_3$ which corresponds to the minimal attenuation required for the first harmonic component.

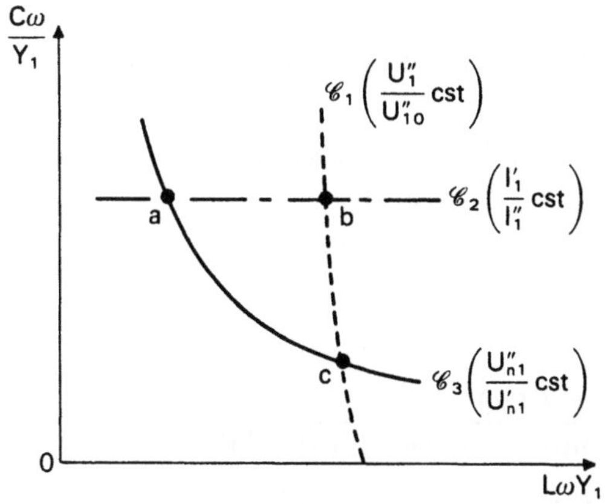

Fig. 2.39

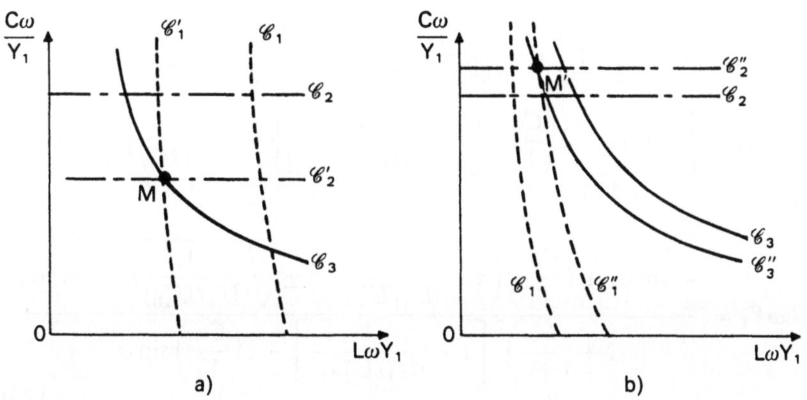

Fig. 2.40

All points located inside the curvilinear triangle abc fulfill the three imposed conditions.

- L and C are normally chosen to correspond to a point M (Fig. 3.40a) on curve $\mathscr{C}_3$ so as not to increase LC and thus the cost of the filter. M is placed so as to take full advantage of LC as concerns the reduction of U''_1/U''_{10} (change from $\mathscr{C}_1$ to $\mathscr{C}'_1$) and reduction of I'_1/I''_1 (change from $\mathscr{C}_2$ to $\mathscr{C}'_2$).
- If there is no point which fulfills the conditions as initially laid down (curves $\mathscr{C}_1, \mathscr{C}_2, \mathscr{C}_3$ of Fig. 2.40b), these conditions must be modified, by changing from $\mathscr{C}_1$ to $\mathscr{C}''_1$, $\mathscr{C}_2$ to $\mathscr{C}''_2$ and $\mathscr{C}_3$ to $\mathscr{C}''_3$. The value of L and C can then be deduced from the coordinates of intersection M' of $\mathscr{C}''_1, \mathscr{C}''_2$ and $\mathscr{C}''_3$.

2.6.4.2 The Three Networks of Curves

By studying the effects of the filter on the fundamental component, it has been shown that such effects depend not only on $L\omega Y_1$ and $C\omega/Y_1$ but also on φ_1.

The characteristics $\mathscr{C}_1$ and $\mathscr{C}_2$ must therefore also be plotted for various φ_1 and especially for the boundary values of this angle, according to the load which is to be supplied.

We will limit the curves to values of $L\omega Y_1$ and $C\omega/Y_1$ below 0.80, i.e. to values which will give reasonably sized filter elements.

- *Curves with constant* U_1''/U_{10}''

Equations (2.83) and (2.84) give

$$\frac{U_1''}{U_{10}''} = \frac{1 - LC\omega^2}{[(1 - LC\omega^2 + L\omega Y_1 \sin\varphi_1)^2 + (L\omega Y_1 \cos\varphi_1)^2]^{1/2}}.$$

For given values of U_1''/U_{10}'' and $C\omega/Y_1$, the value of $L\omega Y_1$ is solution of

$$(L\omega Y_1)^2\left\{1 + \left(\frac{C\omega}{Y_1}\right)^2\left[1 - \frac{1}{(U_1''/U_{10}'')^2}\right] - 2\left(\frac{C\omega}{Y_1}\right)\sin\varphi_1\right\}$$

$$+ L\omega Y_1\left\{2\sin\varphi_1 - 2\left(\frac{C\omega}{Y_1}\right)\left[1 - \frac{1}{(U_1''/U_{10}'')^2}\right]\right\} + 1 - \frac{1}{(U_1''/U_{10}'')^2} = 0$$

giving

$$L\omega Y_1 = \frac{-\sin\varphi_1 + \left(\frac{C\omega}{Y_1}\right)\left(1 - \frac{1}{(U_1''/U_{10}'')^2}\right) \pm \sqrt{\frac{1}{(U_1''/U_{10}'')^2} - \cos^2\varphi_1}}{1 + \left(\frac{C\omega}{Y_1}\right)^2\left[1 - \frac{1}{(U_1''/U_{10}'')^2}\right] - 2\left(\frac{C\omega}{Y_1}\right)\sin\varphi_1}. \tag{2.88}$$

The dashed lines in Fig. 2.41 show curves corresponding to constant U_1''/U_{10}'' successively for

(a) $\varphi_1 = 0$
(b) $\varphi_1 = \pi/6$
(c) $\varphi_1 = \pi/2$
(d) $\varphi_1 = -\pi/6$
(e) $\varphi_1 = -\pi/2$.

- With a purely resistive load (Fig. 2.41a), U_1''/U_{10}'' equal to 1 for zero $L\omega Y_1$ can only decrease when $L\omega Y_1$ increases. The curves plotted for U_1''/U_{10}'' equal successively to 1, 0.95, 0.90, 0.80, 0.70 and 0.60 show the influence of $C\omega/Y_1$ which becomes stronger as U_1''/U_{10}'' decreases.
- As the load becomes more inductive (Fig. 2.41b for $\varphi_1 = \pi/6$; Fig. 2.41c for $\varphi_1 = \pi/2$), the curves at constant U_1''/U_{10}'' become closer together, especially for values of U_1''/U_{10}'' near to unity. For given value of $L\omega Y_1$ and $C\omega/Y_1$ the variation in the fundamental component of output voltage between no load and full load operation is increasingly larger as $|\cos\varphi_1|$ moves away from 1.

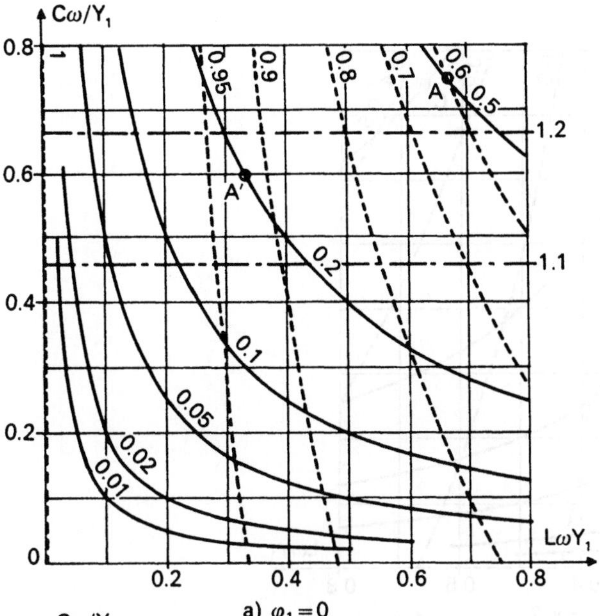

a) $\varphi_1 = 0$

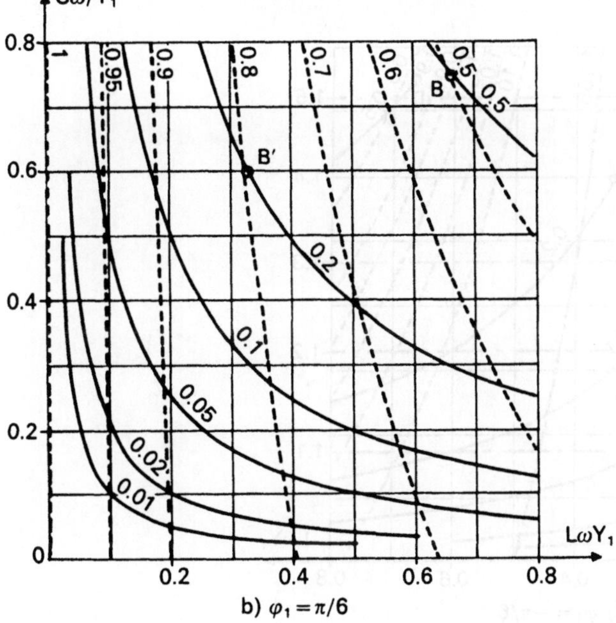

b) $\varphi_1 = \pi/6$

Fig. 2.41

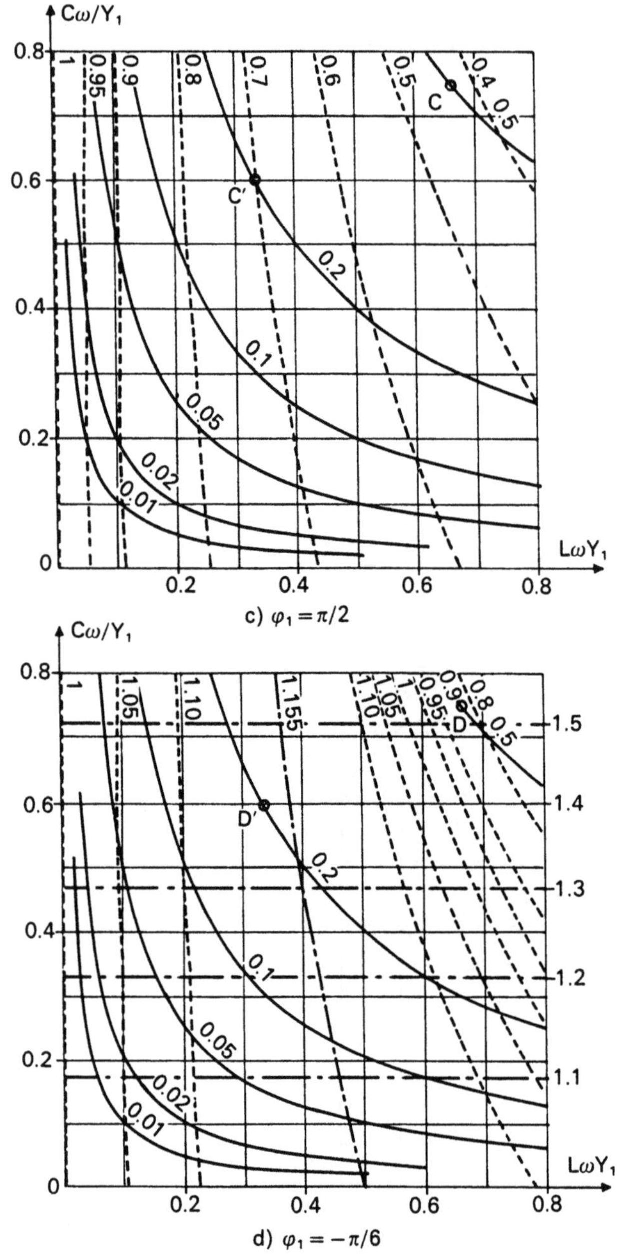

Fig. 2.41 (continued)

– For a less or more capacitive load (Fig. 2.41d for $\varphi_1 = -\pi/6$), when $L\omega Y_1$ increases, U_1''/U_{10}'' first of all increases and reaches a maximum value (in this case, 1.155) equal to $1/\cos\varphi_1$, and then decreases.

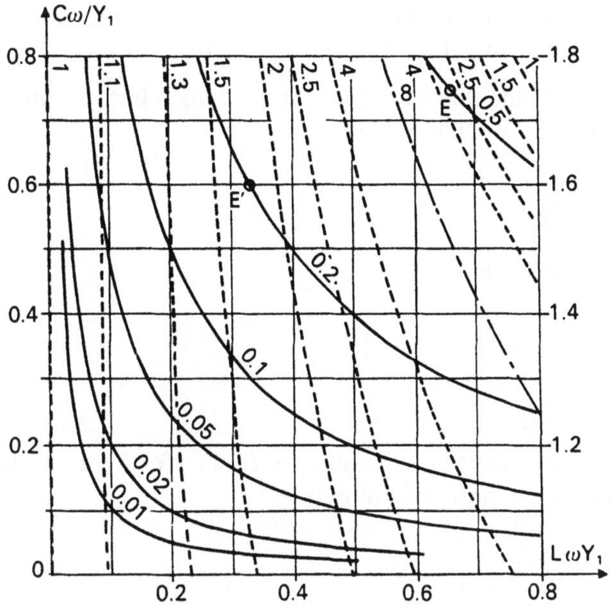

Fig. 2.41 (continued)

As the load becomes more capacitive, the curves with constant U_1''/U_{10}'' come closer together and maximum value of U_1''/U_{10}'' is greater.

For a purely capacitive load (Fig. 2.41e), U_1''/U_{10}'' has a theoretically infinite maximum value. The reduction is the resonance frequency shown by the curve plotted for $Y_n = jG_f$ in Fig. 2.38 can also be found here.

- *Curves with constant I_1'/I_1''.*

Ratio I_1'/I_1'' is given by Eq. (2.85):

$$\frac{I_1'}{I_1''} = \sqrt{\left(1 - \frac{C\omega}{Y_1}\sin\varphi_1\right)^2 + \left(\frac{C\omega}{Y_1}\cos\varphi_1\right)^2}.$$

It depends only on $C\omega/Y_1$ and on φ_1. It is only greater than one in cases where $C\omega/Y_1$ is greater than $2\sin\varphi_1$.

- In the case of inductive loads (positive φ_1), ratio I_1'/I_1'' is less than 1 or only slightly above it, for the normal values of $C\omega/Y_1$ at least.

- For $\varphi_1 = 0$,

$$\frac{I_1'}{I_1''} = \sqrt{1 + \left(\frac{C\omega}{Y_1}\right)^2}.$$

We have plotted – in alternately dashed lines on Fig. 2.41a – the horizontals of equal I'_1/I''_1 which correspond to values 1.10 and 1.20 of this ratio.

- The more capacitive the load, the closer together are the straight lines of equal I'_1/I''_1. For $\varphi_1 = -\pi/6$ (Fig. 2.41d), this gives

$$\frac{I'_1}{I''_1} = \sqrt{\left(1 + \frac{C\omega}{Y_1}\right)^2 + \frac{C\omega}{Y_1}}.$$

For a purely capacitive load (Fig. 2.41e),

$$\frac{I'_1}{I''_1} = 1 + \frac{C\omega}{Y_1}.$$

- *Curves with constant U''_{n_1}/U'_{n_1}.*

A hyperbola corresponds, in the system of axes $L\omega Y_1$, $C\omega/Y_1$, to each value U''_{n_1}/U'_{n_1} required for the harmonic of rank n_1:

$$L\omega Y_1 \frac{C\omega}{Y_1} = \frac{1}{n_1^2}\left[1 + \frac{1}{U''_{n_1}/U'_{n_1}}\right]. \tag{2.89}$$

For example in Fig. 2.41, the different lines show the hyperbolae corresponding to $LC\omega^2$ equals to 0.01, 0.02, 0.05, 0.10, 0.20 and 0.50.

2.6.4.3 Uses and Conclusions

- In the case of single-phase two-switch voltage-source inverters the first non zero harmonic component in the output voltage is of *rank* 3 ($n_1 = 3$).
For the latter to be reduced and not be amplified by filter, Eq. (2.87) shows that $LC\omega^2$ must be greater than 2/9, i.e. than 0.222. Obtaining a ratio U''_3/U'_3 less than 1/10 would require $LC\omega^2$ to be greater than 1.222, thus leading to an extremely costly filter at medium and high power.

If a filter with a reasonable value of $LC\omega^2$ is required, a relatively small reduction of 3rd harmonic has to be accepted.

For example $LC\omega^2 = 0.5$, thus $\omega/\omega_f = 0.707$ and $3\omega/\omega_f = 2.12$, gives

$$U''_3/U'_3 = 1/3.5$$

- Even if the load is always resistive, the performances are poor. With $LC\omega^2 = 0.5$ and $C\omega/Y_1 = 0.75$, the following is obtained (point A in Fig. 2.41a):

$$U''_1/U''_{10} = 0.6 \quad \text{and} \quad I'_1/I''_1 = 1.25.$$

- If the load can be either resistive or inductive, with $\cos\varphi_1$ varying from 1 to $\sqrt{3}/2$ for example, the voltage drop at full load is made even worse. For $LC\omega^2 = 0/5$, $C\omega/Y_1 = 0.75$ and $\cos\varphi_1 = \sqrt{3}/2$ (point B in Fig. 2.41b).

U''_1/U''_{10} drops to 0.493. If U''_1/U''_{10} has to be brought back to 0.6 with $LC\omega^2 = 0.5, C\omega/Y_1$ must be taken with a value of around 1 (a value which is so high that it does not appear on the figure drawn); for $\cos\varphi_1 = 1, I'_1/I''_1$ would reach the value $\sqrt{2}$.

If $\cos\varphi_1$ of the inductive load could go down to around zero, the performances would be even more adversely affected.

– If the load can be either inductive or capacitive, and with $\cos\varphi_1$ taken between 1 and $\sqrt{3}/2$ (with positive or negative φ_1), operation at $\varphi_1 = -\pi/6$ with $LC\omega^2 = 0.5$ and $C\omega/Y_1 = 0.75$ (point D in Fig. 2.41d) would lead to $I'_1/I''_1 = 1.52$.

If φ_1 can take any value between $+\pi/2$ and $-\pi/2$ performances become catastrophic. With the previous values of $LC\omega^2$ and $C\omega/Y_1$, a comparison of points C and E of Figs. 2.42c and e shows that

U''_1/U''_{10} varies from 0.493 to 3,
I'_1/I''_1 varies from 0.25 to 1.75

With $LC\omega^2$ equal to 0.5, the only way to reduce variations in U''_1/U''_{10} would be to give $C\omega/Y_1$ an abnormally high value, but this would also lead to an abnormally high maximum value of I'_1/I''_1.

– When n_1 equals 3 and if φ_1 can have any value, $LC\omega^2$ is often given a value above 1 so that ω/ω_r and $3\omega/\omega_r$ can both be found in the right-hand of the characteristics shown in Fig. 2.35. A value of $LC\omega^2$ equal to 2 can, for example, be used, but this leads to a heavy and costly filter.

- If the first non zero harmonic in the output voltage to be filtered is of *rank* 5 ($n_1 = 5$) and providing that U''_5/U'_5 does not have to be reduced too much, choosing L and C is easier.

With $LC\omega^2 = 0.20$ (i.e. $\omega/\omega_r = 0.447$ and $5\omega/\omega_r = 2.236$) for example, $U''_5/U'_5 = 1/4$.

If $LC\omega^2 = 0.20$ and $C\omega/Y_1 = 0.60$, U''_1/U''_{10} and I'_1/I''_1 can be read using points A′, B′, C′, D′, and E′ in Fig. 2.41a, b, c, d, and e.

- for $\cos\varphi_1 = 1$,
  $U''_1/U''_{10} = 0.923$, $I'_1/I''_1 = 1.17$
- for $\sqrt{3}/2 < \cos\varphi_1 < 1$, with positive φ_1,
  U''_1/U''_{10} varies from 0.79 to 0.92
- for $0 < \cos\varphi_1 < 1$, with positive φ_1,
  U''_1/U''_{10} varies from 0.71 to 0.92.
- for $\sqrt{3}/2 < \cos\varphi_1 < 1$, with positive or negative φ_1,
  U''_1/U''_{10} varies from 0.79 to 1.15,
  I'_1/I''_1 reaches a maximum of 1.40

- for $0 < \cos\varphi_1 < 1$, with positive or negative φ_1,

  U_1''/U_{10}'' varies from 0.71 to 1.71,

  I_1'/I_1'' reaches a maximum of 1.60

The value of $C\omega/Y_1$ can obviously be changed to make it more suitably adapted to the range of values specified of φ_1.

- These two examples show what can be expected of using a simple LC filter to filter an AC voltage.

  For the filter to be relatively cheap and to avoid excessive modification of the fundamental frequency voltage and current, the following must apply:

  $$\omega \ll \omega_f \quad \text{or} \quad LC\omega^2 \ll 1.$$

  For the filter to adequately reduce the first harmonic component, the following must apply:

  $$n_1\omega \gg \omega_f \quad \text{or} \quad LCn_1^2\omega^2 \gg 1.$$

  The double inequality

  $$1 \gg LC\omega^2 \gg \frac{1}{n_1^2}$$

  is only possible if the first non zero harmonic component has a sufficiently high rank.

At $n_1 = 3$, filtering is costly and of little efficiency. At $n_1 = 5$, the cost of such a filter becomes acceptable as long as a rather modest reduction in the first harmonic can be suffered.

This explains the increasingly frequent use of pulse width modulated (PWM) inverters (the following chapter will deal with them). By increasing the rank of the first non zero harmonic PWM makes the filtering much easier.

Moreover, it can be seen that studying a filter cannot be limited to its effect on the harmonics unless $LC\omega^2$ is very weak, with $L\omega Y_1$ and $C\omega/Y_1$ considerably less than 1. The effects of the filter on the fundamental component must also be taken into account.

2.6.5 Notes on Other Filters

The LC filter can obviously be replaced by several LC filters in series or by more intricate filters.

- If only one harmonic component of angular frequency $n_1\omega$ had to be eliminated, this could be done by shunt-connecting a series resonant circuit tuned on this harmonic (Fig. 2.42a), thus such that

  $$L_1 C_1 n_1^2 \omega^2 = 1.$$

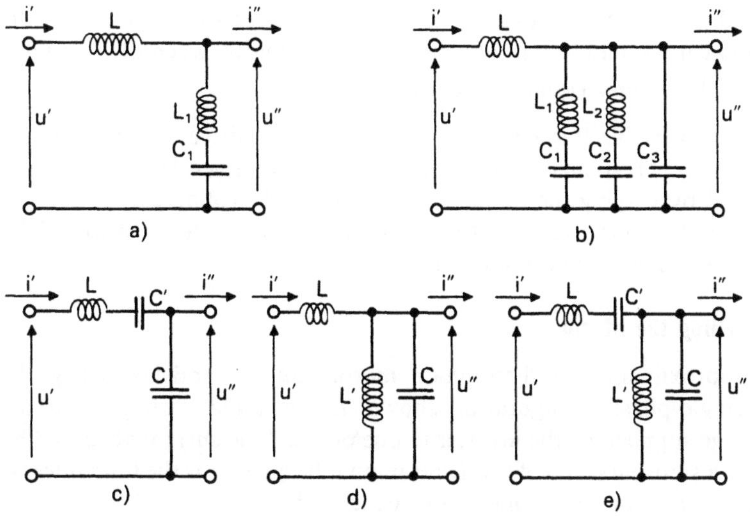

Fig. 2.42

If two important harmonic components had to be eliminated, one of rank n_1 and the other of n_2, this could be done by shunt-connecting two series resonant circuits (Fig. 2.42b), such that

$$L_1 C_1 n_1^2 \omega^2 = 1 \qquad L_2 C_2 n_2^2 \omega^2 = 1$$

and a capacitor C_3 could be added to reduce the remaining harmonics.

- The voltage drop caused by the flow of the fundamental component of i' through inductor L could be suppressed by adding in series to the latter a capacitor C' (Fig. 2.42c), such that

  $$LC'\omega^2 = 1.$$

  The fundamental frequency current taken by capacitor C could be cancelled by connecting in parallel with the latter an inductor L' (Fig. 2.42d) such that

  $$L'C\omega^2 = 1.$$

  Combining these two additions produces the configuration in Fig. 2.42e.

In general, for a given overall cost, these modifications only improve performances if the rank of the first non zero harmonic component is sufficiently high ... but in such cases only a simple LC filter is required.

2.6.6 Notes on Variable Frequency-Supplied Loads

When the output frequency of an inverter varies between a maximum value and a very low fraction of the latter, the output voltages cannot be filtered, at least not by use of a filter with constant elements. For example, if an LC filter was used, when the fundamental angular frequency ω decreases, the various harmonic components would be amplified rather than reduced, when their angular frequency would become equal to $1/\sqrt{LC}$.

Inverters at variable frequency are mainly used to drive synchronous or induction motors. These machines are normally of the three-phase type. Their speed is varied by varying the frequency of their supply voltages.

We shall confine ourselves here to a very approximate determination of the waveforms of the current and the torque.

2.6.6.1 Modelling the Motor

In a first approximation, a three-phase motor can be modelled using the equivalent single-phase configuration shown in Fig. 2.43a. Voltage v' is the stepped voltage supplied by the inverter to one phase, i' the current absorbed by the latter, e is a sinusoidal emf of the same angular frequency as the fundamental of v', and L an inductance assumed to be constant[1].

The first phase-to-neutral voltage v_A delivered by the full-bridge three-phase inverter has the waveforms shown at the top of Fig. 2.22:

$$\begin{aligned} v'_A &= U/3 \quad &&\text{for } 0 < \omega t < \pi/3 \\ &= 2U/3 \quad &&\text{for } \pi/3 < \omega t < 2\pi/3 \\ &= U/3 \quad &&\text{for } 2\pi/3 < \omega t < \pi, \ \ldots \end{aligned}$$

$$v'_A(\omega t + \pi) = -v'_A(\omega t).$$

Its fundamental component has the following rms value:

$$V'_1 = U\sqrt{2}/\pi.$$

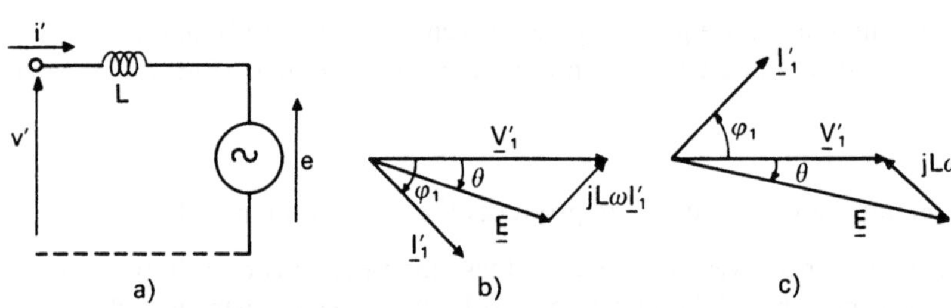

Fig. 2.43

[1] This is the simplest model, typically adopted for an initial study of the synchronous machine. We have presented how to deduce such a model for the induction machine (Vol. 2, Chap. 4, Sect. 5).

The main approximations lie in neglecting the stator resistance and especially in the use of a single inductance L.

For the fundamental frequency variables, the phasor diagram shown in Fig. 2.43b or 2.43c applies, depending on whether E is lower or higher than V'_1. E is lower than V'_1 in the case of an asynchronous motor or an under-excited synchronous motor. E is greater than V'_1 in the case of an over-excited synchronous motor.

If the fundamental component of v'_A is taken to be the phase-angle origin, the following can be written:

$$e_A = kU\sin(\omega t - \theta) \tag{2.90}$$

with

$$k = \frac{E\sqrt{2}}{U} = \frac{E}{V'_1}\frac{2}{\pi}.$$

2.6.6.2 Successive Expressions of the Currents

- Current i'_A in the first phase of the load is the solution of

$$L\omega\frac{di'_A}{d\omega t} = v'_A - e_A.$$

– For $0 < \omega t < \pi/3$,

$$L\omega\frac{di'_A}{d\omega t} = \frac{U}{3} - kU\sin(\omega t - \theta)$$

gives

$$i'_A = \frac{U}{L\omega}\left[\frac{\omega t}{3} + k\cos(\omega t - \theta) + A_1\right].$$

At the end of this period,

$$i'_A(\pi/3) = \frac{U}{L\omega}\left[\frac{\pi}{9} + k\cos\left(\frac{\pi}{3} - \theta\right) + A_1\right].$$

– For $\pi/3 < \omega t < 2\pi/3$

$$L\omega\frac{di'_A}{d\omega t} = \frac{2U}{3} - kU\sin(\omega t - \theta)$$

gives

$$i'_A = \frac{U}{L\omega}\left[\frac{2}{3}\omega t + k\cos(\omega t - \theta) + A_2\right].$$

The fact that i'_A is not subject to any discontinuity in $\omega t = \pi/3$ gives $A_2 = A_1 - (\pi/9)$. Thus

$$i'_A = \frac{U}{L\omega}\left[\frac{2}{3}\omega t + k\cos(\omega t - \theta) + A_1 - \frac{\pi}{9}\right].$$

At the end of this period,

$$i'_A(2\pi/3) = \frac{U}{L\omega}\left[\frac{3\pi}{9} + k\cos\left(\frac{2\pi}{3} - \theta\right) + A_1\right].$$

For $2\pi/3 < \omega t < \pi$,

$$L\omega\frac{di'_A}{d\omega t} = \frac{U}{3} - kU\sin(\omega t - \theta)$$

gives

$$i'_A = \frac{U}{L\omega}\left[\frac{\omega t}{3} + k\cos(\omega t - \theta) + A_3\right]$$

or, as i'_A cannot be discontinuous for $\omega t = 2\pi/3$,

$$i'_A = \frac{U}{L\omega}\left[\frac{\omega t}{3} + k\cos(\omega t - \theta) + A_1 + \frac{\pi}{9}\right].$$

Since $i'_A(\omega t + \pi) = -i'_A(\omega t)$, A_1 can be eliminated by stating

$$i'_A(\pi) = -i'_A(0)$$

which gives

$$A_1 = -2\pi/9.$$

This gives the successive expressions of i'_A during the positive half-cycle voltage v'_A.

$$\begin{aligned}
&0 < \omega t < \frac{\pi}{3}, \quad i'_A = \frac{U}{L\omega}\left[\frac{\omega t}{3} + k\cos(\omega t - \theta) - \frac{2\pi}{9}\right]\\
&\frac{\pi}{3} < \omega t < \frac{2\pi}{3}, \quad i'_A = \frac{U}{L\omega}\left[\frac{2\omega t}{3} + k\cos(\omega t - \theta) - \frac{\pi}{3}\right]\\
&\frac{2\pi}{3} < \omega t < \pi, \quad i'_A = \frac{U}{L\omega}\left[\frac{\omega t}{3} + k\cos(\omega t - \theta) - \frac{\pi}{9}\right].
\end{aligned} \tag{2.91}$$

In the upper part of Fig. 2.44, the waveform of the current i'_A has been plotted for the following case:

$L\omega I'_1 = V'_1/4$

I'_1 lagging on V'_1 with $\cos\varphi_1 = 0.80$;

this corresponds to

$E = 0.8732\, V'_1, \qquad k = 0.556, \qquad \theta = 13°24,$

The waveforms of currents i'_B and i'_C have been added: these are identical to that of i'_A but shifted by $2\pi/3$ and $4\pi/3$, respectively.

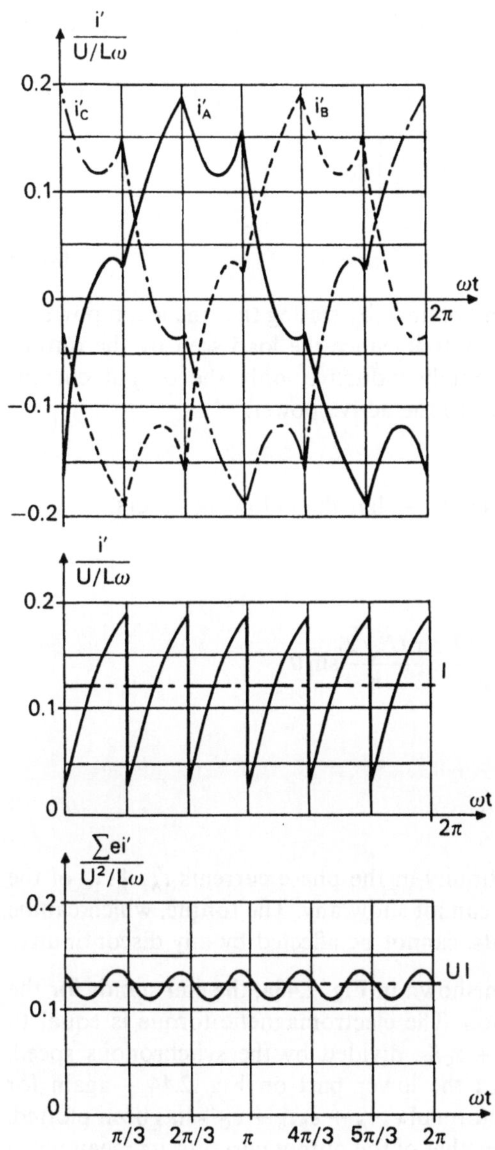

Fig. 2.44

- Current i at the inverter input can be deduced from the output currents. It has a frequency equal to six times that of the output currents or voltages.

  For $\pi/3 < \omega t < 2\pi/3$, in the configuration shown in Fig. 2.21, switches K_1, K'_2 and K'_3 are on; therefore,

$$i = i'_A = \frac{U}{L\omega}\left[\frac{2\omega t}{3} + k\cos(\omega t - \theta) - \frac{\pi}{3}\right].$$

The waveform of i has been plotted for $k = 0.556$ and $\theta = 13°24$ at the centre of Fig. 2.44.

Current i has a mean value of

$$I = \frac{6}{2\pi} \int_{\pi/3}^{2\pi/3} \frac{U}{L\omega} \left[\frac{2\omega t}{3} + k \cos(\omega t - \theta) - \frac{\pi}{3} \right] d\omega t$$

$$I = \frac{3}{\pi} \frac{U}{L\omega} k \sin\theta. \qquad (2.92)$$

The same result could have been obtained by stating that the mean power is the same at the inverter input and output. Since the load seen by the output voltage harmonic components is purely inductive, only the output current fundamental component contributes to the active power:

$$UI = 3V'_1 I'_1 \cos\varphi_1 .$$

The phasor diagrams of Fig. 2.43 show that the following occurs:

$$L\omega I'_1 \cos\varphi_1 = E \sin\theta.$$

Therefore,

$$I = \frac{1}{U} \frac{3}{L\omega} V'_1 E \sin\theta = \frac{1}{U} \frac{3}{L\omega} \frac{U\sqrt{2}}{\pi} k \frac{\pi}{2} \frac{U\sqrt{2}}{\pi} \sin\theta$$

$$= \frac{3}{\pi} \frac{U}{L\omega} k \sin\theta,$$

2.6.6.3 Torque Ripple

Inductances L prevent any discontinuity in the phase currents i'_A, i'_B, i'_C of the motor; similarly the rotor currents cannot show any. The torque, which can be computed from products of currents, cannot be affected by any discontinuity.

In the highly simplified diagram shown in Fig. 2.43a, the emf stands for the electro-mechanical power conversion. The electromagnetic torque is equal to the power converted, $e_A i'_A + e_B i'_B + e_C i'_C$, divided by the synchronous speed. The latter is proportional to ω. In the lower part on Fig. 2.44 – again for $k = 0.556$ and $\theta = 13°24$ – the waveform of $e_A i'_A + e_B i'_B + e_C i'_C$ has been plotted. This power has a frequency six times that of the output currents; its mean value is equal to $3 V'_1 I'_1 \cos\varphi_1$ or to UI, but has a lower ripple than Ui and does not show the same discontinuities at the latter.

2.6.6.4 Harmonic Effects

- In the model which is used, the $h\omega$ frequency harmonics of currents i'_A, i'_B, i'_C have an rms values I'_h related to that V'_h of terms of the same angular frequency of voltages v'_A, v'_B, v'_C by

  $$I'_h = V'_h / Lh\omega.$$

For given values of the supply voltage and of the inverter operating frequency, I'_h is constant. The load variations caused by the motor modify the current fundamental component without affecting the harmonics.

Figure 2.45 shows the waveforms of i'_A and $e_A i'_A + e_B i'_B + e_C i'_C$

(a) for $L\omega I'_1, = 0.5\ V'_1$, with $\cos\varphi_1 = 0.8$ (φ_1 lagging)

$E = 0.806\ V'_1, \quad k = 0.513, \quad \theta = 29°74$

(b) for $L\omega I'_1 = 0.25\ V'_1$, with $\cos\varphi_1 = 0.8$ (φ_1 lagging)

$E = 0.873\ V'_1, \quad k = 0.556, \quad \theta = 13°24$

(c) for $L\omega I'_1 = 0$

$E = V'_1, \quad k = 0.637, \quad \theta = 0.$

A comparision of the three waveforms of current i'_A shows that the differences compared to the fundamental component caused by the harmonics remain unchanged. In (c), there are only these differences since the fundamental is zero.

A comparison of the three waveforms of the sum $e_A i'_A + e_B i'_B + e_C i'_C$ which gives an image of the torque, shows that, by dividing by 2 and then cancelling

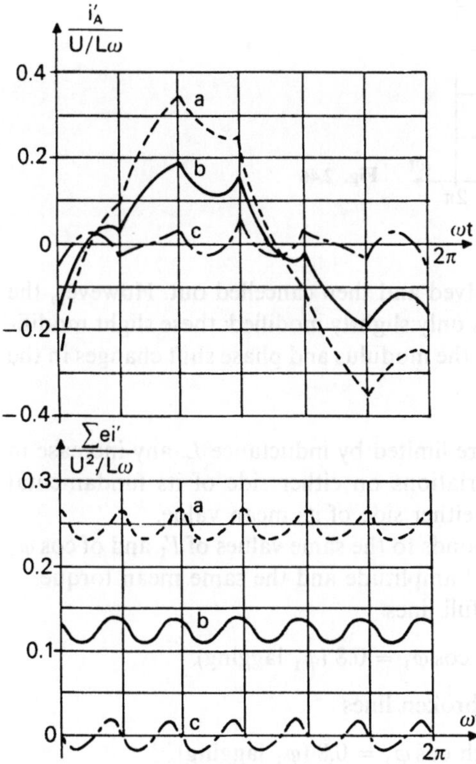

Fig. 2.45

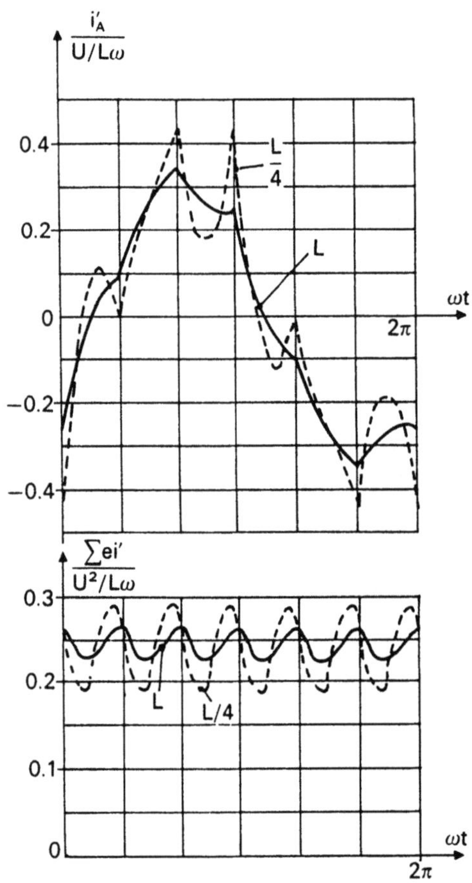

Fig. 2.46

out $I'_1 \cos \varphi_1$, the mean torque is halved and then cancelled out. However, the torque ripple caused by harmonics is only slightly modified: these slight modifications of the ripple are produced by the modulus and phase shift changes in the emf's e_A, e_B, e_C.

- The harmonics and their effects are limited by inductance L: any increase in the latter reduces the current variations on either side of its fundamental component and of the torque on either side of its mean value.

  The curves in Fig. 2.46 corresponds to the same values of I'_1 and of $\cos \varphi_1$ and thus to the same fundamental amplitude and the same mean torque.

  For the waveforms shown in full lines

  $L\omega I'_1 \cos \varphi_1 = 0.50\ V'_1$, with $\cos \varphi_1 = 0.8$ (φ_1 lagging).

  For the waveforms shown in broken lines

  $L'\omega I'_1 \cos \varphi_1 = 0.125\ V'_1$, with $\cos \varphi_1 = 0.8$ (φ_1 lagging).

Chapter 3

Pulse-Width-Modulated Voltage-Source Inverters

In a pulse-width-modulated inverter, each half-cycle of the output voltage is made up of several square pulses of suitable width, instead of only one square-wave as was the case in the previous chapter.

The configurations of the inverters are the same as before but the control of the "switches" is modified: the commutation frequency is higher than that of the output voltage.

Multiplying the number of "pulses" which form each output voltage half-cycle offers two important advantages:

- it pushes the output voltage harmonics towards higher frequencies and thus makes filtering easier,
- it enables the magnitude of the output voltage fundamental component to be varied, even in the case of inverters with only two "switches" per phase.

The increasing use of pulse width modulation is linked to progress made in the field of power semiconductor devices: the increase in the number of commutations would lead to excessive losses if reduction in the losses at each commutation had not been achieved.

The turn-on and turn-off instants of "switches" are usually determined in real time by use of an analogical or numerical control, or by simultaneous use of both techniques.

In order to achieve this, the crossing points between a reference wave and a modulation wave are used.

In some applications, the control instants are calculated and memorised previously; the switches are then controlled by a microprocessor.

The most frequently used configuration in three-phase applications is the six-"switch" bridge; its control and voltage waveforms are directly linked to those of the single-phase half-bridge inverter. In single-phase applications, the full-bridge inverter tends to be used.

This explains the plan adopted for this chapter. The study

- firstly, of the single-phase half-bridge inverter,
- secondly, of the three-phase full-bridge inverter,
- thirdly, of the single-phase full-bridge inverter

will be carried out with direct determination of the switching instants.

Finally, notes on the programmed modulation will be presented.

3.1 Single-Phase Half-Bridge Inverter

3.1.1 "Switch" Control. Waveforms

3.1.1.1 Modulation Principle

The principle of PWM voltage-source inverters can be presented in the same way as for cycloconverters (cf. Vo. 2, Chap. 5). In both cases, the aim is to generate an output voltage which is approximatively sinusoidal, by making its "mean value" vary sinusoidally.

Figure 3.1 indicates the configuration of the single-phase half-bridge inverter and the notations used.

The turn-on instants of the complementary "switches" K_1 and K'_1 are determined by the crossing points between the *reference* wave u'_w, which represents the wanted output voltage (from which the suffix w used as in the cycloconverter study) of frequency f, and the *modulation* wave, or carrier, M.

The carrier, with a frequency f' well above f, is a triangular wave with magnitude $U/2$ (Fig. 3.2):

- the intersections of u'_w with the rising sides of M give the turn-on instants of K'_1 and the beginning of periods where u' equals $-U/2$;
- the intersections of u'_w with the falling sides of M give the turn-on instants of K_1 and the beginning of periods where u' equals $+U/2$.

In order to calculate the mean value of the output voltage u' over a cycle T' of the modulation wave, it is assumed that this period is sufficiently short to

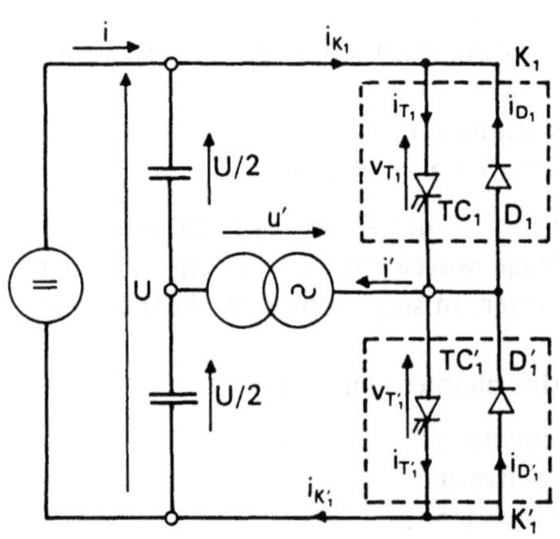

Fig. 3.1

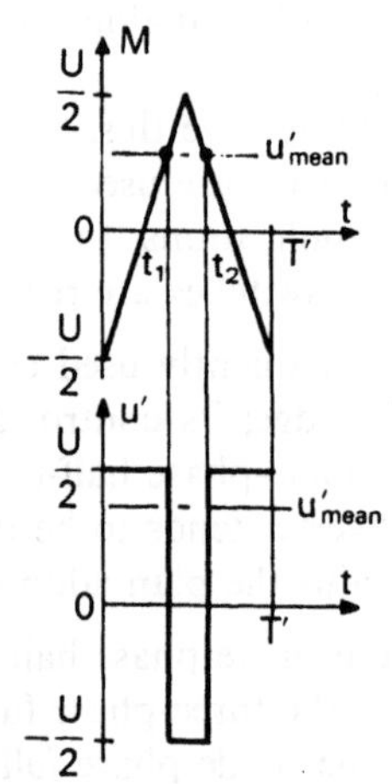

Fig. 3.2

enable the variations in reference voltage u'_w to be ignored throughout its duration.

The rising part of M has the following equation:

$$M = -\frac{U}{2} + 2U\frac{t}{T'}$$

it crosses u'_w for $t = t_1$ such that

$$t_1 = \left(u'_w + \frac{U}{2}\right)\frac{T'}{2U}$$

The falling part of M has the following equation:

$$M = \frac{3U}{2} - 2U\frac{t}{T'}$$

it crosses u'_w for $t = t_2$ such that

$$t_2 = \left(\frac{3U}{2} - u'_w\right)\frac{T'}{2U}$$

The output voltage has the following mean value during this cycle of M:

$$u'_{mean} = \frac{1}{T'}\left[\frac{U}{2}T' - \frac{U}{2}(t_2 - t_1)\right]$$

If t_2 and t_1 are replaced by their values, this gives

$$u'_{mean} = u'_w.$$

If the reference shows a sinusoidal variation, the "mean value" of u' will vary according to the same law.

3.1.1.2 Characterising the Modulation

• If the reference is sinusoidal, the control can be characterized by two parameters:

- *the frequency–modulation ratio m*, equal to the ratio f'/f of the modulation and reference frequencies;
- *the amplitude–modulation ratio r*, equal to the reference voltage magnitude referred to the peak value $U/2$ of the carrier wave.

• The modulation is normally *synchronous*, i.e. f' is an integer multiple of f. Voltage u' is then actually periodic and has a cycle T equal to $1/f$.

However, in certain cases, the modulation is asynchronous, in particular when, at a given modulation frequency f', the reference frequency is varied continuously.

In synchronous modulation, if m is uneven, the negative half-cycle of u' is identical to its positive half-cycle to within a sign. The series expansion of u' comprises only uneven harmonics.

Conversely, if m is even, both even and uneven harmonics and a DC component can be found in the series expansion of u'.

When the inverter has a single-phase output, *uneven values* of m must be chosen.

However, we will deal with the case where m is even, on account of the three-phase full-bridge inverter which will be analyzed next, since, in this configuration, the DC component and some harmonics are suppressed. Furthermore, since certain properties are maintained whether m be even or uneven, the remarks concerning asynchronous modulation will be more easily understood.

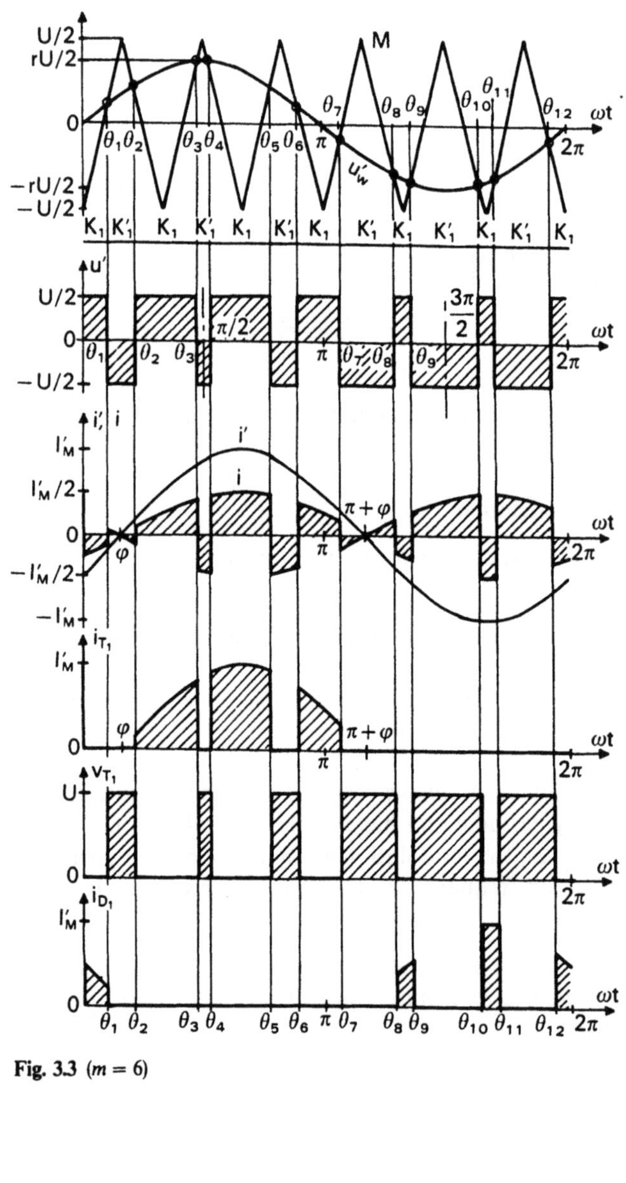

Fig. 3.3 ($m = 6$)

- In synchronous modulation, by a proper phase *locking* of the carrier with respect to the reference, the output voltage can be given a symmetrical waveform.

  If M reaches a maximum or minimum value in the middle of u'_w half-cycles, the half-cycles of u' are symmetrical in relation to their mid-points: the phase lock is then said to be optimal.

3.1.1.3 Waveforms

- The upper part of Figs. 3.3 and 3.4 shows the curve of u' obtained using triangular modulation and sinusoidal reference.

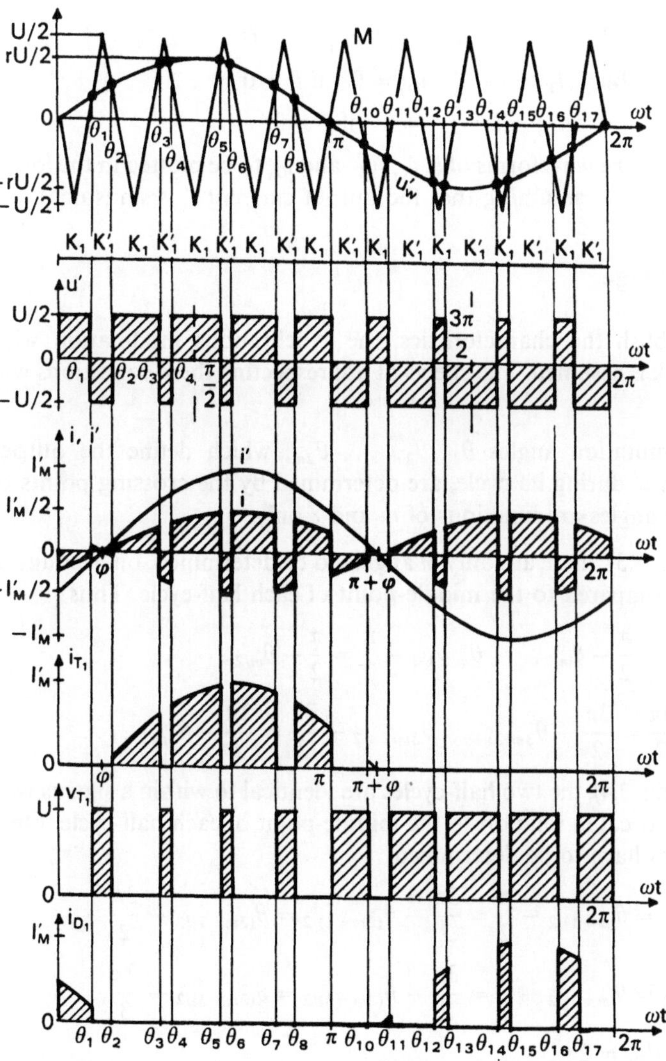

Fig. 3.4 ($m = 9$)

Figure 3.3 corresponds to m equal to 6, and Fig. 3.4 to m equal to 9. Thus, in both cases, the modulation is synchronous; curves have been plotted for optimal phase lock.

- The relations established in the previous chapter between instantaneous values of currents and voltages remain valid:
  - when K_1 is on:

$$u' = U/2, \quad i = i'/2, \quad i_{K'_1} = 0, \quad v_{K'_1} = +U$$

$$i_{K_1} = i' \quad \text{thus} \quad i_{T_1} = i', \quad i_{D_1} = 0 \quad \text{if } i' > 0$$

$$i_{D_1} = -i', \quad i_{T_1} = 0 \quad \text{if } i' < 0$$

  - when K'_1 is on:

$$u' = -U/2, \quad i = -i'/2, \quad i_{K_1} = 0, \quad v_{K_1} = +U$$

$$i_{K'_1} = -i' \quad \text{thus} \quad i_{T'_1} = -i' \quad i_{D'_1} = 0 \quad \text{if } i' < 0$$

$$i_{D'_1} = i', \quad i_{T'_1} = 0 \quad \text{if } i' > 0.$$

This has enabled the waveforms of i', i_{T_1}, v_{T_1} and i_{D_1} to be plotted in the lower part of Figs 3.3 and 3.4, assuming that the output currrent i' is sinusoidal.

3.1.2 Output Voltage

In order to establish the characteristics, the synchronous modulation with optimal phase lock is assumed. The effects of not respecting these conditions will be seen later.

The $2m$ commutation angles $\theta_1, \theta_2, \ldots, \theta_{2m}$, which define the output voltage waveform u' during its cycle, are determined by the crossing points of M and u'_w. These angles are functions of m and r only.

- If m is even (Fig. 3.3), there are only m angles to be determined, on account of the symmetry compared to the middle-point of each half-cycle. Thus

$$\theta_{m/2+1} - \frac{\pi}{2} = \frac{\pi}{2} - \theta_{m/2}, \qquad \theta_{m/2+2} - \frac{\pi}{2} = \frac{\pi}{2} - \theta_{m/2-1} \ldots$$

$$\theta_{3m/2+1} - \frac{3\pi}{2} = \frac{3\pi}{2} - \theta_{3m/2}, \qquad \theta_{3m/2+2} - \frac{3\pi}{2} = \frac{3\pi}{2} - \theta_{3m/2-1} \ldots$$

- If m is uneven (fig. 3.4), the two half-cycles are identical to within a sign as well as being symmetrical in relation to the middle-point of each half-cycle. Only $(m-1)/2$ angles have to be determined.

$$\frac{\pi}{2} - \theta_{(m-1)/2} = \theta_{(m+1)/2} - \frac{\pi}{2} = \frac{3\pi}{2} - \theta_{(3m-1)/2} = \theta_{(3m+1)/2} - \frac{3\pi}{2}$$

$$\frac{\pi}{2} - \theta_{(m-3)/2} = \theta_{(m+3)/2} - \frac{\pi}{2} = \frac{3\pi}{2} - \theta_{(3m-3)/2} = \theta_{(3m+3)/2} - \frac{3\pi}{3}, \ldots$$

$$\theta_m = \pi, \qquad \theta_{2m} = 2\pi.$$

3.1.2.1 Series Expansion

- *If m is uneven*, the Fourier series of u' comprises only uneven harmonics in addition to the fundamental.

  If the zero crossing of u'_w is taken as the zero time, voltage u' can be written as

$$u' = A_1 \sin \omega t + A_3 \sin 3\omega t + A_5 \sin 5\omega t + \ldots + A_n \sin n\omega t + \ldots$$

with

$$A_n = \frac{4}{\pi} \int_0^{\pi/2} u' \sin n\omega t \, d\omega t$$

$$= \frac{4}{\pi} \frac{U}{2} \left[+ \int_0^{\theta_1} \sin n\omega t \, d\omega t - \int_{\theta_1}^{\theta_2} \sin n\omega t \, d\omega t + \ldots \pm \int_{\theta_{m/2-1}}^{\pi/2} \sin n\omega t \, d\omega t \right]$$

giving

$$A_n = \frac{1}{n} \frac{4}{\pi} \frac{U}{2} (1 - 2 \cos n\theta_1 + 2 \cos n\theta_2 - \ldots \pm 2 \cos n\theta_{m/2-1}) \tag{3.1}$$

  The rms value U'_1 of the fundamental component, as well as those U'_3, $U'_5, \ldots, U'_n$ of the different uneven harmonics which form voltage u', are given by

$$U'_1 = \frac{A_1}{\sqrt{2}}, \quad U'_3 = \frac{|A_3|}{\sqrt{2}}, \quad U'_5 = \frac{|A_5|}{\sqrt{2}}, \quad \ldots, \quad U'_n = \frac{|A_n|}{\sqrt{2}}, \quad \ldots$$

- *If m is even* the Fourier series of u' comprises a DC component[1] and even or uneven harmonics, in addition to the fundamental.

  It is convenient to use the mid-point of the positive half-cycle of u'_w as the time origin. The series expansion has, in that case, the following form:

$$u' = U'_0 + B_1 \cos \omega t' + B_2 \cos 2\omega t' + B_3 \cos 3\omega t' + \ldots + B_n \cos n\omega t' + \ldots$$

with

$$B_n = \frac{2}{\pi} \int_0^{\pi} u' \cos n\omega t' \, d\omega t' \quad \text{and} \quad \omega t' = \omega t - \pi/2.$$

  This gives

  - for $m/2$ uneven,

$$B_n = \frac{1}{n} \frac{4}{\pi} \frac{U}{2} (-\sin n\theta'_1 + \sin n\theta'_2 - \ldots + \sin n\theta'_{3m/2}) \tag{3.2}$$

[1] This component is high for m equal to 2, small for m equal to 4, and can be ignored for m equal to or greater than 6.

– for $m/2$ even,

$$B_n = \frac{1}{n}\frac{4}{\pi}\frac{U}{2}(+\sin n\theta'_1 - \sin n\theta'_2 + \ldots - \sin n\theta'_{3m/2}). \tag{3.2'}$$

The rms values of the fundamental and harmonic components are given by

$$U'_1 = \frac{B_1}{\sqrt{2}}, \quad U'_2 = \frac{|B_2|}{\sqrt{2}}, \quad U' = \frac{|B_3|}{\sqrt{2}}, \ldots, \quad U'_n = \frac{|B_n|}{\sqrt{2}} \ldots$$

3.1.2.2 Fundamental Component. Voltage Loss

- As soon as m reaches a sufficient value (equal to or above 6), the fundamental of the output voltage is virtually equal to the reference voltage:

$$U'_1 = \frac{1}{\sqrt{2}} r \frac{U}{2}. \tag{3.3}$$

Figure 3.5 shows the variations, as a function of r, of the $U'_1/U/2\sqrt{2}$ ratio, for m equal to 2, 3 and 5 respectively. At these values of m, this ratio moves furthest away from r.

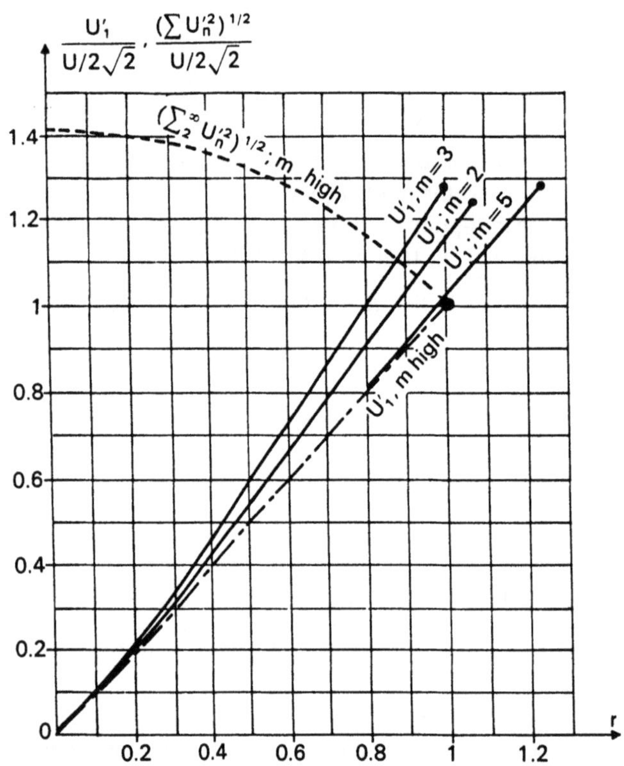

Fig. 3.5

When m has a sufficient value, the value which corresponds to infinite m (straight line segment drawn in dashed line in Fig. 3.5) may be used.

- By acting on r, U'_1 can only be made to rise from zero up to a maximum which corresponds to the disappearance of some pulses of voltage u' (or to the disappearance of some intersections between the reference and modulation waves).

  The maximum value of U'_1 can be compared to the rms value of the fundamental obtained when there is no pulse width modulation, i.e. to

$$\frac{1}{\sqrt{2}}\frac{4}{\pi}\frac{U}{2} = \frac{\sqrt{2}}{\pi}U.$$

  The maximum value of U'_1 is usually less than this value: the PWM produces a "*voltage loss*".

- When m (even or uneven) is very high, the maximum value of r is equal to 1. If r is made equal to 1 in (3.3) and by referring U'_1 to $(\sqrt{2}/\pi)\ U$, the following is obtained:

$$\left(\frac{U'_1}{\sqrt{2}\,U/\pi}\right)_{max} = \frac{\pi}{4} = 0.7854. \tag{3.4}$$

  The "voltage loss" equals 21.46%.

- If m is even, the maximum value of r always remains equal to 1. Equation (3.4) can be used, except in the specific case where m equals 2.

- If m is uneven and such that $(m-1)/2$ is also uneven, the maximum value of r is once more equal to 1. Equation (3.4) can again be used, except in the specific case where m is equal to 3.

- If m is uneven and such that $(m-1)/2$ is even, r can be increased slightly beyond 1. The maximum value of u'_w corresponds to a minimum value of M, and the minimum value of u'_w to a maximum value of M. The value of r may reach $1/\cos(\pi/m)$. However, the greater the value of m, the smaller the difference between $1/\cos(\pi/m)$ and 1, and the nearer the voltage loss is to that given by Eq. (3.4).

- In fact, the "voltage loss" is greater, since each square-pulse of voltage u' must be left with a minimal width. The turn-off of a "switch" cannot be controlled before the end of its turn-on and vice-versa. Now, either the turn-off nor the turn-on are instantaneous.

  For a high value of m, r must be limited to

$$r_{max} = 1 - m\frac{\Delta\theta}{\pi}$$

  if an angular width of $\Delta\theta$ has to be maintained for the pulses which are located at the mid-point of u'_w half-cycles or in the immediate vicinity of these points.

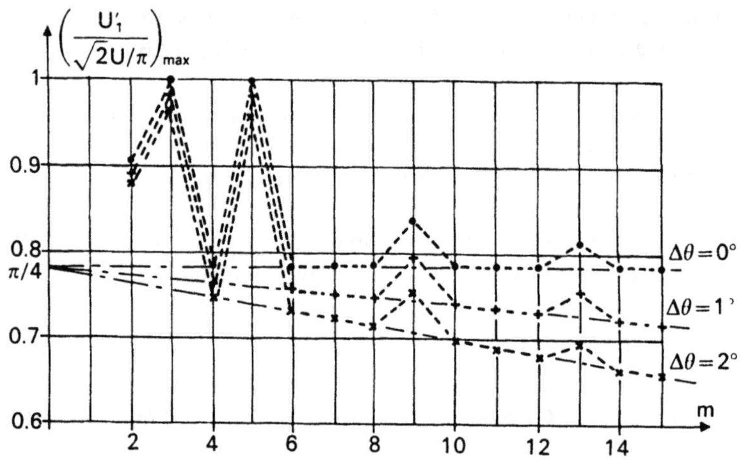

Fig. 3.6

This gives the maximum value of U'_1:

$$\left(\frac{U'_1}{\sqrt{2}U/\pi}\right)_{\max} = \frac{\pi}{4}\left(1 - m\frac{\Delta\theta}{\pi}\right). \tag{3.5}$$

Figure 3.6 shows, for the lower values of m, the normalized value of the maximum of U'_1 for zero $\Delta\theta$, $\Delta\theta$ equal to 1° and $\Delta\theta$ equal to 2°, respectively.

The increase in the "voltage loss" as m increases is one of the reasons for limiting the value of m.

3.1.2.3 Harmonics

Pulse width modulation does not reduce the harmonic content.

This content is very high and virtually independent of m.

Since voltage u' is always equal to $\pm U/2$, its rms value U' is equal to $U/2$. The quadratic sum of the harmonics, computed by

$$\left(\sum_2^\infty U_n'^2\right)^{1/2} = \sqrt{U'^2 - U_1'^2}$$

is, if the value of U'_1 given by (3.3) is taken,

$$\left(\sum_2^\infty U_n'^2\right)^{1/2} = \frac{U}{2}\sqrt{1 - \frac{r^2}{2}}.$$

The dashed line in Fig. 3.5 shows the variations of this sum as a function of r. It is greater than U'_1 and it becomes equal to it only when r is equal to 1.

However, the pulse width modulation enables the harmonics to be moved towards higher frequencies.

- When r is zero, voltage u' is made up of alternative rectangular pulses, all of the same width. Its fundamental frequency is mf; its series expansion contains terms of frequencies mf, $3mf$, $5mf$, ... and of relative values 1, 1/3, 1/5, ...

When r increases, the component at f frequency increases. The components at mf, $3mf$, $5mf$, ... frequencies decrease, but other harmonics appear:

- on either side of the mth harmonic, appears the pair of harmonics of rank $m - 2$ and $m + 2$, whose values increase with r. For high values of this factor, harmonics of rank $m - 4$ and $m + 4$ can no longer be ignored;
- on either side of frequency $2mf$ appears the pair of harmonics of rank $2m - 1$ and $2m + 1$, followed by the pair $2m - 3$, $2m + 3$, ...
- on either side of harmonic of rank $3m$, appears a third "family" of harmonics of rank $3m \pm 2$, $3m \pm 4$, ...

In general, the values of the harmonics decrease as the central frequency of the family increases and, within a family, when they deviate from the central frequency. Therefore, in normal circumstances, only the first two families need to be examined, since the harmonics of the others are smaller and, more importantly, easier to filter.

When m is high, the terms of the first families do not interfere. The initial part of the spectrum is made up of a group of lines which are symmetrical in relation to mf, and then of another group which is symmetrical in relation to $2mf$; both harmonics of a same pair have the same value.

This separation of the various "families" becomes much less clear as m decreases.

- In order to show how *the increase in m shifts the harmonics towards the higher frequencies* and so makes the spectrum clearer, we have drawn, in Fig. 3.7, the spectra which correspond to the maximum value of r, for m equal to 3, 6, 9 and 12, respectively. All the values are referred to $\sqrt{2}U/\pi$.

  For m equal to 3, the spectrum is not as previously described. For m equal to 6, the first family starts to become apparent. For m equal to 9, the first family can clearly be distinguished from the second, but there is some interference between the latter family and the third. The spectrum drawn for m equal to 12 shows that the groups of harmonics centered on mf and $2mf$ can be clearly distinguished and that the harmonics of the same pair (10 and 14, 8 and 16, 23 and 25, 21 and 27, 19 and 29) are equal. For greater values of m, the spectrum is the same and only the gap between the families increases.

- Figure 3.8 shows variations, as a function of r, of the value of the fundamental and of the first harmonics for m equal to 3, to 6, to 9 and to 15, respectively.

  For m equal to 3, when r increases from 0 to 1, the waveform of u' changes from a square-wave AC signal of frequency $3f$ to a signal which has the same waveform but which is of frequency f. In addition to the fundamental all the uneven harmonics are found.

  For m equal to 6, the different harmonics are to be found as foreseen. The values of harmonics which occupy the same position in the spectrum on either

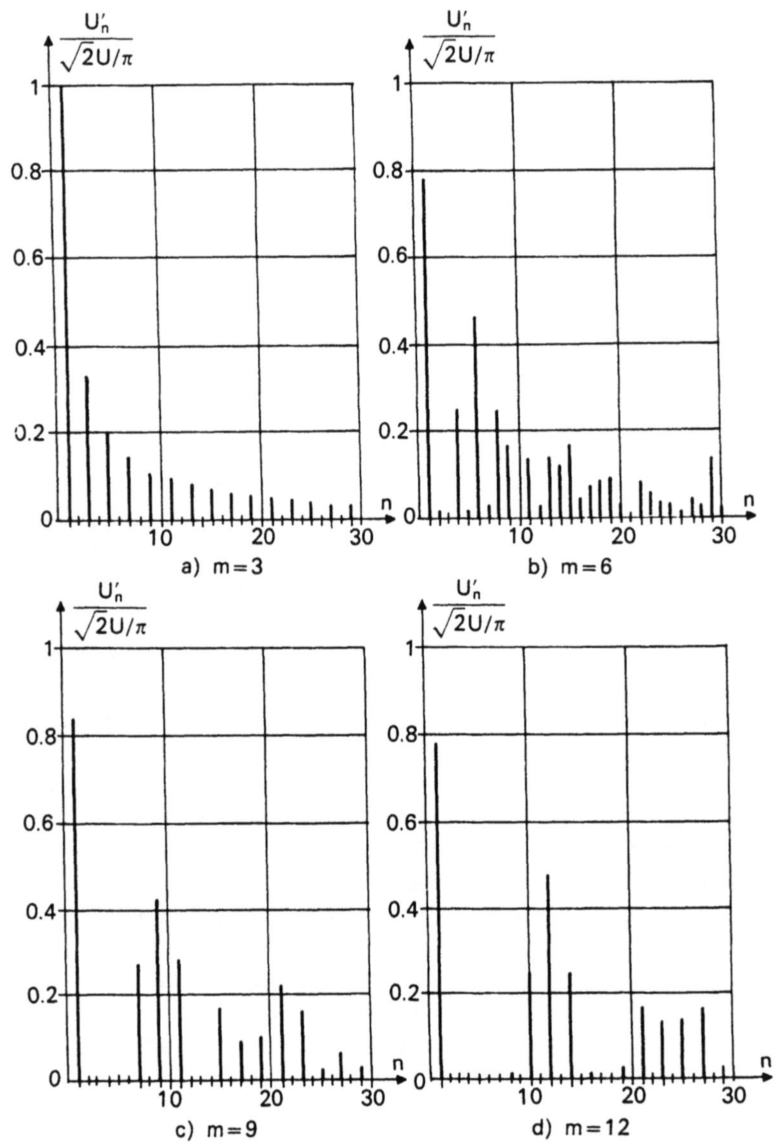

Fig. 3.7

side of mf and $2mf$ show little variation when m changes from 6 to 9. From 15 onwards, the curves no longer show any change and the harmonics of a same pair have values which are virtually identical.

When m is high (equal to or greater than 15), the curves plotted for m equal to 15 can be used whether it be even or uneven. The followings are taken:

U'_1, for the fundamental,
U'_{15}, for the harmonic of rank m,
U'_{13} equal to U'_{17}, for harmonics of rank $m - 2$ and $m + 2$,
U'_{11}, equal to U'_{19}, for harmonics of rank $m - 4$ and $m + 4$,
U'_{29} equal to U'_{31}, for harmonics of rank $2m - 1$ and $2m + 1$,

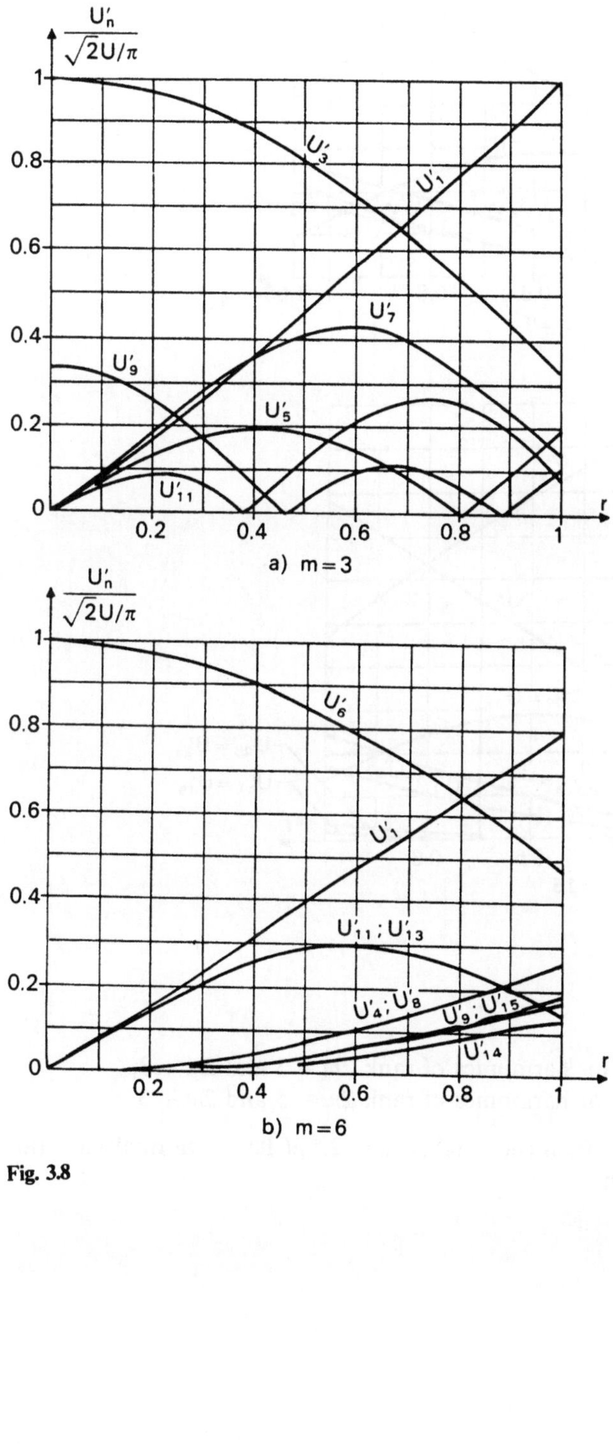

Fig. 3.8

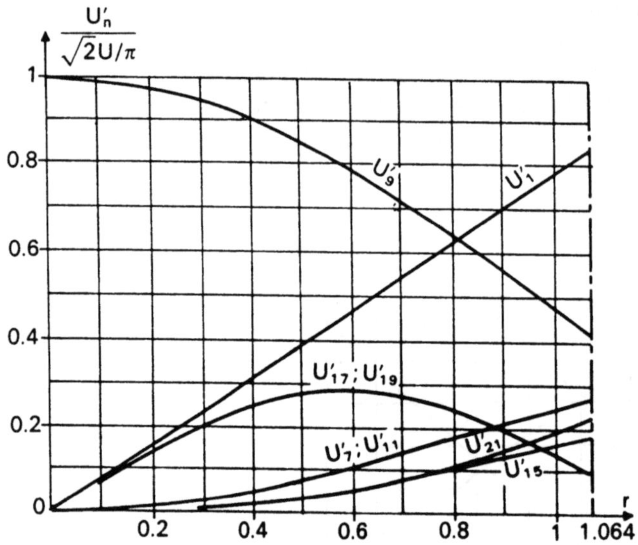

c) m=9

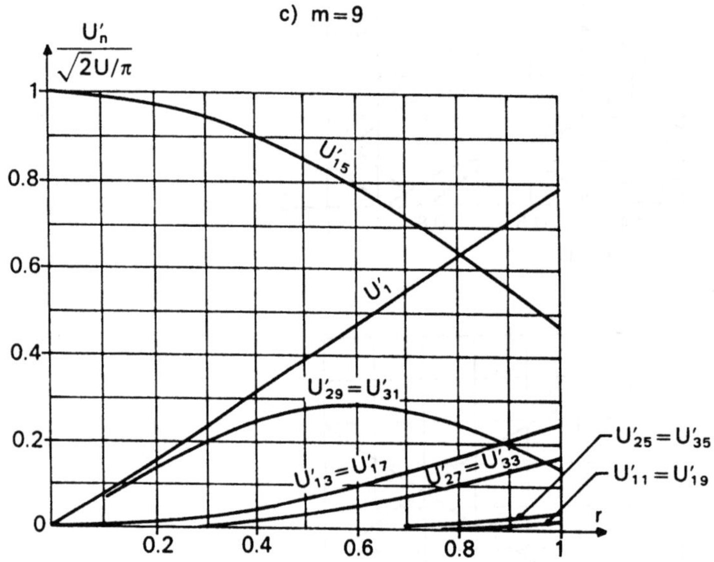

d) m=15

Fig. 3.8 (continued)

U'_{27}, equal to U'_{33}, for harmonics of rank $2m - 3$ and $2m + 3$,
U'_{25}, equal to U'_{35}, for harmonics of rank $2m - 5$ and $2m + 5$.

All the other terms of frequency higher than $2.5mf$ have a normalized value of less than a hundredth.

3.1.3 Currents

In studying currents, the output current i' is assumed to have a sinusoidal waveform:

$$i' = I'\sqrt{2}\sin(\omega t - \varphi).$$

Figures 3.3 and 3.4 provide two examples of the waveform of input current i and of currents i_{T_1} and i_{D_1} in the semiconductor devices.

At a given output current value I', the current values in the semiconductor devices and at the input were dependent on φ when there was no pulse width modulation. With the use of the latter, the values of these currents are dependent on m and r in addition.

3.1.3.1 Current in the Semiconductor Devices

- The *number of commutations* to be carried out, at non-zero current, throughout one cycle T of the output current, changes from 2 to $2m$. During the positive half-cycle of i', TC_1–D'_1 and D'_1–TC_1 commutations occur; during the negative half-cycle of i', TC'_1–D_1 and D_1–TC'_1 commutations occur. These commutations are similar to those found for the series chopper.

  Whatever the value of φ, switches TC_1 and TC'_1 must be turn-on and turn-off controlled devices.

  As soon as the value of m is high enough, a commutation takes place in the immediate vicinity of the mid-point of each half-cycle of i'. The maximum current to be commutated is equal to $I'\sqrt{2}$.

  For the same reason, the maximum value of the current in each semiconductor device is close to the maximum value of i':

$$i_{T\,max} = i_{D\,max} \simeq I'\sqrt{2}. \tag{3.6}$$

- The controlled switch TC_1 and diode D'_1 alternately let the positive half-cycle of current i' flow. Similarly TC'_1 and D_1 let the negative half-cycle of this current flow.

  As in the case where there was no PWM, this gives

$$I_{T\,mean} + I_{D\,mean} = \frac{1}{\pi} I'\sqrt{2}, \qquad I^2_{T\,rms} + I^2_{D\,rms} = \frac{I'^2}{2}$$

  but the pulse width modulation modifies the sharing of the currents between the two types of devices.

For example, in the graph shown in Fig. 3.4, i_{T_1} would be equal to i' from $\omega t = \varphi$ to $\omega t = \pi$, if there were no PWM. The latter leads to the last pulse being continued between $\omega t = \pi$ and $\omega t = \pi + \varphi$, but also brings out three zero-current periods. Without PWM, current i_{D_1} would be equal to $-i'$ for

$0 < \omega t < \varphi$; PWM causes a slight suppression at the end of this period but leads to four additional conduction intervals during period $\pi + \varphi, 2\pi$. PWM reduces the difference between the values of the current in the controlled devices and those of the current in the diodes.

This reduction of the difference increases all the more as the difference is large. Figure 2.4 shows that this corresponds to φ equal to 0 or π; the current in one type of device is then zero.

This reduction becomes much more apparent when r decreases. A decrease in r does lead to enlargement of the periods at negative u' during the positive half-cycle of u' and of the periods at positive u' during the negative half-cycle of u'.

In order to show how these values are brought closer to each other, the variations in the mean values, $I_{T\,mean}$ and $I_{D\,mean}$, and the rms values $I_{T\,rms}$ and $I_{D\,rms}$, of the currents in both types of devices are plotted as a function of r (Fig. 3.9), in the case where the difference is the greatest, i.e. for φ equal to zero. (For φ equal to π, the subscripts D and T need only to be permuted).

These curves were plotted for m equal to 15; they can be used for any sufficiently high value of m.

3.1.3.2 Current Delivered by the Supply

As shown in Fig. 3.3 and 3.4, input current i has a frequency twice that of the output variables. However, during each of its cycles, it is equal to either $+ i'/2$ or $- i'/2$.

– *Its rms value*, as in the case where there was no PWM, is

$$I_{rms} = I'/2. \tag{3.7}$$

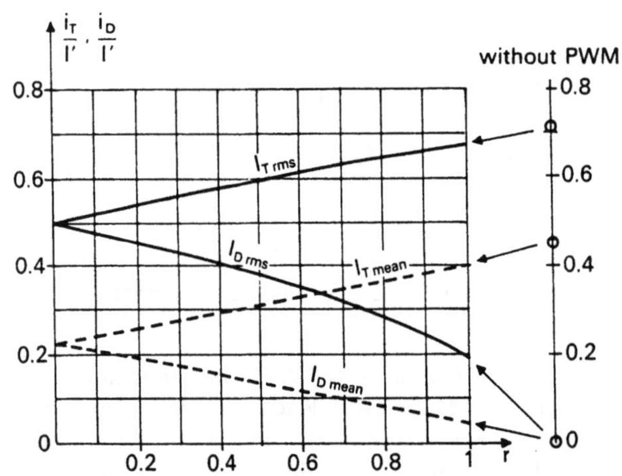

Fig. 3.9

- *Its mean value* I can be deduced from the equality of power at input and output. As i' is assumed to be sinusoidal,

  $$UI = (u'i')_{\text{mean}} = U'_1 I' \cos\varphi.$$

  If the value given by (3.3) is taken for U'_1,

  $$I = \frac{1}{2\sqrt{2}} rI' \cos\varphi. \tag{3.8}$$

  When it reaches its maximum, I is equal to $I'/2\sqrt{2}$ while, without PWM, this value was $I'\sqrt{2}/\pi$. This maximum value is thus divided by $4/\pi$; this is due to the "voltage loss" given by Eq. (3.4).

- *The harmonic ratio*

  $$\tau_i = \frac{1}{I}\sqrt{I_{\text{rms}}^2 - I^2} = \sqrt{\frac{2}{r^2\cos^2\varphi} - 1}$$

  is, for a given $\cos\varphi$, greater than that which was obtained without PWM when r equals 1. Furthermore, it increases when r decreases.

- The value of the different *harmonics* is deduced from the conservation of the instantaneous power:

  $$i = \frac{1}{U}(u'i').$$

  The fundamental component of the output voltage u' produces, in the input current – in addition to the average component – a term at $2f$ frequency. If m is sufficient for there to be no term at $3f$ frequency in u', the 2nd harmonic of the input current is only due to the fundamental of u' and has the following rms value:

  $$I_2 = \frac{U'_1 I'}{\sqrt{2}U}$$

  or, if the value given by (3.3) is taken for U'_1,

  $$I_2 = \frac{1}{4} rI'. \tag{3.9}$$

  It is this harmonic which determines the size of the input filter.

  Harmonics of rank $m - 2$, m, and $m + 2$ of u' correspond in i to those of rank $m - 3, m - 1, m + 1$ and $m + 3$.

  Harmonics of rank $2m - 3, 2m - 1, 2m + 1$ and $2m + 3$ of u' correspond in i to those of rank $2m - 4, 2m - 2, 2m, 2m + 2$ and $2m + 4$ . . .

3.1.4 Remarks on Control Phase Lock and on Asynchronous Modulation

- The optimal *phase lock*, for which the waveforms have been plotted and the characteristics determined, is that which makes the voltage u' fundamental closest to the wanted sine wave u'_w.

If the reference wave is kept as the time origin, i.e.

$$u'_w = r\frac{U}{2}\sin\omega t$$

the voltage u' waveform is modified if the carrier is shifted by an angle δ; the same also applies to input current i.

Angle δ is counted, on the ωt scale, positively when the carrier is lagging the reference wave. Its value may vary from zero to $2\pi/m$.

Figure 3.10 shows for $m = 5$ and $r = 2/3$ the modifications of the waveforms of u' which correspond to the change from $\delta = 0$ to $\delta = \pi/10$, and then to $\delta = \pi/5$.

- With the optimal phase lock, the series expansion of voltage u' for m uneven, had the following form:

$$u' = \sum_{n=1}^{\infty} U'_n\sqrt{2}\sin(n\omega t + \varphi_{n0})$$

with n uneven and φ_{n0} equal to zero or π.

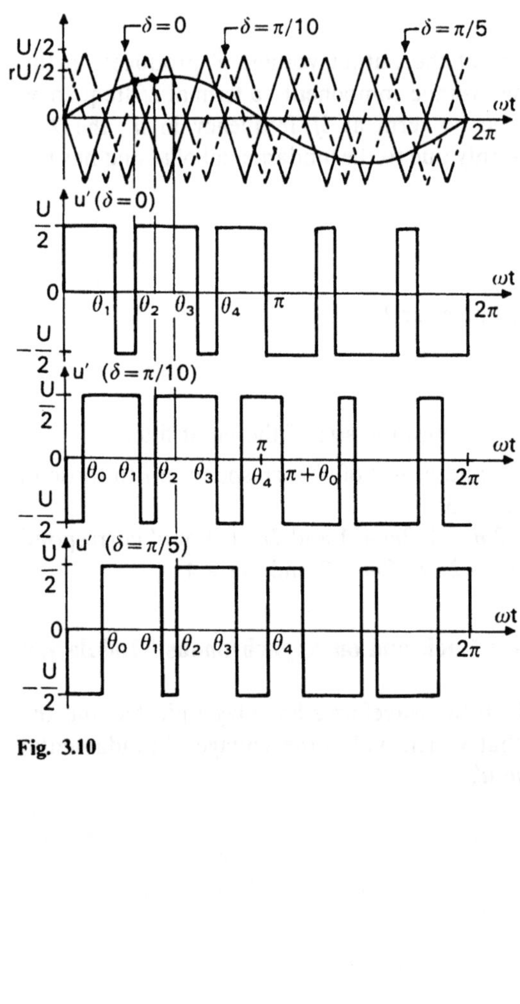

Fig. 3.10

U'_n was calculated over a quarter-cycle using

$$|U'_n|\sqrt{2} = \frac{1}{n}\frac{4}{\pi}\frac{U}{2}(1 - 2\cos n\theta_1 + 2\cos n\theta_2 - \ldots \pm 2\cos n\theta_{m/2-1}).$$

When δ is different from zero, there is no longer a necessary symmetry with respect to the mid-points of the half-cycles; but if m is uneven, the negative half-cycle once more reproduces the positive half-cycle to within a sign: the series expansion then comprises the only terms of uneven rank and has the following form:

$$u' = \sum_{n=1}^{\infty} U'_n\sqrt{2}\sin(n\omega t + \varphi_n)$$

with

$$U'_n\sqrt{2} = \sqrt{A_n^2 + B_n^2}, \qquad \tan\varphi_n = \frac{B_n}{A_n}$$

$$A_n = \frac{1}{n}\frac{4}{\pi}\frac{U}{2}(\cos n\theta_0 - \cos n\theta_1 + \cos n\theta_2 - \ldots \pm \cos n\theta_{m-1})$$

$$B_n = \frac{1}{n}\frac{4}{\pi}\frac{U}{2}(-\sin n\theta_0 + \sin n\theta_1 - \sin n\theta_2 + \ldots \mp \cos n\theta_{m-1})$$

The m (no longer $m/2 - 1$) commutation angles to be determined vary as a function of δ.

- Figure 3.11 shows, for r equal to 0.8 and m equal to 3 and then to 9, the variations of the rms value and phase of the fundamental and the main harmonics, when δ changes from 0 to $2\pi/m$.

• For m equal to 3, the fundamental greatly decreases when moving away from optimal phase lock: this reduction is accompanied by a sharp increase in harmonics 3 and 5. The fundamental phase shows significant variations above and below zero. Angles φ_3 and φ_5 decrease, one of 2π and the other of 4π but in a non-linear manner.

• For m equal to 9, the fundamental and harmonics 9, 7 and 11, 17 and 19 remain virtually constant when δ changes from 0 to $2\pi/9$. Angle φ_1 remains at zero. Angles φ_7, φ_9 and φ_{11} decrease of 2π; angles φ_{17} and φ_{19} of 4π; the differences compared to a linear variation remain less than 0.2°.

- The characteristics of Fig. 3.11 show that the influence of the phase shift is very different according to whether m is high or low.

• *For low values of m*, operation in the optimal phase lock mode is required if the output voltage fundamental has to vary in accordance with r, and if the harmonics are not allowed to take on an abnormal relative importance.

• *For high values of m*, the phase shift of the carrier with respect to the reference has virtually no importance:
  - the fundamental and harmonic rms values are independent of δ;

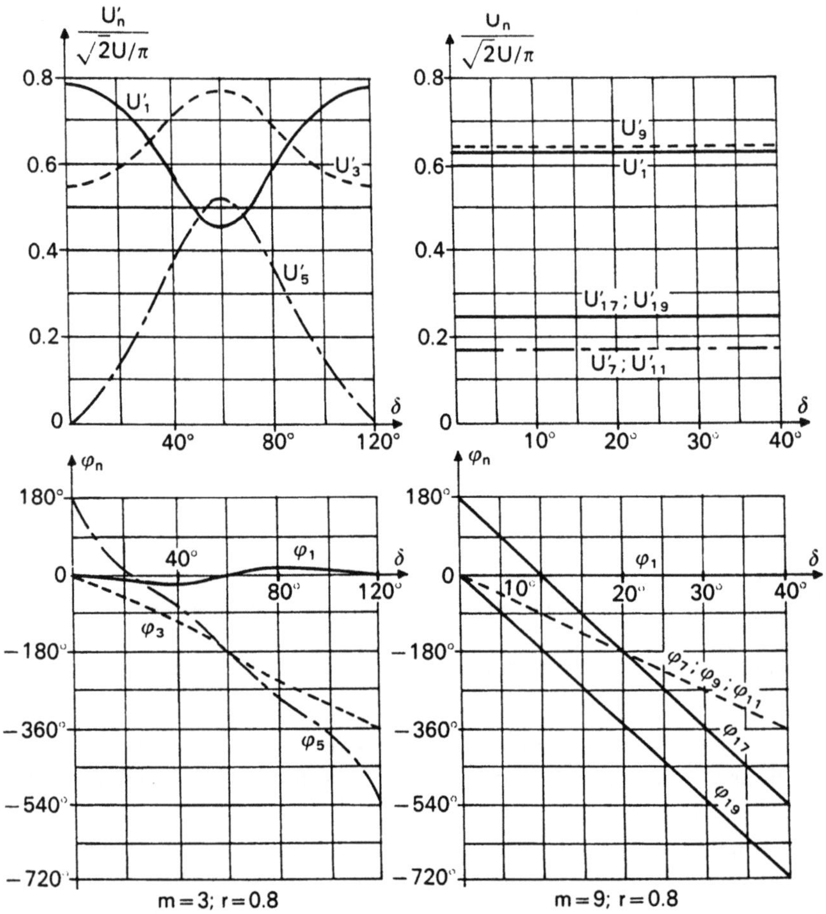

Fig. 3.11

- the phase angles varies proportionately to δ *in the same way for all the terms of a same "family"*:
  - the fundamental remains in phase with the reference,
  - for the harmonics centered on frequency *mf*,

$$\varphi_n = -m\delta$$

  - for the harmonics centered on frequency *2mf*,

$$\varphi_n = -2m\delta$$

  - for the harmonics of the third family,

$$\varphi_n = -3m\delta \ldots$$

- Studying the effect of the phase shift between the carrier and the reference wave enables us to see whether the PWM must be *synchronous* or may be *asynchronous*, i.e. whether the modulation frequency f' must be a whole multiple of the reference frequency or not.

  If Δm denotes the difference between the modulation ratio m, equal to f'/f, and the nearest integer, angle δ varies by $2\pi\,\Delta m/m$ during each cycle, equal to $1/f$, of the reference. This angle varies by $2\pi/m$ in $1/\Delta m$ cycle; it thus varies at frequency $f\,\Delta m$.

– For low values of m, the asynchronous modulation cannot be used. The fundamental of u' and its harmonics, the mean value of i and its harmonics would show wide oscillations at the beat frequency $f\,\Delta m$.

– If the value of m is high enough, the choice of the asynchronous modulation poses no problems. For variable frequency inverters, f' can be kept constant and only frequency f of the reference wave has to be varied, as long as it remains well below f'.

3.2 Three-Phase Full-Bridge Inverter

By grouping together three single-phase half-bridge inverters, the six-switch three-phase inverter shown in Fig. 3.12 is obtained.

When operating in PWM, this inverter is widely used to feed balanced three-phase loads at variable frequency and voltage.

If a sinusoidal reference is adopted, the three-phase inverter provides the same advantages as the single-phase half-bridge inverter.

- possibility of varying the output voltage fundamental,
- repelling the harmonics of these voltages towards higher frequencies.

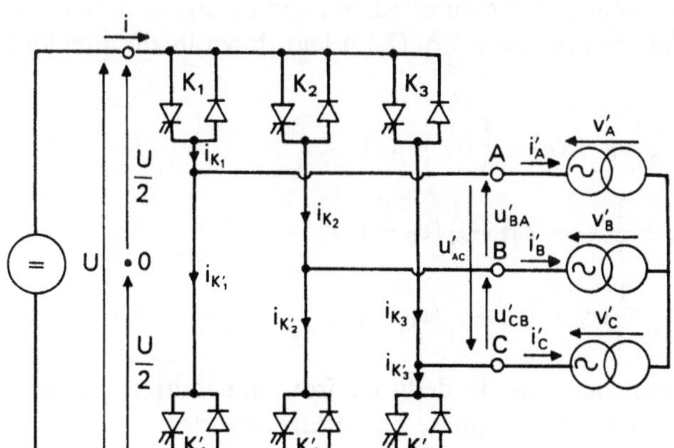

Fig. 3.12

Furthermore, if the load is star-connected without a neutral conductor or is delta-connected, the main harmonics of these voltage can be suppressed.

The reference is often modified in order to reduce the "voltage loss". This also enables certain harmonics which remain in the output voltages to be reduced.

3.2.1 "Switch" Control. Waveforms

- In order to determine the turn-on instant of the complementary switches K_1 and K'_1, K_2 and K'_2, K_3 and K'_3 of each inverter leg, the method described for the single-phase half-bridge inverter should be used:
  - The three reference waves are plotted which represent the three wanted voltages between the three output terminals A, B, C and the (fictitious) mid-point of the DC voltage supply.
  - The intersections with the triangular modulation wave (or carrier) M determine the turn-on instants of the different switches.

- *The voltages* supplied by the half-bridges are:

  $v_A - v_O = U/2$, when K_1 is on;

  $v_A - v_O = -U/2$, when K'_1 is on;

  $v_B - v_O = U/2$, when K_2 is on . . .

  The phase-to-phase voltages between the output terminals can be deduced from the above voltages:

  $$u'_{BA} = (v_A - v_O) - (v_B - v_O)$$

  $$u'_{CB} = (v_B - v_O) - (v_C - v_O).$$

  $$u'_{AC} = (v_C - v_O) - (v_A - v_O).$$

  If the load is balanced and star-connected, voltage v'_A, v'_B, v'_C across the three phases can be determined using Eq. (2.49) laid down in the previous chapter:

  $$v'_A = \frac{2}{3}(v_A - v_O) - \frac{1}{3}(v_B - v_O) - \frac{1}{3}(v_C - v_O)$$

  $$v'_B = -\frac{1}{3}(v_A - v_O) + \frac{2}{3}(v_B - v_O) - \frac{1}{3}(v_C - v_O)$$

  $$v'_C = -\frac{1}{3}(v_A - v_O) - \frac{1}{3}(v_B - v_O) + \frac{2}{3}(v_C - v_O).$$

- *The currents* in the switches can be deduced from the output currents i'_A, i'_B, i'_C as in the case of the single-phase half-bridge inverter:

  $i_{K_1} = i'_A$, $\quad i_{K'_1} = 0$ when K_1 is on

  $i_{K_1} = 0$, $\quad i_{K'_1} = -i'_A$ when K'_1 is on

  $i_{K_2} = i'_B$, $\quad i_{K'_2} = 0$ when K_2 is on . .

Table 3.1. Laying down the voltage and current expressions

K_1	K_2	K_3	$v_A - v_O$	$v_B - v_O$	$v_C - v_O$	u'_{BA}	v'_A	i_{K_1}	i_{K_2}	i_{K_3}	i
ON	ON	ON	$U/2$	$U/2$	$U/2$	0	0	i'_A	i'_B	i'_C	0
ON	OFF	ON	$U/2$	$-U/2$	$U/2$	U	$U/3$	i'_A	0	i'_C	$-i'_B$
ON	ON	OFF	$U/2$	$U/2$	$-U/2$	0	$U/3$	i'_A	i'_B	0	$-i'_C$
ON	OFF	OFF	$U/2$	$-U/2$	$-U/2$	U	$2U/3$	i'_A	0	0	i'_A
OFF	ON	ON	$-U/2$	$U/2$	$U/2$	$-U$	$-2U/3$	0	i'_B	i'_C	$-i'_A$
OFF	OFF	ON	$-U/2$	$-U/2$	$U/2$	0	$-U/3$	0	0	i'_C	i'_C
OFF	ON	OFF	$-U/2$	$U/2$	$-U/2$	$-U$	$-U/3$	0	i'_B	0	i'_B
OFF	OFF	OFF	$-U/2$	$-U/2$	$-U/2$	0	0	0	0	0	0

Current i delivered by the supply is given by

$$i = i_{K_1} + i_{K_2} + i_{K_3}.$$

- Table 3.1 shows the expressions that $u'_{BA}, v'_A, i_{K1}, i_{K2}, i_{K3}$ and i can take depending on the off- or on-state of switches K_1, K_2, K_3.

In synchronous modulation, the *modulation ratio chosen in a multiple of* 3, which enables the three voltages $v_A - v_O, v_B - v_O, v_C - v_O$ to be identical to within a third of their cycle; this identity can then be found with voltages $u'_{BA}, u'_{CB}, u'_{AC}$ or with voltages v'_A, v'_B, v'_C.

Figure 3.13 gives an example of the waveforms obtained with sinusoidal voltage references and with output currents assumed to be sinusoidal. This graph corresponds to

$$m = 6, \qquad r = 0.75, \qquad \varphi = \pi/6.$$

The waveforms of currents in the two devices which build up K_1 and the voltage across them have not been shown; they are the same as in Fig. 3.3.

3.2.2 Sinusoidal Reference

If the reference voltages are sinusoidal, voltages $v_A - v_O$, $v_B - v_O$ and $v_C - v_O$ have, for given values of m and r, the same waveform as that found in single-phase inverters; the value and phase of the fundamental and harmonics of these voltages are those found in single-phase inverters.

The stresses on the semiconductor devices can be deduced from voltage U and the current delivered by each inverter-leg, in the same way as for single-phase inverters.

If the modulation ratio m is multiple of 3, the three voltages $v_A - v_O$, $v_B - v_O$, $v_C - v_O$ are identical to within a third of a cycle. This has two beneficial effects: reduction in the harmonics of output voltages v'_A, v'_B, v'_C and input current i.

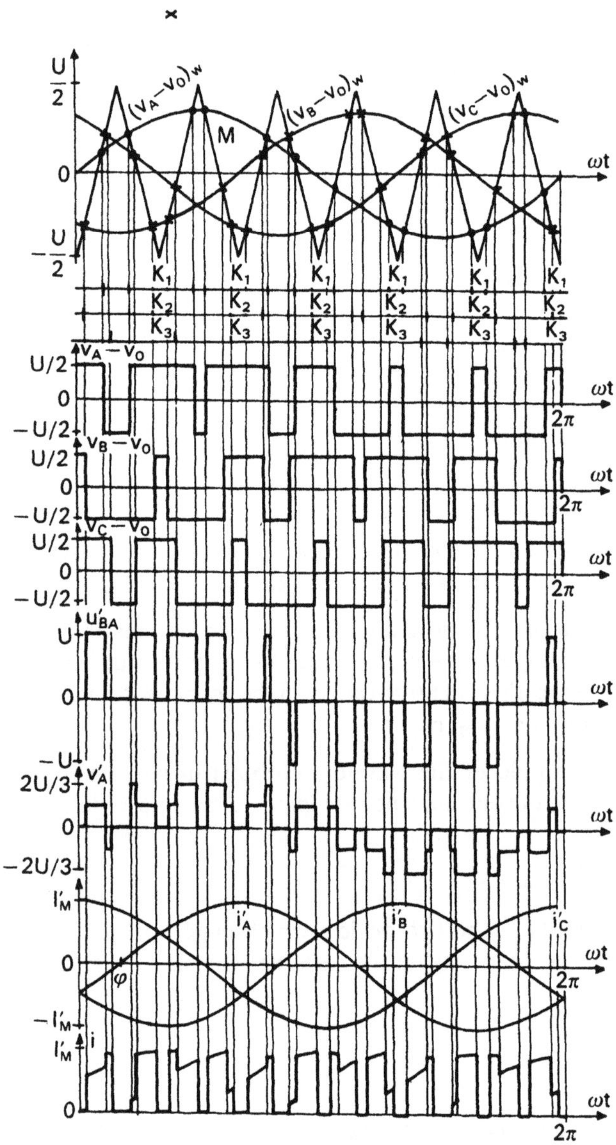

Fig. 3.13 ($m = 6$)

3.2.2.1 Improvement in Output Voltages

Output voltages v'_A, v'_B, v'_C, can be deduced from $v_A - v_0, v_B - v_0, v_C - v_0$, by *suppressing harmonics of rank* 3 *or multiple of* 3 (cf. Chap. 2, Sect. 2.4.1.1).

Since m is a multiple of 3, the following harmonics are suppressed:

- the most important harmonic – of frequency mf, situated in the centre of the first family,

- the harmonics of frequency $2mf - 3f$ and $2mf + 3f$ of the second family,
- the harmonic of frequency $3mf$, situated in the centre of third family, . . .

The only important harmonics of the first two families which remain are those of rank

$$\boxed{\begin{array}{ll} m-2, & m+2 \\ 2m-1, & 2m+1 \end{array}} \tag{3.10}$$

those of rank $m, m-6, m+6, 2m-3, 2m+3$ have been suppressed.

For the fundamental and the remaining harmonics, the characteristics plotted for the single-phase inverter can be used.

3.2.2.2 Improvement in Input Current

If the output currents i'_A, i'_B, i'_C are balanced, input current i is the sum of three currents i_{K1}, i_{K2}, i_{K3} identical to within $T/3$ or $2T/3$.

Each of these currents (i_{K_1} is given by $i_{T_1} - i_{D_1}$ in Figs. 3.3 and 3.4) is the sum of a mean term and of harmonics. When making the sum of these currents,

- the mean terms are added together;
- the harmonics of rank 3 or multiple of 3, forming homopolar systems, can also be added together;
- the other harmonics, forming positive or reverse balanced systems, give sums equal to zero.

A large number of harmonics have thus been suppressed, notably that of frequency $2f$, which causes the most problems.

The conservation of instantaneous power can be used to find the value and rank of the remaining harmonics.

$$Ui = v'_A i'_A + v'_B i'_B + v'_C i'_C \tag{3.11}$$

with

$$i'_A = I'\sqrt{2}\sin(\omega t - \varphi)$$

$$i'_B = I'\sqrt{2}\sin(\omega t - 2\pi/3 - \varphi)$$

$$i'_C = I'\sqrt{2}\sin(\omega t - 4\pi/3 - \varphi).$$

If m is high, according to (3.10) the first terms of the series expansion of v'_A, v'_B, v'_C are:

$$v'_A = V'_1\sqrt{2}\sin\omega t + V'_{m\pm 2}\sqrt{2}\,[\sin(m-2)\omega t + \sin(m+2)\omega t]$$
$$+ V'_{2m\pm 1}\sqrt{2}[-\sin(2m-1)\omega t + \sin(2m+1)\omega t]$$

$$v'_B = V'_1\sqrt{2}\sin\left(\omega t - \frac{2\pi}{3}\right)$$
$$+ V'_{m\pm 2}\sqrt{2}\left[\sin(m-2)\left(\omega t - \frac{2\pi}{3}\right) + \sin(m+2)\left(\omega t - \frac{2\pi}{3}\right)\right]$$
$$+ V'_{2m\pm 1}\sqrt{2}\left[-\sin(2m-1)\left(\omega t - \frac{2\pi}{3}\right) + \sin(2m+1)\left(\omega t - \frac{2\pi}{3}\right)\right]$$
$$v'_C = V'_1\sqrt{2}\sin\left(\omega t - \frac{4\pi}{3}\right)$$
$$+ V'_{m\pm 2}\sqrt{2}\left[\sin(m-2)\left(\omega t - \frac{4\pi}{3}\right) + \sin(m+2)\left(\omega t - \frac{4\pi}{3}\right)\right]$$
$$+ V'_{2m\pm 1}\sqrt{2}\left[-\sin(2m-1)\left(\omega t - \frac{4\pi}{3}\right) + \sin(2m+1)\left(\omega t - \frac{4\pi}{3}\right)\right]$$

with V'_1, $V'_{m\pm 2}$ and $V'_{2m\pm 1}$ having the values given for U'_1, $U'_{m\pm 2}$ and $U'_{2m\pm 1}$ in the study of the single-phase half-bridge inverter.

By introducing the expression of v'_A, v'_B, v'_C, of i'_A, i'_B, i'_C in (3.11) and taking into account that $m-3$, $m+3$ and $2m$ are multiples of 3, the following can be written:

$$Ui = 3V'_1 I' \cos\varphi$$
$$+ 3V'_{m\pm 2} I'\{\cos[(m-3)\omega t + \varphi] - \cos[(m+3)\omega t - \varphi]\}$$
$$+ 3V'_{2m\pm 1} I'[\cos(2m\omega t - \varphi) + \cos(2m\omega t + \varphi)].$$

The only remaining important harmonics are those of rank

$$\boxed{m-3, \quad m+3, \quad 2m.} \tag{3.12}$$

It can be seen that the increase in m makes the filtering (or smoothing) of the input current easier.

The mean value of this current and the rms value of its first important harmonics are given by

$$I = \frac{3V'_1 I'}{U}\cos\varphi = \frac{3}{2\sqrt{2}} rI'\cos\varphi$$
$$I_{m-3} = I_{m+3} = \frac{3}{\sqrt{2}U} V'_{m\pm 2} I' \tag{3.13}$$
$$I_{2m} = \frac{6}{\sqrt{2}U} V'_{m\pm 1} I' \cos\varphi.$$

3.2.3 Injection of Harmonic 3 into the Reference

3.2.3.1 Increasing *r* in the Three-Phase Mode

In three-phase inverters, *the "voltage loss" can be reduced* without decreasing the quality of the output voltage and input currents.

Since the harmonics of rank 3 or multiple of 3 are eliminated from the output voltages, *a harmonic 3 can be added* to the sinusoid of frequency f in order to form the reference wave. Harmonic 3 will be found in voltages $v_A - v_O$, $v_B - v_O$, $v_C - v_O$ related to the fictitious mid-point O of the input, but it disappears in the phase-to-neutral voltages v'_A, v'_B, v'_C and phase-to-phase voltages $v_A - v_B$, $v_B - v_C$, $v_C - v_A$ at the output.

Adding harmonic 3 enables the maximum amplitude of the fundamental to be increased in the reference and, as a result, in the output voltages.

If the reference voltage is sinusoidal, $(v_A - v_O)_w$ equal to v'_{1w} cannot be greater than $U/2$, at least if m is high (Fig. 3.14a).

If the reference $(v_A - v_O)_w$ is the sum of a fundamental component and a sufficiently strong harmonic 3, the maximum values of the reference are no longer to be found at the mid-point of its half-cycles. Without voltage $(v_A - v_O)_w$ going beyond $U/2$, a value of v'_{1w} can be obtained which is greater than $U/2$ (Fig. 3.14b).

The control using a reference voltage with the form

$$(v_A - v_O)_w = \frac{U}{2}(r \sin \omega t + k \sin 3\omega t) \tag{3.14}$$

is called *suboptimal control.*

In order to find the value of k which enables r to reach its maximum value, it is assumed that the waveform of the reference remains always the same by writing

$$(v_A - v_O)_w = \frac{U}{2} r(\sin \omega t + k' \sin 3\omega t).$$

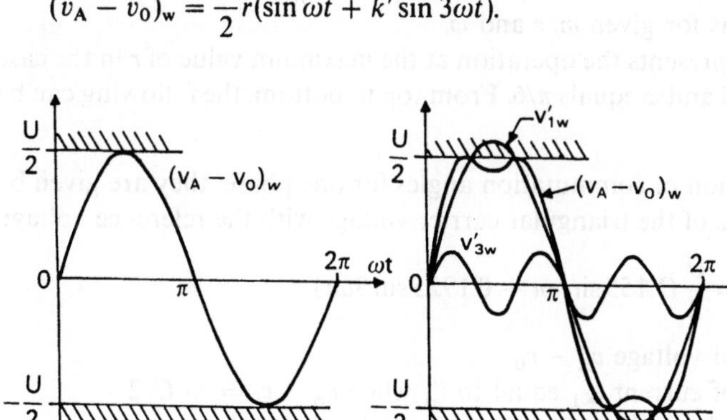

Fig. 3.14

The time derivative of $\sin\omega t + k'\sin 3\omega t$ shows that the first maximum occurs before $t = \pi/2\omega$ if

$$k' > \tfrac{1}{9}.$$

If this condition is fulfilled, the instant of this maximum value is such that

$$\cos^2\omega t = \frac{1}{4}\left(3 - \frac{1}{3k'}\right).$$

The maximum value of r corresponds to the minimum value of the quantity $(\sin\omega t + k'\sin 3\omega t)$ at this instant. The following can thus be found:

$$r_{\max} = \frac{2}{\sqrt{3}} = 1.155$$

for $k' = \dfrac{1}{6}$ or $k = \dfrac{1}{6}\dfrac{2}{\sqrt{3}} = 0.1925.$

The theoretical maximum of the phase-to-neutral voltage fundamental change

from $U/2$, with sinusoidal control
to $1.155U/2$, with suboptimal control.

The voltage loss

$$1 - \left(\frac{V'_1}{\sqrt{2}U/\pi}\right)_{\max}$$

changes from 21.46% to 9.31%.

3.2.3.2 Waveforms. Characteristics

- For given values of m (multiple of 3) and r, the voltage waveforms show little difference from those found with a sinusoidal reference voltage; there is only a slight variation in the width of certain pulses. The same applies to the current waveforms for given m, r and φ.

  Figure 3.15 represents the operation at the maximum value of r in the case where m equals 15 and φ equals $\pi/6$. From top to bottom, the following can be seen:

  - The determination of commutation angles for one phase; they are given by the intersections of the triangular carrier voltage with the reference voltage

  $$(v_A - v_O)_w = \frac{U}{2}(1.155\sin\omega t + 0.1925\sin 3\omega t)$$

  - the waveform of voltage $v_A - v_O$
  - the waveform of current i_{K1} equal to i'_A when $v_A - v_O = +U/2$
  - the waveform of the phase-to-neutral output voltage v'_A, given by

  $$v'_A = \frac{2}{3}(v_A - v_O) - \frac{1}{3}(v_B - v_O) - \frac{1}{3}(v_C - v_O)$$

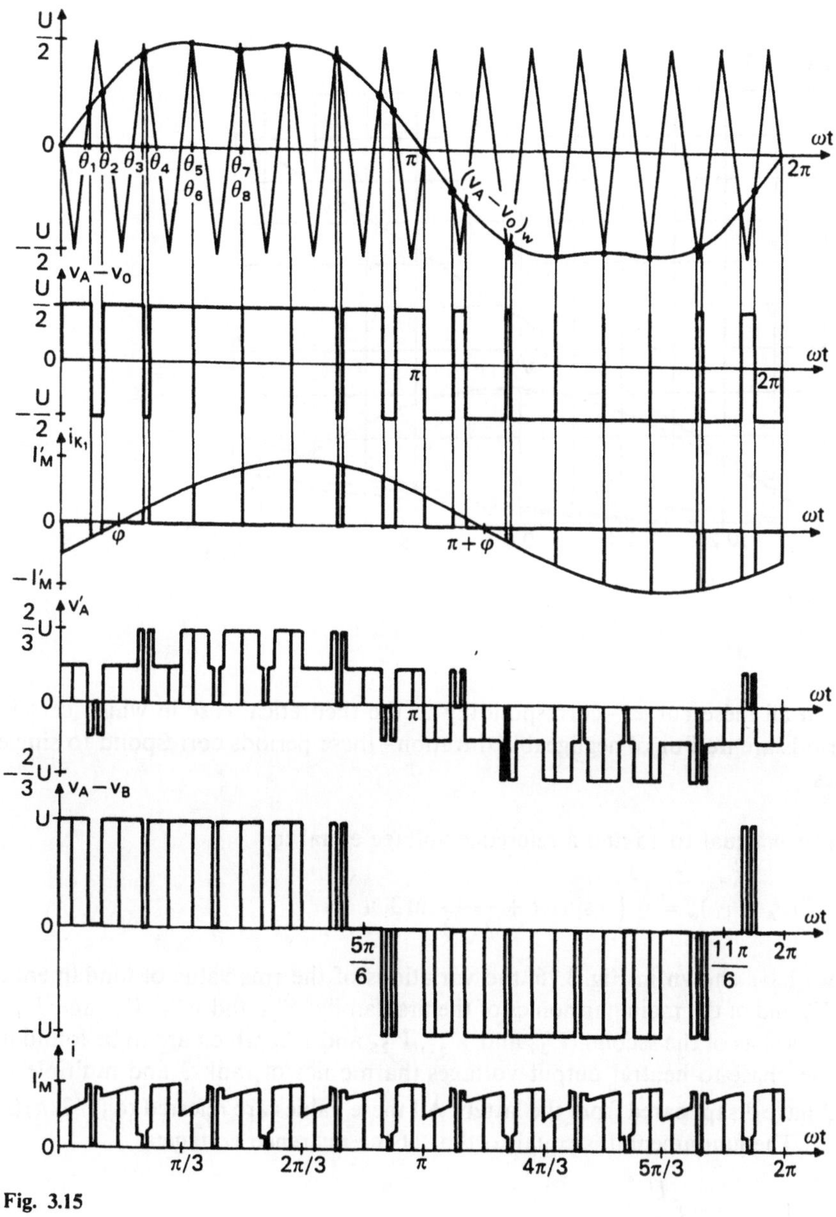

Fig. 3.15

- the waveform of the phase-to-phase output voltage $v_A - v_B$ given by

$$v_A - v_B = (v_A - v_O) - (v_B - v_O) = v'_A - v'_B$$

- finally, the waveform of the current i delivered by the DC supply.

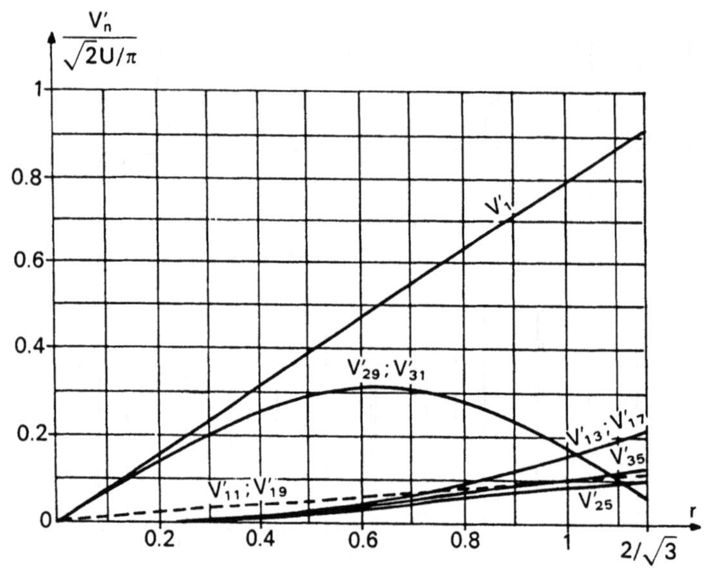

Fig. 3.16

In all these curves – corresponding to the theoretical case in which certain periods are null or of negligeable duration – these periods correspond to single lines.

– For m equal to 15 and a reference voltage equal to

$$(v_A - v_O)_w = \frac{U}{2}\left(r \sin \omega t + \frac{1}{3\sqrt{3}} \sin 3\omega t\right)$$

we have shown in Fig. 3.16, the variations of the rms value of fundamental V'_1 and of the main harmonics of the first family, V'_{13} and V'_{17}, V'_{11} and V'_{19}, as well as of the second, V'_{29} and V'_{31}, V'_{25} and V'_{35} which are to be found in the phase-to-neutral output voltages (harmonics of rank 3 and multiple of 3 have disappeared from the latter). All these values are referred to $(\sqrt{2}/\pi)U$.

The fundamental is equal to that of the reference voltage:

$$V'_1 = \frac{1}{\sqrt{2}} r \frac{U}{2}.$$

A comparison with the harmonics of the same rank given by the sinusoidal reference (Fig. 3.8d) shows that, for a given r,

harmonic 29 and 31 are slightly stronger,
harmonic 13 and 17 are smaller,
harmonic 11 and 19, 25 and 35 are more noticeable.

Moreover, for a same pair of harmonics, there are considerable differences in magnitudes. For these to disappear, the modulation frequency must be high not only compared to f but also to $3f$.

As in the case of the sinusoidal reference voltage, the input current mean value and harmonics can be deduced from the output voltage fundamental and harmonics.

Terms of rank

$1;\ m \pm 2$ and $m \pm 4;\ 2m \pm 1$ and $2m \pm 5$

correspond, at the input, to those of rank

$0;\ m \pm 3;\ 2m$ and $2m \pm 6$.

As the current i harmonic of rank $m - 3$ is proportional to $V'_{m-2} - V'_{m-4}$ and harmonic $m + 3$ proportional to $V'_{m+2} - V'_{m+4}$, the current harmonics of the first family are greatly reduced.

3.2.4 Other Modifications of the Control Law

- Other control laws which use an injection of harmonic 3 into the reference voltage can be found. Some use another relative value of this harmonic or make it vary with the fundamental. For example, the following can be found:

$$(v_A - v_O)_w = r\frac{U}{2}\left(\sin\omega t + \frac{1}{4}\sin 3\omega t\right)$$

– Occasionally, *harmonic* 9 *can be injected togther* with harmonic 3, yielding a reference which has the following form:

$$(v_A - v_O)_w = \frac{U}{2}(r\sin\omega t + k'_3\sin 3\omega t - k'_9\sin 9\omega t)$$

or

$$(v_A - v_O)_w = \frac{U}{2}r\,(\sin\omega t + k'_3\sin 3\omega t - k'_9\sin 9\omega t).$$

Without noticeably decreasing r_{max} if m is high, the harmonics of rank $m \pm 3$ in the input current can be greatly reduced.

– The choice of the reference voltage depends on the importance given to the voltage loss, to the output voltage harmonics and to the input current harmonics.

- Fundamental V'_1 of the output voltages must vary only slightly in cases where the inverter acts as a constant AC voltage supply; to achieve this, the effects of the variation in the load and in the value of the DC voltage must only be compensated. The use of *partial pulse width modulation* is of particular interest in such circumstances.

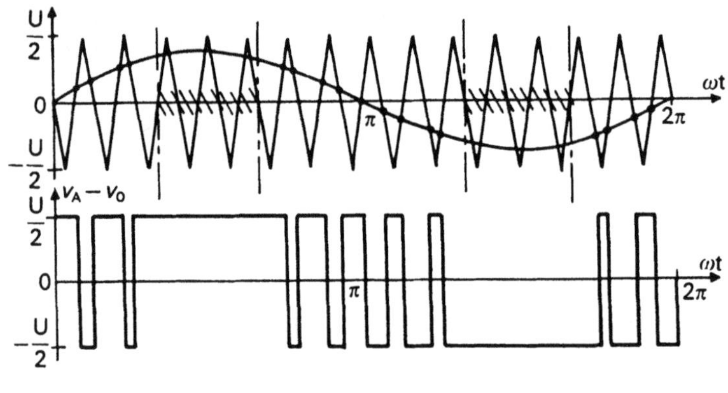

Fig. 3.17

The commutations are suppressed during the central part of each half-cycle in the reference voltage–during the third of the latter, for example (Fig. 3.17).

This requires a slightly more complex control and limits the range of output voltage variations. However it offers two advantages:

- the number of commutations is reduced, which involves a reduction of the commutation losses
- the "voltage loss" is reduced since the periods of very short duration, situated in the middle of the half-cycles, are suppressed.

- The three-phase full-bridge voltage inverter, operating in PWM is widely used in variable-speed drive systems. Frequency f of the fundamental V'_1 of the output voltages must then be made to vary within a wide range; the value of V'_1 must vary more or less proportionately to f. To keep the commutation frequency $2mf$ at a value which is compatible with the characteristics of the semiconductor devices, *the modulation ratio must be modified.*

  Let it be assumed, for example, that the modulation frequency is limited to 600 Hz and that the output frequency must vary between 0 and 100 Hz.

  - For f between 0 and 45 Hz, asynchronous modulation with constant mf equal to 600 Hz is adopted.
  - For f between 45 and 66 Hz, synchronous modulation with m equal to 9 is selected; mf thus goes from 385 to 594 Hz.
  - For f between 66 and 100 Hz, synchronous modulation with m equal to 6 is chosen; mf thus goes from 396 to 600 Hz.

3.3 Single-Phase Full-Bridge Inverter

Figure 3.18 shows the configuration of the single-phase full-bridge voltage-source inverter, fed by a DC voltage U. Its use enables a voltage u' equal to

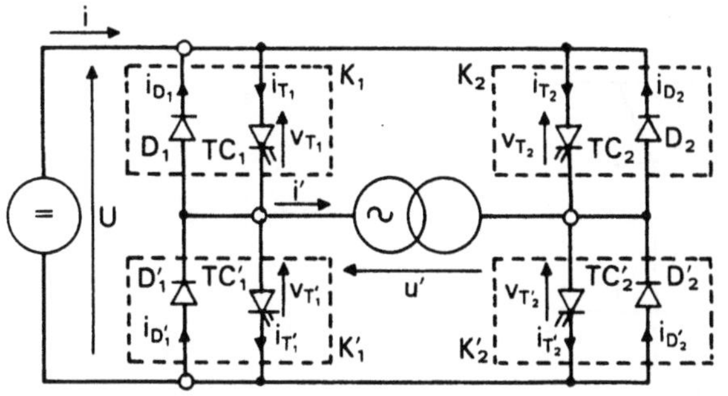

Fig. 3.18

$+ U$, $- U$ or zero to be obtained at the output. The possibility of obtaining periods with zero u' was not available with the single-phase half-bridge inverter and enables the output voltage to be more close to the required sinewave.

The full-bridge inverter is thus most widely used when the inverter does in fact have a single-phase output.

3.3.1 Switch Control. Waveforms

3.3.1.1 Switch Control

The positive half-cycle of u' consists of pulses of amplitude $+ U$ separated by periods at zero u'; the negative half-cycle of u' consists of negative pulses separated by periods at zero u'.

- During the positive half-cycle, K'_2 can be used, for example, as a directing switch by turning it on permanently and by taking K_1 and K'_1 as chopping switches (see Vol. 3, Chap. 6, Sect. 3.1.1).

  When K_1 is on, u' is equal to $+ U$; when K'_1 is on, u' is equal to zero.

  In order for the "mean value" of u' to follow the variations in reference voltage u'_w, a unipolar triangular carrier wave M^+ of amplitude U is used (Fig. 3.19a). The intersection points of the rising parts of M^+ with u'_w give the turn-on points of K'_1 and the beginning of the periods at zero u'; the intersections of falling parts of M^+ with u'_w give the turn-on points of K_1 and the beginning of the periods at u' equal to $+ U$.

  A study of Fig. 3.19a shows that, if there is no variation in u'_w during cycle T'

$$u'_{\text{mean}} = U \frac{t_1}{T'/2} = U \frac{u'_w}{U} = u'_w .$$

- Similarly, for the negative half-cycle of u', K_2 can be constantly on and the chopping performed by controlling alternatively the turn-on of K_1 ($u' = 0$) and that of K'_1 ($u' = - U$).

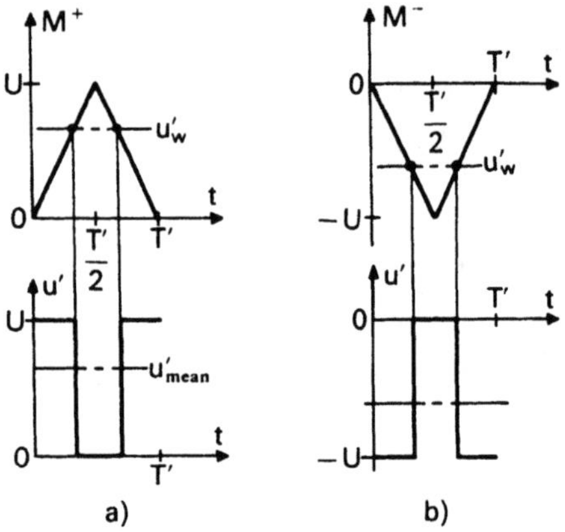

Fig. 3.19

A carrier wave M^- formed of negative triangles is now used (Fig. 3.19b). The intersections of the falling parts of M^- with u'_w determine the turn-on of K_1 and make u' fall to zero; those of the rising parts of M^- with u'_w determine the turn-on of K'_1 and make u' equal to $-U$.

If there is no variation in u'_w during cycle T' of the carrier wave, the following applies once more:

$$u'_{mean} = u'_w.$$

The polarity of the carrier wave must be reversed for every half-cycle of u'_w.

3.3.1.2 Waveforms

For the two half-cycles of u' to be identical to within a sign in synchronous modulation, the modulation frequency f' must be an even multiple of the frequency f of u'.

The optimal phase lock corresponds to the inversion of the carrier at the change in half-cycle of u'_w. Under these conditions, the half-cycles of u' are symmetrical in comparison with their mid-point.

The control can be characterised by

- *the voltage modulation ratio r*, which gives the relative amplitude of the wanted voltage.

  $$u'_w = rU \sin \omega t$$

- and the *frequency modulation ratio m* – the ratio of the modulation frequency f' to the output voltage frequency f, or *the number* of pulses p per half-cycle.

  p and m are linked by

$$p = \frac{m}{2} - 1. \tag{3.15}$$

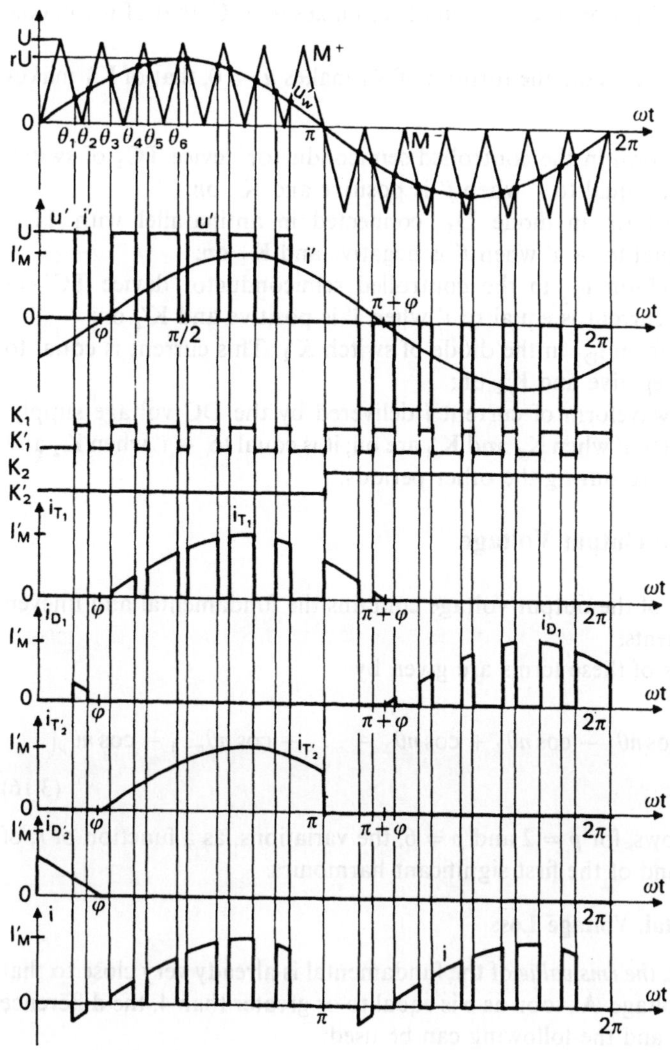

Fig. 3.20 ($m = 14$)

Figure 3.20, plotted for $m = 14$ (therefore $p = 6$) and for $r = 0.8$, shows from top to bottom:

- the determination of the commutation angles; in view of the symmetries, six angles need to be determined $\theta_1, \theta_2, \ldots, \theta_6$;
- the output voltage u' waveform. The output current i' (assumed to be sinusoidal and lagging by angle φ the fundamental of u') is also shown;
- the conduction period of the four switches for the case where K_2 and K'_2 are used as directing switches, and K_1 and K'_1 as chopping switches:

For $0 < \omega t < \pi$, K'_2 is on: the turn-on of K_1 makes $u' = U$, that of K'_1 makes $u' = 0$;
For $\pi < \omega t < 2\pi$, K_2 is on: the turn-on of K_1 makes $u' = 0$, that of K'_1 makes $u' = -U$;

- the current waveform in the controlled semiconductor device TC_1 of switch K_1. Current i_{T1} is equal to i' when i' is positive and K_1 on;
- the current waveform in diode D_1, connected in antiparallel with TC_1. Current i_{D_1} is equal to $-i'$ when i' is negative and K_1 on;
- the current waveform $i_{T'_2}$ in the controlled semiconductor device TC'_2 of switch K'_2. This current is equal to i' when i' is positive and K'_2 on;
- the current waveform $i_{D'_2}$ in the diode of switch K'_2. This current is equal to $-i'$ when i' is negative and K'_2 on;
- and finally, the waveform of current i delivered by the DC voltage supply. Current i is equal to i' when K_1 and K'_2 are on; it is equal to $-i'$ when K_2 and K'_1 are on; it is zero during the other periods.

3.3.2 Study of the Output Voltage

The Fourier series of the output voltage contains the fundamental and uneven harmonic components.

The rms values of these terms are given by

$$U'_n = \frac{1}{n}\frac{2\sqrt{2}}{\pi}|\cos n\theta_1 - \cos n\theta_2 + \cos n\theta_3 - \ldots + \cos n\theta_{p-1} - \cos n\theta_p|. \tag{3.16}$$

Figure 3.21 shows, for $p = 2$ and $p = 6$, the variations, as a function of r, of the fundamental and of the first significant harmonics.

3.3.2.1 Fundamental. Voltage Loss

- For p equal to 2, *the rms value* of the fundamental is already very close to that of the wanted voltage. As soon as p is equal to or greater than 4, the difference can be ignored, and the following can be used:

$$U'_1 = \frac{1}{\sqrt{2}} rU. \tag{3.17}$$

- For the uneven values of p, the voltage regulation ratio r can slightly exceed 1, but the possible excess rapidly decreases as p increases. For even p, the theoretical maximum value of r is always equal to 1. This corresponds to

$$(U'_1)_{\max} = \frac{U}{\sqrt{2}}.$$

The result is a *voltage loss*, i.e. $U'_{1\,\max}$ is lower than the rms value of the fundamental which would be given by a full-bridge inverter producing pulses of a width equal to π (i.e. $2\sqrt{2}U/\pi$).

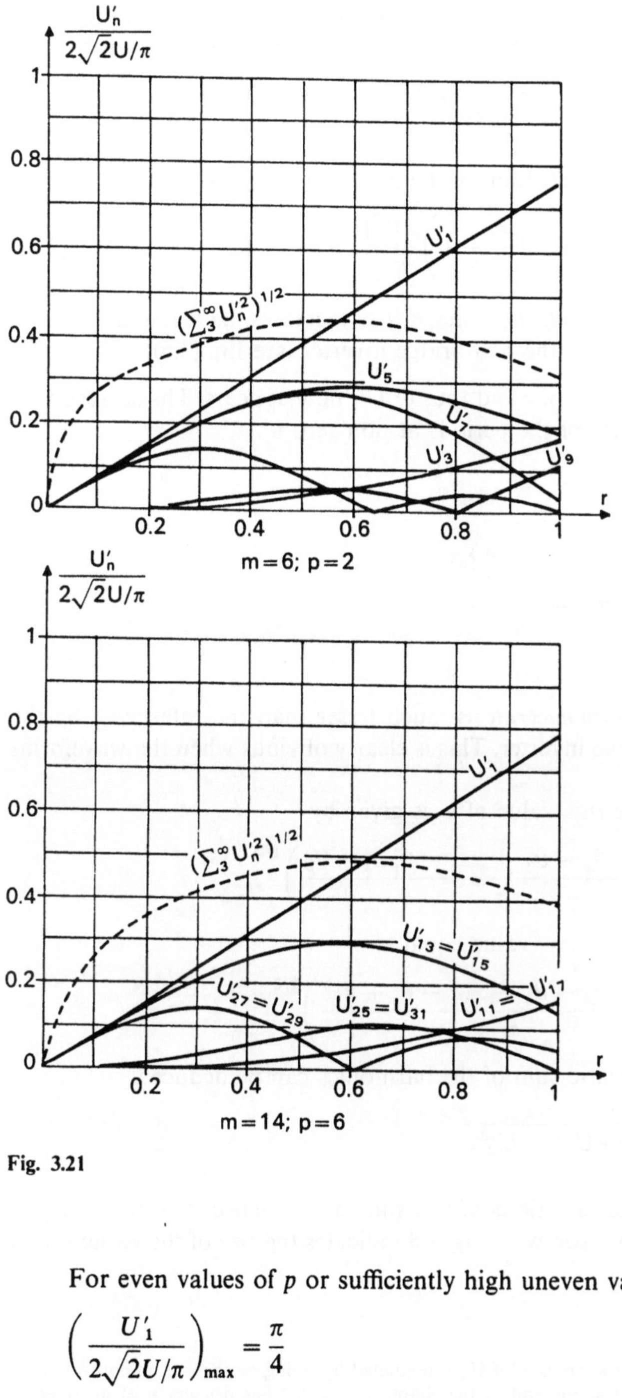

Fig. 3.21

For even values of p or sufficiently high uneven values of p,

$$\left(\frac{U'_1}{2\sqrt{2}U/\pi}\right)_{\max} = \frac{\pi}{4} \tag{3.18}$$

the theoretical voltage loss in 21.46% as in the case of the half-bridge inverter.

- When the minimal width $\Delta\theta$ to be given to each pulse is taken into account, r must be limited to a value of less than 1. If p is high, this is given by

$$r_{\max} = 1 - p\frac{\Delta\theta}{\pi} = 1 - \left(\frac{m}{2} - 1\right)\frac{\Delta\theta}{\pi}.$$

This gives the new expression of the maximum value of U'_1:

$$\left(\frac{U'_1}{2\sqrt{2}U/\pi}\right)_{\max} = \frac{\pi}{4}\left[1 - \left(\frac{m}{2} - 1\right)\frac{\Delta\theta}{\pi}\right]. \tag{3.19}$$

For a given value of m, the voltage loss increases, as a function of $\Delta\theta$, roughly half as fast as for the half-bridge inverter (see Eq. (3.5)).

- However a minimal value of r and thus of U'_1 now appears. Ths corresponds to a minimal width $\Delta\theta'$ of the periods at non zero u'.[1]

$$r_{\min} = p\frac{\Delta\theta'}{\pi}$$

$$\left(\frac{U'_1}{2\sqrt{2}U/\pi}\right)_{\min} = \frac{\pi}{4}\left(\frac{m}{2} - 1\right)\frac{\Delta\theta'}{\pi}.$$

3.3.2.2 Harmonics

- *The output voltage harmonic rate* is much lesser than the rate given by the single-phase half-bridge inverter. This is clearly obvious when the waveforms are observed.

If p is uneven, the rms value of u' is given by

$$U' = U\left(\frac{-\theta_1 + \theta_2 - \theta_3 + \ldots - \theta_{p-1} + \theta_p}{\pi}\right)^{1/2}.$$

If p is even, u' has a rms value equal to

$$U' = U\left(\frac{-\theta_1 + \theta_2 - \theta_3 + \ldots - \theta_p + \pi/2}{\pi}\right)^{1/2}$$

from which the quadratic sum of the harmonics can be deduced:

$$\left(\sum_{n=3}^{\infty} U'^2_n\right)^{1/2} = \sqrt{U'^2 - U'^2_1}.$$

Figure 3.21 shows the variations of this sum, as a function of r, by curves in broken lines. A comparison with Fig. 3.8 indicates the size of the reduction in

[1] This minimum value can be suppressed if U'_1 is regulated by acting on the shift in control of the two half-bridges (cf. Sect. 4.4 at the end of the chapter). But this has drawback of introducing a harmonic of rank m in voltage u'.

the harmonic sum, notably in the case of low and medium values of r. This reduction is caused essentially by the *suppression of harmonic m*.

- The harmonics which are present in the output voltage are grouped together in families as soon as m reaches a sufficiently high value:
  - the family centred on the mf frequency, which comprises notably the pairs $(m-1)f$ and $(m+1)f$, $(m-3)f$ and $(m+3)f$;
  - the family centred on the $2mf$ frequency, which comprises notably the pairs $(2m-1)f$ and $(2m+1)f$, $(2m-3)f$ and $(2m+3)f$;
  - the family centred on $3mf$ . . .

For $p = 2$ or $m = 6$ (upper part of Fig. 3.21), there is no clear separation between the fundamental and the first family, between the latter and the second, . . ; the harmonics of the same pair have noticeably different values.

For $p = 6$ or $m = 14$ (lower part of Fig. 3.21), the value of m is sufficiently high for the families to be clearly separated and for the values belonging to a same pair to be equal.

For high values of p, the following can be written, if only the principal harmonics of the first two families are taken into consideration:

$$\begin{aligned} u' \simeq\ & U'_1\sqrt{2}\sin\omega t \\ & + U'_{m-1}\sqrt{2}[-\sin(m-1)\omega t + \sin(m+1)\omega t] \\ & + U'_{m-3}\sqrt{2}[-\sin(m-3)\omega t + \sin(m+3)\omega t] \\ & + U'_{2m-1}\sqrt{2}[-\sin(2m-1)\omega t + \sin(2m+1)\omega t] \\ & + U'_{2m-3}\sqrt{2}[-\sin(2m-3)\omega t + \sin(2m+3)\omega t] \end{aligned}$$

and the values of harmonics indicated for harmonics 13, 15, 27, 29 for $m = 14$ in Fig. 3.21 can be taken for harmonics $m-1$, $m+1$, $2m-1$, $2m+1$. (The signs shown for harmonics $2m-1$ and $2m+1$ in the above expression of u' corresponds to low values of r. For r greater than 0.6, these signs must be reversed).

3.3.3 Study of the Currents

3.3.3.1 Currents in the Semiconductor Devices

The number of commutations to be carried out at non zero current over each cycle is still equal to $2m$.

The switches which are used for chopping have $2m-2$ commutations to carry out. For given values of amplitude $I'\sqrt{2}$ and phase-shift φ of the output current as well as of the modulation ratio m, the current in the semiconductor devices has the same waveforms and values as for the half-bridge inverter (see Sect. 3.1.2.4)

Switches which operate as directing switches have two non-zero current commutations to carry out for each cycle T. For given values of I' and φ, the commutations, the waveforms and the values of the currents in the semiconductor devices are the same as those shown for the full-bridge inverter operation at zero β without PWM (see Chap. 2, Sect. 2.3.2.2).

In order to balance out the temperature rise of the different switches, the role played by the two half-bridges are sometimes reversed: during one cycle, one of them operates as a chopper and the other as a directing switch; whereas, during the following cycle, the roles are reversed.

3.3.3.2 Input Current

Current i taken from DC voltage supply U has a fundamental frequency of $2f$. The values of its different components can be deduced from the conservation of the instantaneous power between the input and output of the converter.

- Its mean value I is given by

$$I = \frac{U'_1}{U} I' \cos \varphi$$

or, if Eq. (3.14) is taken into account, by

$$I = \frac{1}{\sqrt{2}} r I' \cos \varphi. \tag{3.20}$$

- If the value of m is high enough for U'_3 to be negligible, the term of the input current at $2f$ frequency has the following rms value:

$$I_2 = \frac{U'_1 I'}{\sqrt{2} U} = \frac{1}{2} r I'. \tag{3.21}$$

This term once more determines the size of the input filter.

- The other input current harmonics are linked to the output voltage harmonics as follows:
  - For the first family, the current harmonics of rank

    $m - 4, m - 2, m, m + 2, m + 4$

    correspond to the voltage harmonics of rank

    $m - 3, m - 1, m + 1, m + 3$

  - For the second family, the current harmonics of rank

    $2m - 4, 2m - 2, 2m + 2, 2m + 4$

    correspond to the voltage harmonics of rank

    $2m - 3, 2m - 1, 2m + 1, 2m + 3 \ldots$

If the value m is sufficiently high for both harmonics of the same pair situated on either side of the centre of each family ($m-1$ and $m+1$, $2m-1$ and $2m+1, \ldots$) to have the same value, the input current harmonics of rank m, $2m, \ldots$ disappear.

- *Application to three-phase systems*

The single-phase full-bridge inverter operating in PWM supplies an output voltage which is much easier to filter than that supplied by the half-bridge inverter. This justifies its use in many uninterruptible power supplies.

It is also used in three-phase systems when the load is unbalanced and when the output voltage has to be regulated phase by phase. In addition to the presence of three inverters, each with four switches, a transformer must also be used unless the three phases are galvanically independent.

If the load is balanced and m is multiple of 3 – as in the case of a three-phase full-bridge inverter with six switches, the only harmonics to be found at the input are those whose order is a multiple of 3, i.e. the main input current harmonics have disappeared. They reappear, however, as soon as the load is unbalanced and their values increase as the imbalance grows.

3.3.4 Digital Reference

For the half-bridge inverter as well as for the full-bridge, digital control is generally used:

- the sinusoidal reference is generated by reading tables;
- the modulation is generated by up/down counters.

The reference is usually digitized, i.e. the sinusoid (continuously variable depending on the time) is replaced by discrete values: a different value is taken for each cycle of the carrier – every from $2\pi/m$.

Figure 3.22 shows the determination of commutation instants for a half-bridge inverter when m is equal to 9 (compare with the upper part of Fig. 3.4).

Figure 3.23 shows this determination for a full-bridge inverter when m is equal to 14 (compare with the upper part of Fig. 3.20).

In the latter case, for example,

$$\theta_1 = \frac{2\pi}{m} - \frac{\pi}{m} r \sin \frac{2\pi}{m}, \qquad \theta_2 = \frac{2\pi}{m} + \frac{\pi}{m} r \sin \frac{2\pi}{m}$$

$$\theta_3 = \frac{4\pi}{m} - \frac{\pi}{m} r \sin \frac{4\pi}{m}, \qquad \theta_4 = \frac{4\pi}{m} + \frac{\pi}{m} r \sin \frac{4\pi}{m}$$

$$\theta_5 = \frac{6\pi}{m} - \frac{\pi}{m} r \sin \frac{6\pi}{m}, \qquad \theta_6 = \frac{6\pi}{m} + \frac{\pi}{m} r \sin \frac{6\pi}{m} \cdots$$

As soon as the value of m is sufficiently high, there is virtually no difference between the characteristics obtained with the determination of commutation

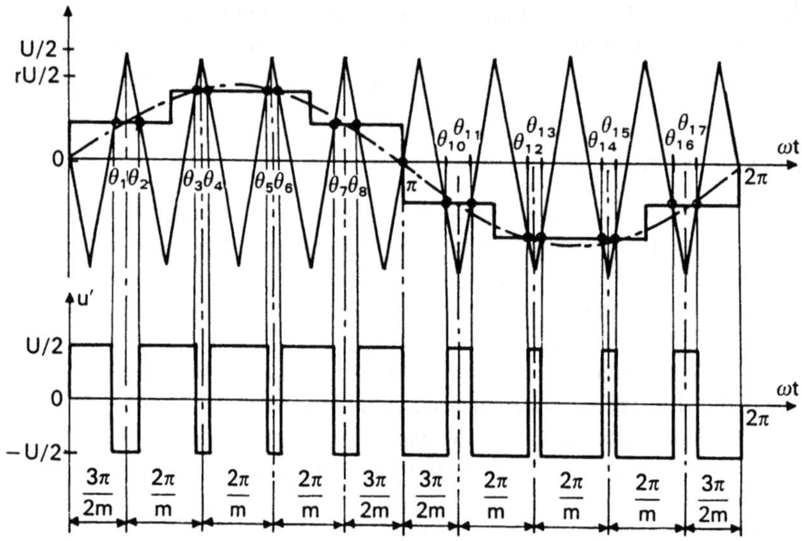

Fig. 3.22 ($m = 9$)

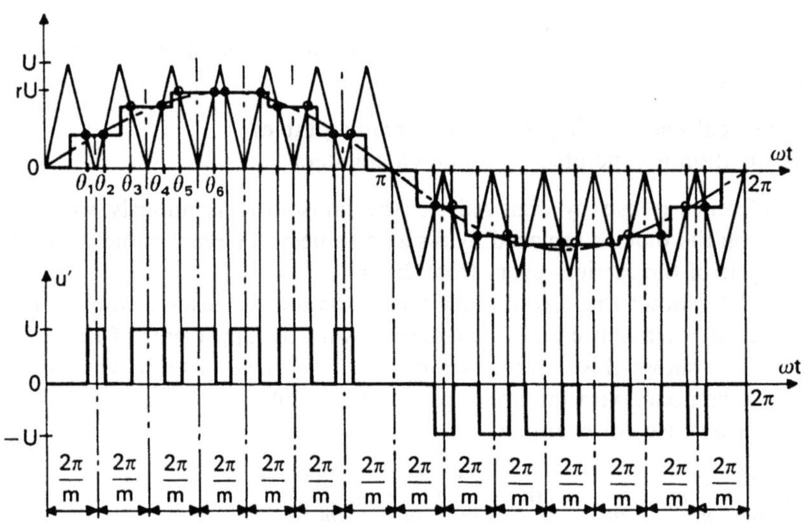

Fig. 3.23 ($m = 14$)

angles using the sinusoidal reference voltage or using the digitized reference voltage.

(The latter offers the advantage of making the calculation of the characteristics easier.)

3.4 Computed Modulation

Instead of determining the commutation angles in real time, the control sequences can be previously computed and memorised. These stored sequences are then used to control the inverter switches.

The determination of angles corresponding to the sequences can be performed using the same laws as those used for the real-time control. The performance obtained using the latter is in fact relatively satisfactory.

However, since the angles are determined off line, other laws can be adopted:

- at the beginning of the development of PWM control, the aim was to *eliminate the first harmonics* of the output voltage;
- today, the nature of the output filter is normally taken into account and the aim is to *minimise the weighted harmonic rate.*

Other laws may naturally be adapted to the needs linked to specific applications.

3.4.1 Characterising the Output Voltage

The notion of frequency–modulation ratio can no longer be used as such, since the angles are calculated on other bases than the intersection of waves of different frequencies.

3.4.1.1 Half-Bridge Inverter

The output voltage is compared to that which would be obtained if there were no pulse width modulation (upper part of Fig. 3.24): if u' were equal to $+U/2$ throughout the positive half-cycle, to $-U/2$ throughout the negative half-cycle, it would have a fundamental with the following rms value:

$$U'_{10} = \frac{\sqrt{2}}{\pi} U .$$

The waveform can be characterised by the *number of chops per half-cycle c*, i.e. the number of periods at u' equal to $-U/2$ during the positive half-cycle and at u' equal to $+U/2$ during the negative half-cycle.

The rms value can be characterized by the *relative value of the fundamental* U'_1/U'_{10}.

Figure 3.24 initially shows a wave with three chops and then a wave with four chops. The identical nature – to within a sign – of the positive and negative half-cycles and the symmetry of each half-cycle compared to its centre-point mean that *c angles* are sufficient to determine the widths of all the pulses.

3.4.1.2 Full-Bridge Inverter

For the full-bridge inverter, the output voltage is characterised by

- *the number of pulses per half-cycle p,*

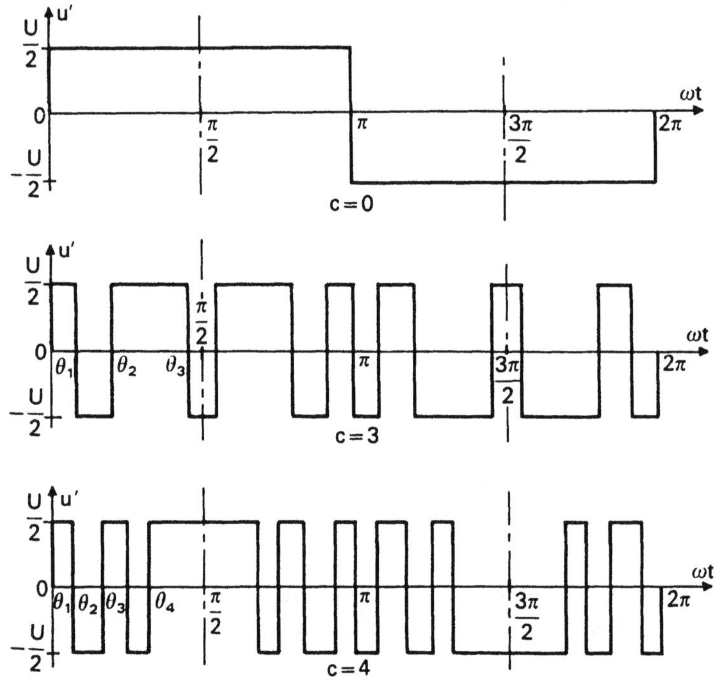

Fig. 3.24

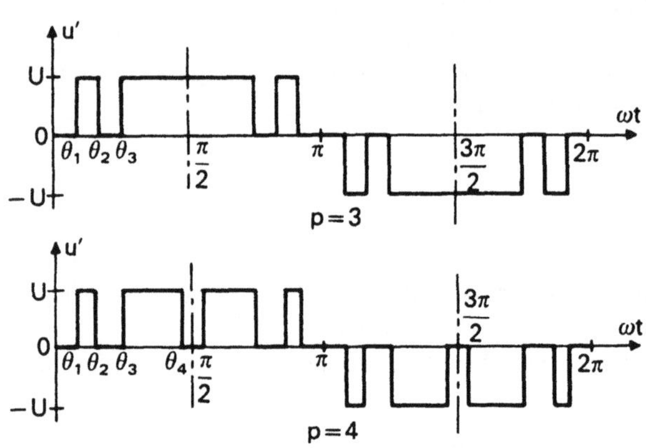

Fig. 3.25

- *the rms value U'_1 of the fundamental referred to the value U'_{10}* equal to $2\sqrt{2}\,U/\pi$ which would be obtained if u' was made of a single pulse with an angular width of π, during each half-cycle.

Whether p is uneven or even (Fig. 3.25) p *angles* are enough to determine the width of all the pulses.

3.4.2 Eliminating the First Harmonics of the Output Voltage

3.4.2.1 Single-Phase Half-Bridge Inverter

Whether c be even or uneven, the rms value of the fundamental and those of the various uneven harmonics of voltage u' are once more given by the equation

$$\boxed{U'_n = \left| U'_{10}\frac{2}{n}\left(\frac{1}{2} - \cos n\theta_1 + \cos n\theta_2 - \cos n\theta_3 + \dots + (-1)^{c-1}\cos n\theta_c\right)\right|} \tag{3.22}$$

with $n = 1, 3, 5, 7, \dots$
and $0 < \theta_1 < \theta_2 < \dots < \theta_c < \pi/2$.
In order to determine the size of the c chops, c angles are available $\theta_1, \theta_2, \theta_3, \dots, \theta_c$.

- *If the output voltage is to be varied*, the c angles are calculated by solving a system of c equations with c unknowns. The first equation gives the value U'_1 required for the fundamental; the remaining $c - 1$ correspond to the *elimination of the $c - 1$ first uneven harmonics*:

$$\begin{aligned}
&\frac{1}{2} - \cos\theta_1 + \cos\theta_2 - \cos\theta_3 + \dots \pm \cos\theta_c = \frac{U'_1}{2U'_{10}}\\
&\frac{1}{2} - \cos 3\theta_1 + \cos 3\theta_2 - \cos 3\theta_3 + \dots \pm \cos 3\theta_c = 0\\
&\frac{1}{2} - \cos 5\theta_1 + \cos 5\theta_2 - \cos 5\theta_3 + \dots \pm \cos 5\theta_c = 0\\
&\cdots\cdots\cdots\cdots\cdots\cdots\cdots\cdots\\
&\frac{1}{2} - \cos(2c-1)\theta_1 + \cos(2c-1)\theta_2 - \dots \pm \cos(2c-1)\theta_c = 0.
\end{aligned} \tag{3.23}$$

The first remaining harmonic is that of order $2c + 1$. There are $4c + 2$ commutations per cycle.

For example, for c equal to 4, Fig. 3.28 shows the variations of angles $\theta_1, \theta_2, \theta_3, \theta_4$ as a function of U'_1/U'_{10}, which enable harmonics 3, 5 and 7 to be cancelled out.

As shown in Fig. 3.29b, for low and medium values of U'_1/U'_{10}, harmonic 9 has a very high value.

- The problem of *voltage loss* once more occurs. The theoretical maximum value of U'_1 is less than U'_{10}.

  The theoretical maximum value is reached

  - in the case of uneven c, when the width of the chop in the middle of each half-cycle falls to zero, or when θ_c is equal to $\pi/2$;

- in the case of even c, when the first chop begins as soon as the half-cycle starts, in i.e., when θ_1 is equal to zero.

If the minimal width to be given to each chop is taken into account, this leads to the maximum value of U'_1 being further reduced; this reduction increases with an increase in c.

- If the aim is *no longer to vary the output voltage*, the angles are computed for the theoretical maximum value of U'_1/U'_{10}.

– Starting with an uneven number $(c + 1)$ of chops per half-cycle – which eliminates the first c uneven first harmonics, U'_1/U'_{10} is taken in such a way that the chop at the centre of each half-cycle is suppressed.
This leaves c chops per half-cycle,
c angles to be calculated,
$4c + 2$ commutations per cycle,
harmonics of rank greater than or equal to $2c + 3$.

– Alternatively, starting with an even number c of chops per half-cycle – which eliminates the first $c - 1$ uneven harmonics, U'_1/U'_{10} is taken such that θ_1 is equal to zero. In other words, period $[0, \theta_1]$ disappears.
This leaves c chops per half-cycle,
$c - 1$ angles to be calculated,
$4c - 2$ commutations per cycle,
harmonics of rank greater than or equal to $2c + 1$.

An even or uneven number of uneven harmonics can thus be eliminated: the number of harmonics which are eliminated increases proportionally to the number of commutations.

Figure 3.26 shows the waveforms which correspond, firstly to the elimination of harmonic 3, then to that of harmonics 3 and 5, then to that of harmonics 3, 5, 7 and 9.

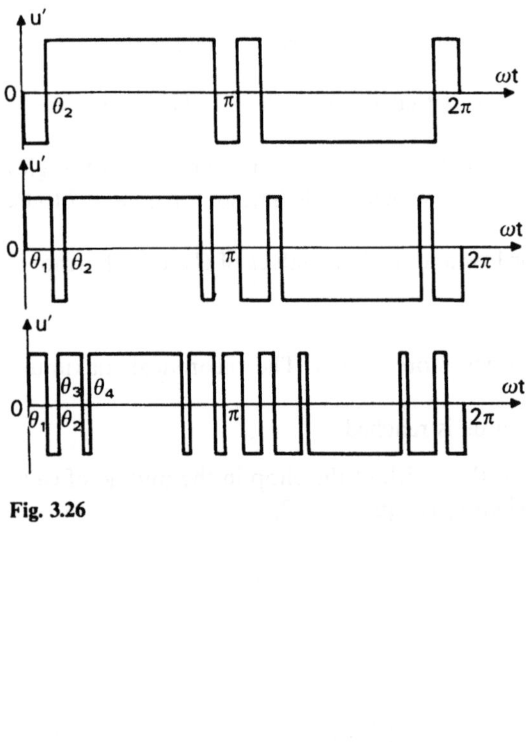

Fig. 3.26

In the first case,

$\theta_1 = 0, \qquad \theta_2 = 20°$

$u' \simeq U'_{10}\sqrt{2}(0.88 \sin \omega t + 0 - 0.27 \sin 5\omega t - 0.36 \sin 7\omega t \ldots).$

In the second case,

$\theta_1 = 23°6, \qquad \theta_2 = 33°3$

$u' \simeq U'_{10}\sqrt{2}(0.84 \sin \omega t + 0 + 0 + 0.25 \sin 7\omega t + 0.41 \sin 9\omega t \ldots).$

In the third case,

$\theta_1 = 15°9, \qquad \theta_2 = 24°8, \qquad \theta_3 = 46°9, \qquad \theta_4 = 50°4$

$u' \simeq U'_{10}\sqrt{2}(0.80 \sin \omega t + 0 + 0 + 0 + 0 + 0.27 \sin 11\omega t + 0.45 \sin 13\omega t \ldots).$

3.4.2.2 Three-Phase Full-Bridge Inverter

In the output voltages of the three-phase full-bridge inverter, the harmonics of rank 3 or multiple of 3, which existed in the half-bridge voltages, disappear.

There is thus no need to eliminate these harmonics. The conditions which concerned this elimination are replaced by that of the following uneven harmonics.

- *If the value of the output voltages has to be varied*, the c angles which define the c chops of each half-cycle are calculated in such a way as to cancel out the first $c - 1$ uneven harmonics apart from those of rank 3 or multiple of 3.

  The following system of equations therefore replaces system (3.23):

$$
\begin{aligned}
&\frac{1}{2} - \cos\theta_1 + \cos\theta_2 - \cos\theta_3 + \ldots \pm \cos\theta_c = U'_1/2U'_{10} \\
&\frac{1}{2} - \cos 5\theta_1 + \cos 5\theta_2 - \cos 5\theta_3 + \ldots \pm \cos 5\theta_c = 0 \\
&\frac{1}{2} - \cos 7\theta_1 + \cos 7\theta_2 - \cos 7\theta_3 + \ldots \pm \cos 7\theta_c = 0 \\
&\frac{1}{2} - \cos 11\theta_1 + \cos 11\theta_2 - \cos 11\theta_3 + \ldots \pm \cos 11\theta_c = 0. \\
&- -
\end{aligned}
\tag{3.24}
$$

With c equal successively to $2, 3, 4, \ldots$, harmonic 5, harmonics 5 and 7, harmonics $5, 7, 11, \ldots$ are suppressed.

The number of commutations per cycle is once more equal to $4c + 2$.

Figure 3.30 shows for c equal to 4, the variations in angles $\theta_1, \theta_2, \theta_3$ and θ_4 as a function of U'_1/U'_{10}.

Figure 3.31b indicates, for this value of c, the variations in the first harmonics (other than those of rank 3 or multiple of 3) which remain in the voltage supplied by each half-bridge, i.e. harmonics of rank 13, 17, 19.

The problem remains of the voltage loss and of the increase in its theoretical value when each chop is given the minimum value which it can technically have.

- *If the output voltage value is not to be varied*, the situation described in the previous paragraph for single-phase operation can be found again.

  With c chops per half-cycle, the following is possible:

  - either to eliminate c harmonics, if there are $4c + 2$ commutations per cycle.
  - or to eliminate $c - 1$ harmonics, if there are $4c - 2$ commutations per cycle.

For example, by using waveforms similar to those of Fig. 3.26 but with different values for the angles, harmonic 5 can be eliminated, followed by harmonics 5 and 7 and then harmonics 5, 7, 11, 13.

In the first case,

$\theta_1 = 0, \quad \theta_2 = 12°$

$$v_A - v_O \simeq U'_{10}\sqrt{2}\,[0.955 \sin \omega t + (0.21 \sin 3\omega t) + 0 - 0.11 \sin 7\omega t$$
$$- (0.18 \sin 9\omega t) - 0.33 \sin 11\omega t - 0.22 \sin 13\omega t \ldots].$$

In the second case,

$\theta_1 = 16°25, \quad \theta_2 = 22°05$

$$v_A - v_O \simeq U'_{10}\sqrt{2}\,[0.935 \sin \omega t + (0.16 \sin 3\omega t) + 0 + 0 + (0.08 \sin 9\omega t)$$
$$+ 0.19 \sin 11\omega t - 0.25 \sin 13\omega t + (0.24 \sin 15\omega t) \ldots].$$

In the third case,

$\theta_1 = 9°8, \quad \theta_2 = 15°1, \quad \theta_3 = 85°2, \quad \theta_4 = 86°4$

$$v_A - v_O \simeq U'_{10}\sqrt{2}\,[0.92 \sin \omega t + (0.26 \sin 3\omega t) + 0 + 0 - (0.08 \sin 9\omega t)$$
$$+ 0 + 0 + (0.11 \sin 15\omega t) + 0.13 \sin 17\omega t + 0.19 \sin 19\omega t \ldots].$$

3.4.2.3 Single-Phase Full-Bridge Inverter

Whether the number p of pulses per half-cycle be even or uneven, the rms values of the fundamental and of the different uneven harmonics of output voltage u' are given by

$$\boxed{U'_n = \left| U'_{10}\frac{1}{n}[\cos n\theta_1 - \cos n\theta_2 + \cos n\theta_3 - \ldots - (-1)^p \cos n\theta_p] \right|.} \qquad (3.25)$$

p angles are available in order to determine the width of pulses.

- *If the value of the output voltage has to be varied*, the first $p - 1$ uneven harmonics can be cancelled out by solving the following system:

$$\begin{aligned}
&\cos \theta_1 - \cos \theta_2 + \cos \theta_3 - \ldots \pm \cos \theta_p = U'_1/U'_{10} \\
&\cos 3\theta_1 - \cos 3\theta_2 + \cos 3\theta_3 - \ldots \pm \cos 3\theta_p = 0 \\
&\cos 5\theta_1 - \cos 5\theta_2 + \cos 5\theta_3 - \ldots \pm \cos 5\theta_p = 0 \\
&\ldots\ldots\ldots\ldots\ldots\ldots \\
&\cos(2p-1)\theta_1 + \cos(2p-1)\theta_2 + \ldots \pm \cos(2p-1)\theta_p = 0.
\end{aligned} \qquad (3.26)$$

The harmonic of rank $2p + 1$ is the first remaining harmonic. There are $4p + 2$ commutations at non-zero current per cycle, shared between the two half-bridges.

For p equal to 4, for example, Fig. 3.32 shows the variation of angles $\theta_1, \theta_2, \theta_3, \theta_4$ as a function of U'_1/U'_{10}, when harmonics 3, 5 and 7 are eliminated.

Figure 3.33b gives the variations as a function of U'_1/U'_{10} of the remaining first harmonics – those of rank 9, 11, 13 and 15.

- The theoretical maximum value of U'_1/U'_{10} is reached:
  - for even p, when θ_p equals $\pi/2$, i.e. when the period between the two pulses situated on either side of the centre-point of each half-cycle disappears;
  - for uneven p, when θ_1 is equal to zero, i.e. when the periods at zero u' at either end of the half-cycles disappear.

The corresponding voltage loss increases when the minimal duration to be allowed between two successive pulses is taken into account. Moreover, as it has already been shown (cf. Sect. 3.3.2.1), the minimal pulse duration imposes a minimal value on ratio U'_1/U'_{10}.

- *By no longer varying the output voltage value* and operating at the theoretical limit, the following can be obtained with an uneven number p of pulses per half-cycle:
  - suppression of p harmonics – as a result there are $4p + 2$ commutations per cycle;
  - suppression of $p - 1$ harmonics – as a result there are $4p - 2$ commutations per cycle.

Figure 3.27 gives, for example, the waveforms which correspond to the elimination of harmonic 3, followed by that of harmonics 3 and 5, and finally that of harmonics, 3, 5 and 7.

In the first case,

$p = 1, \quad \theta_1 = 30°$

$$u' \simeq U'_{10}\sqrt{2}(0.865 \sin \omega t + 0 - 0.17 \sin 5\omega t - 0.12 \sin 7\omega t + 0 + \ldots).$$

In the second case,

$p = 3, \quad \theta_1 = 0, \quad \theta_2 = 17°9, \quad \theta_3 = 38°$

$$u' \simeq U'_{10}\sqrt{2}(0.835 \sin \omega t + 0 + 0 + 0.21 \sin 7\omega t + 0.32 \sin 11\omega t + \ldots).$$

In the third case,

$p = 3, \quad \theta_1 = 22°7, \qquad \theta_2 = 37°85, \qquad \theta_3 = 46°8$

$$u' \simeq U'_{10}\sqrt{2}(0.815 \sin \omega t + 0 + 0 + 0 - 0.15 \sin 9\omega t - 0.16 \sin 11\omega t + \ldots).$$

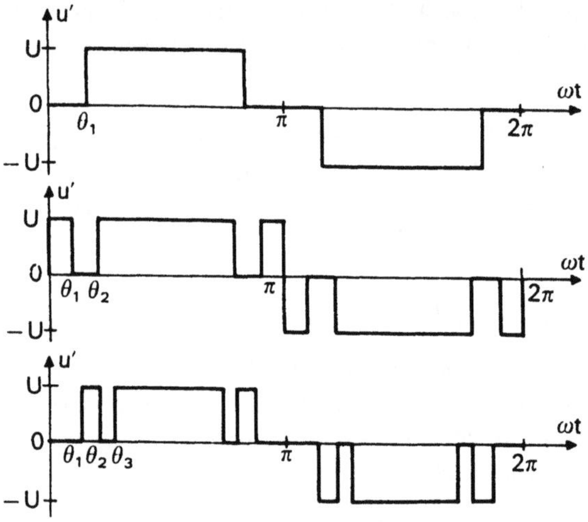

Fig. 3.27

3.4.3 Minimising the Weighted Harmonic Rate

When the commutation angles are calculated so as to eliminate the first uneven harmonics, the first non-eliminated harmonics show relatively high values.

A harmonic normally causes more problems when its frequency is lower. This leads to the idea of giving a different "weight" to various harmonics and of calculating the angles in such a way as to minimise the weighted harmonic rate.

This weighting must take into account the nature of the load connected at the inverter output of the filter placed between the inverter and the load.

3.4.3.1 First-Order "Filter"

When the inverter supplies the load directly, the aim is usually to supply the latter with a current as close to a sinusoid as possible. This is especially the case with variable frequency and voltage supplied motors.

As the impedance presented by the loads to the harmonics is often of the *R–L* type, if *L* is assumed to be constant and $Ln\omega$ high with respect to *R*, the impedance is proportional to the harmonic rank *n*. In that case, the load behaves like a first-order filter and the harmonic current of rank *n* is proportional to U'_n/n. This leads to the idea of giving each harmonic a weight which is inversely proportional to its rank.

- For full- or half-bridge single-phase inverters, the angles are calculated in such a way as to minimise the rate of harmonics weighted in this manner:

$$\frac{1}{U'_1}\left[\sum_{k=1}^{\infty}\left(\frac{U'_{2k+1}}{2k+1}\right)^2\right]^{1/2}.$$

In other words, for each value of U'_1/U'_{10} the quadratic sum of the weighted harmonics is minimised or, what gives the same result, the square of this sum:

$$\sum_{k=1}^{\infty}\left(\frac{U'_{2k+1}}{2k+1}\right)^2 .$$

- Since, in the case of the three-phase full-bridge inverter, the harmonics of rank 3 or multiple 3 disappear in the output voltages, the angles are calculated in order to minimise the following sum in the voltages supplied by the half-bridges:

$$\sum_{k=1}^{\infty}\left(\frac{U'_{6k\pm 1}}{6k\pm 1}\right)^2 .$$

3.4.3.2 Second-Order Filter

For fixed frequency inverters which have to supply a variable load, a filter is usually placed at the inverter output so as to make the filtered voltage as sinusoidal as possible.

If the filter is simply made up of a series inductor followed by a shunt capacitor (see Chap. 2, Section 2.6), the attenuation of voltage harmonics is clearly proportional to the square of their frequency. Each harmonic is thus given a weight which is inversely proportional to the square of its rank.

- For single-phase output inverters, the commutation angles are calculated in such a way as to minimise the sum

$$\sum_{k=1}^{\infty}\left[\frac{U'_{2k+1}}{(2k+1)^2}\right]^2 .$$

- For three-phase output inverters, the sum to be minimised is

$$\sum_{k=1}^{\infty}\left[\frac{U'_{6k\pm 1}}{(6k\pm 1)^2}\right]^2 .$$

- *Remark*

It can easily be imagined that a great number of control laws are possible, as long as the angles are computed off line.

If another type of filter is used, another weighting law can be adopted.

The attenuation curve of the filter can be exactly taken into account.

For each value of U'_1/U'_{10}, there are c or p variables. The value of U'_1/U'_{10} imposes one condition. Next, the harmonics which are closest to the filter cut-off frequency can be eliminated. Finally, the remaining variables can be used to minimise the remaining harmonics suitably weighted or to respect a limiting gauge of the filter output voltage harmonics or of the inverter input current harmonics.

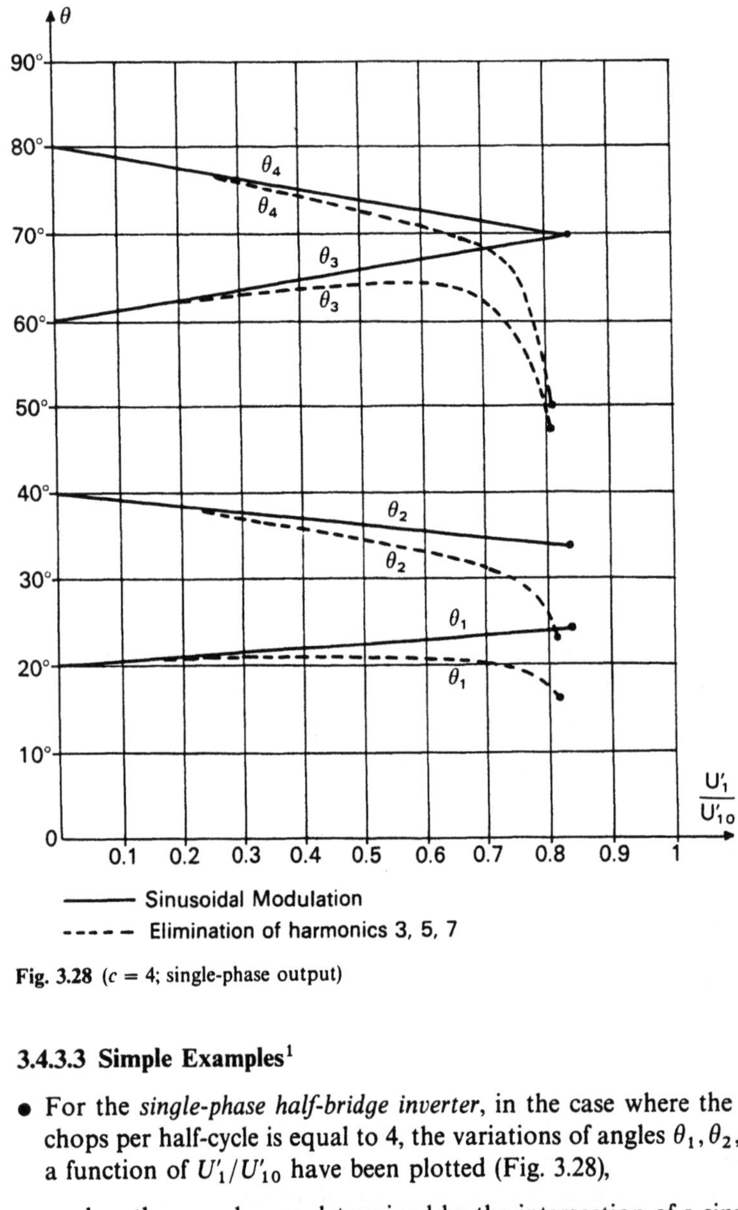

Fig. 3.28 ($c = 4$; single-phase output)

3.4.3.3 Simple Examples[1]

- For the *single-phase half-bridge inverter*, in the case where the number c of chops per half-cycle is equal to 4, the variations of angles $\theta_1, \theta_2, \theta_3$ and θ_4 as a function of U'_1/U'_{10} have been plotted (Fig. 3.28),
  - when these angles are determined by the intersection of a sinusoidal reference voltage with a triangular carrier voltage (the frequency–modulation ratio corresponding to 4 chops per half-cycle is equal to 9),
  - when the angles are computed so as to eliminate harmonics 3, 5 and 7.

[1] For plotting Figs. 3.27 to 3.33, we have used the results of computations carried out by M. El Khamlichi Drissi.

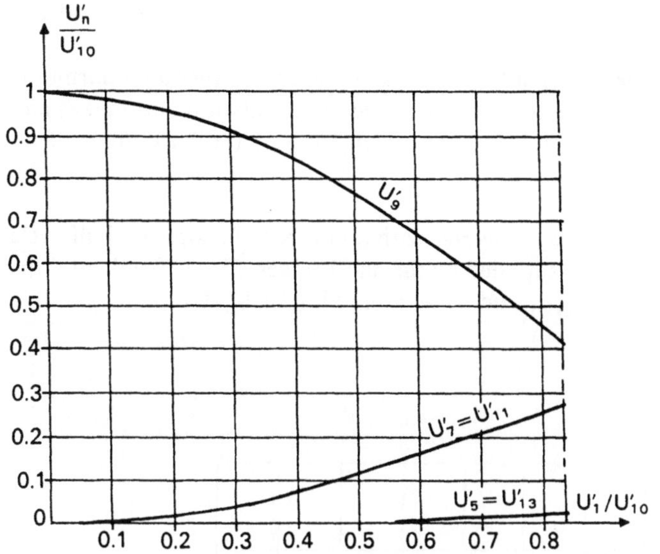

a) Sinusoidal Modulation

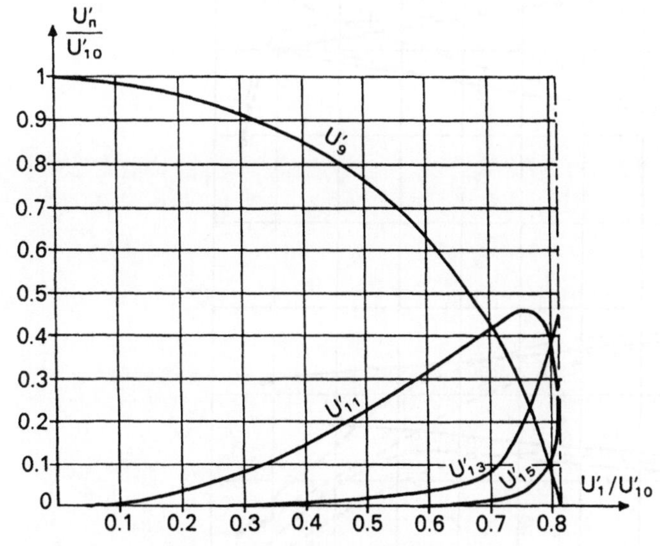

b) Elimination of harmonics 3, 5, 7

Fig. 3.29 ($c = 4$; single-phase output)

Figure 3.29 shows the variations of the first harmonics of voltage u' as function of U'_1/U'_{10} for these two control laws.

The computed modulation suppresses harmonic 7 but greatly increases harmonic 11. It is only if the regulation range of U'_1 is limited to the immediate proximity of its maximum value that it can show higher performance since it enables a low harmonic 9 to be obtained.

For low and average values of U'_1/U'_{10}, the presence of a very high harmonic 9 is to be noted; this is inherent in the principle of modulation and makes any adoption of a computed modulation with weighting of the harmonics of no interest.

- For the *three-phase full-bridge inverter* with a number of chops per half-cycle c equal to 4 in the voltages supplied by the half-bridges, Fig. 3.30 indicates the variations in angles $\theta_1, \theta_2, \theta_3$ and θ_4 as a function of U'_1/U'_{10},

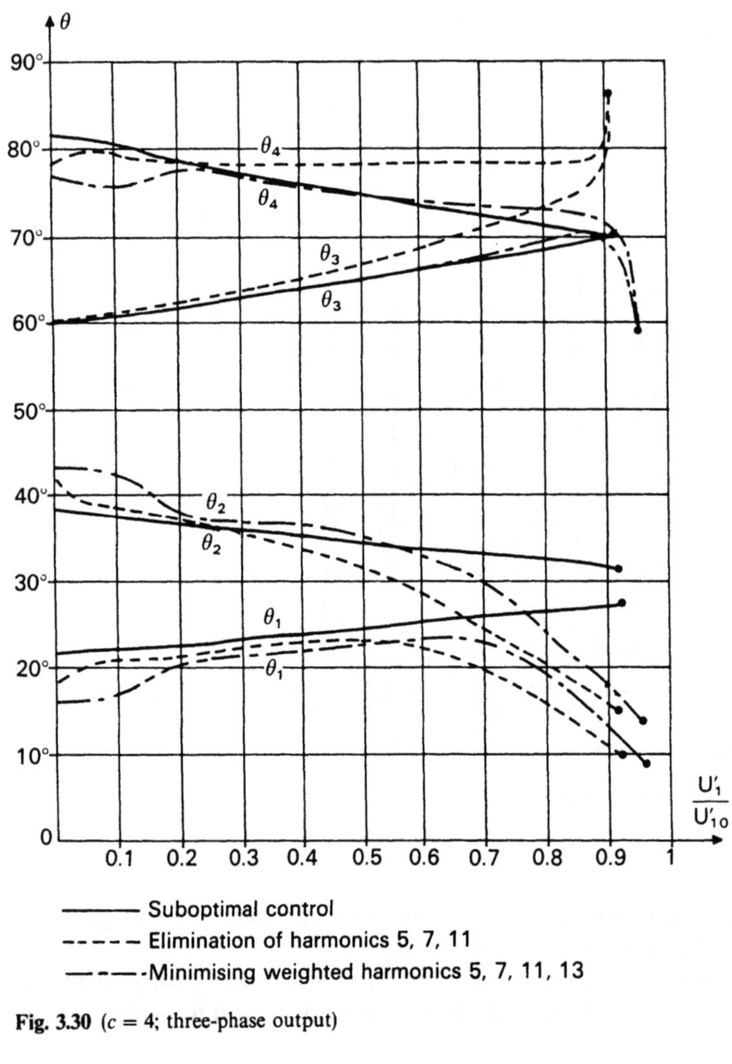

Fig. 3.30 ($c = 4$; three-phase output)

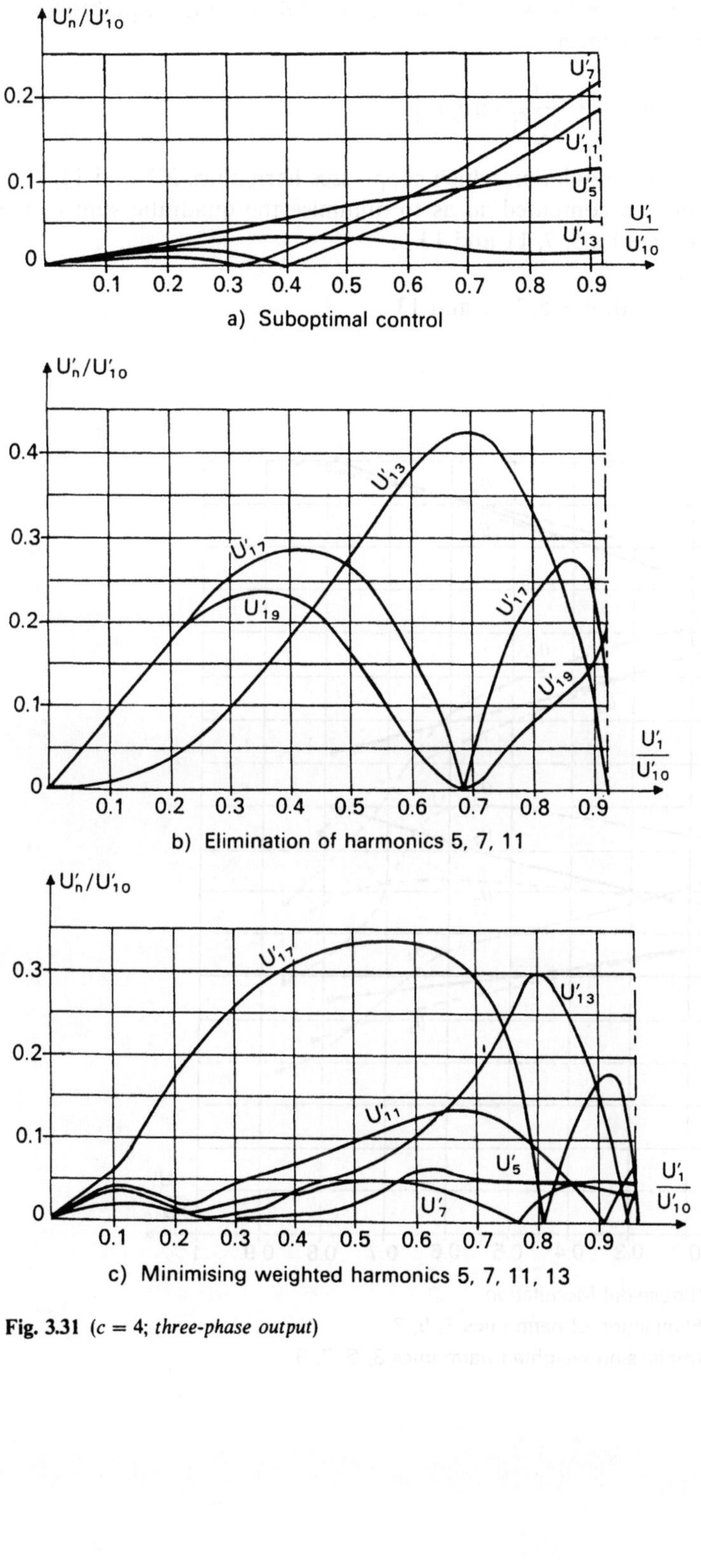

Fig. 3.31 (*c* = 4; *three-phase output*)

- for a suboptimal control with a frequency-modulation ratio equal to 9 and a reference whose form is

$$u'_w = \frac{U}{2}\left(r \sin \omega t + \frac{1}{3\sqrt{3}} \sin 3\omega t\right)$$

- for a computed modulation which suppresses harmonics 5, 7 and 11,
- for a modulation computed so as to minimise the quadratic sum of the weighted harmonics 5, 7, 11 and 13

$$\sum_n \left(\frac{U'_n}{n}\right)^2 \quad \text{with } n = 5, 7, 11 \text{ and } 13$$

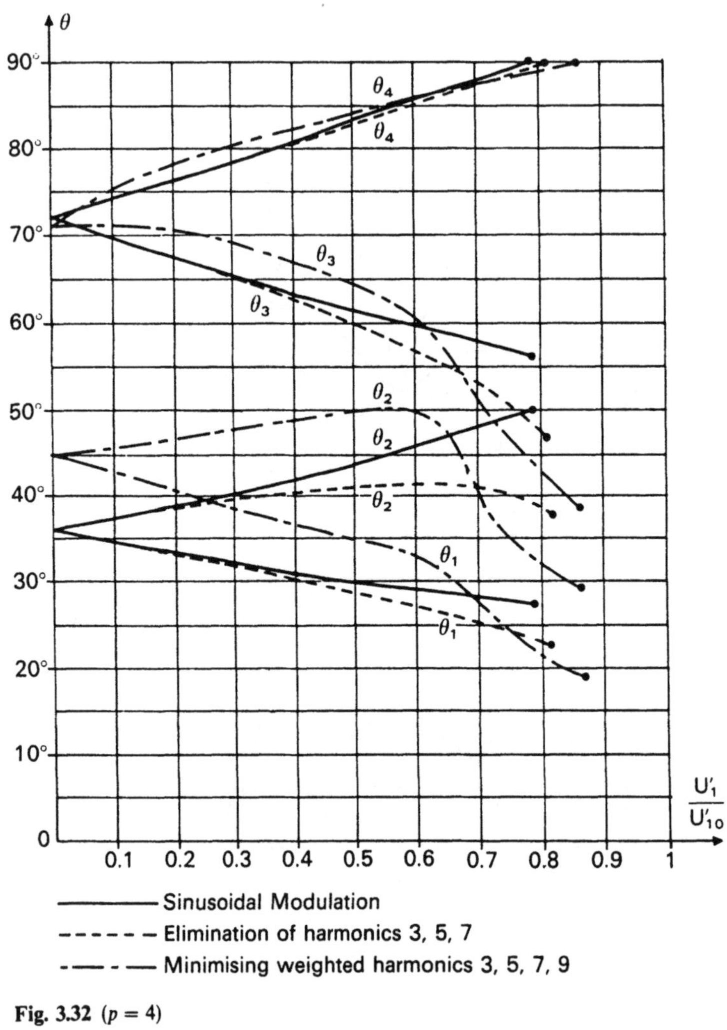

Fig. 3.32 ($p = 4$)

Figure 3.31 shows the variations of the first harmonics (other than those of rank 3 or multiple of 3) of voltage u' supplied by each half-bridge.

When harmonics 5, 7 and 11 are suppressed, the first harmonics which follow – in this case those of rank 13, 17 and 19 – show very high values.

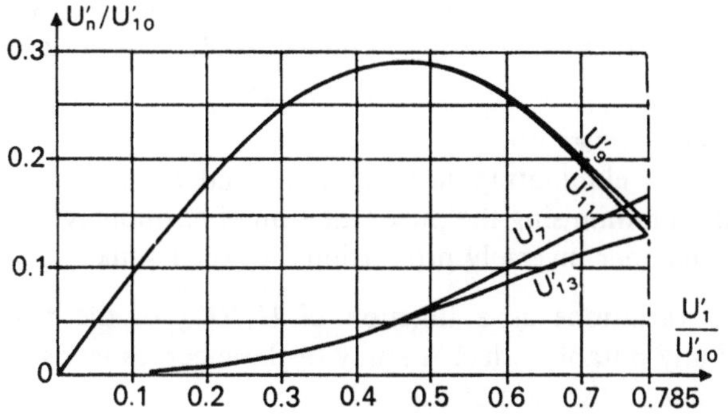

a) Sinusoidal Modulation

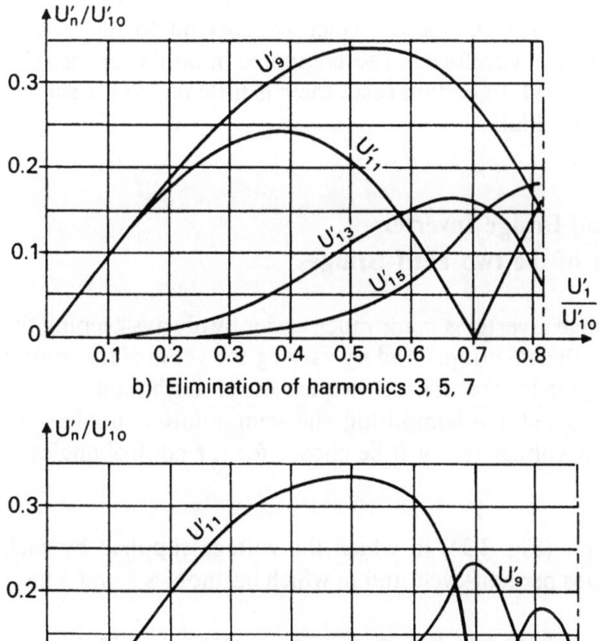

b) Elimination of harmonics 3, 5, 7

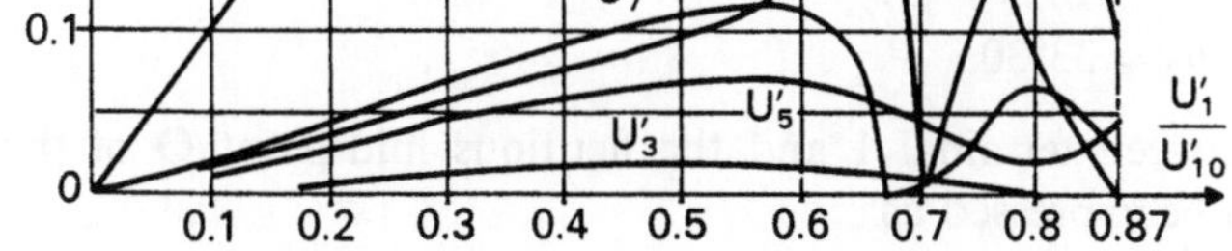

c) Minimising weighted harmonics 3, 5, 7, 9

Fig. 3.33 ($p = 4$)

By allowing reduced relative values of harmonics 5, 7 and 11, the third control law enables harmonic 13 in particular to be greatly reduced. But the first harmonic following those taken into account in determining the angle variation law – in this case harmonic 17 – shows a high relative value.

- For the *single-phase full-bridge inverter* and a number p of pulses per half-cycle equal to 4, Fig. 3.32 shows the variations of angles $\theta_1, \theta_2, \theta_3$ and θ_4 as a function of U'_1/U'_{10}
    - for a sinusoidal modulation ($m = 10$),
    - for a computed modulation eliminating harmonics 3, 5 and 7,
    - for computed modulation minimising the quadratic sum of harmonics 3, 5, 7 and 9, weighted in a manner inversely proportional to their rank.

  Figure 3.33 shows the variations as a function of U'_1/U'_{10} of the first harmonics of voltage u' of any notable value. A study of these curves leads to remarks similar to those made in the case of Fig. 3.31.

- Whatever the inverter and the number of chops or pulses per half-cycle, if the commutation angles are given the values which correspond to the criteria which are imposed on the waveforms, the computed modulation enables performances to be improved. In certain cases there is little improvement but in others it is more substantial.

3.4.4 Control of the Full-Bridge Inverter by Shifting the Voltage of the two Half-Bridges

The control of the full-bridge inverter is made much easier by always keeping the same control angles for both half-bridges and by varying the value of the output voltage by shifting adequately the control sequences of both half-bridges. Whatever the criteria adopted for computing the commutation angles, the values which give the least voltage loss will be chosen for the control angles of the two half-bridges.

– Take the simple example (Fig. 3.34) in which the voltage supplied by each half-bridge has two chops per half-cycle and in which harmonics 3 and 5 have to be suppressed.

  Switches K_1 and K'_1 of the first half-bridge are controlled so that

  $$\theta_1 = 23°62, \qquad \theta_2 = 33°30.$$

  The voltage between terminal 1 and the fictitious mid-point O of the voltage supply can be expressed as

  $$v_1 - v_0 = \frac{4}{\pi}\frac{U}{2}(0.839 \sin \omega t + 0.248 \sin 7\omega t + 0.408 \sin 9\omega t \ldots).$$

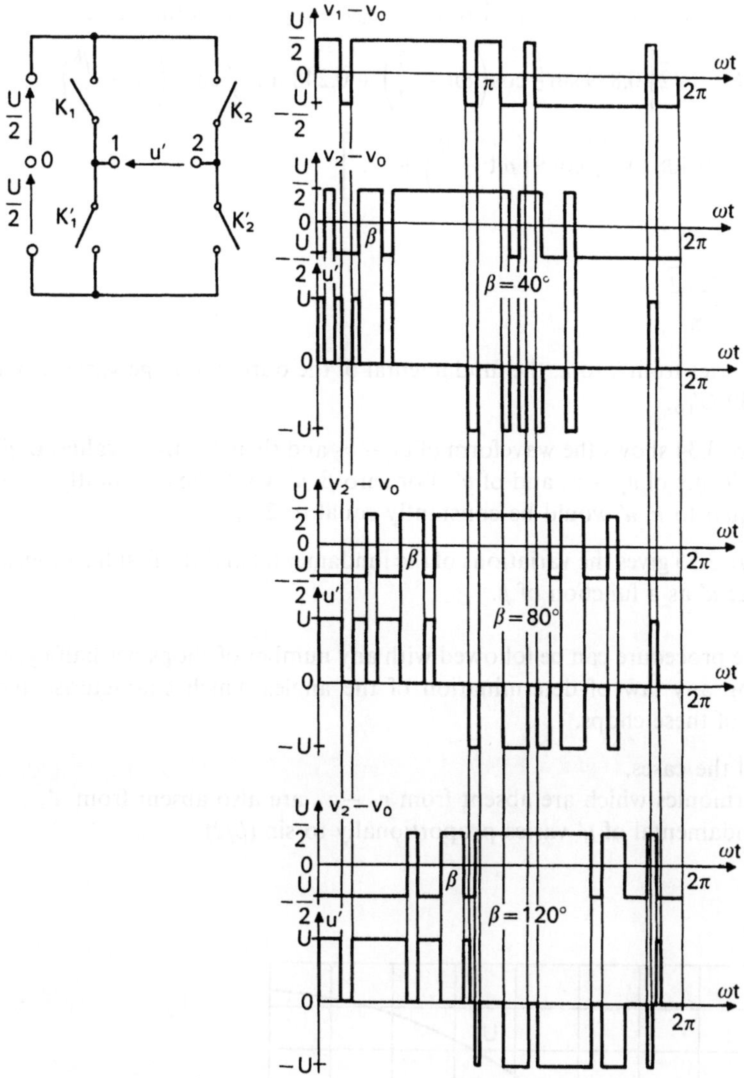

Fig. 3.34

Switches K_2 and K'_2 of the second half-bridge are controlled by following the same sequence but by shifting the latter by an angle β.

The voltage between the second output terminal 2 and the mid-point O is expressed as

$$v_2 - v_0 = \frac{4}{\pi}\frac{U}{2}(0.839 \sin(\omega t - \beta) + 0.248 \sin 7(\omega t - \beta)$$

$$+ 0.408 \sin 9(\omega t - \beta) \ldots).$$

The output voltage u', equal to $(v_1 - v_0) - (v_2 - v_0)$ is thus given by

$$u' = U'_{10}\sqrt{2}\left[0.839 \sin\frac{\beta}{2}\cos\left(\omega t - \frac{\beta}{2}\right) + 0.248 \sin\frac{7\beta}{2}\cos 7\left(\omega t - \frac{\beta}{2}\right)\right.$$
$$\left. + 0.408 \sin\frac{9\beta}{2}\cos 9\left(\omega t - \frac{\beta}{2}\right) + \dots\right]$$

with

$$U'_{10} = \frac{2\sqrt{2}}{\pi}U.$$

If β varies from 0 to π, the fundamental of the output voltage varies from 0 to 0.839 U'_{10}.

Figure 3.34 shows the waveform of $v_1 - v_0$ and then, for three values of β, the waveforms of $v_2 - v_0$ and of u'. For zero β, u' would be constantly zero. For β equal to π, u' would be constantly equal to $2(v_1 - v_0)$.

Figure 3.35 gives the variations of the fundamental and the first harmonics of voltage u' as a function of β.

– The same procedure can be followed with any number of chops per half-cycle and using any law of determination of the angles which characterise the position of these chops.

In all the cases,

- the harmonics which are absent from $v_1 - v_0$ are also absent from u',
- the fundamental of u' varies proportionally to sin $(\beta/2)$,

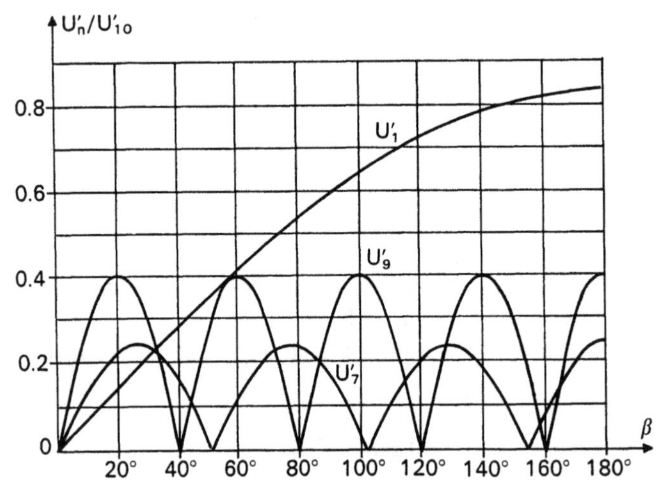

Fig. 3.35

- the harmonic of rank n which is present in u' varies proportionally to $\sin n\ (\beta/2)$.

  The number of commutations per cycle is either $2(4c + 2)$ or $2\ (4c - 2)$.

- The control which uses the shifting of the voltages of both half-bridges has the disadvantage of increasing the number of commutations. In order to suppress x uneven harmonics, $8x + 2$ commutations per cycle are needed instead of $4\ (x + 1) + 2$.

  This control offers the advantage of keeping constant the angular width of each chop which has to be taken into account in determining the maximum voltage loss. Moreoever the problem of the minimal voltage value is no longer relevant.

  In particular, control is made much easier since variations in the output voltage value can be obtained simply by shifting the control sequences of the two half-bridges. If the value and the frequency of the output voltage must both be variable, thcsc two variables can be acted on separately.

Bibliography

The bibliography concerning both choppers and inverters is to be found in the third volume of this series. Specific information can be found on publications concerning

- *semiconductor devices*, at the end of Chap. 2,
- *forced turn-off of thyristors*, at the end of Chap. 5,
- *snubbers*, at the end of the appendix of this volume.

In the case of voltage-source inverters, PWM control has replaced control with only one switch conduction period per cycle in most applications. Most publications thus deal with PWM controlled inverters.

- The articles on *voltage-source Inverters without PWM* can be broadly divided into two groups:
  - The first, which deals with *classically structured* inverters [1–11], concentrates mainly on improving the thyristor turn-off procedures, on modelling and controlling converters taking the presence of filters into account.
  - The second group [12–23] concerns the study of *variations*: "quasi-sinusoidal" inverters as presented in Appendix B of this volume; inverters with a number of outputs other than one or three; inverters using specific connections of the output transformer; inverters which involve the motor windings, which they supply, in their operating principle.

- *Pulse-width-modulated voltage-source inverters* have given rise to a great many publications. The most frequently studied modulations are in open loop, i.e. the turn-on and turn-off instants of the switches are determined from the frequency and amplitude of the required voltage.

- It is difficult to separate the articles dealing with *sinusoidal modulation* or its variations from those dealing with *pre-programmed modulation* [24–64] since they often deal with them together. Many authors have looked for the control laws enabling voltage loss or the number of commutations to be reduced, to compensate for the effects of any fluctuations in the DC input voltage, or to ensure a continuous variation in the output voltage frequency ... ,

- We have separated out the articles on the *modulation fitted to the induction motor* [65–73] supplied by the inverter. Instead of concentrating solely on the quality of the output voltage, the characteristics of the machine and the effects of the harmonic currents in the windings are taken into account. Commutation instants are determined so as to reduce the most troublesome harmonic currents and to reduce the losses, the torque pulsations and the noise which they cause.

  These studies have close links with those concerning the effects on the machines of being supplied by an inverter (see Ref. [87–107], at the end of Chap. 5).

- In the *delta modulation* (also known as current-hysteresis control or closed-loop modulation [74–83]), the turn-on and turn-off of inverter switches is controlled by observing the output current.

  Output current i' is compared to its fundamental i'_f. When the gap between i' and i'_f reaches a certain positive threshold, the beginning of a negative voltage square-wave is under control; when $i' - i'_f$ reaches the negative threshold, the comparator controls the beginning of a positive square-wave.

  This modulation is very similar to that used in active filters (see Ref. [21–28] in the bibliography at the end of the appendices) and requires the output current fundamental to be calculated at each instant.

- The other publications [84–105] on PWM voltage-source inverters deal mainly with the following themes:
  - thyristor turn-off circuits when the latter are used in converters with a relatively high commutation frequency,
  - variations on classical types of structure, notably inverters with output other that single- or three-phase and multilevel inverters using the mid-point of the DC supply,
  - characteristics relating to currents,
  - the consequences of the limited precision with which the commutation angles are determined, and of the delay between the actual turn-on and turn-off of the switches and the instant when commutations are controlled.

- *Uninterruptible power supplies* [106–115] are the main field of application for voltage-source inverters with constant output frequency. Since powered loads are often non-linear (particularly in the case of computers), the apparent impedance of the assembly formed by the inverter supply–inverter–output filter, seen from the output terminals of this filter, must be made as low as possible. PWM closed-loop techniques are thus often used; in this case, the filter output voltage fundamental is compared to the reference voltage.

- Voltage-source inverters with variable output voltage frequency and amplitude are mainly used to produce *AC variable-speed drives*.

- Firstly, we have noted some articles [116–125] which present *AC drives as a whole* – whether they be synchronous or asynchronous motors, or supplied by voltage-source inverters or current-source inverters. A lengthy bibliography is to be found in certain articles.

- Drives with a voltage-source inverter feeding a *synchronous motor* [126–137] are studied, mainly on account of the increasingly frequent use of permanent magnet machines, at low- or medium-power.

  These machines have an emf with a waveform nearer to a trapezoidal shape than a sinusoid; there are also differences between their reactances and those of the wound rotor.

  Most articles deal with modelling the machine-converter assembly (with or without PWM) with the aim of establishing steady-state characteristics and the study of dynamic behaviour.

- Articles on the *asynchronous motor* supplied with a voltage-source inverter [138–156] deal mainly with the problem of control. Progress in the use of microcomputers has led to a change from regulations based on the flux, the current, the rotor frequency ... , to vector control. The latter enables the asynchronous motor to be controlled like a DC motor.

  In this group of references, where nearly all the inverters are considered to be pulse width modulated, there are some articles which deal with special diagrams and with drives using motors with more than three phases.

  Articles dealing with the effects of supplying the motor by non-sinusoidal voltages are noted in the bibliography annexed to Chap. 5.

Voltage-Source Inverters Without PWM

– With Conventional Diagrams

1. Kuo SL (1974) Half-bridge inverter for DC power conversion, *IEEE Trans. Ind. Electron. Control Instrum.*, **21**(4), 249–253.
2. Nikravesh KY, Hoft RG (1975) Lyapounov stability analysis of thyristor inverter system, *IEEE Trans. Ind. Appl.*, **11**(2), 158–164.
3. Palaniappan R, Abrol RK, Datta SK, Vithayathil J (1977) Modified Mc Murray inverter with pulse forming network commutation circuits, *IEEE Trans. Ind. Electron. Control Instrum.*, **24**(1), 66–73.
4. Wilson JA (1978) The forced-commutated inverter as a regenerative rectifier, *IEEE Trans. Ind. Appl.*, **14**(4), 335–340.
5. Mapes JS, Bose BK (1978) Linearization of the transfer characteristic of a phase-controlled converter under discontinuous conduction, *IEEE Trans. Ind. Appl.*, **14**(6), 559–564.
6. Green RM, Boys JT (1982) A reconsideration of the design of the modified Mc Murray inverter circuit, *IEEE Trans. Ind. Appl.*, **18**(2), 152–162.
7. Lee FC, Wilson TG (1982) Modeling and analysis of several classes of self-oscillating converters. Part 1: State plane representations, *IEEE Trans. Circ. Syst.*, **29**(6) 355–365. Part 2: Model extension, classification and duality relationship, *ibid.*, 366–374.
8. Bowes SR, Clare J. Clements RR (1982) Transient performance of inverter systems, *Proc. Inst. Electr. Eng., Part B*, **129**(6), 301–314.
9. Ziogas PD (1983) A complementary current impulse commutated thyristor inverter, *IEEE Trans. Ind. Electron.*, **30**(1), 29–34.

10. Zach FC, Kaiser KH, Faschang LJ (1989) A new Mc Murray type inverter with asymmetric SCR's, *IEEE Trans. Power Electron.*, **4**(2), 272–278.
11. Kumar BSA, Mirbod, A. (1989) A variable-voltage current-commutated three-phase inverter, *IEEE Trans. Ind. Appl.*, **25**(3), 419–426.

– Variations

12. Bedford RE, Nene V (1970) Analysis and performance of a three-phase ring inverter, *IEEE Trans. Ind. Gen. Appl.*, **6**(5), 488–496.
13. Yair A, Ben-Uri J (1971) New 3-phase invertor with three thyristors, *Proc. Inst. Electr. Eng.*, **118**(7), 901–905.
14. Lissack T, Appelbaum J (1977) Stepped voltage DC/AC inverter. Part 1: Analysis, *IEEE Trans. Ind. Electron. Control Instrum.*, **24**(2), 192–199. Part 2: Optimal construction waveforms by means of discrete levels, *ibid.*, **24**(3), 273–276.
15. Bose BK (1978) A single-phase self-oscillating thyristor inverter, *IEEE Trans. Ind. Appl.*, **14**(2), 123–127.
16. Martinez S, Aldana F (1978) Current-source double DC-side forced commutated inverter, *IEEE Trans. Ind. Appl.*, **14**(6), 581–593.
17. Tez ES, Clarke J, Smith IR (1982) Inverter regulation by transformer tap changing, *Proc. Inst. Electr. Eng., Part B*, **129**(6), 315–322.
18. Williams BW (1982) High reliability 3-phase variable-frequency inverter, *Proc. Inst. Electr. Eng., Part B*, **129**(6), 353–354.
19. Mhango LMC, Creighton GK (1984) Novel two-phase inverter-fed induction-motor drive, *Proc. Inst. Electr. Eng., Part B*, **131**(3), 99–104.
20. Oguchi K, Ohta M. (1984) An improved three-phase multistepped-voltage inverter combined with a single-phase inverter through switching devices, *IEEE Trans. Ind. Appl.*, **20**(3), 655–666.
21. Bruning, GW (1986) A new high-voltage oscillator, *IEEE Trans. Ind. Electron.*, **33**(2), 171–175.
22. Asumadu JA, Hoft RG (1989) Microprocessor-based sinusoidal waveform synthesis using Walsh and related orthogonal functions, *IEEE Trans. Power Electron.*, **4**(2), 234–241.
23. Smith IR, Creighton GK, Mhango LMC (1989) Analysis and performance of a novel two-phase drive for fan and water-pumping applications, *IEEE Trans. Ind. Electron.*, **36**(4), 530–538.

PWM Voltage-Source Inverters

– Sinusoidal Modulation. Preprogrammed Modulation

24. Jackson SP (1970) Multiple pulse modulation in static inverters reduces selected output harmonics and provides smooth adjustement of fundamentals, *IEEE Trans. Ind. Gen. Appl.*, **6**(4), 357–360.
25. Dewan SB, Rosenberg SA (1970) Output voltage in three-phase pulsewidth-modulated inverters, *IEEE Trans. Ind. Gen. Appl.*, **6**(6), 570–579.
26. Patel HS, Hoft RG (1973) Generalized techniques of harmonic elimination and voltage control in thyristor inverters. Part 1: Harmonic elimination, *IEEE Trans. Ind. Appl.*, **9**(3), 310–317.
27. Patel HS, Hoft RG (1974) Generalized techniques of harmonic elimination and voltage control in thyristor inverters. Part 2: Voltage control techniques, *IEEE Trans. Ind. Appl.*, **10**(5), 666–673.
28. Bowes SR, Bird BM (1975) New approach to the analysis and synthesis of modulation processes in power convertors, *Proc. Inst. Electr. Eng.*, **122**(5), 507–513.
29. Jones NB, Stevens RTA, Readman MC (1975) Simulations and mathematical models for a pulse-width-modulated controller, *Proc. Inst. Electr. Eng.* **122**(5), 561–564.
30. Bowes SR (1975) New sinusoidal pulsewidth-modulated invertor, *Proc. Inst. Electr. Eng.*, **122**(11), 1279–1285.
31. Bansal SC, Rao UM (1978) Evaluation of PWM inverter schemes, *Proc. Inst. Electr. Eng.*, **125**(4), 328–335.
32. Pitel IJ, Talukdar SN, Wood P (1980) Characterization of programmed-waveform pulse width modulation, *IEEE Trans. Ind. Appl.*, **16**(5), 707–715.
33. Grant DA (1981) Technique for pulse dropping in pulse-width modulated inverters, *Proc. Inst. Electr. Eng., Part B*, **128**(1), 67–72.

34. Sriraghavan SM, Pradhan BD, Revankar GN (1981) Three-phase pulse-amplitude and width-modulated systems, *Proc. Inst. Electr. Eng., Part B*, **128**(3), 167–171.
35. Grant DA, Seidner R (1981) Ratio changing in pulse-width modulated inverters, *Proc. Inst. Electr. Eng. Part B*, **128**(5), 243–248.
36. Mazucchelli M, Puglisi L, Sciutto G (1981) PWM systems in power converters: an extension of the "subharmonic" method, *IEEE Trans. Ind. Electron. Control Instrum.*, **28**(4), 315–322.
37. Bowes SR, Mount MJ (1981) Microprocessor control of PWM inverters, *Proc. Inst. Electr. Eng., Part B*, **128**(6), 293–305.
38. Bowes SR, Clements RR (1982) Computer-aided design of PWM inverters, *Proc. Inst. Electr. Eng., Part B*, **129**(1), 1–17.
39. Green RM, Boys JT (1982) Implantation of pulsewidth modulated inverters modulation strategies, *IEEE Trans. Ind. Appl.* **18**(2), 138–145.
40. Green RM, Boys JT (1982) PWM sequence selection and optimization: a novel approach, *IEEE Trans. Ind. Appl.*, **18**(2), 146–151.
41. Grant D, Seidner R (1982) Technique for pulse elimination in pulsewidth-modulation inverters with no waveform discontinuity, *Proc. Inst. Electr. Eng., Part B*, **129**(4), 205–210.
42. Varnovitsky M (1983) A microcomputer-based control signal generator for a three-phase switching power inverter, *IEEE Trans. Ind. Appl.*, **19**(2), 228–234.
43. Bose B, Sutherland HA (1983) A high-performance pulsewidth modulation for an inverter-fed drive system using a microcomputer, *IEEE Trans. Ind. Appl.*, **19**(2), 235–243.
44. Sen PC, Premchandran G (1984) Improved PWM control strategy for inverters and induction motor drives, *IEEE Trans. Ind. Electron.*, **31**(1), 43–50.
45. Chin TH, Nakano M, Fuwa Y (1984) New PWM technique using a triangular carrier wave of saturable amplitude, *IEEE Trans. Ind. Appl.*, **20**(3), 643–650.
46. Houldsworth JA, Grant DA (1984) The use of harmonic distortion to increase the output of the three-phase PWM inverter, *IEEE Trans. Ind. Appl.*, **20**(5), 1224–1228.
47. Bowes SR, Midoun A (1985) Suboptimal switching strategies for microprocessor-controlled PWM inverter drives, *Proc. Inst. Electr. Eng., Part B*, **132**(3), 133–148.
48. Boys JT, Walton SJ (1985) A loss minimised sinusoidal PWM inverter, *Proc. Inst. Electr. Eng., Part B*, **132**(5), 260–268.
49. Lee JY, Sun YY (1986) Adaptative harmonic control in PWM inverters with fluctuating input voltage, *IEEE Trans. Ind. Electron.*, **33**(1), 92–98.
50. Taniguchi K, Irie H (1986) Trapezoidal modulating signal for three-phase PWM inverter, *IEEE Trans. Ind. Electron.*, **33**(2), 193–200.
51. Bowes SR, Midoun A (1986) New PWM switching strategy for microprocessor controlled inverter drives, *Proc. Inst. Electr. Eng., Part B*, **133**(4), 237–254.
52. Mangal M, De G (1987) Novel control strategy for sinusoidal PWM inverters, *IEEE Trans. Ind. Appl.*, **23**(3), 561–566.
53. Murai Y, Ohashi K, Hosono I (1987) New PWM method for fully digitized inverters, *IEEE Trans. Ind. Appl.*, **23**(5), 887–893.
54. Bellini A, De Santis A, Figalli G, Ulivi G (1987) Application of pulse ratio modulation to fixed-frequency sinusoidal tracking inverters, *IEEE Trans. Power Electron.*, **2**(4), 337–342.
55. Van den broeck HW, Skudelny HC, Stanke GV (1988) Analysis and realization of a pulsewidth modulator based on voltage space vectors, *IEEE Trans. Ind. Appl.*, **24**(1), 142–150.
56. Boost MA, Ziogas PD (1988) State-of-the-art carrier PWM techniques: a critical evaluation, *IEEE Trans. Ind. Appl.*, **24**(2), 271–280.
57. Lee JY, Sun YY (1988) Novel sinusoidal pulsewidth modulation schemes for voltage-source inverters with fluctuating input voltage, *IEEE Trans. Ind. Electron.*, **35**(2), 284–294.
58. Hamman J, Van der merwe FS (1988) Voltage harmonics generated by voltage-fed inverters using PWM natural sampling, *IEEE Trans. Power Electron.*, **3**(3), 297–302.
59. Taniguchi K, Ogino Y, Irie H (1988) PWM technique for power Mosfet inverter, *IEEE Trans. Power Electron.*, **3**(3), 328–334.
60. Enjeti P, Lindsay JF, Ziogas PD, Rashid MH (1988) New current control scheme for PWM inverters, *Proc. Inst. Electr. Eng., Part B*, **135**(4), 172–179.
61. Haneyoshi T, Kawamura A, Hoft RG (1988) Waveform compensation of PWM inverters with cyclic fluctuation loads, *IEEE Trans. Ind. Appl.*, **24**(4), 582–589.
62. Bowes SR, Midoun A (1988) Microprocessor implementation of new optimal PWM switching strategies, *Proc. Inst. Electr. Eng., Part B*, **135**(5), 269–280.
63. Zach FC (1989) Comparison of voltage and current spectra using optimized forms of inverter pulsewidth modulations, *IEEE Trans. Power Electron.*, **4**(1), 64–72.

64. Trzynadlowski AT (1989) Nonsinusoidal modulating functions for three-phase inverters, *IEEE Trans. Power Electron.*, **4**(3), 331–338.

– Modulation for Feeding Induction Motor

65. Buja GS, Indri GB (1977) Optimal pulsewidth modulation for feeding AC motor, *IEEE Trans. Ind. Appl.*, **13**(1), 38–44.
66. Kliman GB, Plunkett AB (1979) Development of a modulation strategy for a PWM inverter drive, *IEEE Trans. Ind. Appl.*, **15**(1), 72–79.
67. De Buck F, Gistelink P, De Backer D (1983) Loss-optimal PWM waveforms for variable-speed induction-motor drives, *Proc. Inst. Electr. Eng., Part B*, **130**(5), 310–320.
68. Park, MH, Sul SK (1984) Microprocessor-based optimal-efficiency drive of an induction motor, *IEEE Trans. Ind. Electron.*, **31**(1), 69–73.
69. Nagase H, Masuda Y, Ohnishi K, Ninomya H, Koike T (1984) High-performance induction motor drive system using a PWM inverter, *IEEE Trans. Ind. Appl.*, **20**(6), 1482–1489.
70. Takahashi I, Mochikawa H (1985) A new control of PWM inverter waveform for minimum loss operation of an induction motor drive, *IEEE Trans. Ind. Appl.*, **21**(3), 580–587.
71. Zach FC, Ertl H (1985) Efficiency optimal control for AC drives with PWM inverters, *IEEE Trans. Ind. Appl.*, **21**(4), 987–1000.
72. Takahashi I, Mochikawa H (1986) Optimum PWM waveforms of an inverter for decreasing acoustic noise of an induction motor, *IEEE Trans. Ind. Appl.*, **22**(5), 828–834.
73. Thorborg K, Nystrom A (1988) Staircase PWM: An uncomplicated and efficient technique for AC motor drives, *IEEE Trans. Power Electron.*, **3**(4), 391–398.

– Delta Modulation

74. Kernick A, Stechschulte DL, Shireman DW (1977) Static inverter with synchronous output waveform synthetized by time-optimal-response feedback, *IEEE Trans. Ind. Electron. Control Instrum.*, **24**(4), 297–305.
75. Ziogas PD (1981) The Delta modulation technique in static PWM inverters, *IEEE Trans. Ind. Appl.*, **17**(2), 199–204.
76. Mc Murray W (1984) Modulation of the chopping frequency in DC choppers and PWM inverters having current-hysteresis controllers, *IEEE Trans. Ind. Appl.*, **20**(4), 763–768.
77. Kawamura A, Hoft RG (1984) Instantaneous feedback controlled PWM inverter with adaptive hysteresis, *IEEE Trans. Ind. Appl.*, **20**(4), 769–775.
78. Rahman MA, Quaicoe JE, Choudhury MA (1987) Performance analysis of delta modulated PWM inverters, *IEEE Trans. Power Electron.*, **2**(3), 227–233.
79. Kulkarni AB, Nixon JW, Nishimoto M, Ooi BT (1987) Transient tests on a voltage-regulated controlled-current PWM converter, *IEEE Trans. Ind. Electron.*, **34**(3), 319–324.
80. Gokhale KP, Kawamura A, Hoft RG (1987) Deadbeat microprocessor control of PWM inverter for sinusoidal output waveform synthesis, *IEEE Trans. Ind. Appl.*, **23**(5), 901–910.
81. Kawamura A, Haneyoshi T, Hoft RG (1987) Deadbeat controlled PWM inverter with parameter estimation using only voltage sensor, *IEEE Trans. Power Electron.*, **3**(2), 118–125
82. Christiansen CF, Valla MI, Rivetta CH (1988) A synchronisation technique for static delta-modulated PWM inverters, *IEEE Trans. Ind. Electron.*, **35**(4) 502–507.
83. Salvatore L, Trotta A (1988) Flat-top window for PWM waveform processing via DFT, *Proc. Inst. Electr. Eng., Part B*, **135**(6), 346–361.

– Realization. Characteristics

84. Dewan SB, Caceres J (1980) An input current-commutated three-phase PWM inverter, *IEEE Trans. Ind. Appl.*, **16**(4), 445–451.
85. Sriraghavan SM, Pradhan BD, Revankar GN (1981) Three-phase pulse-amplitude and width-modulated inverter system, *Proc. Inst. Electr. Eng., Part B*, **128**(3), 167–171.
86. Pitel IJ (1981) Spectral errors in the application of pulse-width modulated waveforms, *IEEE Trans. Ind. Appl.*, **17**(3), 289–295.
87. Nabae A, Yakahaschi I, Akagi H (1981) A new neutral point clamped PWM inverter, *IEEE Trans. Ind. Appl.*, **17**(5), 518–522.

88. Bhagwat PD, Stefanovic VR (1982) A versatile commutation circuit for PWM inverters, *IEEE Trans. Ind. Appl.*, **18**(2), 127–137.
89. Ziogas PD, Photiadis PND (1983) An exact input analysis of ideal static PWM inverter, *IEEE Trans. Ind. Appl.*, **19**(2), 281–295.
90. Bhagwat PD, Stefanovic VR (1983) Generalized structure of a multilevel PWM inverter, *IEEE Trans. Ind. Appl.*, **19**(6), 1057–1069.
91. Bhagwat PD, Stefanovic VR (1984) Some new aspects in the design of PWM inverters, *IEEE Trans. Ind. Appl.*, **20**(4), 776–784.
92. Abbas MA, Christen R, Jahns TM (1984) Six-phase voltage source inverter driven induction motor, *IEEE Trans. Ind. Appl.*, **20**(5), 1251–1259.
93. Grant DA, Stevens M, Houldsworth JA (1985) The effect of word length on the harmonic content of microprocessor-based PWM waveform generators, *IEEE Trans. Ind. Appl.*, **21**(1), 218–225.
94. Klingshirn E (1985) Harmonic filters for six-phase and other multiphase motors on voltage source inverter, *IEEE Trans. Ind. Appl.*, **21**(3), 588–594.
95. Gierse G, Schuermann W (1986) Microprocessor control for two magnetically coupled three-phase PWM inverters, *IEEE Trans. Power Electron.*, **1**(3), 141–147.
96. Evans PD, Hill-Cottingham RJ (1986) DC link current in PWM inverters, *Proc. Inst. Electr. Eng., Part B*, **133**(4), 217–224.
97. Alexa D (1987) Static frequency convertor to supplying an asynchronous two-phase motor, *Proc. Inst. Electr. Eng., Part B*, **134**(1), 57–60.
98. Evans PD, Close PR (1987) Harmonic distortion in PWM inverter output waveforms, *Proc. Inst. Electr. Eng., Part B*, **134**(4), 224–232.
99. Murai Y, Watanabe T, Iwasaki H (1987) Waveform distortion and correction circuit for PWM inverters with switching lag-time, *IEEE Trans. Ind. Appl.*, **23**(5), 881–886.
100. Manias S, Wiechmann EP, Ziogas PD (1987) Effects of switching angle phase shift on PWM techniques, *IEEE Trans. Ind. Electron.*, **34**(4), 463–470.
101. Van Den Broeck HW, Skudelny HC (1988) Analytical analysis of the harmonic effects of a PWM AC drive, *IEEE Trans. Power Electron.*, **3**(2), 216–233.
102. Alolah AI, Hulley LN, Shepherd W (1988) A three-phase neutral point clamped inverter for motor control, *IEEE Trans. Power Electron.*, **3**(4), 399–405.
103. Andria G, Dell'Aquila A, Salvatore L (1989) Analysis of distorted unbalanced waveforms in inverter drives, *IEEE Trans. Power Electron.*, **4**(2), 298–310.
104. Mestha LK, Evans PD (1989) Analysis of on-state losses in PWM inverters, *Proc. Inst. Electr. Eng., Part B*, **136**(4), 189–195.
105. Green TC, Williams BW (1989) Derivation of motor line-current waveforms from the DC-link current of an inverter, *Proc. Inst. Electr. Eng., Part B*, **136**(4), 196–204.

Uninterruptible Power Supplies

106. Peterson WV, Yohman EJ (1970) Pulse width modulated inverters for uninterruptible AC power systems, *IEEE Trans. Ind Electron. Control Instrum.*, **17**(4), 339–345.
107. Kusko A, Gilmoro FE (1970) Application of static uninterruptible power systems to computer loads, *IEEE Trans. Ind. Gen. Appl.*, **6**(4), 330–336.
108. Chauprade R (1977) Inverters for uninterruptible power supplies, *IEEE Trans. Ind. Appl.*, **13**(4), 281–297.
109. Srighavan SM, Revankar GN, Pradhan BD (1978) Voltampere ratings of components of pulsewidth-modulated inverter system, *IEEE Trans. Ind. Electron. Control Instrum.*, **25**(3), 278–284.
110. Kawamura A, Chuarayapratip R, Haneyoshi T (1988) Deabeat control of PWM inverter with modified pulse patterns for uninterruptible power supply, *IEEE Trans. Ind. Electron.*, **35**(2), 295–300.
111. Holtz J, Lotzkat W, Werner KH (1988) A high-power multitransistor-inverter uninterruptible power supply system, *IEEE Trans. Power Electron.*, **3**(3), 278–285.
112. Maire J (1989) Alimentation Sans Interruption de moins de 3 KVA: des performances et une qualité mieux controlées, *Rev. Gén. Electr.*, **98**(2), 23–38.
113. Cheron Y, Elisabelar C, Foch H (1989) Face aux impératifs des Alimentations Sans Interruption: Modulation de Largeur d'Impulsions ou résonance, *Rev. Gén. Electr.*, **98**(2), 41–46.

114. Boost MA, Ziogas PD (1989) Towards a zero-output impedance UPS system, *IEEE Trans. Ind. Appl.*, **25**(3), 408–418.
115. Martinez S, Castro M, Antoranz R, Aldana F (1989) Off-line uninterruptible power supply with zero transfer time using integrated magnetics, *IEEE Trans. Ind. Electron.*, **36**(3), 441–445.

Variable-Speed Drives with Voltage-Source Inverters

– AC Drives. General Studies

116. Novotny DW (1976) Steady state performance of inverter fed induction machines by means of time domain complex variables, *IEEE Trans. Power Appar. Syst.*, **95(3)**, 927–935.
117. Pierrat L (1978) Pourquoi la fréquence variable par variateurs statiques, *Rev. Gén. Electr.*, **87**(9), 679–685.
118. Bose BK (1982) Adjustable speed AC drives. A technology status review, *Proc. IEEE*, **70**(2), 116–138.
119. Jones BL, Brown JE (1984) Electrical variable-speed drives, *Proc. Inst. Electr. Eng., Part A*, **131**(7), 516–558.
120. Van Wyck JD, Skudelny HC, Muller-Hellmann A (1986) Power electronics, control of the electromechanical energy conversion process and some applications, *Proc. Inst. Electr. Eng., Part B*, **133**(6), 369–399.
121. Jarc DA, Novotny DW (1987) A graphical approach to AC drive classification, *IEEE Trans. Ind. Appl.*, **23**(6), 1029–1035.
122. Lipo TA (1988) Recent progress in the development of solid-state AC motor drives, *IEEE Trans. Power Electron.*, **3**(2), 105–117.
123. Dote Y (1988) Application of modern control techniques to motor control, *Proc. IEEE*, **76**(4), 438–454.
124. Leonhard W (1988) Adjustable-speed AC drives, *Proc. IEEE*, **76**(4), 455–471.
125. Bowes SR, Clare JC (1988) Computer-aided design of power-electronic variable-speed drives, *Proc. Inst. Electr. Eng., Part B*, **135**(5), 240–260.

– Voltage-Source Inverter Fed Synchronous Motors

126. Lajoie-Mazenc M, Villanueva C, Hector J (1985) Study and implementation of hysteresis controlled inverter on a permanent magnet synchronous machine, *IEEE Trans. Ind. Appl.*, **21**(2), 408–415.
127. Funabiki S, Himei T (1985) Estimation of torque pulsation due to the behaviour of a converter and an inverter in a brushless DC-drive system, *Proc. Inst. Electr. Eng., Part B*, **132**(4), 215–222.
128. Nehl TW, Demerdash NA, Fouad FA (1985) Impact of winding inductances and other parameters on the design and performance of brushless DC motors, *IEEE Trans. Power Appar. Syst.*, **104**(8), 2206–2213.
129. Nehl TW, Demerdash NA, Hijazi TM, Mc Hale TL (1985) Automatic formulation of models for simulation of the dynamic performance of electronically commutated DC machines, *IEEE Trans. Power Appar. Syst.*, **104**(8), 2214–2222.
130. Graham ML, Gruber S (1985) Simulation study of an electronically commutated DC machine, *IEEE Trans. Ind. Electron.*, **32**(4), 399–405.
131. Bolton HR, Liu YD, Mallinson NM (1986) Investigation into a class of brushless DC motor with quasisquare voltages and currents, *Proc. Inst. Electr. Eng., Part B*, **133**(2), 103–111.
132. Namuduri C, Sen PC (1987) Digital simulation of an inverter-fed self-controlled synchronous motor, *IEEE Trans. Ind. Electron.*, **34**(2), 205–215.
133. Krause PC, Nucera RR, Krefta RJ, Wasynczuk O (1987) Analysis of a permanent magnet synchronous machine supplied from a 180° inverter with phase control, *IEEE Trans. Energy Conv.*, **2**(3), 423–431.
134. Hashimoto H, Yamamoto H, Yanagisawa S, Harashima F (1988) Brushless servo motor control using variable structure approach, *IEEE Trans. Ind. Appl.*, **24**(10), 160–170.
135. Pillay P, Krishnan R (1989) Modeling, simulation and analysis of permanent-magnet rotor drives. Part I: The permanent-magnet synchronous motor drive, *IEEE Trans. Ind. Appl.*, **25**(2) 265–273. Part 2: The brushless DC motor drive, *ibid.*, 274–279.
136. Le-Huy H, Dessaint LA (1989) An adaptive current control scheme for PWM synchronous motor drives: analysis and simulation *IEEE Trans. Power Electron.*, **4**(4), 486–495.

137. Nucera RR, Sudhoff SD, Krause, PC (1989) Computation of steady-state performance of an electronically commutated motor, *IEEE Trans. Ind. Appl.*, **25**(6), 1110 1117.

– Voltage-Source Inverter Fed Asynchronous Motors

138. Plunkett AB, Lipo TA (1976) New methods of induction motor torque regulation, *IEEE Trans. Ind. Appl.*, **12**(1), 47–55.
139. Foch H (1978) Alimentation en tension d'une machine asynchrone d'induction à fréquence variable, *Rev. Gén. Electr.*, **87**(4), 271–276.
140. Bellini A, Figalli G (1979) On the selection of the commutating instants in induction motor drives, *IEEE Trans. Ind. Appl.*, **15**(5), 501–506.
141. Bowes SR, Davies T (1985) Microprocessor-based development system for PWM variable-speed drives, *Proc. Inst. Electr. Eng., Part B*, **132**(1), 18–45.
142. Ho EYY, Sen PC (1986) Digital simulator of PWM induction motor drives for transient and steady-state performance, *IEEE Trans. Ind. Electron.*, **33**(1), 66–67.
143. Ahmed MM, Taufiq JA, Goodman CJ, Lockwood M (1986) Electrical instability in a voltage source inverter-fed induction motor drive at constant speed, *Proc. Inst. Electr. Eng., Part B*, **133**(4), 299–307.
144. Takahashi I, Nogushi T (1986) A new quick-response and high-efficiency control strategy of an induction motor, *IEEE Trans. Ind. Appl.*, **22**(5), 820–827.
145. Rajashekara KS, Rajagopalan V, Sevigny A, Vithayathil J (1987) DC link filter design considerations in three-phase voltage source inverter-fed induction motor drive system, *IEEE Trans. Ind. Appl.*, **23**(4), 673–680.
146. Kohlmeier H, Niermeyer O, Schroder DF (1987) Highly dynamic four-quadrant AC motor drive with improved power factor and on-line optimized pulse pattern with PROMC, *IEEE Trans. Ind. Appl.*, **23**(6), 1001–1009.
147. Boys JT (1988) Novel current sensor for PWM AC drives, *Proc. Inst. Electr. Eng., Part B*, **135**(1), 27–32.
148. Ooi BT, Dixon JW, Kulkarni AB, Nishimoto M (1988) An integrated AC drive system using a controlled-current PWM rectifier/inverter link, *IEEE Trans. Power Electron.*, **3**(1), 64–71.
149. Pavithran KN, Parimelalagan R, Krishnamurthy MR (1988) Studies on inverter-fed five-phase induction motor drive *IEEE Trans. Power Electron.*, **3**(2), 224–235.
150. Ho, EYY (1988) Decoupling control of induction motor drives, *IEEE Trans. Ind. Electron.*, **35**(2), 253–262.
151. Boys JT, Walton SJ (1988) Scalar control: an alternative AC drive philosophy, *Proc. Inst. Electr. Eng., Part B*, **135**(3), 151–158.
152. Depenbrock M (1988) Direct self-control of inverter-fed induction machine, *IEEE Trans. Power Electron.*, **3**(4), 420–429.
153. Underland TM, Mohan N (1988) Overmodulation and loss considerations in high-frequency modulated transistorized induction motor drives, *IEEE Trans. Power Electron.*, **3**(4), 447–452.
154. Ueda R, Sonoda T, Takata S (1989) Experimental results and their simplified analysis on instability problems in PWM inverter induction motor drives, *IEEE Trans. Ind. Appl.*, **25**(1), 86–95.
155. Takahashi I, Ohmori Y (1989) High-performance direct torque control of an induction motor, *IEEE Trans. Ind. Appl.*, **25**(2), 257–264.
156. Nishimura T, Nakaoka M, Maruhashi T (1989) Reduction of vibration and acoustic noise in induction motor driven by three-phase PWM AC chopper using static induction transistors, *IEEE Trans. Power Electron.*, **4**(3), 313–318.

Chapter 4

Commutations in Voltage-Source Inverters

Voltage-source inverters use current-reversible switches formed by associating controlled semiconductor devices and antiparallel connected diodes.

A classical thyristor with a turn-off circuit can be used as a controlled device[1], but the following are used increasingly frequently:

- bipolar transistors for medium currents,
- GTO thyristors for high currents,
- field-effect transistors or MOSEFET's for low power applications.

In the previous volume, the structure and the switching behaviour of these devices were studied (Chap. 2), as well as, in the case of the choppers, the snubbers associated to them (Appendix).

The commutations of these three types of device will now be studied in the case of the voltage-source inverters when fitted with appropriate snubbers.

In general, these devices require a series inductive snubber, in order to limit the di/dt at turn-on, and a parallel capacitive snubber, to limit the dv/dt at turn-off. After each commutation, the energy stored in the snubber inductors and capacitors is usually dissipated in resistances. Auxiliary diodes are needed to send the currents correctly to the snubber circuits.

4.1 Commutation of the Bipolar Transistor in Full-Wave Inverters

The problems which are induced by commutations in the voltage-source inverters vary according to whether PWM is used or not.

If PWM is not used, the commutation problems are the same as those encountered in the case of choppers with certain occasional simplifications.

In the usual case where the inverter feeds a load which always remains inductive (Fig. 4.1), the transistors are turned on at zero current i' and thus require protection against dv/dt only.

---

[1] The turn-off circuits of classical thyristors were studied in Vol. 3, Chap. 5, together with the chopper structures. Appendix C provides extra information on the voltage-source inverters.

In the case of current-source inverters, it is advisable not to separate the commutations of the operation as a whole. These inverters are studied in the following chapter.

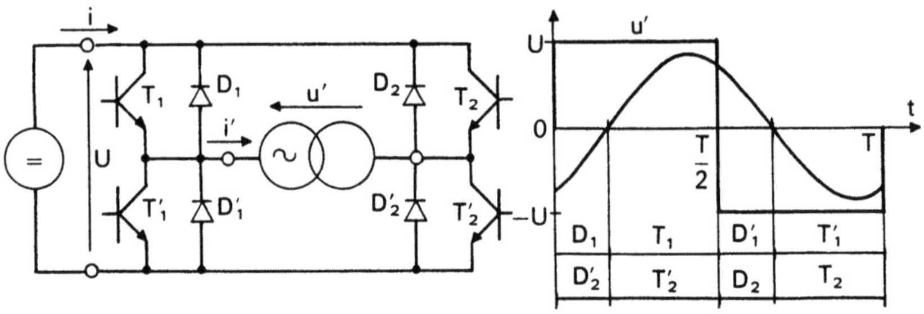

Fig. 4.1

4.1.1 Commutation Process with RC Snubber

In order to protect the transistors from excessive dv/dt, each one is fitted with an *R–C–D* network (Fig. 4.2a) or with a simple *R–C* network (Fig. 4.2b).

For the case in which *R–C* circuits are used, we shall follow the commutation between a transistor and a diode, T_1 and D'_1 for example. This commutation can be divided into three phases; at the top of Fig. 4.3, the thick black lines mark the elements which are involved during each stage; below, the waveforms of v_1 and v'_1, of i_{T_1} and v_{T_1}, and of $i_{D'_1}$ and $v_{D'_1}$ have been plotted.

The commutation is assumed to be sufficiently fast to enable current i' to remain almost constant throughout its duration. The value of current i' during the commutation is referred to as I_{OFF}.

- *Phase 1* ($0 < t < t_f$)

If the forward voltage drop of the conducting transistor is much smaller than DC voltage U, the voltages across the capacitors have the following initial values:

$$v_{10} = 0, \qquad v'_{10} = U.$$

It is stated that current i_{T_1} in the transistor decreases linearly from I_{OFF} to zero in a time t_f when the base control signal is removed.

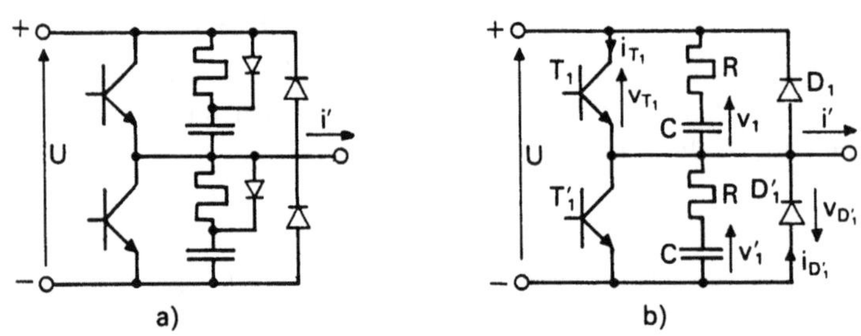

Fig. 4.2

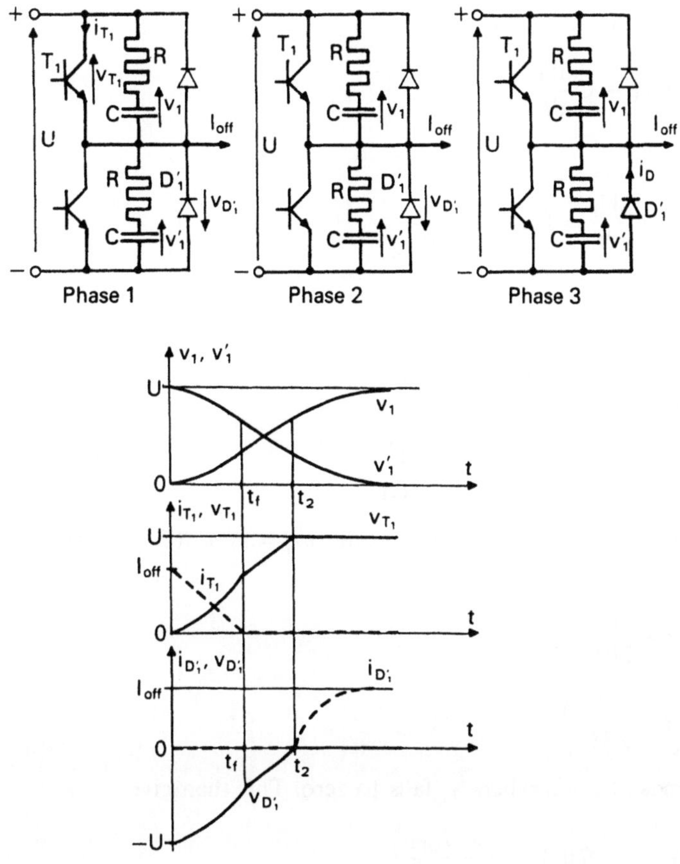

Fig. 4.3

During this period,

$$U = v_1 + RC\frac{dv_1}{dt} + v_1' + RC\frac{dv_1'}{dt}$$

$$= (v_1 + v_1') + RC\frac{d(v_1 + v_1')}{dt}.$$

The sum $(v_1 + v_1')$ thus has the form $U + A\exp(-t/RC)$. However, in this case, A is zero since $v_{10} + v_{10}' = U$. This thus gives:

$$v_1 + v_1' = U$$

which leads to:

$$C\frac{dv_1}{dt} = -C\frac{dv_1'}{dt}.$$

The relation between the currents:

$$i_{T_1} + C\frac{dv_1}{dt} = I_{OFF} + C\frac{dv_1'}{dt}$$

can be written as:

$$I_{OFF}\left(1 - \frac{t}{t_f}\right) + C\frac{dv_1}{dt} = I_{OFF} - C\frac{dv_1}{dt}$$

giving:

$$2C\frac{dv_1}{dt} = I_{OFF}\frac{t}{t_f}.$$

The following can be deduced:

$$v_1 = I_{OFF}\frac{t^2}{4Ct_f} \quad \text{and} \quad v_1' = U - I_{OFF}\frac{t^2}{4Ct_f}. \tag{4.1}$$

Voltages v_{T_1} and $v_{D_1'}$ are deduced from $v_{T_1} = v_1 + RC(dv_1/dt)$ and $v_{D_1'} = v_{T_1} - U$, giving:

$$v_{T_1} = \frac{RI_{OFF}}{2t_f}\left(t + \frac{t^2}{2RC}\right)$$

$$v_{D_1'} = \frac{RI_{OFF}}{2t_f}\left(t + \frac{t^2}{2RC}\right) - U.$$

This first phase ends at $t = t_f$ when i_{T_1} falls to zero. This then gives:

$$v_1(t_f) = \frac{I_{OFF}}{4C}t_f\,, \qquad v_1'(t_f) = U - \frac{I_{OFF}}{4C}t_f.$$

- *Phase 2* ($t_f < t < t_2$)

The role of capacitors is to slow down the voltage rise across the transistors during the latter's turn-off. v_{T_1} is thus usually less than U and $v_{D_1'}$ negative when T_1 is turned off. Diode D_1' is still in the blocking state.

As in phase 1, the following is valid:

$$U = v_1 + RC\frac{dv_1}{dt} + v_1' + RC\frac{dv_1'}{dt}$$

giving once again:

$$v_1 + v_1' = U \quad \text{and} \quad C\frac{dv_1}{dt} = -C\frac{dv_1'}{dt}.$$

However the current equation must now be written as:

$$C\frac{dv_1}{dt} = C\frac{dv_1'}{dt} + I_{OFF}.$$

From $C\,\mathrm{d}v_1/\mathrm{d}t = I_{\mathrm{OFF}}/2$, the following can be deduced:

$$v_1 = \frac{I_{\mathrm{OFF}}}{2C}\left(t - \frac{t_{\mathrm{f}}}{2}\right)$$

$$v_1' = U - v_1 = U - \frac{I_{\mathrm{OFF}}}{2C}\left(t - \frac{t_{\mathrm{f}}}{2}\right). \tag{4.2}$$

The voltages across T_1 and D_1' are given by:

$$v_{\mathrm{T}_1} = v_1 + RC\frac{\mathrm{d}v_1}{\mathrm{d}t}, \qquad v_{\mathrm{D}_1'} = v_{\mathrm{T}_1} - U$$

they can thus be expressed as:

$$v_{\mathrm{T}_1} = \frac{I_{\mathrm{OFF}}}{2C}\left(t - \frac{t_{\mathrm{f}}}{2} + RC\right)$$

$$v_{\mathrm{D}_1'} = \frac{I_{\mathrm{OFF}}}{2C}\left(t - \frac{t_{\mathrm{f}}}{2} + RC\right) - U.$$

This phase ends for $t = t_2$ when $v_{\mathrm{D}_1'}$ equals zero. The value of t_2 is thus:

$$t_2 = \frac{2CU}{I_{\mathrm{OFF}}} + \frac{t_{\mathrm{f}}}{2} - RC$$

and the values of v_1 and v_1' at that instant are:

$$v_1(t_2) = U - \frac{1}{2}RI_{\mathrm{OFF}}, \qquad v_1'(t_2) = \frac{1}{2}RI_{\mathrm{OFF}}.$$

- *Phase 3* ($t > t_2$)

For $t > t_2$, diode D_1' conducts but a current continues to flow in the two RC circuits. As was the case during phases 1 and 2, the following can be written:

$$v_1 + v_1' = U \text{ and } C\frac{\mathrm{d}v_1}{\mathrm{d}t} = -C\frac{\mathrm{d}v_1'}{\mathrm{d}t}.$$

However, since D_1' is conducting, this now gives:

$$U = v_1 + RC\frac{\mathrm{d}v_1}{\mathrm{d}t}.$$

Taking into account the value of $v_1(t_2)$, the following can be deduced:

$$v_1 = U - \frac{RI_{\mathrm{OFF}}}{2}\exp\left(\frac{t_2 - t}{RC}\right) = U - v_1'. \tag{4.3}$$

Current $i_{\mathrm{D}_1'}$, computed from

$$i_{\mathrm{D}_1'} = I_{\mathrm{OFF}} + C\frac{\mathrm{d}v_1'}{\mathrm{d}t} - C\frac{\mathrm{d}v_1}{\mathrm{d}t}$$

can be expressed as:

$$i_{D'_1} = I_{OFF}\left[1 - \exp\left(\frac{t_2 - t}{RC}\right)\right].$$

Remark. The value of time constant RC must be considerably smaller than the minimum conduction time of T_1 in order to allow v_1, v'_1 and $i_{D'_1}$ to reach their steady-state values.

4.1.2 Computing the Losses

4.1.2.1 Power Dissipated in T_1 During Commutations

Figure 4.3 shows that, with the assumptions made, power dissipation in T_1 only occurs during phase 1. If f is the operating frequency of the inverter,

$$P_T = f\int_0^{t_f} v_{T_1} i_{T_1}\,dt = f\int_0^{t_f} \frac{RI_{OFF}}{2t_f}\left(t + \frac{t^2}{2RC}\right) I_{OFF}\left(1 - \frac{t}{t_f}\right)dt$$

$$P_T = f\frac{RI_{OFF}^2}{12}\left(t_f + \frac{t_f^2}{4RC}\right). \tag{4.4}$$

Remarks

- This expression under-estimates the losses to a certain extent since:
  - during the storage time, $V_{CE\,sat}$ shows a slight increase;
  - after the fall-time of current i_{T_1}, the leakage collector current of the transistor remains, for some time, higher than its normal value in the blocking state.
- Conduction losses must be added to commutation losses. The former can be deduced from the rms and average values of current i_{T_1} given in Chap. 2.2.

4.1.2.2 Power Dissipated in the Resistances During Commutations

The current in the resistance of the RC circuit shunt connected across the terminals of T_1 is equal to $C(dv_1/dt)$. From (4.1), (4.2) and (4.3) the expressions of this current during the three periods can be deduced:

$$\frac{I_{OFF}}{2}\frac{t}{t_f}, \quad \frac{I_{OFF}}{2}, \quad \frac{I_{OFF}}{2}\exp\left(\frac{t_2 - t}{RC}\right).$$

Giving the power dissipated in this resistance:

$$P_R = fR\left[\int_0^{t_f} \frac{I_{OFF}^2}{4t_f^2}t^2\,dt + \int_{t_f}^{t_2} \frac{I_{OFF}^2}{4}\,dt + \int_{t_2}^{\infty} \frac{I_{OFF}^2}{4}\exp\left(\frac{2(t_2 - t)}{RC}\right)dt\right]$$

$$P_R = f\frac{RI_{OFF}^2}{4}\left(\frac{t_f}{3} + t_2 - t_f + \frac{RC}{2}\right).$$

If the value of t_2 is taken into account, this gives:

$$P_R = f\,R\,\frac{I_{OFF}^2}{4}\left(\frac{2CU}{I_{OFF}} - \frac{t_f}{6} - \frac{RC}{2}\right). \tag{4.5}$$

The losses in the resistance of the RC circuit shunt connected across the terminals of transistor T'_1 or of diode D'_1 have the same value, since the current in this resistance is equal to $C(dv'_1/dt)$ and the following always occurs.

$$C\,dv_1/dt = -\,C\,dv'_1/dt.$$

4.2 Commutations of the Bipolar Transistor in PWM Inverters

In PWM inverters, each controlled switch is opened and closed several times during a same half-cycle of output current i'. Figure 4.4 indicates the inverter configuration and shows the waveform of voltage $v_1 - v_0$ between terminal 1 and the fictitious midpoint 0 of the voltage supply. Transistor T_1 delivers current i' when the latter is positive and $v_1 - v_0$ is equal to $+U/2$; diode D'_1 is conducting during the remainder of the positive half-cycle of i'.

Almost all commutations take place at non-zero current i' and each transistor must be protected against both excessive di/dt and excessive dv/dt. Each half-bridge presents the configuration shown in Fig. 4.5: the transistors are protected against dv/dt by R, C, D_C circuits and against di/dt by r, L, D_L circuits.

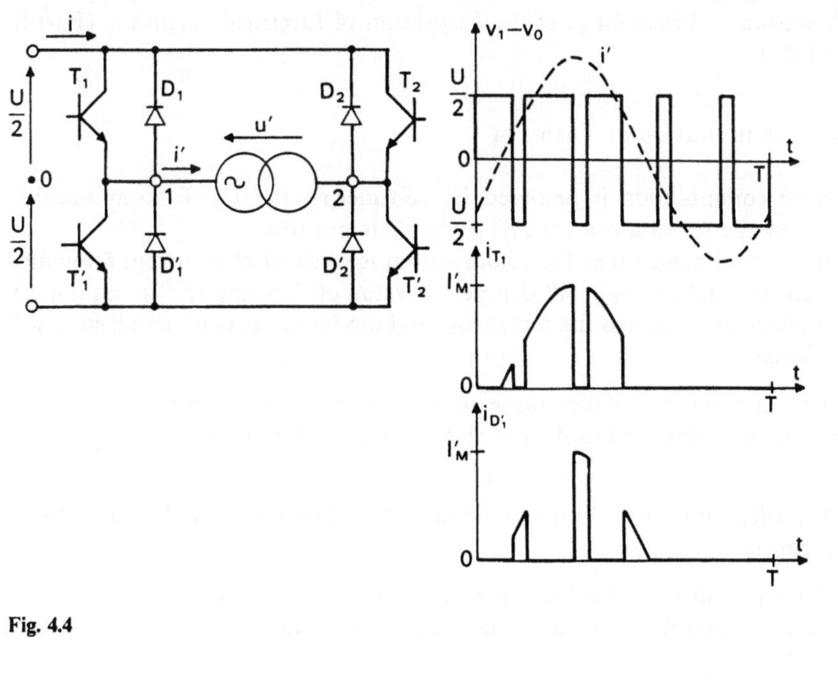

Fig. 4.4

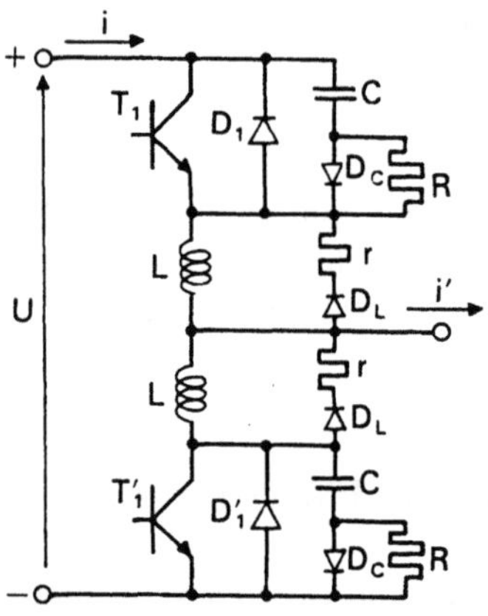

Fig. 4.5

A sufficiently high resistance R is associated to the capacitive snubber in order, firstly, to obtain a time constant RC which is large compared with the switching time of the transistor at turn-on and, secondly, to reduce the initial peak of current U/R. On the other hand, the resistance r associated to the inductive snubber must have a low value in order to limit the peak voltage ri produced by the interruption of current i in the inductance L.

Finally, the time constants RC and L/r are usually close so that the recovery times of the snubbers at turn-on and turn-off are more or less the same.

In studying the commutations, we have used the results published by Evans and Mestah in Proceedings of the Institution of Electrical Engineers (Part B, July 1988).

4.2.1 Commutation at Turn-On

Turn-on commutation is analysed by considering the D'_1–T_1 commutation caused by the turn-on control of T_1 when i' is positive.

It will be assumed that the commutation is so short that current i' remains constant throughout; I_{ON} will denote the value of i' during the commutation. The voltage drops across the transistors and diodes in the conducting state will be ignored.

This commutation takes place in four stages: the elements involved are shown in thick lines in Fig. 4.6 for each stage as well as for the initial and final states.

We will follow the evolution of the variables which cannot suffer discontinuities, namely:

- voltage v_1 and v'_1 across the capacitors of the same capacitance C,
- currents i_1 and i'_1 in the inductances of the same value L.

The expressions of these four variables enable the current and voltage in any branch of the circuit to be obtained at each instant.

At the initial instant, when only D'_1 is conducting:

$$v_1 = U, \qquad v'_1 = 0, \qquad i_1 = 0, \qquad i'_1 = -I_{ON}.$$

Figure 4.7 shows the variations of v_1, of v'_1, of i_1, of i'_1, of v_{T_1} and i_{T_1}, of $v_{D'_1}$ and $i_{D'_1}$.

- *Phase 1 ($0 < t < t_1$): Simultaneous conduction of T_1 and D'_1*

When transistor T_1 becomes conducting, the fall-time of voltage v_{T_1} is ignored.

Capacitor C_1 discharges into R_1 via T_1 (part a in the equivalent circuit of Fig. 4.8) while current i_1 begins to rise (part b in Fig. 4.8).

- The first circuit gives:

$$v_1 = U \exp(-t/RC) \tag{4.6}$$

this equation will remain valid throughout the four phases.

- The second circuit gives:

$$U = L\frac{di_1}{dt} + L\frac{di'_1}{dt}, \quad \text{with } i_1 = i'_1 + I_{ON}$$

thus

$$\frac{di_1}{dt} = \frac{di'_1}{dt}.$$

From $U = 2L\, di_1/dt$ it can be deduced that:

$$i_1 = \frac{U}{2L}t, \qquad i'_1 = \frac{U}{2L}t - I_{ON}. \tag{4.7}$$

The currents in the semiconductor devices, given by:

$$i_{D'_1} = -i'_1 \quad \text{and} \quad i_{T_1} = i_1 - C\frac{dv_1}{dt}$$

are expressed as:

$$i_{D'_1} = I_{ON} - \frac{U}{2L}t, \qquad i_{T_1} = U\left[\frac{t}{2L} + \frac{1}{R}\exp(-t/RC)\right].$$

- Diode D'_1 does not block instantaneously when i'_1 becomes positive; it continues to conduct and the voltage across it can be ignored until the reverse current has reached its maximum value I_{RR}. Phase 1 thus ends at instant t_1,

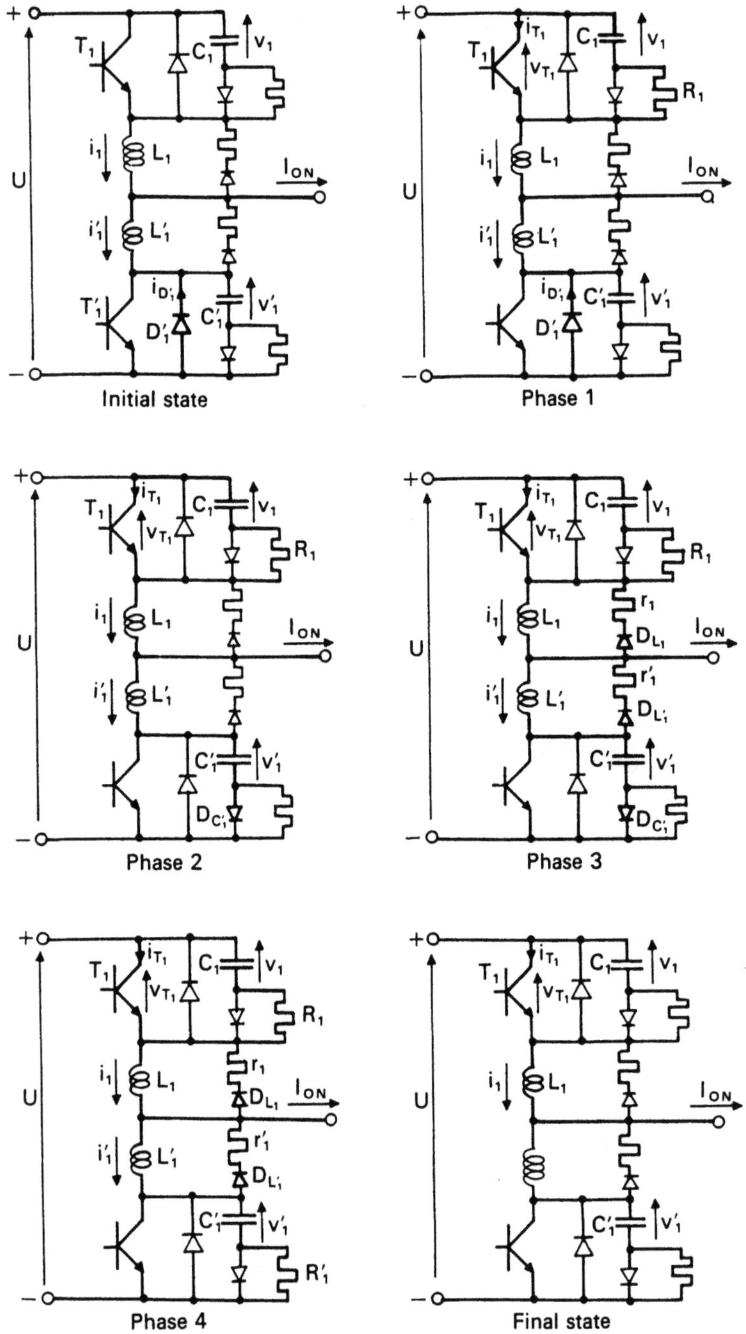

Fig. 4.6. Commutation $D'_1 \rightarrow T_1$

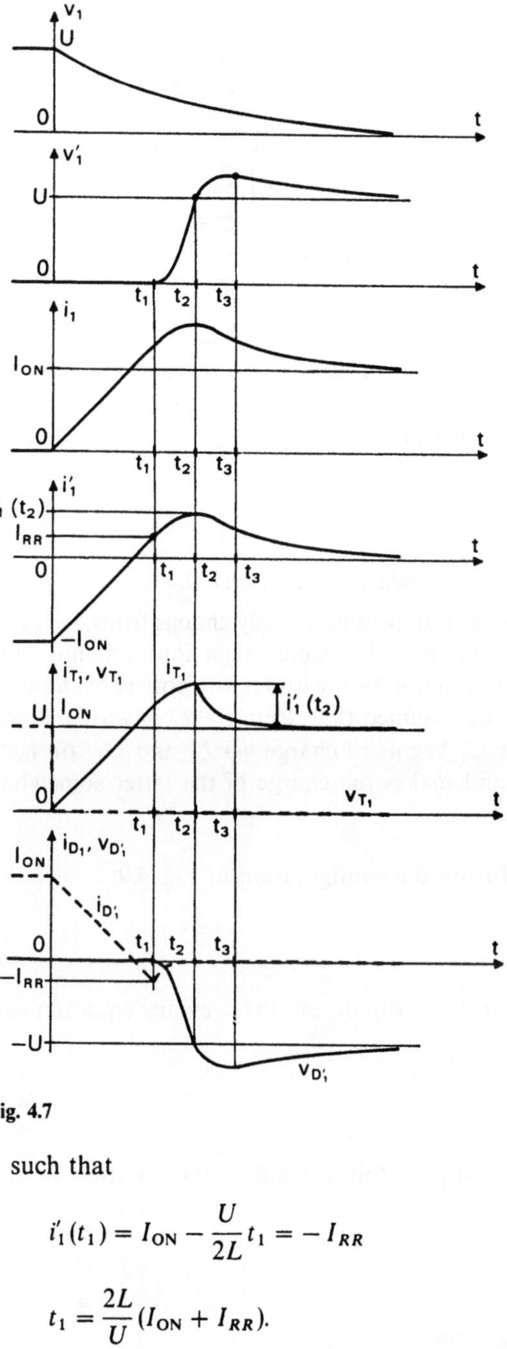

Fig. 4.7

such that

$$i'_1(t_1) = I_{\mathrm{ON}} - \frac{U}{2L} t_1 = -I_{RR}$$

$$t_1 = \frac{2L}{U}(I_{\mathrm{ON}} + I_{RR}).$$

The value of I_{RR} depends on the value of the current to be switched off I_{ON} and on the decrease speed of $-i'_1$, imposed by U and L.

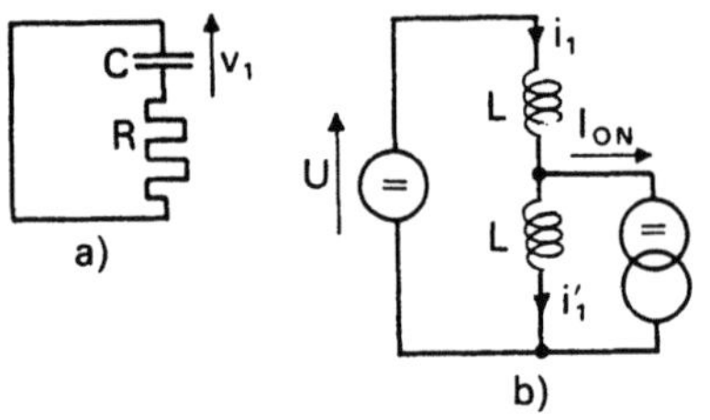

Fig. 4.8

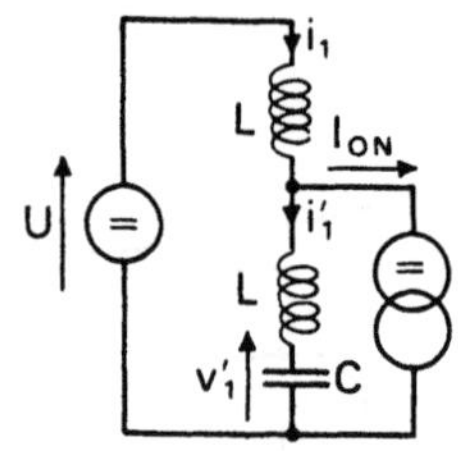

Fig. 4.9

At the end of the first period, there is thus:

$$v_1(t_1) = U \exp\left(-\frac{t_1}{RC}\right), \qquad v'_1(t_1) = 0,$$

$$i_1(t_1) = I_{ON} + I_{RR}, \qquad i'_1(t_1) = I_{RR}.$$

- *Phase 2* ($t_1 < t < t_2$): *Simultaneous conduction of* T_1 *and* $D_{C'_1}$

When D'_1 blocks, current $i_{D'_1}$ does not instantaneously change from $-I_{RR}$ to zero. However if the disappearance of $-I_{RR}$ were taken into account, this would complicate the study of the commutation without any noticeable modification of the results. It will thus be assumed that, from $t = t_1$ onwards, diode D'_1 is blocked and that capacitor C'_1 begins to charge via L'_1 and $D_{C'_1}$ (in fact, $i_{D'_1}$ disappears via $D_{C'_1}$ and C'_1, and makes the charge of the latter somewhat slower).

– The following can be deduced from the configuration in Fig. 4.9:

$$U = L\frac{di_1}{dt} + L\frac{di'_1}{dt} + v'_1.$$

Since $i_1 = i'_1 + I_{ON}$, therefore $di_1/dt = di'_1/dt$ and the previous equation can be written:

$$v'_1 = U - 2L\frac{di'_1}{dt}. \tag{4.8}$$

Since current i'_1 is equal to $C\,dv'_1/dt$, this current is the solution of the differential equation

$$2LC\frac{d^2 i'_1}{dt^2} + i'_1 = 0.$$

The solution has the following form

$$i'_1 = \frac{1}{\omega_2}\frac{di'_1}{dt}(t_1)\sin\omega_2(t - t_1) + i'_1(t_1)\cos\omega_2(t - t_1)$$

with

$$\omega_2 = \frac{1}{\sqrt{2LC}}$$

$$i'_1(t_1) = I_{RR}, \qquad \frac{di'_1}{dt}(t_1) = \frac{U - v'_1(t_1)}{2L} = \frac{U}{2L}$$

giving the expression of i'_1:

$$i'_1 = \frac{U}{2L\omega_2} \sin \omega_2 (t - t_1) + I_{RR} \cos \omega_2 (t - t_1) \tag{4.9}$$

and that of v'_1:

$$v'_1 = U - 2L \frac{di'_1}{dt} = U[1 - \cos \omega_2 (t - t_1)] + 2L\omega_2 I_{RR} \sin \omega_2 (t - t_1). \tag{4.10}$$

Currents i_1 and i_{T_1} can be deduced from i'_1 by

$$i_1 = I_{ON} + i'_1$$

$$i_{T_1} = i_1 - C \frac{dv_1}{dt} = i_1 + \frac{U}{R} \exp\left(-\frac{t}{RC} \right).$$

- Since $di_1/dt = di'_1/dt$, the voltages across both inductors are equal. When voltage v'_1 becomes greater than U, Eq. (4.8) shows that $L\, di'_1/dt$ becomes negative and that diodes D_{L_1} and $D_{L'_1}$ start conducting. This occurs for $t = t_2$, such that:

$$U[1 - \cos \omega_2 (t_2 - t_1)] + 2L\omega_2 I_{RR} \sin \omega_2 (t_2 - t_1) = U$$

$$t_2 = t_1 + \frac{1}{\omega_2} \operatorname{Arc\,tan} \frac{U}{2L\omega_2 I_{RR}}$$

$$= \frac{2L}{U}(I_{ON} + I_{RR}) + \sqrt{2LC} \operatorname{Arc\,tan} \frac{U}{\sqrt{\frac{2L}{C}} I_{RR}}.$$

At the end of this stage, on account of the value of t_2, the following is obtained:

$$v'_1(t_2) = U, \quad i'_1(t_2) = \sqrt{\frac{C}{2L} U^2 + I_{RR}^2}, \quad i_1(t_2) = I_{ON} + i'_1(t_2).$$

- The discharge currents of C_1 and C'_1 are added to I_{ON}, leading to an *over-current* in the transistor. Current i_{T_1} has its maximum value at the end of phase 2 and is then equal to:

$$i_{T_1 \max} = i_1(t_2) + \frac{U}{R} \exp\left(-\frac{t_2}{RC} \right)$$

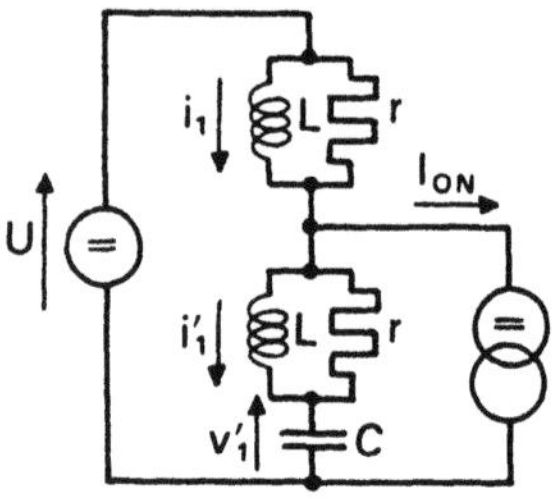

Fig. 4.10

$$i_{T_1\,max} = I_{ON} + \sqrt{\frac{C}{2L}U^2 + I_{RR}^2} + \frac{U}{R}\exp\left(-\frac{t_2}{RC}\right). \tag{4.11}$$

- *Phase 3* ($t_2 < t < t_3$): *Simultaneous conduction of* T_1, $D_{C'_1}$, D_{L_1} *and* $D_{L'_1}$

– The equivalent circuit for computing i_1, i'_1 and v'_1 is now that shown in Fig. 4.10.

The following can be written:

$$i_1 + \frac{L}{r}\frac{di_1}{dt} = i'_1 + \frac{L}{r}\frac{di'_1}{dt} + I_{ON}$$

$$\frac{L}{r}\frac{d(i_1 - i'_1)}{dt} + (i_1 - i'_1) = I_{ON}.$$

The difference $(i_1 - i'_1)$ thus has the form $I_{ON} + A\exp[-r(t - t_2)/L]$; but the integration constant A is zero since $i_1(t_2) - i'_1(t_2)$ is equal to I_{ON}. In spite of D_{L_1} and $D_{L'_1}$ conducting, this once more gives

$$i_1 - i'_1 = I_{ON} \tag{4.12}$$

– Furthermore, since $di_1/dt = di'_1/dt$,

$$v'_1 = U - 2L\frac{di'_1}{dt} \tag{4.13}$$

and

$$i'_1 + \frac{L}{r}\frac{di'_1}{dt} = C\frac{dv'_1}{dt} = -2LC\frac{d^2i'_1}{dt^2}$$

This provides the differential equation giving i'_1:

$$2LC\frac{d^2i'_1}{dt^2} + \frac{L}{r}\frac{di'_1}{dt} + i'_1 = 0$$

Denoting

$$\omega_3 = \frac{1}{\sqrt{2LC}} \quad \text{and} \quad \alpha_3 = \frac{1}{4rC}$$

current i'_1 has the following expression[1]:

- If $\omega_3 > \alpha_3$

$$i'_1 = \frac{i'_1(t_2)}{\sin \varphi_3} \exp[-\alpha_3(t - t_2)] \sin[\beta_3(t - t_2) + \varphi_3] \tag{4.14}$$

with $\beta_3 = \sqrt{\omega_3^2 - \alpha_3^2}$ and $\tan \varphi_3 = \dfrac{\beta_3}{\alpha_3}$

- If $\omega_3 < \alpha_3$

$$i'_1 = \frac{i'_1(t_2)}{\mathrm{sh}\, \varphi_3} \exp[-\alpha_3(t - t_2)] \,\mathrm{sh}[\beta_3(t - t_2) + \varphi_3] \tag{4.14'}$$

with $\beta_3 = \sqrt{\alpha_3^2 - \omega_3^2}$ and $\mathrm{th}\, \varphi_3 = \dfrac{\beta_3}{\alpha_3}$

- Current i_1 can be deduced from i'_1 using Eq. (4.12). Equation (4.13) enables the expression of v'_1 to be laid down; that of charge current $i_{C'_1}$ of capacitor C'_1 can be deduced from v'_1 by.

$$i_{C'_1} = C\, \mathrm{d}v'_1/\mathrm{d}t$$

The following is thus obtained:

$$v'_1 = U + 2L\omega_3 \frac{i'_1(t_2)}{\sin \varphi_3} \exp[-\alpha_3(t - t_2)] \sin \beta_3(t - t_2) \tag{4.15}$$

$$i_{C'_1} = -2LC\omega_3^2 \frac{i'_1(t_2)}{\sin \varphi_3} \exp[-\alpha_3(t - t_2)] \sin[\beta_3(t - t_2) - \varphi_3]$$

- Phase 3 ends when $i_{C'_1}$ changes its polarity, leading to the turn-off of $D_{C'_1}$. When $\mathrm{d}v'_1/\mathrm{d}t$ falls to zero, this gives:

$$t_3 = t_2 + \frac{\varphi_3}{\beta_3}$$

At the end of this period,

$$i'_1(t_3) = \frac{i'_1(t_2)}{\sin \varphi_3} \exp[-\alpha_3(t_3 - t_2)] \sin[\beta_3(t_3 - t_2) + \varphi_3]$$

$$= \frac{1}{2r} \sqrt{U^2 + \frac{2L}{C} I_{RR}^2} \exp\left(\frac{-\alpha_3 \varphi_3}{\beta_3}\right)$$

$$i_1(t_3) = I_{ON} + i'_1(t_3)$$

$$v'_1(t_3) = U + 2r i'_1(t_3)$$

[1] In the following section, only the case where ω_3 is greater than α_3 will be studied. In the opposite case ($\omega_3 < \alpha_3$), the trigonometrical functions merely have to be replaced by the hyperbolic functions.

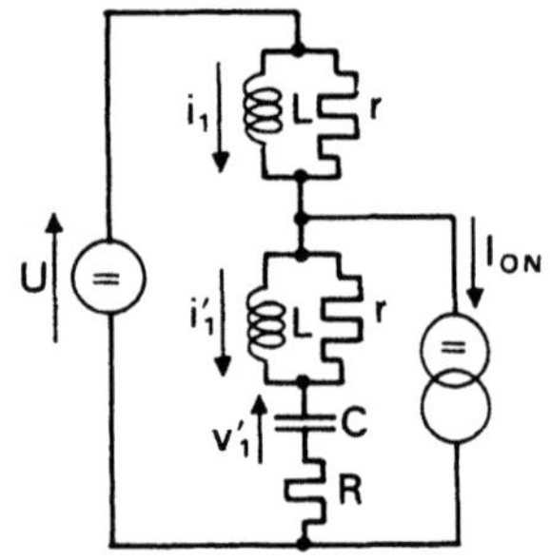

Fig. 4.11

- *Phase 4 ($t > t_3$): Simultaneous conduction of T_1, D_{L1} and $D_{L'1}$*

The circuit enabling i_1, i'_1 and v'_1 to be computed is now that of Fig. 4.11.

- As in phase 3, from the identity:

$$i_1 + \frac{L}{r}\frac{di_1}{dt} = i'_1 + \frac{L}{r}\frac{di'_1}{dt} + I_{ON}, \quad \text{with } i_1(t_3) - i'_1(t_3) = I_{ON}$$

the following can once more be deduced:

$$i_1 - i'_1 = I_{ON}$$

- From

$$U = 2L\frac{di'_1}{dt} + v'_1 + RC\frac{dv'_1}{dt},$$

with

$$i'_1 + \frac{L}{r}\frac{di'_1}{dt} = C\frac{dv'_1}{dt}$$

the following can be obtained, after derivation,

$$LC\left(2 + \frac{R}{r}\right)\frac{d^2 i'_1}{dt^2} + \left(RC + \frac{L}{r}\right)\frac{di'_1}{dt} + i'_1 = 0$$

Denoting

$$\omega_4 = \frac{1}{\sqrt{LC\left(2 + \dfrac{R}{r}\right)}}, \qquad \alpha_4 = \frac{1}{2}\frac{RC + \dfrac{L}{r}}{LC\left(2 + \dfrac{R}{r}\right)}$$

the expression of i'_1 can be obtained:

- If $\omega_4 > \alpha_4$

$$i'_1 = \frac{i'_1(t_3)}{\sin\varphi_4}\exp[-\alpha_4(t - t_3)]\sin[\beta_4(t - t_3) + \varphi_4] \tag{4.16}$$

with

$$\beta_4 = \sqrt{\omega_4^2 - \alpha_4^2}, \qquad \tan\varphi_4 = \frac{\beta_4}{\alpha_4 - \dfrac{r}{L}}$$

- If $\omega_4 < \alpha_4$

$$i'_1 = \frac{i'_1(t_3)}{\operatorname{sh}\varphi_4}\exp[-\alpha_4(t-t_3)]\operatorname{sh}[\beta_4(t-t_3)+\varphi_4] \tag{4.16'}$$

with

$$\beta_4 = \sqrt{\alpha_4^2 - \omega_4^2}, \qquad \operatorname{th}\varphi_4 = \frac{\beta_4}{\alpha_4 - \dfrac{r}{L}}$$

Calculating v'_1 by means of

$$v'_1 = U - Ri'_1 - L\left(2 + \frac{R}{r}\right)\frac{di'_1}{dt}$$

gives, for the case where α_4 is less than ω_4

$$v'_1 = U - Ri'_1\left[1 + \frac{L}{R}\left(2 + \frac{R}{r}\right)(-\alpha_4 + \beta_4 \operatorname{cotan}[\beta_4(t-t_3)+\varphi_4])\right] \tag{4.17}$$

– When t tends towards infinity, i'_1 tends towards zero,
i_1 tends towards I_{ON},
and v'_1 tends towards U;

transistor T_1 is the only conducting device and commutation is ended.

4.2.2 Commutation at Turn-Off

Turn-off commutations are studied by considering the switching from T_1 to D'_1, caused by the suppression of the control signal on the base of T_1, when current i' is positive. I_{OFF} denotes the value of i' – assumed to be constant during the commutation.

This commutation occurs in 4 phases, represented in Fig. 4.12. Figure 4.13 gives the waveforms of v_1, v'_1, i_1, i'_1, v_{T_1} and i_{T_1}, $v_{D'_1}$ and $i_{D'_1}$.

The initial conditions are:

$$v_1(0) = 0, \quad v'_1(0) = U, \qquad i_1(0) = I_{OFF}, \qquad i'_1(0) = 0.$$

- *Phase 1* ($0 < t < t_f$): *Simultaneous conduction of* T_1, D_{C_1}, D_{L_1} *and* $D_{L'_1}$

When the control signal of T_1 disappears, current i_{T_1} does not suddenly revert to zero. It will be assumed, as for the commutation without PWM, that

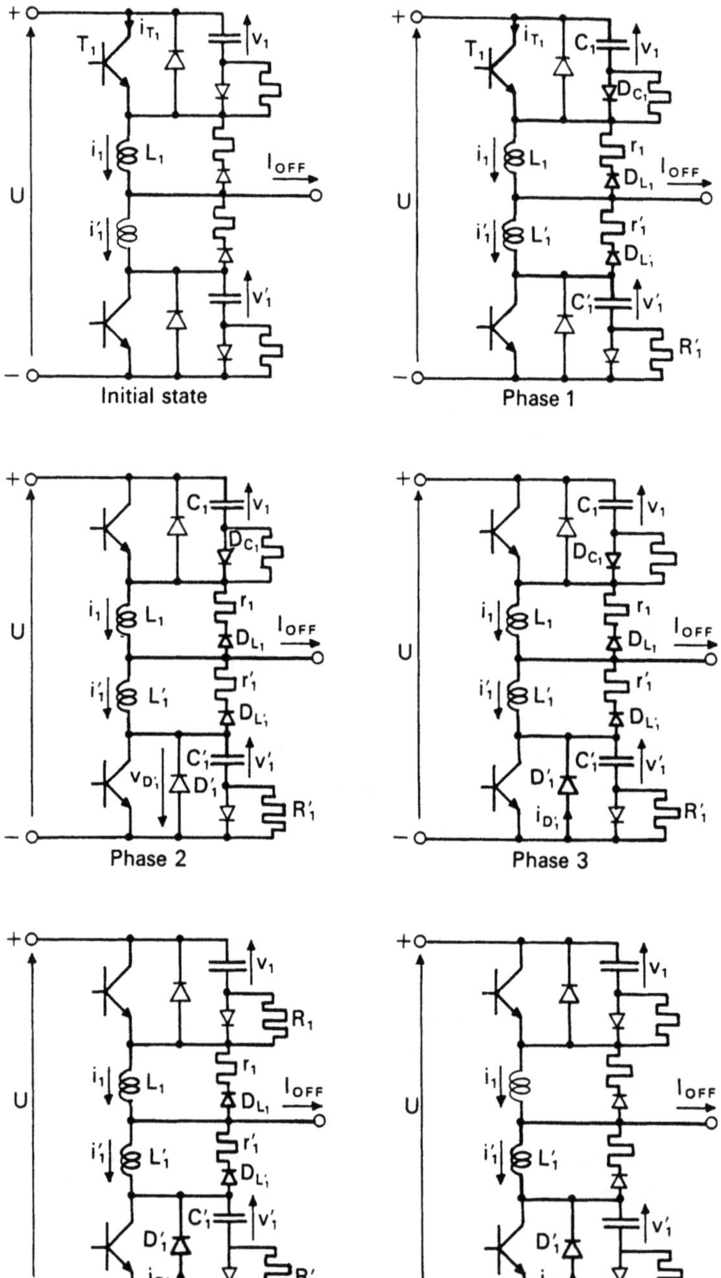

Fig. 4.12. Commutation $T_1 \rightarrow D'_1$

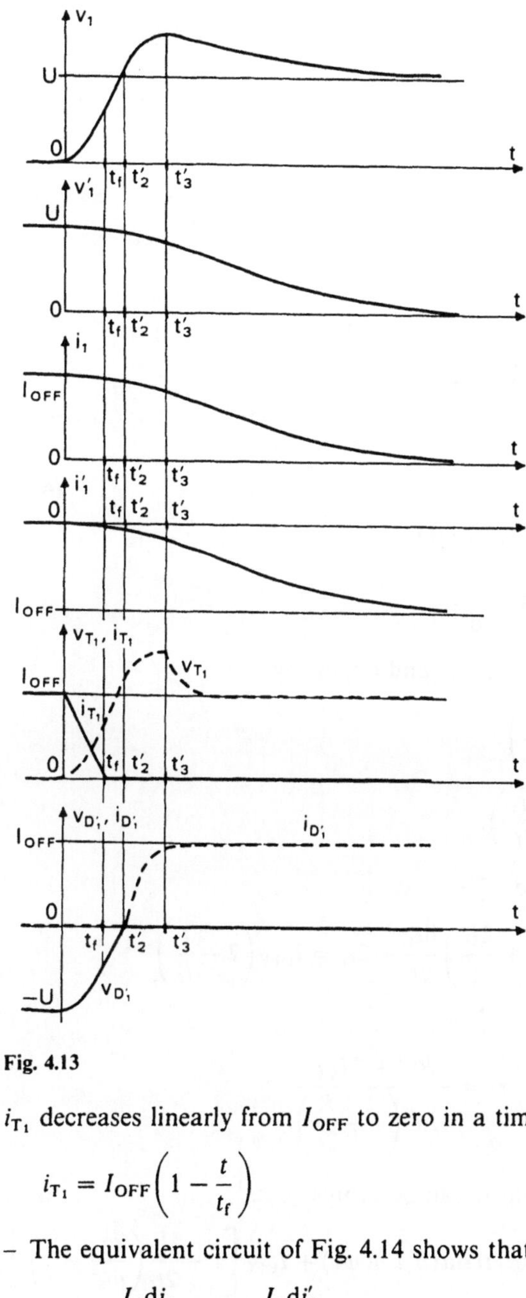

Fig. 4.13

i_{T_1} decreases linearly from I_{OFF} to zero in a time t_f.

$$i_{T_1} = I_{OFF}\left(1 - \frac{t}{t_f}\right)$$

– The equivalent circuit of Fig. 4.14 shows that there is the following

$$i_1 + \frac{L}{r}\frac{di_1}{dt} = i'_1 + \frac{L}{r}\frac{di'_1}{dt} + I_{OFF}$$

$$(i_1 - i'_1) + \frac{L}{r}\frac{d}{dt}(i_1 - i'_1) = I_{OFF}, \quad \text{with } i_1(0) - i'_1(0) = I_{OFF}$$

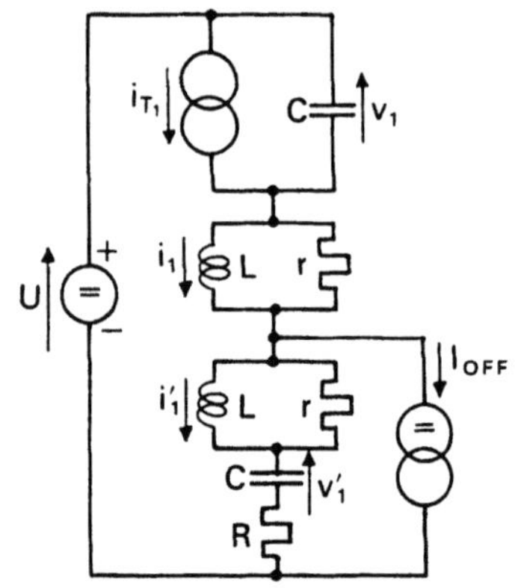

Fig. 4.14

giving:

$$i_1 - i'_1 = I_{OFF} \tag{4.18}$$

The voltage equation can be written:

$$U = v_1 + 2L\frac{di_1}{dt} + v'_1 + RC\frac{dv'_1}{dt}$$

By deriving and replacing dv_1/dt and dv'_1/dt by:

$$\frac{dv_1}{dt} = \frac{1}{C}\left(i_1 + \frac{L}{r}\frac{di_1}{dt} - i_{T_1}\right)$$

$$\frac{dv'_1}{dt} = \frac{1}{C}\left(i_1 - I_{OFF} + \frac{L}{r}\frac{di_1}{dt}\right),$$

the following can be obtained

$$LC\left(2 + \frac{R}{r}\right)\frac{d^2 i_1}{dt^2} + \left(RC + \frac{2L}{r}\right)\frac{di_1}{dt} + 2i_1 = I_{OFF}\left(2 - \frac{t}{t_f}\right)$$

Denoting

$$\omega'_1 = \frac{1}{\sqrt{LC\left(1 + \dfrac{R}{2r}\right)}}, \qquad \alpha'_1 = \frac{RC + 2L/r}{4LC\left(1 + \dfrac{R}{2r}\right)}$$

if ω'_1 is greater than α'_1 current i_1 can be expressed as[1]:

$$i_1 = -\frac{\alpha'_1}{\omega_1'^2 t_f}\frac{I_{OFF}}{\sin\varphi'_1}\exp(-\alpha'_1 t)\sin(\beta'_1 t + \varphi'_1) + I_{OFF}\left[1 + \frac{1}{2t_f}\left(\frac{2\alpha'_1}{\omega_1'^2} - t\right)\right]$$

---

[1] As previously stated, the case in which α'_1 is greater than ω'_1 will not be studied. The expressions which correspond to ω'_1 lower than α'_1 are obtained by replacing the trigonometrical functions by the hyperbolic ones.

with

$$\beta'_1 = \sqrt{\omega_1'^2 - \alpha_1'^2}, \qquad \tan\varphi'_1 = \frac{2\alpha'_1\beta'_1}{2\alpha_1'^2 - \omega_1'^2}. \tag{4.19}$$

– The following two equations are available to calculate v_1 and v'_1:

• $$C\frac{dv_1}{dt} - C\frac{dv'_1}{dt} = \left(i_1 + \frac{L}{r}\frac{di_1}{dt} - i_{T_1}\right) - \left(i'_1 + \frac{L}{r}\frac{di'_1}{dt}\right) = I_{OFF}\frac{t}{t_f}$$

thus

$$v_1 - v'_1 = \frac{I_{OFF}}{Ct_f}\frac{t^2}{2} + v_1(0) - v'_1(0) = \frac{I_{OFF}}{2Ct_f}t^2 - U$$

• $$v_1 + v'_1 = U - 2L\frac{di_1}{dt} - RC\frac{dv'_1}{dt} = U - 2L\frac{di_1}{dt} - R\left(i'_1 + \frac{L}{r}\frac{di'_1}{dt}\right)$$

thus

$$v_1 + v'_1 = U - R(i_1 - I_{OFF}) - 2L\left(1 + \frac{R}{2r}\right)\frac{di_1}{dt}.$$

From this, the following is deduced:

$$v_1 = \frac{1}{2}\left[\frac{I_{OFF}}{2Ct_f}t^2 - R(i_1 - I_{OFF}) - 2L\left(1 + \frac{R}{2r}\right)\frac{di_1}{dt}\right] \tag{4.20}$$

$$v'_1 = \frac{1}{2}\left[2U - \frac{I_{OFF}}{2Ct_f}t^2 - R(i_1 - I_{OFF}) - 2L\left(1 + \frac{R}{2r}\right)\frac{di_1}{dt}\right]. \tag{4.21}$$

– The first phase ends at $t = t_f$ when i_{T_1} falls to zero. This then gives:

$$i_1(t_f) = -\frac{\alpha'_1}{\omega_1'^2 t_f}\frac{I_{OFF}}{\sin\varphi'_1}\exp(-\alpha'_1 t_f)\sin(\beta'_1 t_f + \varphi'_1) + \frac{I_{OFF}}{2}\left(1 + \frac{2\alpha'_1}{\omega_1'^2 t_f}\right)$$

$$i'_1(t_f) = i_1(t_f) - I_{OFF}$$

$$v_1(t_f) = \frac{1}{2}\left[\frac{I_{OFF}}{2C}t_f - Ri'_1(t_f) - 2L\left(1 + \frac{R}{2r}\right)\frac{di_1}{dt}(t_f)\right]$$

$$v'_1(t_f) = U + v_1(t_f) - \frac{I_{OFF}}{2C}t_f$$

It should be noted that, during this first phase, the variations of i_1, i'_1 and v'_1 are very slight. It is possible to take

$$i_1(t_f) \simeq I_{OFF}, \qquad i'_1(t_f) \simeq 0, \qquad v'_1(t_f) \simeq U, \qquad v_1(t_f) \simeq \frac{I_{OFF}}{2C}t_f$$

• *Phase 2 ($t_f < t < t'_2$): Simultaneous conduction of D_{C_1}, D_{L_1} and $D_{L'_1}$*

The equivalent circuit is that shown in Fig. 4.14 except that i_{T_1} is now zero. The equations are the same as those of the first period, with, however, $i_{T_1} = 0$.

Current i_1 is solution of

$$LC\left(2+\frac{R}{r}\right)\frac{d^2 i_1}{dt^2}+\left(RC+\frac{2L}{r}\right)\frac{di_1}{dt}+2i_1=I_{OFF}$$

Giving, if $\omega'_1 > \alpha'_1$,

$$\begin{aligned} i_1 &= \frac{i_1(t_f)-I_{OFF}/2}{\sin\varphi'_2}\exp[-\alpha'_1(t-t_f)]\sin[\beta'_1(t-t_f)+\varphi'_2]+\frac{I_{OFF}}{2} \\ &\simeq \frac{I_{OFF}}{2}\left[1+\frac{\exp(-\alpha'_1(t-t_f))]}{\sin\varphi'_2}\sin[\beta'_1(t-t_f)+\varphi'_2]\right] \end{aligned} \tag{4.22}$$

with

$$\tan\varphi'_2=\beta'_1\frac{i_1(t_f)-I_{OFF}/2}{\alpha'_1\left[i_1(t_f)-\dfrac{I_{OFF}}{2}\right]+\dfrac{di_1}{dt}(t_f)}\simeq\frac{\beta'_1}{\alpha'_1+\dfrac{2}{I_{OFF}}\dfrac{di_1}{dt}(t_f)}$$

- Current i'_1 is once more given by

$$i'_1=i_1-I_{OFF}$$

Voltages v_1 and v'_1 are computed from

$$C\frac{d}{dt}(v_1-v'_1)=i_1-i'_1=I_{OFF}$$

$$v_1+v'_1=U-Ri'_1-2L\left(1+\frac{R}{2r}\right)\frac{di_1}{dt}$$

Giving

$$v_1=\frac{1}{2}\left[\frac{I_{OFF}}{C}\left(t-\frac{t_f}{2}\right)-Ri'_1-2L\left(1+\frac{R}{2r}\right)\frac{di_1}{dt}\right] \tag{4.23}$$

$$v'_1=\frac{1}{2}\left[2U-\frac{I_{OFF}}{C}\left(t-\frac{t_f}{2}\right)-Ri'_1-2L\left(1+\frac{R}{2r}\right)\frac{di_1}{dt}\right] \tag{4.24}$$

- This phase ends, for $t=t'_2$, when voltage $v_{D'_1}$ across diode D'_1 becomes positive and this diode thus starts conducting. The value of t'_2 is such that

$$v_1(t'_2)+2L\frac{di_1}{dt}(t'_2)=U$$

or

$$\frac{I_{OFF}}{2C}\left(t'_2-\frac{t_f}{2}\right)-\frac{Ri'_1(t'_2)}{2}+L\left(1-\frac{R}{2r}\right)\frac{di_1}{dt}(t'_2)=U$$

i'_1 and di_1/dt can be replaced by their values in $t=t'_2$, but this does not simplify matters. There is no analytical solution to this equation and the value of t'_2 must be computed by a numerical method. At the end of the second phase, the values of the variables are noted as follows:

$$i_1(t'_2), \qquad i'_1(t'_2), \qquad v_1(t'_2), \quad v'_1(t'_2).$$

- *Phase 3* ($t'_2 < t < t'_3$): *Simultaneous conduction of* D_{C_1}, D_{L_1}, $D_{L'_1}$ *and* D'_1

Figure 4.15 provides the equivalent circuits – one for the computation of i_1, i'_1 and v_1, and the other for v'_1.

– The second circuit gives:

$$v'_1 = v'_1(t'_2)\exp\left(-\frac{t-t'_2}{RC}\right) \tag{4.25}$$

– The first circuit shows that, as in the previous phases and for the same reasons, the following occurs:

$$i'_1 = i_1 - I_{OFF} \quad \text{and} \quad \frac{di_1}{dt} = \frac{di'_1}{dt}$$

Since

$$U = v_1 + 2L\frac{di_1}{dt} \quad \text{with} \quad C\frac{dv_1}{dt} = i_1 + \frac{L}{r}\frac{di_1}{dt},$$

current i_1 is solution of the differential equation:

$$2LC\frac{d^2i_1}{dt^2} + \frac{L}{r}\frac{di_1}{dt} + i_1 = 0$$

By using the parameters ω_3, α_3 and β_3 defined when studying phase 3 of the commutation at turn-on, this gives:

if $\omega_3 > \alpha_3$

$$i_1 = \frac{i_1(t'_2)}{\sin\varphi'_3}\exp[-\alpha_3(t-t'_2)]\sin[\beta_3(t-t'_2)+\varphi'_3] \tag{4.26}$$

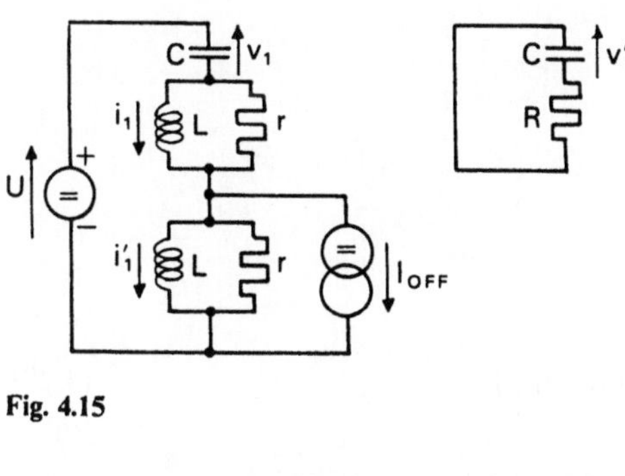

Fig. 4.15

with

$$\tan \varphi_3' = \frac{\beta_3 i_1(t_2')}{\alpha_3 i_1(t_2') + \frac{di_1}{dt}(t_2')}$$

- From $v_1 = U - 2L\frac{di_1}{dt}$, it can be deduced that:

$$v_1 = U - 2Li_1(-\alpha_3 + \beta_3 \operatorname{cotan}[\beta_3(t - t_2') + \varphi_3']) \tag{4.27}$$

- This phase ends when the current in diode D_{C_1} changes polarity, leading to this diode being blocked. This occurs for $t = t_3'$, such that:

$$i_1(t_3') + \frac{L}{r}\frac{di_1}{dt}(t_3') = 0$$

giving:

$$t_3' = t_2' + \frac{1}{\beta_3}\left[\operatorname{Arctan}\left(\frac{\beta_3}{\alpha_3 - (r/L)}\right) - \varphi_3'\right]$$

At the end of the third period,

$$i_1(t_3') = \beta_3 \frac{L}{r} \frac{i_1(t_2')}{\sin \varphi_3'} \exp[-\alpha_3(t_3' - t_2')]$$

$$i_1'(t_3') = i_1(t_3') - I_{OFF}$$

$$v_1(t_3') = U + 2ri_1(t_3')$$

- It is at the end of the third phase that voltage v_{T_1} across transistor T_1 has its maximum value. The *peak voltage* is given by

$$v_{T_1 \max} = v_1(t_3') = U + 2\beta_3 L \frac{i_1(t_2')}{\sin \varphi_3'} \exp[-\alpha_3(t_3' - t_2')]$$

An approximate value of this maximum may be obtained by assuming that current i_1' is still equal to I_{OFF} for $t = t_2'$. This then gives

$$\tan \varphi_3' \simeq \frac{\beta_3}{\alpha_3} = \varphi_3,$$

yielding

$$t_3' - t_2' \simeq \frac{\varphi_3}{\beta_3}$$

and

$$v_{T_1 \max} \simeq U + \sqrt{\frac{2L}{C}} I_{OFF} \exp(-\alpha_3 \varphi_3/\beta_3) \tag{4.28}$$

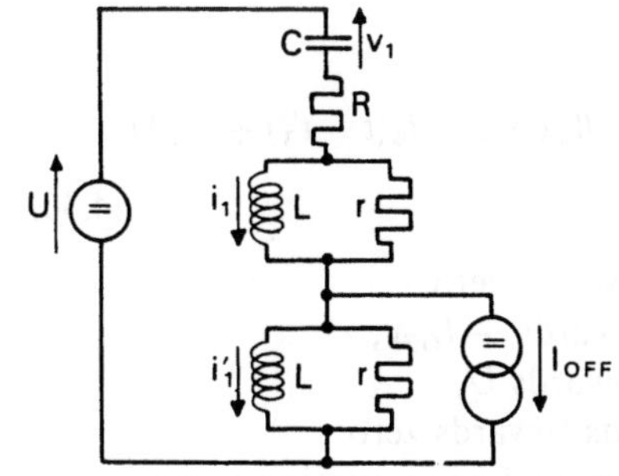

Fig. 4.16

- *Phase 4 ($t > t'_3$): Simultaneous conduction of D_{L_1}, $D_{L'_1}$ and D'_1*

The equivalent circuit for computing i_1, i'_1 and v_1 is that shown in Fig. 4.16.

- As in the previous phases, the following is obtained:

$$i'_1 = i_1 - I_{OFF} \quad \text{and} \quad \frac{di_1}{dt} = \frac{di'_1}{dt}$$

- From

$$U = v_1 + RC\frac{dv_1}{dt} + 2L\frac{di_1}{dt}$$

and

$$C\frac{dv_1}{dt} = i_1 + \frac{L}{r}\frac{di_1}{dt}$$

the following can be deduced:

$$2LC\left(1 + \frac{R}{2r}\right)\frac{d^2i_1}{dt^2} + \left(RC + \frac{L}{r}\right)\frac{di_1}{dt} + i_1 = 0$$

Using the parameters ω_4, α_4, β_4 and φ_4 as defined during phase 4 of the turn-on commutation, the following can be obtained:

if $\omega_4 > \alpha_4$,

$$i_1 = \frac{i_1(t'_3)}{\sin\varphi_4}\exp[-\alpha_4(t - t'_3)]\sin[\beta_4(t - t'_3) + \varphi_4] \tag{4.29}$$

The expression of v_1 can be calculated by

$$v_1 = U - RC\frac{dv_1}{dt} - 2L\frac{di_1}{dt} = U - Ri_1 - 2L\left(1 + \frac{R}{2r}\right)\frac{di_1}{dt}$$

which gives:

$$v_1 = U - Ri_1 - 2L\left(1 + \frac{R}{2r}\right)i_1(-\alpha_4 + \beta_4 \operatorname{cotan}[\beta_4(t - t'_3) + \varphi_4]) \tag{4.30}$$

– When t tends towards infinity, i_1 tends towards zero,
i'_1 tends towards $-I_{OFF}$,
v_1 tends towards U,
and v'_1 tends towards zero;
diode D'_1 is now the only conducting device and commutation is over.

4.2.3 Power Dissipated in the Resistances

The energy dissipated in the snubber resistances during each commutation is evaluated first of all. Next, the average power dissipated in these elements is obtained by taking into account all the commutations which must be carried out per output voltage cycle.

4.2.3.1 Energy Dissipated in R_1

- *During the turn-on commutation* $D'_1 \to T_1$, the capacitor discharges into resistance R_1 via transistor T_1. If the voltage drop in the latter is ignored, the energy dissipated in R_1 has the following value:

$$W_{R_1\,ON} = \frac{1}{2}CU^2;$$

it does not depend on the switched current.

- *During the turn-off commutation* $T_1 \to D'_1$, resistance R_1 is only crossed by a current during phase 4. This current can be expressed as:

$$C\frac{dv_1}{dt} = i_1 + \frac{L}{r}\frac{di_1}{dt}$$

with i_1 being given by Eq. (4.29).

$$C\frac{dv_1}{dt} = \frac{i_1(t'_3)}{\sin\varphi_4}\left(\left(1 - \alpha_4\frac{L}{r}\right)\sin[\beta_4(t - t'_3) + \varphi_4] + \beta_4\frac{L}{r}\cos[\beta_4(t - t'_3) + \varphi_4]\right)\exp[-\alpha_4(t - t'_3)]$$

By developing the sine and cosine and taking into account the value of $\tan\varphi_4$, this gives, after simplifying:

$$C\frac{dv_1}{dt} = -\beta_4\frac{L}{r}\frac{i_1(t'_3)}{\sin^2\varphi_4}\exp[-\alpha_4(t - t'_3)]\sin\beta_4(t - t'_3)$$

If current $C\,\mathrm{d}v_1/\mathrm{d}t$ is admitted to have reached a negligible value before the next commutation, the energy dissipated in R_1 is:

$$W_{R_1\,\mathrm{OFF}} = \int_{t'_3}^{\infty} R\left(C\frac{\mathrm{d}v_1}{\mathrm{d}t}\right)^2 \mathrm{d}t$$

$$= \int_0^{\infty} R\left(\beta_4 \frac{L}{r}\frac{i_1(t'_3)}{\sin^2\varphi_4}\right)^2 \exp(-2\alpha_4 t)\sin^2\beta_4 t\,\mathrm{d}t$$

$$= \frac{R}{4}\left(\beta_4\frac{L}{r}\frac{i_1(t'_3)}{\sin^2\varphi_4}\right)^2 \frac{\beta_4^2}{\alpha_4\omega_4^2}$$

$$= \frac{R}{4}\frac{r^2}{L^2} i_1^2(t'_3)\frac{1}{\alpha_4\omega_4^2}\left[\omega_4^2\frac{L^2}{r^2} - 2\alpha_4\frac{L}{r} + 1\right]^2.$$

Denoting:

$$\tau_C = RC \quad \text{and} \quad \tau_L = \frac{L}{r} \tag{4.31}$$

therefore:

$$\alpha_4 = \frac{\tau_C + \tau_L}{2\tau_C\tau_L\left(1 + 2\dfrac{r}{L}\right)}, \qquad \omega_4^2 = \frac{1}{\tau_C\tau_L\left(1 + 2\dfrac{r}{R}\right)}.$$

By transferring this into the expression of $W_{R_1\,\mathrm{OFF}}$, the latter becomes:

$$W_{R_1\,\mathrm{OFF}} = 2\frac{r^2}{R}\frac{\tau_C^2}{\tau_C + \tau_L} i_1^2(t'_3).$$

If the same simplifying hypothesis is made as in the computation of the voltage peak across T_1, this gives:

$$i_1(t'_3) \simeq \sqrt{\frac{2L}{C}}\frac{I_{\mathrm{OFF}}}{2r}\exp(-\alpha_3\varphi_3/\beta_3)$$

thus:

$$W_{R_1\,\mathrm{OFF}} = L\frac{1}{1 + \dfrac{\tau_L}{\tau_C}} I_{\mathrm{OFF}}^2 \exp(-2\alpha_3\varphi_3/\beta_3).$$

Time constants τ_L and τ_C have frequently very close values so that the recovery time of the snubbers at turn-on and turn-off are similar. In this case, the following can be written:

$$W_{R_1\,\mathrm{OFF}} \simeq \frac{1}{2} L I_{\mathrm{OFF}}^2 \exp(-2\alpha_3\varphi_3/\beta_3).$$

The energy dissipated in R_1 is proportional to the energy stored in inductance L during the commutation.

4.2.3.2 Energy Dissipated in R'_1

- *During the turn-on commutation* $D'_1 \rightarrow T_1$, resistance R'_1 is crossed by the discharge current of C'_1 throughout phase 4.

  This current is equal to:

$$C\frac{dv'_1}{dt} = i'_1 + \frac{L}{r}\frac{di'_1}{dt}$$

or, taking (4.16) into account,

$$C\frac{dv'_1}{dt} = -\beta_4 \frac{L}{r}\frac{i'_1(t_3)}{\sin^2\varphi_4}\exp[-\alpha_4(t-t_3)]\sin\beta_4(t-t_3)$$

which is an expression similar to that previously obtained for $C(dv_1/dt)$.

Computing $W_{R_2\,\mathrm{ON}}$ by

$$W_{R_2\,\mathrm{ON}} = \int_{t_3}^{\infty} R\left(C\frac{dv'_1}{dt}\right)^2 dt$$

similarly gives:

$$W_{R_2\,\mathrm{ON}} = \frac{2r^2}{R}\frac{\tau_C^2}{\tau_C+\tau_L} i'^2_1(t_3).$$

Since

$$i'_1(t_3) = \frac{1}{2r}\sqrt{U^2 + \frac{2L}{C}I_{RR}^2}\exp\left(\frac{-\alpha_3\varphi_3}{\beta_3}\right)$$

$$W_{R_2\,\mathrm{ON}} = \frac{1}{2}C\frac{1}{1+\tau_L/\tau_C}\left(U^2 + \frac{2L}{C}I_{RR}^2\right)\exp\left(\frac{-2\alpha_3\varphi_3}{\beta_3}\right).$$

If τ_C is virtually the same as τ_L,

$$W_{R_2\,\mathrm{ON}} \simeq \frac{1}{4}C\frac{2L}{C}\left(\frac{C}{2L}U^2 + I_{RR}^2\right)\exp\left(-\frac{2\alpha_3\varphi_3}{\beta_3}\right)$$

$$\simeq \frac{1}{2}Li'^2_1(t_2)\exp\left(-\frac{2\alpha_3\varphi_3}{\beta_3}\right).$$

In this equation $i'_1(t_2)$ denotes the maximum value of current i'_1 during the turn-on commutation, this value being independent on the current i' which is to be switched.

- *During the turn-off commutation* $T_1 \rightarrow D'_1$ current flows through resistance R'_1 during the four phases. The energy dissipated in R'_1 during phases 1 and 2 cannot be calculated in any simple way; but, during these two periods, the current in R_1, equal to $C\,dv'_1/dt$ is small since v'_1 shows little variation. The energy dissipated in R'_1 can be seen as corresponding to the discharge in this resistance of capacitor C'_1 – initially charged at U – via diode D'_1 throughout phases 3 and 4.

$$W_{R_2\,\mathrm{OFF}} \simeq \frac{1}{2} C U^2.$$

4.2.3.3 Energy Dissipated in r_1 and r'_1

The same current flows through the two resistances r_1 and r'_1 throughout all the phases of turn-on and turn-off commutations. Only the energy dissipated in one of them needs thus to be calculated.

- *During the turn-on commutation* $D'_1 \rightarrow T_1$, resistance r_1 and r'_1 are only crossed by a current during phases 2 and 3, thus giving:

$$W_{r\,\mathrm{ON}} = \int_{t_2}^{t_3} r \left(\frac{L}{r} \frac{\mathrm{d}i'_1}{\mathrm{d}t} \right)^2 \mathrm{d}t + \int_{t_3}^{\infty} r \left(\frac{L}{r} \frac{\mathrm{d}i'_1}{\mathrm{d}t} \right)^2 \mathrm{d}t.$$

Taking into account the expressions (4.14) and (4.16) of i'_1, the following is obtained:

$$W_{r\,\mathrm{ON}} = \frac{1}{2} L i'^2_1(t_2) \left[1 - \frac{\tau_C}{\tau_C + \tau_L} \exp\left(-\frac{2\alpha_3 \varphi_3}{\beta_3} \right) \right].$$

If the time constants τ_C and τ_L are assumed to be equal,

$$W_{r\,\mathrm{ON}} \simeq \frac{1}{2} L i'^2_1(t_2) \left[1 - \frac{1}{2} \exp\left(-\frac{2\alpha_3 \varphi_3}{\beta_3} \right) \right]$$

with $i'_1(t_2)$ equal to $\sqrt{(C/2L)U^2 + I^2_{RR}}$ and therefore independent of I_{ON}.

- *During the turn-off commutation* $T_1 \rightarrow D'_1$, resistances r_1 and r'_1 are crossed by a current throughout the duration of the commutation. However, as in the case of the calculation of the energy dissipated in R'_1 when T_1 was turned off, the variations in i_1 and i'_1 during phases 1 and 2 can be ignored, thus giving:

$$W_{r\,\mathrm{OFF}} \simeq \int_{t'_2}^{t'_3} r \left(\frac{L}{r} \frac{\mathrm{d}i_1}{\mathrm{d}t} \right)^2 \mathrm{d}t + \int_{t'_3}^{\infty} r \left(\frac{L}{r} \frac{\mathrm{d}i'_1}{\mathrm{d}t} \right)^2 \mathrm{d}t.$$

If it is assumed that $\varphi'_3 \simeq \varphi_3$ and $i'_1(t'_2) \simeq I_{\mathrm{OFF}}$, the following can be deduced from Eqs. (4.26) and (4.29):

$$W_{r\,\mathrm{OFF}} \simeq \frac{1}{2} L I^2_{\mathrm{OFF}} \left[1 - \frac{\tau_C}{\tau_C + \tau_L} \exp\left(-\frac{2\alpha_3 \varphi_3}{\beta_3} \right) \right].$$

If τ_C equals τ_L, the result is that

$$W_{r\,\mathrm{OFF}} \simeq \frac{1}{2} L I^2_{\mathrm{OFF}} \left[1 - \frac{1}{2} \exp\left(-\frac{2\alpha_3 \varphi_3}{\beta_3} \right) \right].$$

Moreover, the exponential term is often much less than 1, and, as a first approximation, the following is valid:

$$W_{r\,\mathrm{OFF}} \simeq \frac{1}{2} L I_{\mathrm{OFF}}^2.$$

4.2.3.4 Power Dissipated in the Snubber Resistances

In making an approximate calculation of the power dissipated in resistances R_1 and R'_1, r_1 and r'_1, the simplified expressions of the energies obtained by assuming τ_C equal to τ_L are used.

If f is the frequency of the inverter output current i' and m the frequency-modulation ratio, the number of commutations per cycle $1/f$ is equal to $2m$ (see Chap. 3, §3.1.3):

$m/2$ commutations $D'_1 \rightarrow T_1$
$m/2$ commutations $T_1 \rightarrow D'_1$
$m/2$ commutations $D_1 \rightarrow T'_1$
$m/2$ commutations $T'_1 \rightarrow D_1$

- In order to calculate the power P_R dissipated in resistance R of a *voltage snubber*, it can be noted that, on account of the symmetry of the half-bridge, the energy dissipated in R_1 during a commutation $T_1 \rightarrow D'_1$ is equal to the energy dissipated in R'_1 during a commutation $T'_1 \rightarrow D_1$, if $|i'|$ is the same.

  The power dissipated in each resistance R is given by

$$P_R = f \sum_{k=1}^{m/2} (W_{R_1\,\mathrm{ON}} + W_{R'_1\,\mathrm{ON}} + W_{R_1\,\mathrm{OFF}} + W_{R'_1\,\mathrm{OFF}})$$

  if $1, 2, \ldots, k, \ldots, m/2$ are used to denote the $m/2$ commutations which follow each other during a half-cycle.

  Since only energy $W_{R_1\,\mathrm{OFF}}$ depends on the switched current,

$$P_R = f\left[\frac{m}{2}(W_{R_1\,\mathrm{ON}} + W_{R'_1\,\mathrm{ON}} + W_{R'_1\,\mathrm{OFF}}) + \sum_{k=1}^{m/2} W_{R_1\,\mathrm{OFF}}\right].$$

  If i' has the form $i' = I'\sqrt{2}\sin(\omega t - \varphi)$,

$$\sum_{k=1}^{m/2} W_{R_1\,\mathrm{OFF}} = \frac{1}{2} L\left[\exp\left(-\frac{2\alpha_3\varphi_3}{\beta_3}\right)\right] \sum_{k=1}^{m/2} [I'\sqrt{2}\sin(\omega t_k - \varphi)]^2.$$

  In case where m is high, the mean quadratic value of current i' can be taken as the mean quadratic value of this current at the instants of commutation:

$$\sum_{k=1}^{m/2} W_{R_1\,\mathrm{OFF}} = \frac{1}{2} L\left[\exp\left(-\frac{2\alpha_3\varphi_3}{\beta_3}\right)\right] \frac{m}{2} I'^2.$$

If the other energies are replaced by their previously calculated values, the following is obtained:

$$P_R = f\frac{m}{2}\left[\frac{1}{2}CU^2 + \frac{1}{2}Li_1'^2(t_2)\exp\left(-\frac{2\alpha_3\varphi_3}{\beta_3}\right) + \frac{1}{2}CU^2 + LI'^2\exp\left(-\frac{2\alpha_2\varphi_3}{\beta_3}\right)\right]$$

$$P_R = \frac{mf}{2}\left[CU^2 + \frac{1}{2}L[I'^2 + i_1'^2(t_2)]\exp\left(-\frac{2\alpha_3\varphi_3}{\beta_3}\right)\right]. \tag{4.32}$$

- In computing the power P_r dissipated in resistance r of a *current snubber*, we proceed as previously. Furthermore, resistances r_1 and r_1' are always crossed by equal current, thus giving:

$$P_r = f\sum_{k=1}^{m/2} 2(W_{r\mathrm{ON}} + W_{r\mathrm{OFF}}) = f\left(mW_{r\mathrm{ON}} + 2\sum_{k=1}^{m/2} W_{r\mathrm{OFF}}\right)$$

since only energy $W_{r\mathrm{OFF}}$ depends on the value of switched current i'.

If the value I'^2 is again taken as the mean value of i'^2 during commutations, this gives:

$$P_r = \frac{mf}{2}[I'^2 + i_1'^2(t_2)]\left[1 - \frac{1}{2}\exp\left(-\frac{2\alpha_3\varphi_3}{\beta_3}\right)\right]. \tag{4.33}$$

- *Remark.* The term $\exp(-2\alpha_3\varphi_3/\beta_3)$ is found in the expression of P_R and P_r. β_3 is given by

$$\beta_3 = \sqrt{\omega_3^2 - \alpha_3^2}$$

Since

$$\omega_3^2 - \alpha_3^2 = \frac{1}{2LC} - \frac{1}{16r^2C^2} = \frac{R}{2r\tau_L\tau_C} - \frac{R^2}{16r^2\tau_C^2} = \frac{R}{2r\tau_C}\left(\frac{1}{\tau_L} - \frac{R}{8r\tau_C}\right)$$

$\omega_3^2 - \alpha_3^2$ is positive for $8r/R > \tau_L/\tau_C$ or, if it is assumed that τ_L and τ_C are equal, for $8r/R > 1$.

For $\tau_L = \tau_C$ and $8r/R > 1$,

$$\exp\left(-\frac{2\alpha_3\varphi_3}{\beta_3}\right) = \exp\left(\frac{-2}{\sqrt{(8r/R) - 1}}\arctan\sqrt{\frac{8r}{R} - 1}\right).$$

If $\tau_L = \tau_C$ and $8r/R$ is < 1,

$$\exp\left(-\frac{2\alpha_3\varphi_3}{\beta_3}\right) = \exp\left(\frac{-2}{\sqrt{1 - (8r/R)}}\operatorname{arcth}\sqrt{1 - \frac{8r}{R}}\right).$$

Figure 4.17 shows how the value of $(1/2)\exp(-2\alpha_3\varphi_3/\beta_3)$, which appears in P_R, and that of $[1 - (1/2)\exp(-2\alpha_3\varphi_3/\beta_3)]$, which appears in P_r, vary depending on r/R.

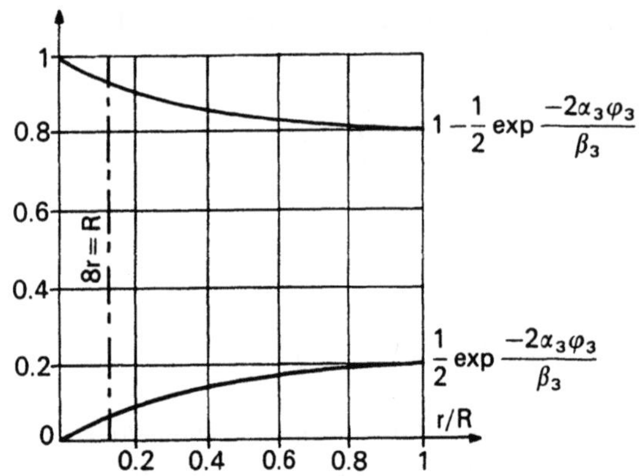

Fig. 4.17

From Eqs. (4.32) and (4.33), it can be seen that at given values of mf, U, I', C and L, the sum of P_R and P_r is independent of R and r.

When r/R is low,

there is little difference between P_R and $\frac{mf}{2}CU^2$

there is little difference between P_r and $\frac{mf}{2}L[I'^2 + i_1'^2(t_2)]$.

It can be observed that the voltage peak which appears across the capacitors of the voltage snubber has no effect on the power dissipated in R.

On the contrary, the overcurrent in the inductances of the current snubber plays a role via the term $i_1'^2(t_2)$ which is equal to $\sqrt{(C/2L)U^2 + I_{RR}^2}$. It is thus important to minimise this current first of all.

4.2.3.5 Switching Losses in Transistors

The fall-time of voltage v_{T_1} across transistor T_1 during its turn-on has been ignored. It has already been shown (Vol. 3, Appendix, §2) that the energy dissipated at turn-on in a transistor protected by a snubber was very low. Only the energy dissipated at turn-off will thus be taken into account.

- At turn-off commutation (see Figs. 4.2 and 4.13), current i_{T_1} in transistor T_1 only differs from zero during phase 1. During this phase, it has been seen that voltages v_1 and v_1' were given by

$$v_1 = \frac{1}{2}\left[\frac{I_{OFF}}{2Ct_f}t^2 - Ri_1' - 2L\left(1 + \frac{R}{2r}\right)\frac{di_1}{dt}\right]$$

$$v_1' = U - \frac{1}{2}\left[\frac{I_{OFF}}{2Ct_f}t^2 + Ri_1' + 2L\left(1 + \frac{R}{2r}\right)\frac{di_1}{dt}\right]$$

and that there was little variation in voltage v_1'.

If it is assumed that v_1 shows little variation from U, this is equivalent to making the following approximation:

$$\frac{I_{OFF}}{2Ct_f}t^2 \simeq -Ri_1' - 2L\left(1+\frac{R}{2r}\right)\frac{di_1}{dt}$$

which leads to:

$$v_1 \simeq \frac{I_{OFF}}{2Ct_f}t^2.$$

As, during phase 1, voltage v_{T_1} across the transistor is equal to v_1, the power dissipated at each turn-off is:

$$W_{T\,OFF} = \int_0^{t_f} \frac{I_{OFF}}{2Ct_f}t^2 I_{OFF}\left(1-\frac{t}{t_f}\right)dt = \frac{I_{OFF}^2 t_f^2}{24C}$$

- Each transistor carries out $m/2$ turn-offs per output current cycle. If the mean quadratic value of i' is again taken as the mean quadratic value of the current I_{OFF} at turn-off, the commutation losses for each transistor are given by

$$P_T = mf\frac{t_f^2}{48C}I'^2. \qquad (4.34)$$

4.2.3.6 Choice of Snubber Elements

This choice is the result of a compromise between limiting the overcurrent and overvoltage in the transistors, when commutation occurs, and between limiting the power dissipated in the transistors during these commutations.

- The maximum value of the current in a transistor is obtained during the turn-on; *overcurrent* Δi, equal to $i_{T\,max} - I_{ON}$, is given by Eq. (4.11) and has the following value:

$$\Delta i = \sqrt{\frac{C}{2L}U^2 + I_{RR}^2} + \frac{U}{R}\exp\left(-\frac{t_2}{RC}\right),$$

with

$$t_2 = \frac{2L}{U}(I_{ON} + I_{RR}) + \sqrt{2LC}\,\mathrm{Arctan}\frac{U}{\sqrt{\frac{2L}{C}}I_{RR}}.$$

If the reverse current in the diode is ignored, Δi becomes:

$$\Delta i = U\sqrt{\frac{C}{2L}} + \frac{U}{R}\exp\left[-\frac{1}{RC}\left(\frac{2L}{U}I_{ON} + \frac{\pi}{2}\sqrt{2LC}\right)\right].$$

The overcurrent has its maximum value for the minimum value of I_{ON}. However, since Δi is added to current i', Δi must be calculated for the maximum value of i', i.e. $I'\sqrt{2}$, since this is the moment at which the current peak in the transistors is at its highest.

Denoting:

$$k_1 = \sqrt{\frac{C}{2L}}\frac{U}{I'}, \qquad k_2 = \frac{r}{R} \quad \text{and} \quad \tau = \frac{L}{r} = RC$$

the overcurrent is given by

$$\Delta i = k_1 I' + k_1\sqrt{2k_2}I' \exp\left[-\frac{1}{\tau}\left(\frac{\tau\sqrt{2k_1}}{k_1 I'}I'\sqrt{2} + \frac{\pi}{2}\tau\sqrt{2k_2}\right)\right].$$

This results in:

$$\frac{\Delta i}{I'\sqrt{2}} = \frac{k_1}{\sqrt{2}}\left[1 + \sqrt{2k_2}\exp\left(-\sqrt{2k_2}\left(\frac{\sqrt{2}}{k_1} + \frac{\pi}{2}\right)\right)\right].$$

- The maximum voltage value across a transistor appears during the turn-off. The *overvoltage* is given by Eq. (4.28) and has the following value:

$$\Delta v = \sqrt{\frac{2L}{C}}\, I_{OFF} \exp(-\alpha_3\varphi_3/\beta_3)$$

which can be put in the following form:

$$\Delta v = \sqrt{\frac{2L}{C}}\, I_{OFF} \exp\left(-\frac{1}{\sqrt{1-(8r/R)}}\operatorname{Arcth}\sqrt{1-\frac{8r}{R}}\right)$$

when $8r/R$ is less than 1, as was shown when calculating the power dissipated in the snubber resistances.

The maximum overvoltage corresponds to the maximum value of I_{OFF}, i.e. $I'\sqrt{2}$. Its value referred to U is then:

$$\frac{\Delta v}{U} = \frac{\sqrt{2}}{k_1}\exp\left(-\frac{1}{\sqrt{1-8k_2}}\operatorname{Arcth}\sqrt{1-8k_2}\right).$$

- Figure 4.18 shows the variations of $\Delta i/I\sqrt{2}$ and $\Delta v/U$ as a function of k_1 and k_2.

Having chosen k_1 and k_2 in order to limit Δi and Δv, the snubber components can be deduced by means of the following equations:

$$R = \frac{U}{I'k_1\sqrt{2k_2}}, \qquad r = k_2 R, \qquad C = \frac{\tau}{R} = \frac{I'}{U}\tau k_1\sqrt{2k_2}, \qquad L = \tau r,$$

where the value of τ must be high as compared to the current fall time ($\tau \simeq 10t_f$).

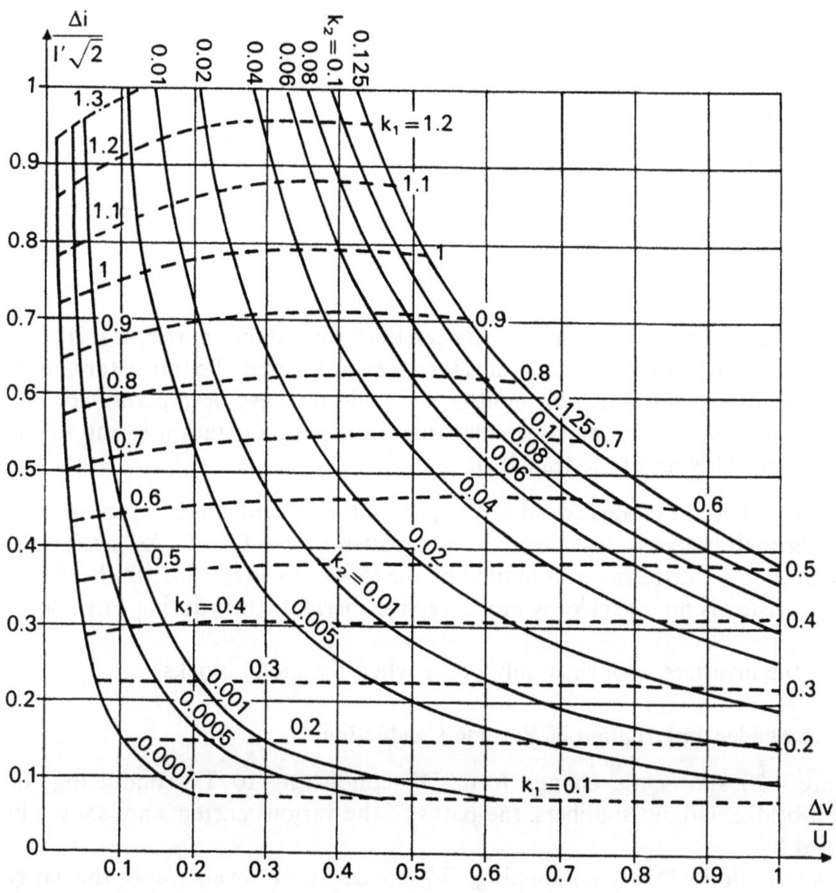

Fig. 4.18

It can be noted that k_1 and k_2 – and thus the value of C – must be small in order to reduce Δi and Δv. But the power dissipated in the transistors is inversely proportional to C.

The choice of k_1 and k_2 results from a compromise, leading to the adoption of values considerably higher than those used in the case of choppers (Vol. 3, Appendix) for C and especially for L.

4.2.4 Remarks on Reverse Conduction in the Transistor

4.2.4.1 Description of the Phenomenon

When the diode connected in reverse parallel across a transistor is conducting, on account of the forward voltage drop in the diode, the transistor emitter is at a higher potential than that of the collector (Fig. 4.19).

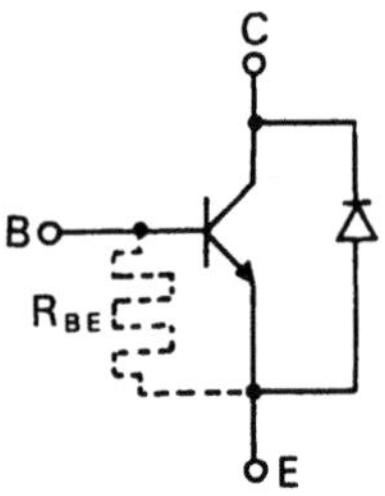

Fig. 4.19

If a resistance R_{BE} is connected between the base and the emitter, as it is often the case, the collector-base junction is then forward biased: the transistor acts as an NPN transitor in which the emitter and collector have been permuted. The current gain is however much smaller than in normal operation owing to the dissymetrical layout of the transistor.

Thus, when the diode is conducting, part of the current which is supposed to flow through it goes in fact through the transistor from E to C. As the reverse voltage across the transistor is limited by the forward voltage drop in the diode, the transistor is not affected by any excessive thermal stress during this operation.

In the inverters, problems only occur when the diode blocks.

4.2.4.2 Sudden Interruption of Reverse Conduction

Figure 4.20 shows the change from D'_1 conducting to T_1 conducting for a half-bridge without snubbers: the paths of the various currents are shown in dashed lines.

When diode D'_1 is conducting, T'_1 conducts in reverse and the stray capacitances of junctions C–B and B–E are charged at low voltages and biased as shown in Fig. 4.20a.

When the $D'_1 \rightarrow T_1$ commutation is over (Fig. 4.20c), the voltages across the stray capacitances have changed polarities: the one which appears across junction C–B is close to U since the B–E junction is forward-biased.

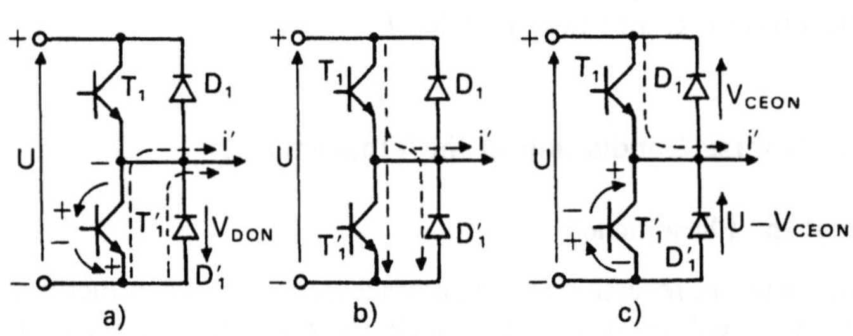

Fig. 4.20

As a result, during the commutation (Fig. 4.20b), transistor T_1 is crossed by a current which is the sum of

- the current i' supplied to the load at the instant of commutation,
- the reverse recovery current in diode D'_1,
- the reverse recovery current of transistor T'_1 which is used to charge the stray capacitance of junction C–B.

The faster the commutation, the higher the currents in D'_1 and T'_1: the overcurrent in T_1 can be very high and cause excessive commutation losses at turn-on.

4.2.4.3 Protection against Overcurrent Due to Reverse Conduction

The inductance of the snubber which protects from excessive di/dt can have a size which enables this overcurrent to be limited. Other solutions enables the reverse conduction to be suppressed.

- A diode can be series-connected with the transistor (Fig. 4.21). However, in medium- or high-power applications, the resulting extra losses make this solution of limited use.
- The most effective solution consists in giving a negative bias to the B–E junction of the transistor during the conduction phases of the diode which is connected in antiparallel across it.
  - It should be noted that, in a voltage-source inverter without PWM, this problem does not arise when feeding an inductive load (Fig. 4.22a), as the transistors begin conducting when current i' goes through zero. When feeding a capacitive load (Fig. 4.22b), the reverse conduction of T_1 is to be avoided when D_1 is conducting and that of T'_1 when D'_1 is conducting. It is

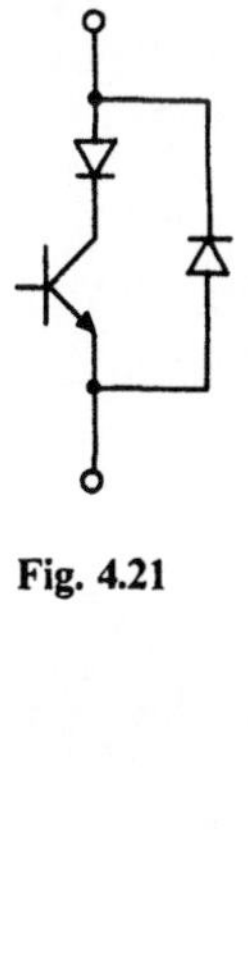

Fig. 4.21

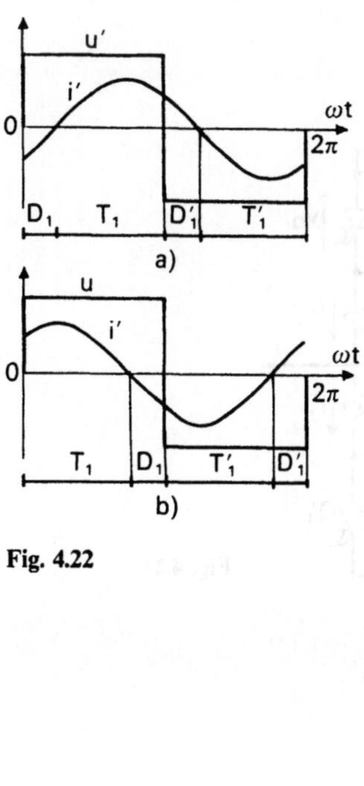

Fig. 4.22

thus necessary to detect the zero crossing instant of current i' so as to know the exact moment at which to begin applying the negative bias v_{BE}.
- In PWM inverters, the problem arises whatever the nature of the load (see Fig. 4.4). This solution further complicates the control of the bases.

4.3 GTO Thyristor Commutations

The problems posed by the GTO thyristor commutation are virtually the same as those for the bipolar transistor. However, as the power involved is considerably higher, it is desirable to reduce the number, size and the losses of the elements in the snubber circuits limiting di/dt and dv/dt.

Figure 4.23 shows some of the configurations used in designing a phase-leg of a GTO voltage-source inverter:

- the first configuration (Fig. 4.23a) is identical to that which was studied in detail with the bipolar transistor: each GTO has its voltage snubber and

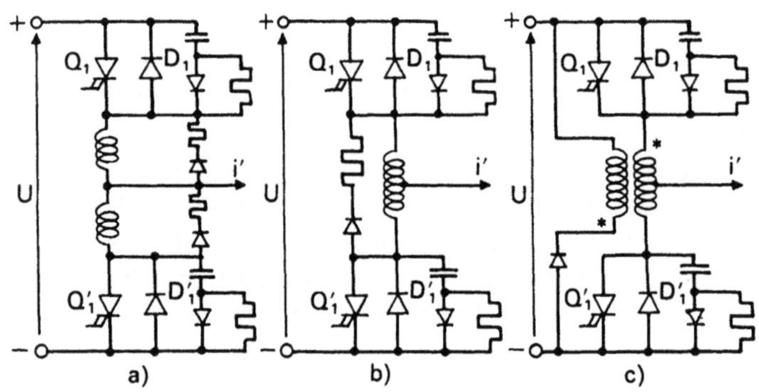

Fig. 4.23

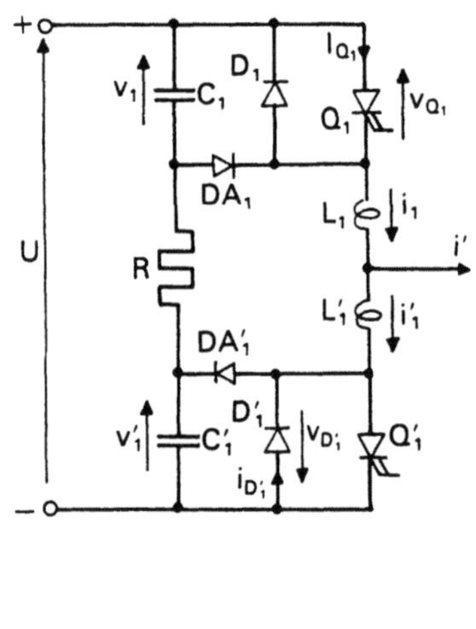

Fig. 4.24

current snubber; each snubber has its discharge resistance and its diverting diode.

- the two snubbers can be brought together around a centre-tapped inductor with a single resistance and diode (Fig. 4.23b).
- if a secondary winding is coupled to this centre-tapped inductor and is then connected to the supply via a diode (Fig. 4.23c), a part of the snubber inductance energy can be recovered.

A closer study will be made of the operation of the circuit proposed by W. McMurray in IEEE Transaction on Power Electronics, July 1987; its diagram is shown in Fig. 4.24. It uses only one resistance R, two diverting diodes DA_1 and DA'_1, two capacitors C_1 and C'_1 of the same capacitance C and a centre-tapped inductor, each half of which (L_1 and L'_1) has an inductance $L/2$.

It should be noted that L_1 and L'_1 can be coupled, thus enabling the same value of L to be obtained with a reduced size.

4.3.1 Commutation at Turn-On

The transfer from D'_1 to Q_1 of a positive current i' – assumed to be constant and equal to I_{ON} during the whole commutation process – will be followed. This commutation occurs in three phases.

In the initial state, current I_{ON} flows through D'_1 and L'_1; capacitors C_1 and C'_1 are charged respectively at $v_1 = U$ and $v'_1 = 0$, and the currents in the inductances are $i_1 = 0$ and $i'_1 = -I_{ON}$.

The configuration of the active part of the circuit in the initial state, during each of the three operating phases and in the final state, is represented in Fig. 4.25.

Figure 4.26 gives the waveforms of $v_1, v'_1, i_1, i'_1, v_{Q_1}$ and $i_{Q_1}, v_{D'_1}$ and $i_{D'_1}$.

- *Phase 1 ($0<t<t_1$): Q_1 and D'_1 conducting simultaneously*

Q_1 receives its firing pulse in $t = 0$; the time taken by the voltage across its terminals to fall from U to 0 is ignored.

- Starting from $t = 0$, there is:

$$U = \frac{L}{2}\frac{di}{dt} + \frac{L}{2}\frac{di'_1}{dt} \quad \text{and} \quad i_1 = I_{ON} + i'.$$

Thus

$$\frac{di_1}{dt} = \frac{di'_1}{dt} = \frac{U}{L}$$

and

$$i_1 = \frac{U}{L}t \; ; \qquad i'_1 = \frac{U}{L}t - I_{ON}. \tag{4.35}$$

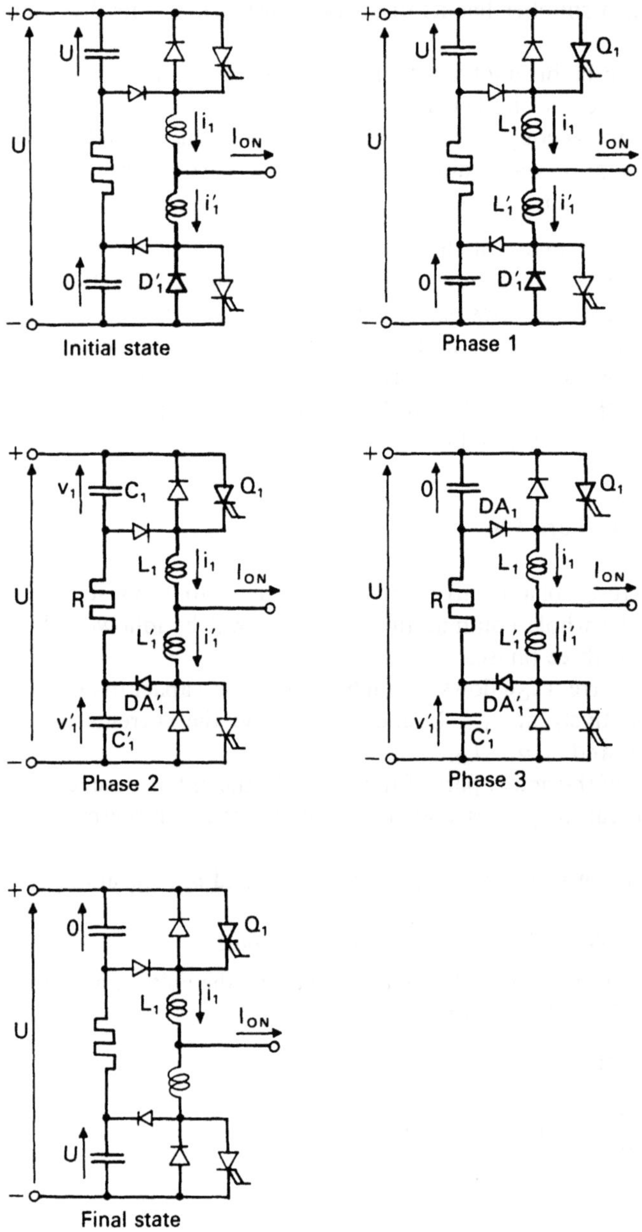

Fig. 4.25. Commutation $D'_1 \rightarrow Q_1$

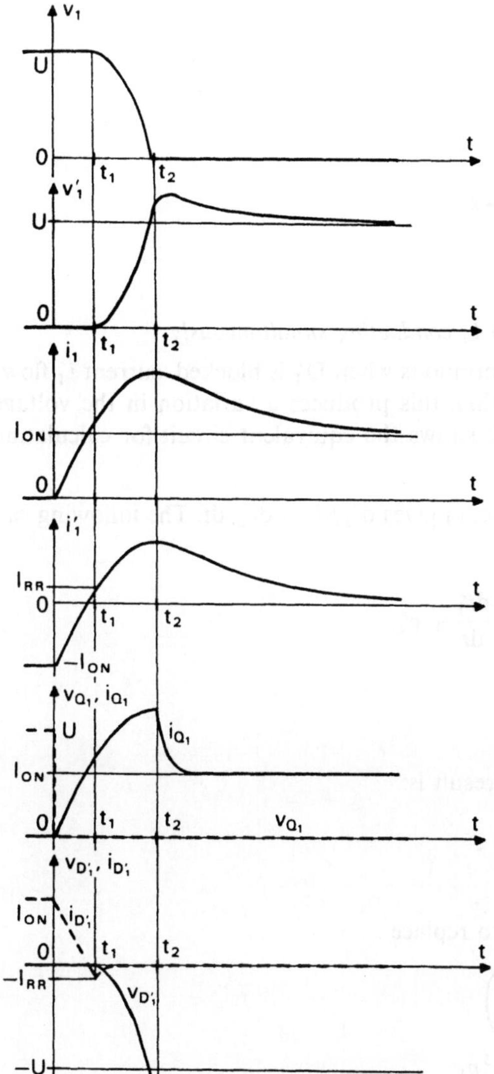

Fig. 4.26

– This phase ends, for $t = t_1$, when i'_1 equals I_{RR}, the maximum value of the reverse current of diode D'_1.

$$t_1 = \frac{L}{U}(I_{RR} + I_{ON}).$$

At instant $t = t_1$,

$$v_1(t_1) = U, \qquad v'_1(t_1) = 0, \qquad i_1(t_1) = I_{ON} + I_{RR}, \qquad i'_1(t_1) = I_{RR}.$$

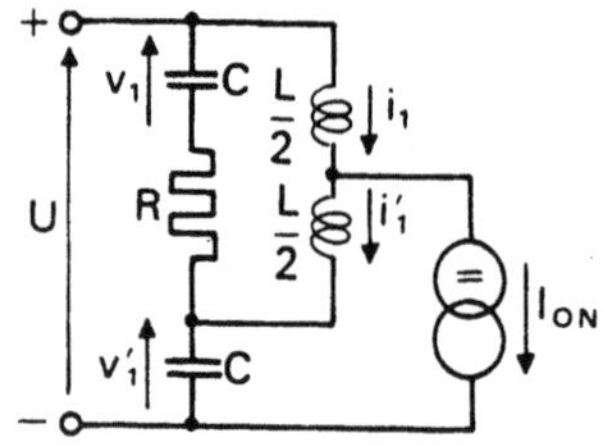

Fig. 4.27

- *Phase 2 ($t_1<t<t_2$): Q_1 and DA'_1 conducting simultaneously*

As current i'_1 cannot be discontinuous when D'_1 is blocked, current i'_1 flows via DA'_1. When DA'_1 is conducting, this produces a variation in the voltage across the capacitors. Figure 4.27 shows the equivalent circuit for calculating v_1, v'_1, i_1 and i'_1.

– Since $i_1 = i'_1 + I_{ON}$, this once again gives $di_1/dt = di'_1/dt$. The following can be written:

$$U = v_1 + RC\frac{dv_1}{dt} + v'_1 = L\frac{di'_1}{dt} + v'_1$$

$$C\frac{dv_1}{dt} + i'_1 = C\frac{dv'_1}{dt}.$$

By deriving this equation, the result is:

$$C\frac{d^2v_1}{dt^2} + \frac{di'_1}{dt} = C\frac{d^2v'_1}{dt^2}.$$

In this equation it is possible to replace

$$\frac{di'_1}{dt} \text{ by } \frac{1}{L}\left(v_1 + RC\frac{dv_1}{dt}\right)$$

$$\frac{d^2v'_1}{dt^2} \text{ by } -\frac{d^2v_1}{dt^2} - RC\frac{d^3v_1}{dt^3}.$$

Leading to the differential equation giving v'_1:

$$RLC^2\frac{d^3v_1}{dt^3} + 2LC\frac{d^2v_1}{dt^2} + RC\frac{dv_1}{dt} + v_1 = 0.$$

The solution has the following form:

$$v_1 = A\sin(\omega_1 t + \varphi_1)\exp(\alpha_1 t) + B\exp(\beta_1 t).$$

– For a simpler evaluation of v_1, v'_1 and i_1 during this phase, R can be assumed as equal to zero.

Hence:

$$U = v_1 + v'_1 = L\frac{di'_1}{dt} + v'_1$$

$$C\frac{dv_1}{dt} + i'_1 = C\frac{dv'_1}{dt}$$

which gives

$$2LC\frac{d^2v_1}{dt^2} + v_1 = 0.$$

The solution has the form:

$$v_1 = \frac{1}{\omega_2}\frac{dv_1}{dt}(t_1)\sin\omega_2(t-t_1) + v_1(t_1)\cos\omega_2(t-t_1)$$

with

$$\omega_2 = \frac{1}{\sqrt{2LC}}$$

Since

$$\frac{dv_1}{dt}(t_1) = -\frac{i'_1(t_1)}{2C} = -\frac{I_{RR}}{2C} \quad \text{and} \quad v_1(t_1) = U$$

$$v_1 = -\frac{I_{RR}}{2C\omega_2}\sin[\omega_2(t-t_1)] + U\cos[\omega_2(t-t_1)] \tag{4.36}$$

$$i'_1 = -2C\frac{dv_1}{dt} = I_{RR}\cos[\omega_2(t-t_1)] + 2C\omega_2 U\sin[\omega_2(t-t_1)] \tag{4.37}$$

$$i_1 = i'_1 + I_{ON}, \qquad v'_1 = U - v_1. \tag{4.38}$$

- This period ends for $t = t_2$, when v_1 reverses its polarity, leading to the start of conduction of diode DA_1.

  If I_{RR} is very small compared to $2C\omega_2 U$,

  $$t_2 \simeq t_1 + \frac{\pi}{2\omega_2}.$$

  At the end of this phase,

  $$v_1(t_2) = 0, \qquad v'_1(t_2) \simeq U,$$

  $$i'_1(t_2) \simeq 2C\omega_2 U, \qquad i_1(t_2) \simeq I_{ON} + 2C\omega_2 U.$$

  At instant $t = t_2$, the current in thyristor Q_1, equal to i_1, reaches its maximum value:

  $$i_{Q_1\,max} \simeq I_{ON} + \sqrt{\frac{2C}{L}}\,U.$$

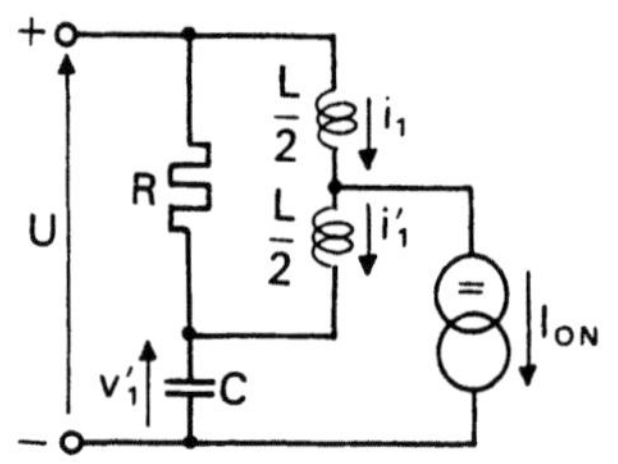

Fig. 4.28

• *Phase 3 ($t > t_2$): Q_l, DA_l and DA'_l conducting simultaneously*

Figure 4.28 then shows the equivalent configuration; it enables the following to be written:

$$U = L\frac{di'_1}{dt} + v'_1$$

and

$$\frac{L}{R}\frac{di'_1}{dt} + i'_1 = C\frac{dv'_1}{dt}.$$

Giving:

$$LC\frac{d^2 i'_1}{dt^2} + \frac{L}{R}\frac{di'_1}{dt} + i'_1 = 0.$$

Denoting $\omega_3 = 1/\sqrt{LC}$ and $\alpha_3 = 1/2RC$, current i'_1 can be expressed as:

• if $\omega_3 > \alpha_3$

$$i'_1 = \frac{i'_1(t_2)}{\sin\varphi_3}\exp[-\alpha_3(t - t_2)]\sin[\beta_3(t - t_2) + \varphi_3] \tag{4.39}$$

with

$$\beta_3 = \sqrt{\omega_3^2 - \alpha_3^2} \quad \text{and} \quad \tan\varphi_3 = \frac{\beta_3 i'_1(t_2)}{\alpha_3 i'_1(t_2) + \frac{di'_1}{dt}(t_2)}$$

• if $\omega_3 < \alpha_3$

$$i'_1 = \frac{i'_1(t_2)}{\text{sh}\,\varphi_3}\exp[-\alpha_3(t - t_2)]\,\text{sh}[\beta_3(t - t_2) + \varphi_3] \tag{4.39'}$$

with

$$\beta_3 = \sqrt{\alpha_3^2 - \omega_3^2} \quad \text{and} \quad \text{th}\,\varphi_3 = \frac{\beta_3 i'_1(t_2)}{\alpha_3 i'_1(t_2) + \frac{di'_1}{dt}(t_2)}.$$

When t tends towards infinity, i'_1 tends towards zero,
i_1 tends towards I_{ON},
and v'_1 tends towards U;
the GTO thyristor Q_1 is the only remaining conducting device: commutation is over.

4.3.2 Commutation at Turn-Off

This section describes the transfer from Q_1 to D'_1 of a current i', assumed to be constant and equal to I_{OFF} throughout the commutation.

In the initial state, the current in the load flows through Q_1 and L_1. Voltages v_1 and v'_1 across the capacitors C_1 and C'_1 have the values 0 and U respectively, and currents i_1 and i'_1 in inductances L_1 and L'_1 have the values I_{OFF} and 0.

The parts of the circuit concerned by the successive commutation phases are shown in thick lines on Fig. 4.29. Figure 4.30 gives the waveforms of $v_1, v'_1, i_1, i'_1, v_{Q_1}$ and $i_{Q_1}, v_{D'_1}$ and $i_{D'_1}$.

- *Phase 1 ($0<t<t_f$): Q_1 and DA_1 conducting simultaneously*

It is assumed that, from the instant when a turn-off control signal is applied to the gate of the GTO thyristor Q_1, current i_{Q_1} decreases linearly from I_{OFF} to 0 in time t_f (the tail current will be ignored).

– The equivalent circuit in Fig. 4.31 enables the following to be written:

$$U = v_1 + RC\frac{dv'_1}{dt} + v'_1$$

and

$$C\frac{dv_1}{dt} + I_{OFF}\left(1 - \frac{t}{t_f}\right) = C\frac{dv'_1}{dt} + I_{OFF}.$$

Hence:

$$C\frac{dv_1}{dt} - I_{OFF}\frac{t}{t_f} = C\frac{dv'_1}{dt}.$$

Since the current in resistance R is equal to $C\,dv'_1/dt$, the following can be deduced from the preceding equations:

$$\frac{RC}{2}\frac{di_R}{dt} + i_R = -\frac{1}{2}I_{OFF}\frac{t}{t_f}.$$

The solution has the form:

$$i_R = A_1\exp\left(-\frac{2t}{RC}\right) + \frac{I_{OFF}}{2t_f}\left(\frac{RC}{2} - t\right).$$

From $i_R(0) = (1/R)[U - v_1(0) - v'_1(0)] = 0$, the value of A_1 can be deduced.

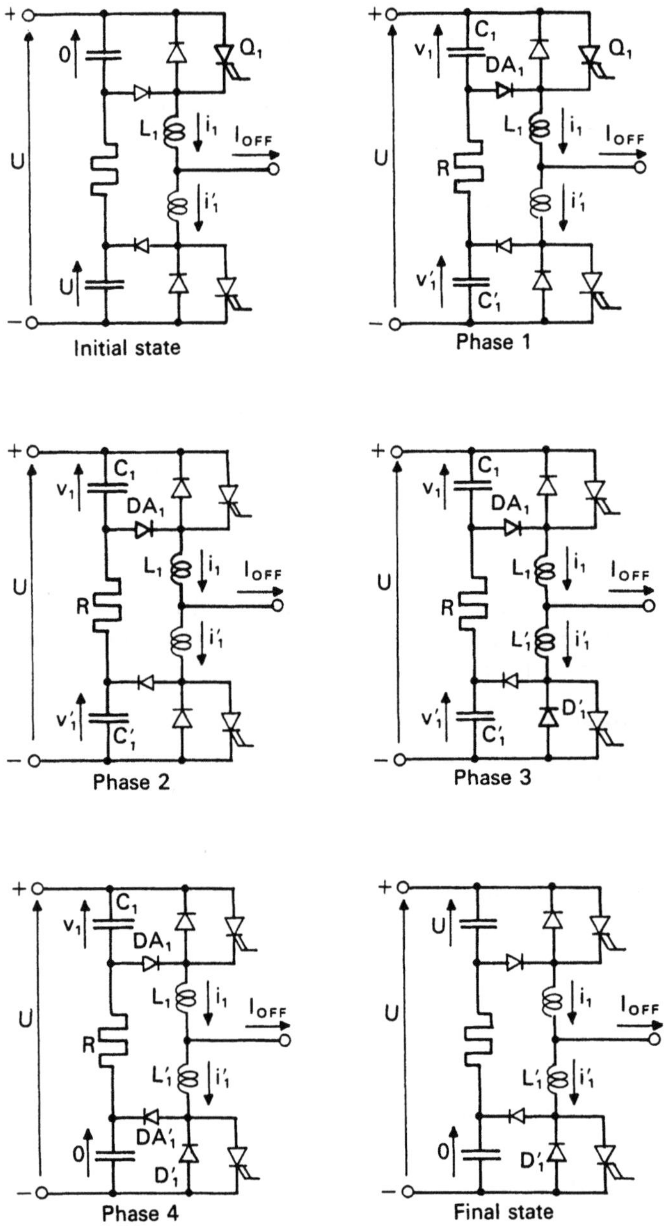

Fig. 4.29. Commutation $Q_1 \rightarrow D'_1$

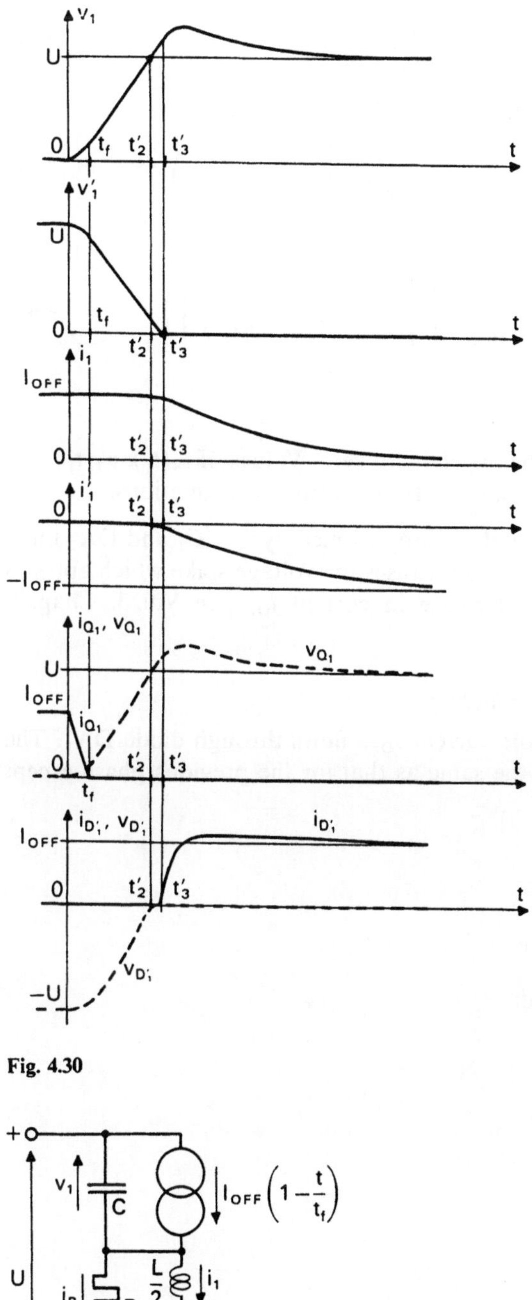

Fig. 4.30

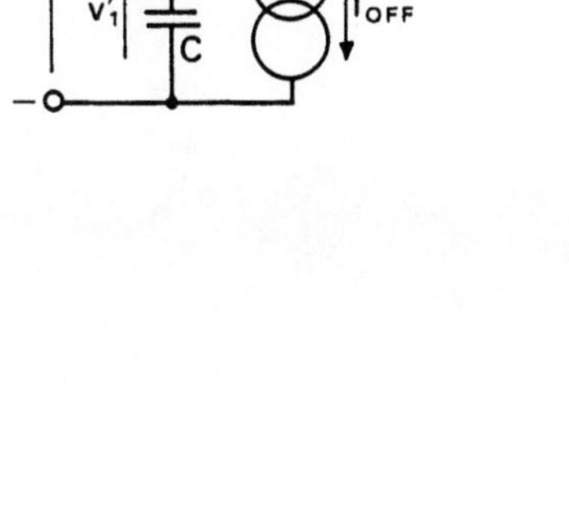

Fig. 4.31

$$i_R = \frac{1}{2} I_{OFF}\left(\frac{RC}{2t_f}\left[1 - \exp\left(-\frac{2t}{RC}\right)\right] - \frac{t}{t_f}\right). \tag{4.40}$$

- Voltage v'_1 can be calculated by $v'_1 = (1/C)\int i_R\,dt$, giving:

$$v'_1 = U - \frac{R I_{OFF}}{4}\left(\frac{RC}{2t_f}\left[1 - \exp\left(-\frac{2t}{RC}\right)\right] - \frac{t}{t_f} + \frac{t^2}{RCt_f}\right) \tag{4.41}$$

and v_1 by $v_1 = U - Ri_R - v'_1$, giving:

$$v_1 = v_{Q_1} = \frac{R I_{OFF}}{4}\left(\frac{t}{t_f} + \frac{t^2}{RCt_f} - \frac{RC}{2t_f}\left[1 - \exp\left(-\frac{2t}{RC}\right)\right]\right) \tag{4.42}$$

During this period,

$$i_1 = I_{OFF}, \qquad i'_1 = 0.$$

- This phase ends for $t = t_f$ when i_{Q_1} falls to zero. Values of $i_R(t_f)$, $v'_1(t_f)$ and $v_1(t_f)$ can be obtained by making $t = t_f$ in the previous equations.
- *Remark*: The stray inductance of the loop formed by Q_1, C_1 and DA_1 must be reduced as much as possible, as it causes the voltage spike which appears across the GTO during the decrease in current i_{Q_1} (see Vol. 3, Chap. 2, §2.5.2.2).

- *Phase 2 ($t_f < t < t'_2$): DA_1 conducting*

When Q_1 turns off, the whole current I_{OFF} flows through diode DA_1. The equivalent circuit (Fig. 4.32) is the same as that for the previous phase except that i_{Q1} is now zero.

- From $v_1 + Ri_R + v'_1 = U$,

with

$$i_R = C\frac{dv'_1}{dt} = C\frac{dv_1}{dt} - I_{OFF}$$

the following can be deduced:

$$\frac{RC}{2}\frac{di_R}{dt} + i_R = -\frac{I_{OFF}}{2}.$$

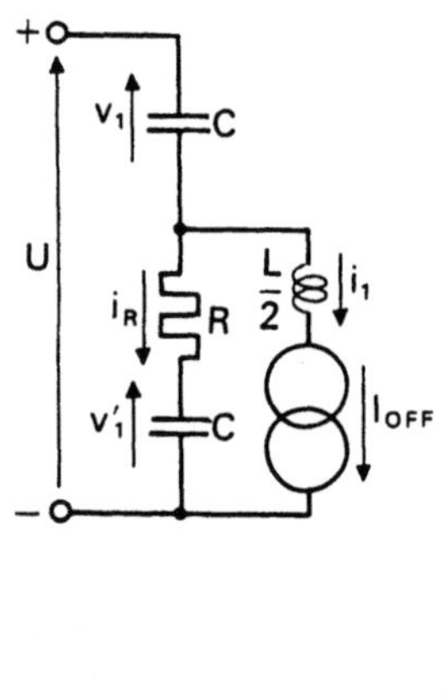

Fig. 4.32

Current i_R can be expressed as:

$$i_R = A_2 \exp\left[-\frac{2(t-t_f)}{RC}\right] - \frac{I_{OFF}}{2} \tag{4.43}$$

with

$$A_2 = i_R(t_f) + \frac{I_{OFF}}{2} = \frac{RCI_{OFF}}{4t_f}\left[1 - \exp\left(-\frac{2t_f}{RC}\right)\right].$$

- Similarly, v'_1 can be calculated by $v'_1 = (1/C)\int i_R \, dt$, giving:

$$v'_1 = -\frac{R}{2} A_2 \exp\left[-\frac{2(t-t_f)}{RC}\right] - \frac{I_{OFF}}{2C}(t-t_f) + B_2 \tag{4.44}$$

with

$$B_2 = v'_1(t_f) + \frac{R}{2} A_2 = U + \frac{RI_{OFF}}{4}\left(1 - \frac{t_f}{RC}\right).$$

Voltage v_1 can be deduced from v'_1 and i_R:

$$v_1 = U - v'_1 - Ri_R$$

$$= U + \frac{RI_{OFF}}{2} - B_2 + \frac{I_{OFF}}{2C}(t-t_f) - \frac{R}{2} A_2 \exp\left[-\frac{2(t-t_f)}{RC}\right]. \tag{4.45}$$

- This phase ends, for $t = t'_2$, when voltage v_1 reaches value U, which leads to the start of diode D'_1 conducting.

  The equation which gives t'_2, i.e.

$$R\frac{I_{OFF}}{2}\left(1 + \frac{t'_2 - t_f}{2RC}\right) = B_2 + \frac{R}{2} A_2 \exp\left[\frac{-2(t'_2 - t_f)}{RC}\right]$$

  cannot be solved analytically. However, if RC is small compared to $t'_2 - t_f$, it can be assumed that current i_R remains close to $-I_{OFF}/2$ throughout phase 2. This gives:

$$C\,dv_1/dt \simeq +I_{OFF}/2$$

$$t'_2 - t_f \simeq \frac{I_{OFF}}{2C}[U - v_1(t_f)]$$

  and

$$v'_1(t'_2) \simeq \frac{RI_{OFF}}{2}.$$

- *Phase 3 ($t'_2 < t < t'_3$): DA_1 and D'_1 conducting simultaneously*

  The equivalent configuration, shown in Fig. 4.33, is the same as that which was obtained for phase 2 of the $D'_1 \rightarrow Q_1$ commutation (Fig. 4.27), provided that i_1 and i'_1, and v_1 and v'_1 are permuted and that I_{ON} is replaced by I_{OFF}.

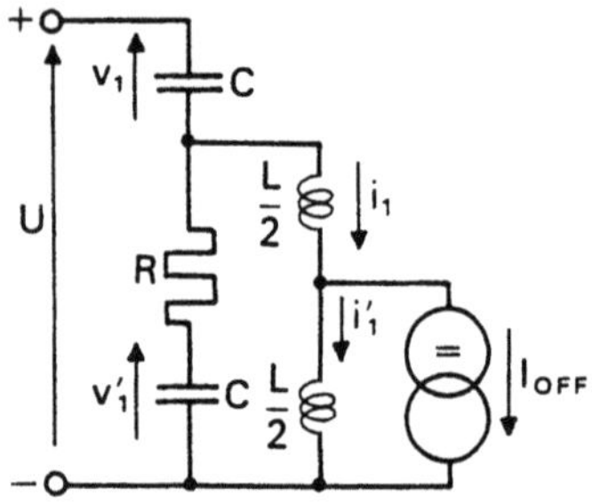

Fig. 4.33

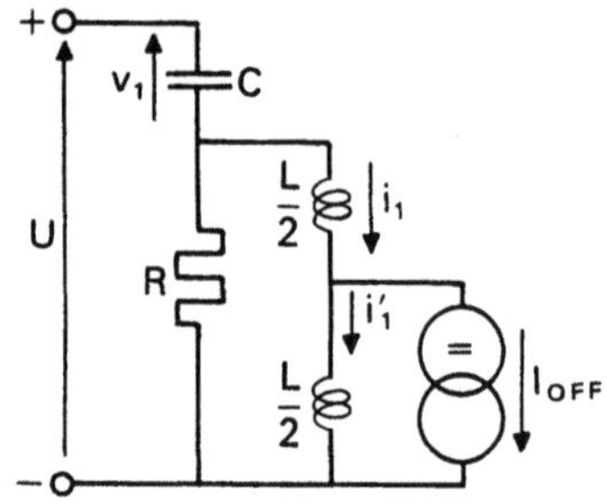

Fig. 4.34

- Voltage v'_1 is thus given by the following equation:

$$RLC^2\frac{d^3v'_1}{dt^3} + 2LC\frac{d^2v'_1}{dt^2} + RC\frac{dv'_1}{dt} + v'_1 = 0.$$

Its solution has the following form

$$v'_1 = A'\sin(\omega_1 t + \varphi_1)\exp(\alpha_1 t) + B'\exp(\beta_1 t).$$

- This phase ends, for $t = t_3$, when, with voltage v_1 falling to zero, diode DA'_1 begins conducting.

Period $t'_3 - t'_2$ is short; it can be assumed that i_R remains once more close to $-I_{OFF}/2$.

$$t'_3 - t'_2 \simeq \frac{2Cv'_1(t'_2)}{I_{OFF}} \simeq RC.$$

At the end of this period, $i_{D'1}$ is still very small and i_1 slightly less than I_{OFF}.

- *Phase 4 ($t>t'_3$): DA_1, D'_1 and DA'_1 conducting simultaneously*

Once again, the equivalent circuit (Fig. 4.34) is similar to that obtained for phase 3 of the $D'_1 \to Q_1$ commutation (see Fig. 4.26).

- By denoting $\omega_3 = 1/\sqrt{LC}$ and $\alpha_3 = 1/2RC$, the following expression of i_1 is obtained:

  - if $\omega_3 > \alpha_3$,

$$i_1 = \frac{i_1(t'_3)}{\sin\varphi'_3}\exp[-\alpha_3(t - t'_3)]\sin[\beta_3(t - t'_3) + \varphi'_3] \tag{4.46}$$

with

$$\beta_3 = \sqrt{\omega_3^2 - \alpha_3^2} \quad \text{and} \quad \tan\varphi'_3 = \frac{\beta_3 i_1(t'_3)}{\alpha_3 i_1(t'_3) + \dfrac{di_1}{dt}(t'_3)}$$

- if $\omega_3 < \alpha_3$,

$$i_1 = \frac{i_1(t'_3)}{\operatorname{sh}\varphi'_3}\exp[-\alpha_3(t-t'_3)]\operatorname{sh}[\beta_3(t-t'_3)+\varphi'_3] \tag{4.46'}$$

with

$$\beta_3 = \sqrt{\alpha_3^2 - \omega_3^2} \quad \text{and} \quad \operatorname{th}\varphi'_3 = \frac{\beta_3 i_1(t'_3)}{\alpha_3 i_1(t'_3) + \dfrac{\mathrm{d}i_1}{\mathrm{d}t}(t'_3)}.$$

From i_1, current i'_1 and voltage v_1 can be deduced:

$$i'_1 = i_1 - I_{\mathrm{OFF}}, \qquad v_1 = U - L\frac{\mathrm{d}i_1}{\mathrm{d}t}.$$

- When t tends towards infinity, i_1 tends towards zero,
  i'_1 tends towards $-I_{\mathrm{OFF}}$,
  and v_1 tends towards U;
  only diode D'_1 is still conducting and commutation is over.

4.3.3 Power Dissipated in the Resistance

4.3.3.1 Energy Dissipated During the $D_1 \to Q_1$ Commutation

A current flows through the resistance during phases 2 and 3 of this commutation.

- For phase 2, we use the simplified expression of v_1 and ignore I_{RR} compared to $2C\omega_2 U$, which gives:

$$i_R = C\frac{\mathrm{d}v_1}{\mathrm{d}t} = -C\omega_2 U \sin\omega_2(t-t_1).$$

The energy dissipated in R during this phase thus has the following value:

$$W_{R\,\mathrm{ON}\,2} = \int_{t_1}^{t_2} R\left(C\frac{\mathrm{d}v_1}{\mathrm{d}t}\right)^2 \mathrm{d}t.$$

Since $t_2 = t_1 + \pi/2\omega_2$,

$$W_{R\,\mathrm{ON}\,2} = \int_0^{\pi/2} RC^2\omega U^2 \sin^2\omega t \,\mathrm{d}\omega t$$

$$= \frac{\pi R}{4}\sqrt{\frac{C}{2L}}CU^2.$$

- During phase 3, current i_R can be expressed as:

$$i_R = \frac{1}{R}(U - v'_1) = \frac{1}{R}\left[U - \left(U - L\frac{\mathrm{d}i'_1}{\mathrm{d}t}\right)\right] = \frac{L}{R}\frac{\mathrm{d}i'_1}{\mathrm{d}t}.$$

By using Eq. (4.39) which gives i'_1 the following can be deduced:

$$i_R = -\frac{L\omega_3}{R}\frac{i'_1(t_2)}{\sin\varphi_3}\exp[-\alpha_3(t-t_2)]\sin\beta_3(t-t_2).$$

This gives the energy dissipated in R during phase 3:

$$W_{R\,\mathrm{ON}\,3} = \int_0^\infty R\left[\frac{L\omega_3}{R}\frac{i'_1(t_2)}{\sin\varphi_3}\exp(-\alpha_3 t)\sin\beta_3 t\right]^2 dt.$$

The calculation is carried out in the same way as the calculation for resistances R_1 and R'_1 in bipolar transistor snubbers. This gives:

$$W_{R\,\mathrm{ON}\,3} = \frac{R}{2}\left[\frac{L\omega_3}{R}\frac{i'_1(t_2)}{\sin\varphi_3}\right]^2\frac{\beta_3^2}{2\alpha_3\omega_3^2}$$

or, taking expressions of ω_3, α_3 and β_3 into account,

$$W_{R\,\mathrm{ON}\,3} = \frac{1}{2}Li'^2_1(t_2).$$

This energy is directly linked to the overcurrent in the thyristor during the ON-commutation, since

$$i_{Q\,\max} = I_{\mathrm{ON}} + i'_1(t_2).$$

$W_{R\,\mathrm{ON}\,3}$ should therefore be minimised.

If, for $i'_1(t_2)$, the approximate value $2C\omega_2 U$ is taken,

$$W_{R\,\mathrm{ON}\,3} = CU^2$$

and the total energy dissipated in R is

$$W_{R\,\mathrm{ON}} = W_{R\,\mathrm{ON}\,2} + W_{R\,\mathrm{ON}\,3} = CU^2\left(\frac{\pi R}{4}\sqrt{\frac{C}{2L}}+1\right).$$

4.3.3.2 Energy Dissipated During the $Q_1 \rightarrow D'_1$ Commutation

A current flows through resistance R throughout the OFF-commutation

- During phase 2, it has been seen that, if a relatively low value of time constant RC was chosen, current i_R remained virtually constant:

$$i_R \simeq -\frac{I_{\mathrm{OFF}}}{2}.$$

In order to make the calculation more simple, phases 1, 2 and 3 can be brought together in a single phase during which capacitor C'_1 discharges at a constant current from U to zero. This phase would last:

$$t'_3 = \frac{2CU}{I_{\mathrm{OFF}}}$$

and the corresponding energy dissipated in R would be:

$$W_{R\,\mathrm{OFF}\,3} = \int_0^{t'_3} R\frac{I^2_{\mathrm{OFF}}}{4}\mathrm{d}t = \frac{1}{2}RCUI_{\mathrm{OFF}}.$$

- The calculation of the energy dissipated in R during phase 4 is carried out as for phase 3 of the $D'_1 \rightarrow Q_1$ commutation. Similarly, the following result is obtained:

$$W_{R\,\mathrm{OFF}\,4} = \frac{1}{2}Li_1^2(t'_3).$$

Current i_1, which had remained constant during phases 1 and 2, has decreased only very slightly during phase 3. The following could thus be taken:

$$i_1(t'_3) \simeq I_{\mathrm{OFF}}, \qquad W_{R\,\mathrm{OFF}\,4} \simeq \frac{1}{2}LI^2_{\mathrm{OFF}}.$$

- The total energy dissipated in R during the $Q_1 \rightarrow D'_1$ commutation is thus equal to:

$$W_{R\,\mathrm{OFF}} = \frac{1}{2}LI^2_{\mathrm{OFF}} + \frac{1}{2}RCUI_{\mathrm{OFF}}.$$

there appears a term which is directly proportional to time constant RC which must therefore be minimised.

4.3.3.3 Power Dissipated in the Resistance

Resistance R is involved in all commutations concerning the half-bridge of which it forms a part.

- If the inverter is operating with only one pulse per alternance, there are only two commutations with non-zero current per cycle $1/f$ of the output values. Chapter 2 gives the value of current i' at the commutation instants.
- If the inverter is operating is Pulse Width Modulation with a modulation ratio equal to m, resistance is involved in $2m$ commutations per cycle $1/f$. This gives the dissipated power:

$$P_R = f\left(\sum_{k=1}^{m} W_{R\,\mathrm{ON}} + \sum_{k=1}^{m} W_{R\,\mathrm{OFF}}\right).$$

Since only $W_{R\,\mathrm{OFF}}$ depends on the switched current,

$$P_R = mfW_{R\,\mathrm{ON}} + f\sum_{k=1}^{m} W_{R\,\mathrm{OFF}}.$$

The following can be written

$$\sum_{k=1}^{m} W_{R\,\mathrm{OFF}} = \frac{1}{2}L\sum_{k=1}^{m} I^2_{\mathrm{OFF}} + \frac{1}{2}RCU\sum_{k=1}^{m} I_{\mathrm{OFF}}$$

and the mean value of i'^2 can be taken as the mean value of I^2_{OFF} and that of $|i'|$ as the mean value of I_{OFF}

As a result this leads to:

$$P_R = mf\left[CU^2\left(1 + \frac{\pi R}{4}\sqrt{\frac{C}{2L}}\right) + \frac{1}{2}LI'^2 + \frac{\sqrt{2}}{\pi}RCUI'\right]. \tag{4.47}$$

- *Remarks*

- Terms CU^2 and LI'^2 can be found in the expression of P_R, as in the case of the bipolar transistor [see Eq. (4.55)]. However, if resistance R is small, both capacitors may be involved simultaneously in order to limit the rate of rise of the voltage across the GTO at turn-off; the capacitance of the capacitors and the associated energy CU^2 can thus be approximately halved.

  Furthermore, since, at Q_1 turn-on, current I_{ON} is completely transferred from D'_1 to Q_1 before the discharge of capacitors starts, the discharge current is limited by inductance L, enabling a low value to be kept for R.

- The commutation losses in the GTO thyristor can be ignored when it is turned on.

  At turn-off, the preceding study only takes into account the GTO losses during phase 1 of the current fall, since the tail current has been ignored. In fact the major part of the GTO commutation losses occurs during the period when this current exists and voltage v_{Q_1} is rising.

  Figure 4.35 shows the variations in the instantaneous power $v_{Q_1}i_{Q_1}$ lost in GTO thyristor Q_1 during the OFF-commutation.

  In order to reduce these losses, the voltage rise which, as an initial approximation, is inversely proportional to C, must be slowed down.

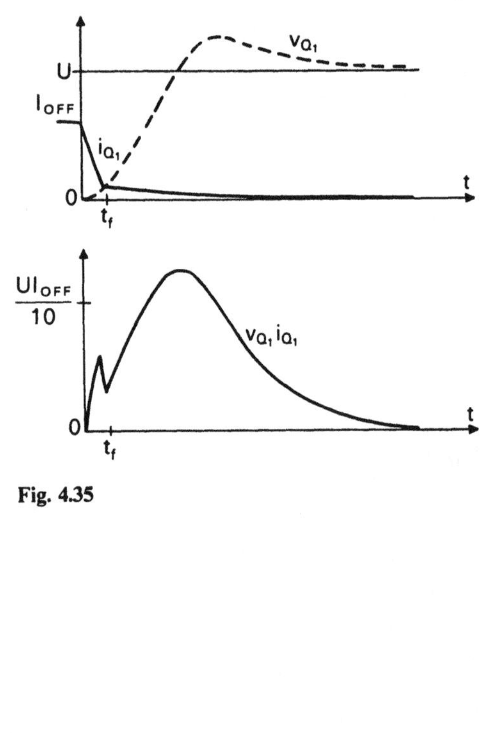

Fig. 4.35

4.4 Field Effect Transistor Commutations

The maximum current and voltage values which the field-effect transistor can sustain limit its use to low-power inverters, requiring no protection circuits against di/dt or dv/dt.

The main problem of field-effect transistor inverters is limited to their commutation speed which requires the use of sufficiently fast *diodes*.

4.4.1 ON-Commutation

In the diagram of a half-bridge inverter shown in Fig. 4.36, l indicates the stray inductances of the connections. Next to this diagram, we have drawn the waveforms of current i_{T_1} and voltage v_{T_1}, of current $i_{D'_1}$ and voltage $v_{D'_1}$ during the $D'_1 \rightarrow T_1$ commutation. It is assumed that current i' remains constant during the commutation and its value is denoted by I_{ON}.

Initially, current i_{T_1} rises from zero to I_{ON} under the effect of the gate control. Voltage v_{T_1} only decreases in relation to U by the value of the voltage drop in the stray inductance; thus voltage fall is constant if i_{T_1} increases linearly.

When i_{T_1} reaches value I_{ON}, voltage v_{T_1} gradually decreases from a value close to U down to zero; but the voltage across diode D'_1 remains very small on

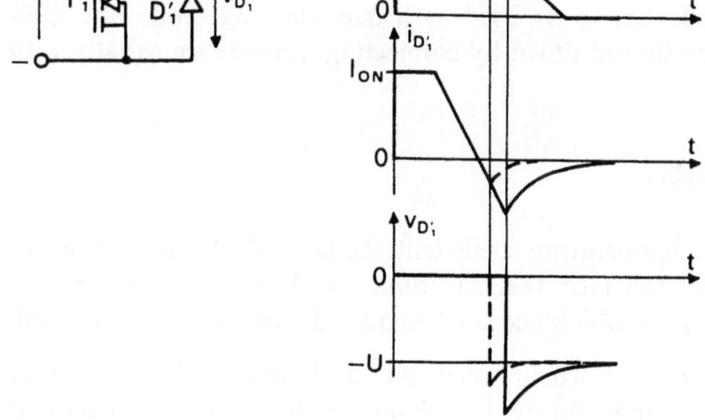

Fig. 4.36

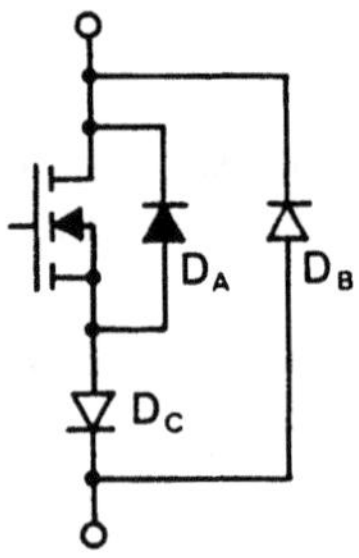

Fig. 4.37

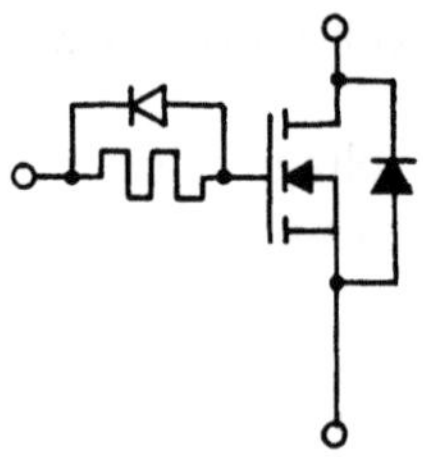

Fig. 4.38

account of the reverse recovery current. Only when this reverse current reaches its maximum value does voltage $v_{D'_1}$ become negative.

- If diode D'_1 is fast (dashed lines), this maximum value is low, thus limiting the overcurrent in the transistor, since

$$i_{T_1} = I_{ON} - i_{D'_1},$$

as well as the voltage surge across D'_1 and T'_1, since

$$v_{D'_1} = v_{T_1} - \left(U + 2l\frac{di_{D'_1}}{dt}\right).$$

- If diode D'_1 is slow (full lines), the voltage surge in transistor T_1 and the surge across T'_1 and D'_1 may be much higher.

- *Remark.* Field-effect transistors have an integrated diode in antiparallel; this body diode is inherent in the manufacturing process and is not a fast diode.

- If a fast diode D_B in antiparallel is connected across the transistor (Fig. 4.37), it may be necessary to prevent the body diode from conducting by series-connecting with the transistor a diode with a low voltage drop.

- If the integrated diode has to be used as a free-wheeling diode, the ON-commutation must be slowed down by connecting a resistance serially with the gate (Fig. 4.38).

4.4.2 OFF-Commutation

As a result of its positive temperature coefficient, the field-effect transistor is not subject to second breakdown or thermal runaway. It is therefore able to withstand the voltage surges which occur when an inductive load is turned off.

However, to avoid breakdown between gate and source which could be caused by capacitive couplings during drain-source voltage surges, a transient

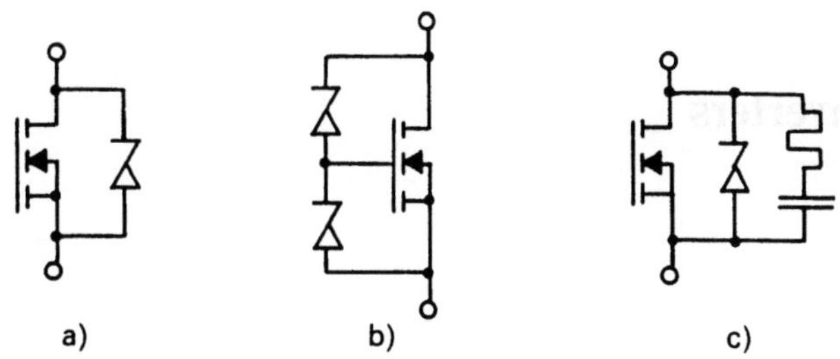

Fig. 4.39

overvoltage protection can be installed between drain and source or even between drain and gate, and gate and source (Fig. 4.39a and b).

In cases where the connection stray inductance would cause excessively high voltage surges, each transistor must be protected by an *RC* circuit (Fig. 4.39c).

Chapter 5

Current-Source Inverters

– In order to generate a single-phase AC current from a DC current supply, the following can be used:

- either a two-switch inverter connected to a centre-tapped transformer (push-pull assembly),
- or a full-bridge inverter with four switches.

To generate a three-phase current system, a six-switch bridge should normally be used.

As suggested in Chap. 1, current-source inverters (also known as current commutators) are essentially used to supply medium- and high-power motors in variable-speed-drive systems. We thus propose only a summary study of single-phase inverters and will devote the major part of this chapter to three-phase inverters and to the multi-phase configurations derived from the latter.

– In the study of the operating principle of each current-source inverter, the DC current supply is assumed to be perfect, i.e. supplying a DC current whose ripple can be ignored. The load is equally assumed to be perfect, i.e. acting as an ideal single- or multi-phase AC voltage source.

The phase shift φ between the voltage across the terminals of each phase and the fundamental component of the current wave which is flowing through it will be taken as positive if the current is lagging the voltage. A leading phase shift will thus correspond to a negative value of φ.

– By studying, firstly, the single-phase inverters and then the full-bridge three-phase inverter, we will examine the full wave operation and then Pulse Width Modulation.

We will then show how the inductive nature of the load can be compensated for during commutations and describe the most common types of configuration used.

Finally, we will make some remarks concerning multi-phase inverters with a number of phases greater than 3 and the problem of filtering the inverter input current.

5.1 Single-Phase Inverters

5.1.1 Inverters with a Centre-Tapped Transformer

Figure 5.1 shows the configuration of a single-phase current-source inverter with two controlled switches TC_1 and TC'_1 and a centre-tapped transformer. The current I supply has one terminal connected to the centre point O; the other is connected either to A by the flow of TC_1 or to B by the flow of TC'_1.

If the transformer is assumed to be perfect and the magnetizing ampere-turns are ignored, this gives:

$$v_1 = v'_1 = (n_1/2n_2)u' \tag{5.1}$$

$$(n_1/2)(i_{T_1} - i_{T'_1}) = n_2 i', \tag{5.2}$$

In the case of this inverter, our study will be confined to full-wave operation.

5.1.1.1 Expressions. Waveforms

For $0 < t < T/2$, TC_1 is on and TC'_1 off:

$$u = v_1 = v'_1 = (n_1/2n_2)u' = (n_1/2n_2)U'\sqrt{2}\sin(\omega t + \varphi)$$

$$v_{T_1} = 0, \qquad v_{T'_1} = u + v'_1 = (n_1/n_2)U'\sqrt{2}\sin(\omega t + \varphi)$$

$$i_{T_1} = I, \qquad i' = (n_1/2n_2)I, \qquad i_{T'_1} = 0.$$

For $T/2 < t < T$, TC_1 is off and TC'_1 on:

$$u = -v'_1 = -v_1 = -(n_1/2n_2)u' = -(n_1/2n_2)U'\sqrt{2}\sin(\omega t + \varphi)$$

$$v_{T_1} = u - v_1 = -(n_1/n_2)U'\sqrt{2}\sin(\omega t + \varphi) \qquad v_{T'_1} = 0$$

$$i_{T_1} = 0, \qquad i_{T'_1} = I, \qquad i' = -(n_1/2n_2)I.$$

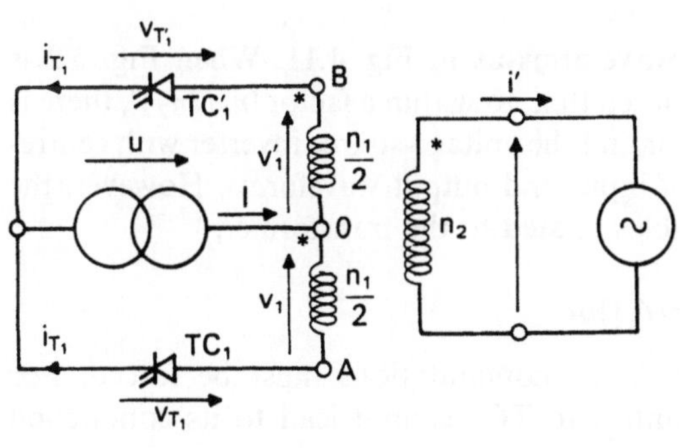

Fig. 5.1

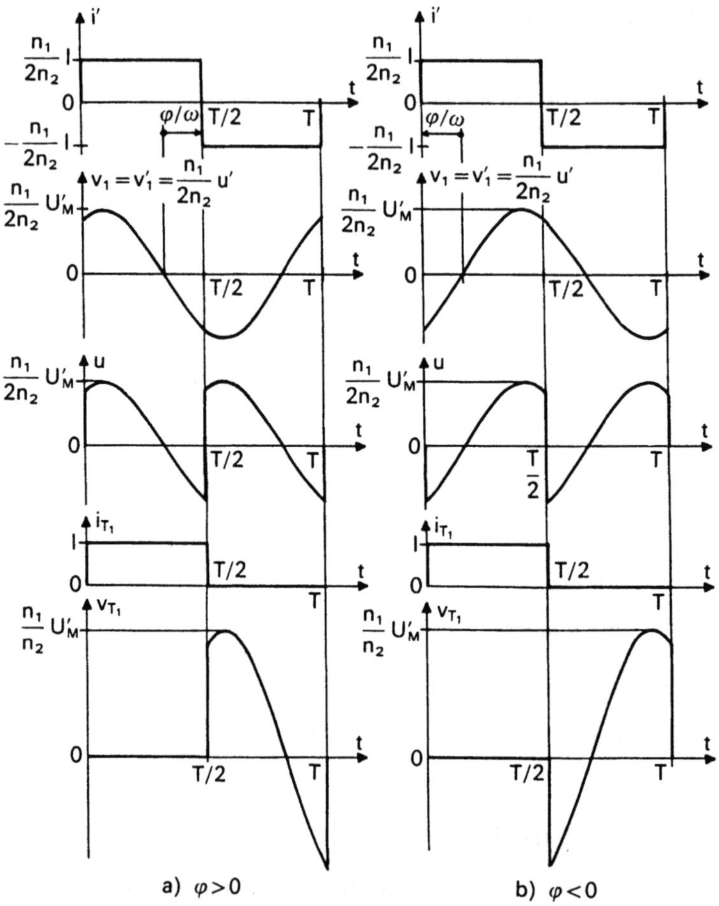

Fig. 5.2

Figure 5.2 gives the waveforms of i', of v_1, and u', of u, of i_{T_1} and v_{T_1}

(a) for a lagging shift ($\varphi > 0$),
(b) for a leading shift ($\varphi < 0$).

This gives exactly the same waveforms as in Fig. 1.11. When Fig. 5.2 is compared with Fig. 2.2, it can be noted that, to within a factor $(n_1/2n_2)^2$, there is perfect duality between this inverter and the voltage-source inverter with centre-tapped transformer at the level of input and output waveforms. However, the same does not apply to the variables related to the transformer.

5.1.1.2 Commutations to be Carried Out

– When feeding an *inductive load*, the commutations must be forced. For $t = T/2$, applying a turn-on control to TC_1' cannot lead to its conduction since it is reverse-biased. The forced interruption of the TC_1 flow causes current I to be transferred from TC_1 towards TC'_1. Turning off TC_1 leads to a positive voltage appearing across it, equal to:

$$v_{T_1} = -(n_1/n_2)U'\sqrt{2}\sin(\pi + \varphi) = (n_1/n_2)U'\sqrt{2}\sin\varphi, \qquad \text{with } \varphi > 0.$$

- When feeding a *capacitive load*, the commutations are natural. For $t = T/2$, TC'_1 has a positive voltage across it; it is fired by the turn-on control which is applied at that instant, leading to TC_1, which is subject to a negative voltage, being turned off[1].

5.1.1.3 Doubling the Voltage

On account of the centre-tapped transformer, switches TC_1 and TC'_1 – when turned off – are subject to a voltage which is twice that which appears across the load at that moment.

5.1.1.4 Magnetizing Current

Flux ϕ in the magnetic circuit of the transformer is linked to voltage u' by

$$n_2 \frac{d\phi}{dt} = u'.$$

The flux has the sinusoidal waveform shown in the upper part of Fig. 5.3. If the material of the magnetic circuit were of constant permeability, the magnetizing current i_μ would also be sinusoidal. Owing to saturation, it has the form shown in the diagram.

The magnetizing current modifies the output current i' of the inverter since

$$\text{for } 0 < t < T/2, \qquad i' = (n_1/2n_2)I - i_\mu$$

$$\text{for } T/2 < t < T, \qquad i' = -(n_1/2n_2)I - i_\mu.$$

In the lower part of Fig. 5.3, the modified waveforms of i' are shown, firstly for an inductive load and then for a capacitive load.

5.1.1.5 Characteristics

- If the magnetizing ampere-turns are ignored, the output current i' is an AC square wave with an amplitude $(n_1/2n_2)I$.

[1] The operating mode in the case of a capacitive load corresponds to that described for the controlled rectifiers in Vol. 1, Chap. 7. Any differences concerning the waveforms and the power signs arise from the change in the relative reference directions of u and i, of u' and i'. In the study of rectifiers, we chose source-type reference directions (see Vol. 3, Chap. 1, §2) for the circuit connected to the AC side and load-type for the circuit connected to the DC side.

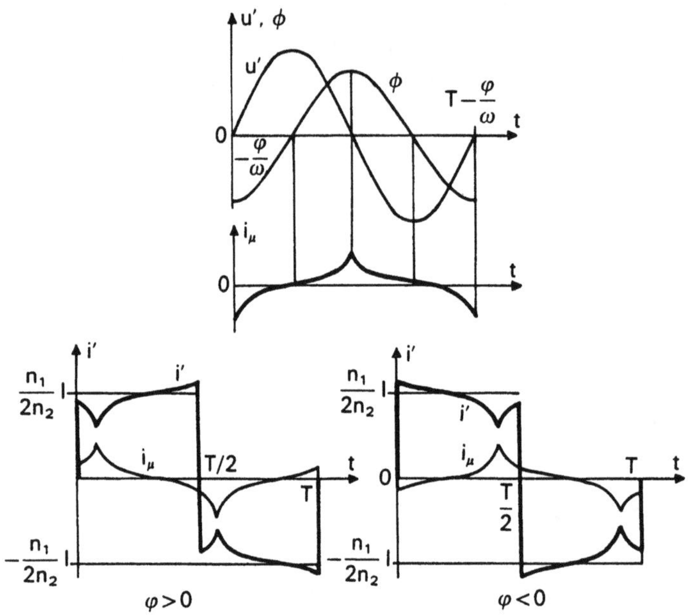

Fig. 5.3

Its series expansion comprises all the uneven harmonics, the amplitude of each being inversely proportional to its rank:

$$i' = \frac{2}{\pi}\frac{n_1}{n_2} I\left(\sin \omega t + \frac{1}{3}\sin 3\omega t + \frac{1}{5}\sin 5\omega t + \cdots\right). \tag{5.3}$$

The rms value of the fundamental is

$$I'_1 = \frac{\sqrt{2}}{\pi}\frac{n_1}{n_2} I \tag{5.4}$$

and the harmonic rate is equal to

$$\tau_{i'} = 0.483. \tag{5.5}$$

- *The output voltage* is assumed to be a sinusoidal wave of rms value U'. As in the case of the half-bridge voltage-source inverter, the series expansion of the instantaneous power shows AC terms of angular frequency 2ω, 4ω, 6ω, ..., in addition with the constant term.

- The apparent power of the *transformer* secondary winding is equal to

$$U'(n_1/2n_2)I. \tag{5.6}$$

The apparent power of the primary is $\sqrt{2}$ higher.

- *The current in the semiconductor devices* has a peak value equal to I, an average value equal to $I/2$, and a rms value of $I/\sqrt{2}$.

  The voltage across these devices is now positive, now negative; its extreme values are

  $$(n_1/n_2)U'\sqrt{2} \quad \text{and} \quad -(n_1/n_2)U'\sqrt{2}.$$

- *The input voltage* has a frequency which is twice that of the inverter operating frequency. For $0 < \omega t < \pi$, it can be expressed as:

  $$u = (n_1/2n_2)U'\sqrt{2}\sin(\omega t + \varphi).$$

  From which the following can be deduced:

  - its mean value:

  $$U = \frac{2}{\pi}\frac{n_1}{2n_2}U'\sqrt{2}\cos\varphi \tag{5.7}$$

  - its rms value:

  $$U_{\text{rms}} = (n_1/2n_2)\,U' \tag{5.8}$$

  - its harmonic rate:

  $$\tau_u = \frac{\pi}{2\sqrt{2}\cos\varphi}\sqrt{1 - \frac{8}{\pi^2}\cos^2\varphi}. \tag{5.9}$$

  In addition to its average value U, the series expansion of voltage u comprises AC terms of 2ω, 4ω, $\ldots$, angular frequency. The rms value of the term with $2k\omega$ angular frequency is:

  $$U_{2k} = \frac{n_1}{2n_2}U'\frac{4|\cos\varphi|}{\pi(4k^2-1)}\sqrt{1 + 4k^2\tan^2\varphi}. \tag{5.10}$$

  This rms value is minimum for zero φ and maximum for φ equal to $\pm\pi/2$.

  $$U_{2k\,\text{min}} = \frac{n_1}{2n_2}U'\frac{4}{\pi(4k^2-1)}$$

  $$U_{2k\,\text{max}} = \frac{n_1}{2n_2}U'\frac{8k}{\pi(4k^2-1)}.$$

- As a result of the duality which exists between the current-source inverter and the voltage-source inveter as far as the input and output waveforms are concerned, Fig. 2.5 in Chap. 2 gives the variations, for the current-source inverter, of the mean value U of the input voltage and of the rms value U_h of its first harmonics, as a function of φ, providing that the following substitutions are made:

$I/(2n_2/n_1)I$, replaced by $u/(n_1/2n_2)U'$

$I_h/(2n_2/n_1)I'$ replaced by $U_h/(n_1/2n_2)U'$.

5.1.2 Full-Bridge Inverter with One Square-Wave Pulse per Alternance

The full-bridge structure was used in Chap. 1 to present the current-source inverters. Fig. 5.4 shows again the diagram of this full-bridge single-phase current-source inverter as well as the notations used.

Firstly, we will study the operating mode and the characteristics for the case of a single pulse per alternance. We will then go on to examine the Pulse Width Modulation control.

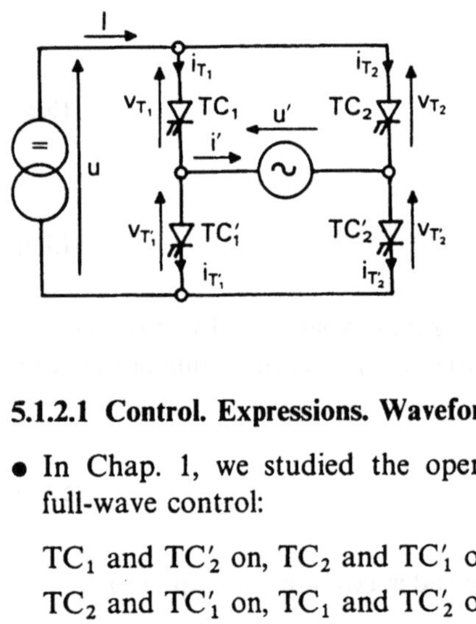

Fig. 5.4

5.1.2.1 Control. Expressions. Waveforms

- In Chap. 1, we studied the operating mode by considering the case of full-wave control:

  TC_1 and TC'_2 on, TC_2 and TC'_1 off, for $0 < \omega t < \pi$
  TC_2 and TC'_1 on, TC_1 and TC'_2 off, for $\pi < \omega t < 2\pi$.

  If the controls of TC_1 and TC_2, on the one hand, and TC'_1 and TC'_2 on the other have to remain complementary in order to avoid putting the DC current source in open circuit, the controls of TC_1 and TC'_2 (and, consequently, of TC_2 and TC'_1) can be shifted between themselves by an angle β other than zero:

  - TC_1 is on, for $0 < \omega t < \pi$
  - TC'_2 is on, for $\beta < \omega t < \pi + \beta$
  - TC_2 is on, for $\pi < \omega t < 2\pi$
  - TC'_1 is on, for $\pi + \beta < \omega t < 2\pi + \beta$.

- Four configurations per operating cycle must thus be considered:
  - for $0 < \omega t < \beta$, TC_1, and TC'_1 are on, TC'_2 and TC_2 off: the current source is short-circuited by TC_1 and TC'_1, the voltage source being in open circuit:

$$u = 0, \qquad i' = 0$$

$$i_{T_1} = I, \quad i_{T'_1} = I, \quad i_{T_2} = 0, \quad i_{T'_2} = 0$$

$$v_{T_1} = 0, \quad v_{T'_1} = 0, \quad v_{T_2} = u', \quad v_{T'_2} = -u'.$$

  - for $\beta < \omega t < \pi$, TC_1 and TC'_2 are on, TC_2 and TC'_1 off: the current source is connected to the voltage source via TC_1 and TC'_2:

$$u = u', \qquad i' = I$$

$$i_{T_1} = I, \quad i_{T'_1} = 0, \quad i_{T_2} = 0, \quad i_{T'_2} = I$$

$$v_{T_1} = 0, \quad v_{T'_1} = u', \quad v_{T_2} = u', \quad v_{T'_2} = 0.$$

  - for $\pi < \omega t < \pi + \beta$, TC_2 and TC'_2 are on, TC_1 and TC'_1 off:

$$u = 0, \qquad i' = 0$$

$$i_{T_1} = 0, \quad i_{T'_1} = 0, \quad i_{T_2} = I, \quad i_{T'_2} = I$$

$$v_{T_1} = -u', \quad v_{T'_1} = u', \quad v_{T_2} = 0, \quad v_{T'_2} = 0.$$

  - for $\pi + \beta < \omega t < 2\pi$, TC_2 and TC'_1 are on, TC_1 and TC'_2 off;

$$u = -u', \qquad i' = -I$$

$$i_{T_1} = 0, \quad i_{T'_1} = I, \quad i_{T_2} = I, \quad i_{T'_2} = 0$$

$$v_{T_1} = -u', \quad v_{T'_1} = 0, \quad v_{T_2} = 0, \quad v_{T'_2} = -u'.$$

- Current i' is equal to $+I$ during period $[\beta, \pi]$, to $-I$ during period $(\pi + \beta, 2\pi)$ and zero during the remainder of the cycle. Its fundamental is given by

  $$I'_1\sqrt{2}\sin(\omega t - \beta/2).$$

  Voltage u' can thus be expressed as:

  $$u' = U'\sqrt{2}\sin(\omega t - \beta/2 + \varphi).$$

In Fig. 5.5, below the semiconductor device conduction diagrams, the waveforms of i', u', u, i_{T_1}, v_{T_1}, $i_{T'_1}$ and $v_{T'_1}$ have been plotted for the case of an inductive load ($0 < \varphi < 180°$)

(a) for $\beta = 120°$, $\varphi = 30°$

(b) for $\beta = 120°$, $\varphi = 90°$.

The nature of the commutations (forced F or natural N) has also been noted.

Figure 5.6 shows the same waveforms, in the case of a capacitive load ($0 > \varphi > -180°$)

(a) for $\beta = 120°$, $\varphi = -90°$

(b) for $\beta = 120°$, $\varphi = -150°$.

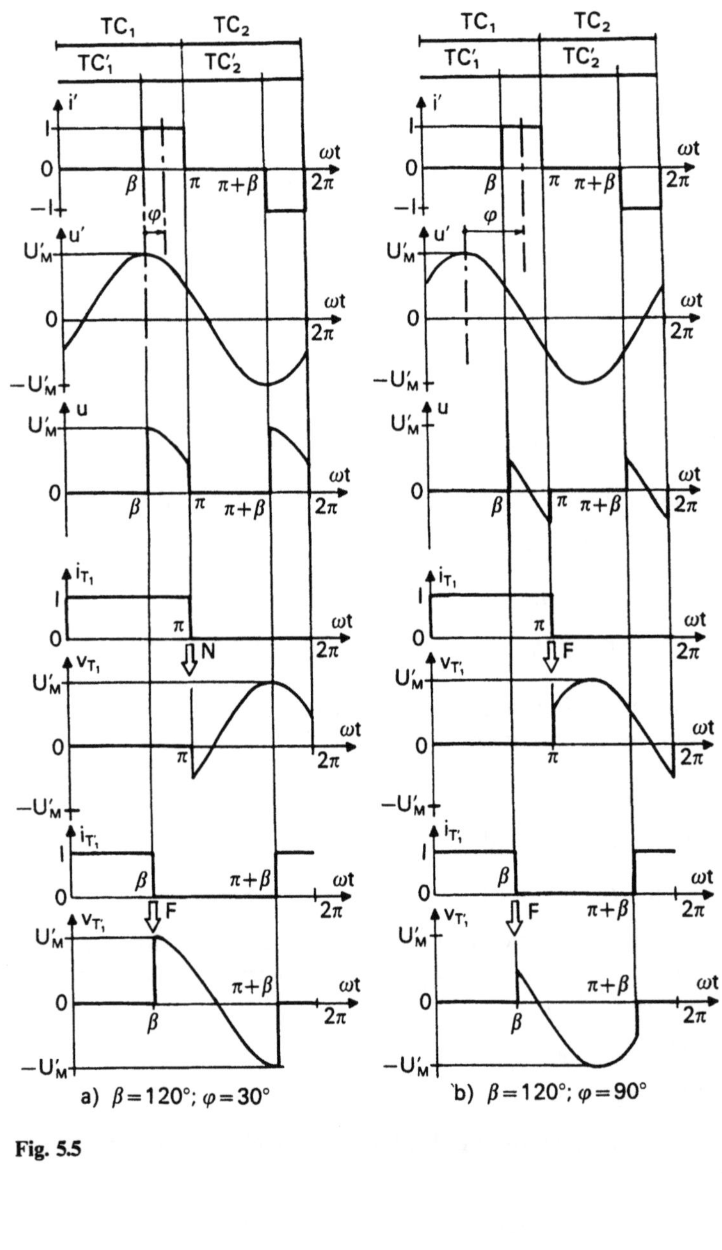

Fig. 5.5

The waveforms of i_{T_2}, v_{T_2}, $i_{T'_2}$ and $v_{T'_2}$ can be directly deduced from those plotted, since the following are still valid:

$$i_{T_2}(\omega t + \pi) = i_{T_1}(\omega t), \qquad v_{T_2}(\omega t + \pi) = v_{T_1}(\omega t)$$

$$i_{T'_2}(\omega t + \pi) = i_{T'_1}(\omega t), \qquad v_{T'_2}(\omega t + \pi) = v_{T'_1}(\omega t).$$

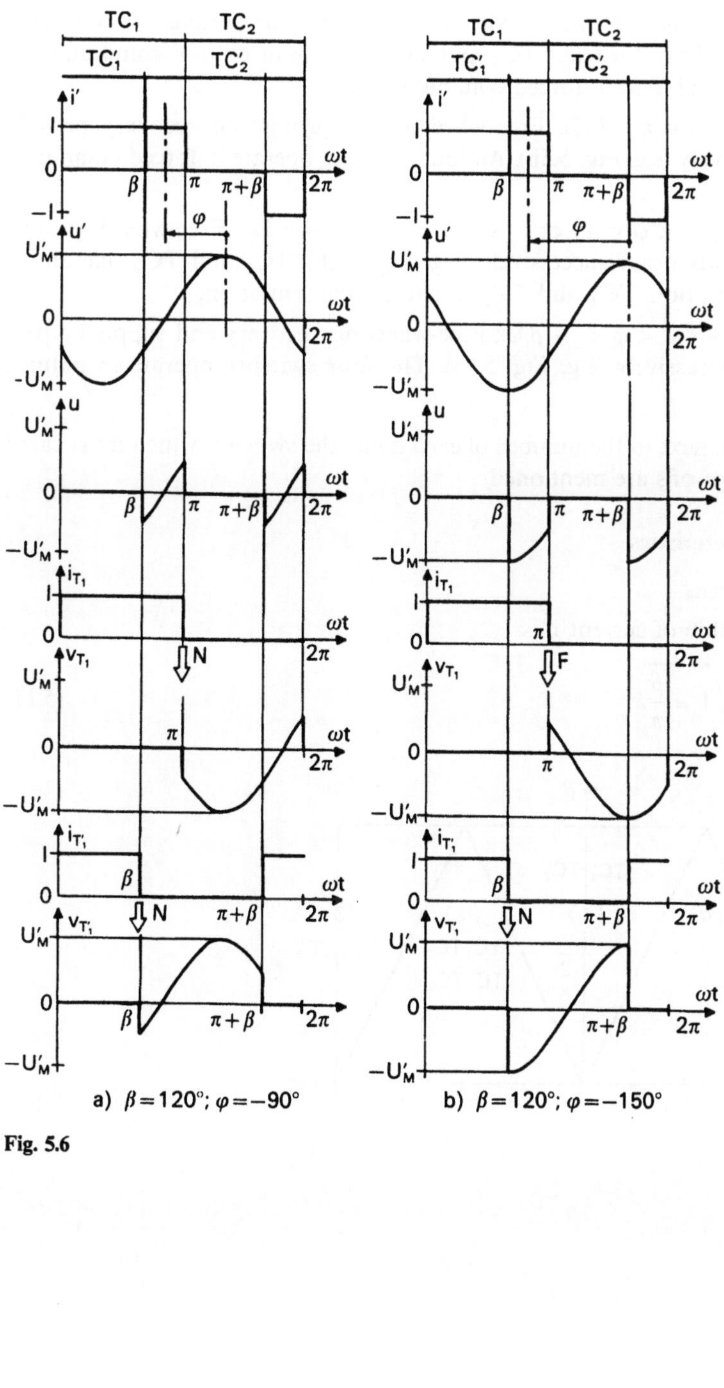

Fig. 5.6

By comparing Figs. 5.5 and 5.6 with Figs. 2.11 and 2.12, it can be observed that, as was noted in Chap. 1, there is perfect duality between the current-source inverter and the voltage-source inverter in the case of the full-bridge single-phase configuration.

5.1.2.2 Commutations to be Carried Out

The nature of the commutations to be carried out depends on φ and β. In the (φ, β) plane, 4 zones are to be distinguished (Fig. 5.7):

Zone 1 ($-\beta/2 < \varphi < \beta/2$). Free-wheeling and supplying periods occur successively (e.g Fig. 5.5a). Switches TC_1 and TC_2 operate in natural commutation, switches TC'_1 and TC'_2 in forced commutation.

Zone 2 ($\beta/2 < \varphi < \pi - \beta/2$). Free-wheeling, supplying and recovery periods occur successively (e.g. Fig. 5.5b). All four switches operate in forced commutation.

Zone 3 ($\pi - \beta/2 < \varphi < \pi$ or $-\pi < \varphi < -\pi + \beta/2$). Free-wheeling and recovery periods occur successively (e.g. Fig. 5.6b). TC_1 and TC_2 operate in forced commutation, TC'_1 and TC'_2 in natural commutation.

Zone 4 ($-\pi + \beta/2 < \varphi < -\beta/2$); Free-wheeling, recovery and supplying periods occur successively (e.g. Fig. 5.6a). The four switches operate in natural commutation.

In Fig. 5.6, next to the number of each zone, the switches which must carry out forced turn-offs are mentioned.

5.1.2.3 Characteristics

- *Output current*

The rms value of current i' is:

$$I' = I\sqrt{1 - \frac{\beta}{\pi}}. \tag{5.11}$$

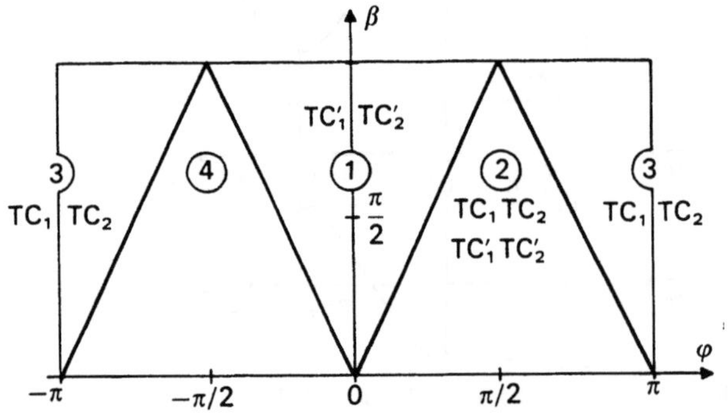

Fig. 5.7

The rms value of its fundamental component is:

$$I'_1 = I\frac{2\sqrt{2}}{\pi}\cos\beta/2. \tag{5.12}$$

The Fourier series comprises all the harmonics of uneven rank. The rms value of the harmonic of rank $2k + 1$ is equal to

$$I'_{2k+1} = \frac{I}{2k+1}\frac{2\sqrt{2}}{\pi}\cos(2k+1)\beta/2. \tag{5.13}$$

The curves in Fig. 2.15, which show the variations of U'/U, U'_1/U, U'_h/U and $(\sum U'^2_h)^{1/2}/U$, as a function of β, for the voltage-source inverter, also give the variations of I'/I, I'_1/I, I'_h/I and $(\sum I^2_h)^{1/2}/I$ for the current-source inverter.

- *Semiconductor devices*

Whatever the values of φ and of β, each semiconductor device delivers a current equal to I during half the cycle. This gives the peak, mean and rms values of the current in the switches.

$$i_{\mathrm{T\,max}} = I$$

$$I_{\mathrm{T\,mean}} = I/2$$

$$I_{\mathrm{T\,rms}} = I/\sqrt{2}.$$

The voltage across the semiconductor devices when they are in the blocked state is, at times, positive and, at others, negative. Its value is between $U'\sqrt{2}$ and $-U'\sqrt{2}$.

- *Input voltage*

The input voltage u cycle is half of that of the output voltage u' cycle. The mean value of voltage u is:

$$U = (2/\pi)\,U'\sqrt{2}\cos\varphi\cos(\beta/2) \tag{5.14}$$

and its rms value:

$$U_{\mathrm{rms}} = U'\sqrt{\frac{\pi-\beta}{\pi} + \frac{\sin\beta\cos 2\varphi}{\pi}}. \tag{5.15}$$

Owing to the duality between the two types of full-bridge inverter,

- Figure 2.17 gives the variations of U/U'
- Figure 2.18 the variations of U_{rms}/U'
- Figure 2.19 the variations of $(\sum U^2_h)^{1/2}/U'$, U_2/U', U_4/U' and U_6/U'

as a function of β for various values of $\cos\varphi$.

5.1.3 PWM-Controlled Full-Bridge Inverter

Filtering the output current i' can be made easier by forming each half-cycle of several square pulses of suitable widths.

The positive half-cycle is made up of pulses of amplitude $+I$ separated by periods at zero i'. The negative half-cycle is made up of pulses of amplitude $-I$ separated by periods at zero i'.

TC_1 and TC_2 must be turned on to obtain i' equal to $+I$. TC_2 and TC'_1 must be turned on to obtain i' equal to $-I$. In order to obtain a zero current i', the current source must be short-circuited by the simultaneous conduction of TC_1 and TC'_1 or of TC_2 and TC'_2.

5.1.3.1 Sinusoidal Modulation

Figure 5.8 shows an example of the i' waveform plotting. The commutation instants are given by the intersections of a sinusoidal reference and a triangular carrier ($m = 8$). We have indicated the sequence of switch turn-on which minimises the number of commutations.

It should be pointed out that, within each half-cycle, the modulation is carried out by commutations between two switches connected to the same terminal of the DC supply (TC_1 and TC_2 or TC'_1 and TC'_2). In the voltage-source inverter, the modulation was obtained by commutations between two switches connected to the same terminal of the load (K_1 and K'_1 or K_2 and K'_2).

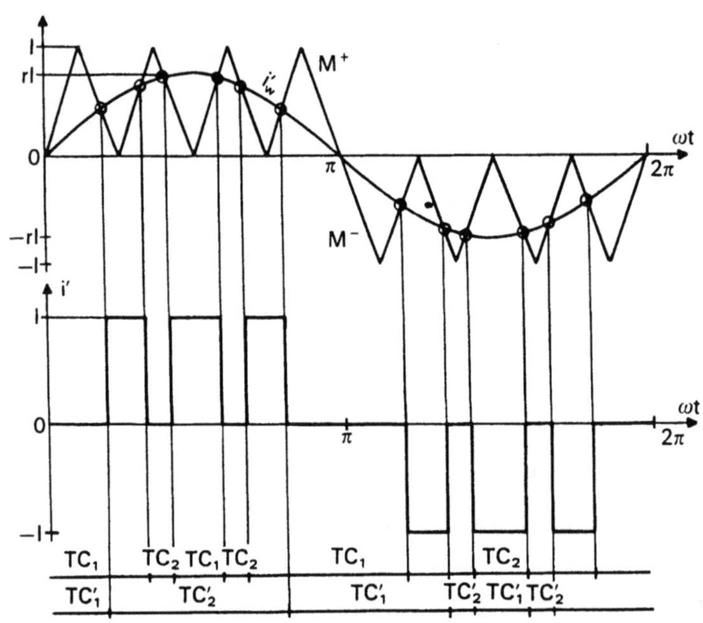

Fig. 5.8

The results of the harmonic analysis of the output voltage, as carried out for the full-bridge voltage-source inverter in Chap. 3, apply to the current-source inverter output current.

5.1.3.2 Computed Modulation

The calculation of angles $\theta_1, \theta_2, \cdots, \theta_p$ which characterise the p pulses forming each half-cycle of current i', can be carried out as in the case of the full-bridge voltage-source inverter.

The results obtained in Chap. 3 can be directly applied, as long as they are applied to i' and not to u'.

The switch control can be deduced from the waveform of current i' as shown in Fig. 5.8.

5.1.4 Remarks on PWM for Current-Source Inverters

5.1.4.1 The Usual Role of PWM

In a current-source inverter the generator connected to the input of the DC side normally imposes the value of the input current.

This is usually obtained by placing an intermediate converter and a smoothing inductance between the supply itself and the inverter input, and by using a current I regulation loop (Fig. 5.9).

This procedure is used since the usual supplies (AC mains, battery, DC current catenary) have static characteristics corresponding to voltage supplies even if their internal inductances, which oppose sudden variations of the current they deliver, means that they are current sources as far as commutations are concerned (see Vol. 3, Chap. 2, §2.2).

When the energy is supplied by the AC mains, current I is regulated by controlling the rectifier (Fig. 5.9a).

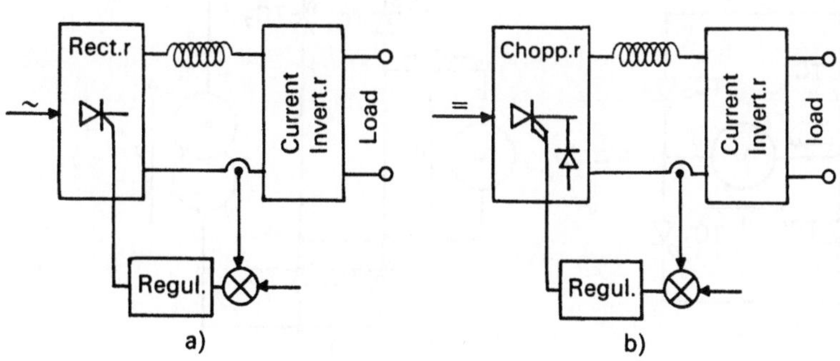

Fig. 5.9

When the supply is a catenary, this regulation is carried out by controlling the chopper (Fig. 5.9b).

The use of a regulation loop to control the input current I enables the latter's value to be varied. This explains why, in the case of current-source inverters, Pulse Width Modulation is *only used to improve the output current waveform* in the majority of cases. The ahead converter regulates the amplitude.

This is thus an important difference between these configurations and the volage-source inverters where, if a DC supply is available, it is very often connected to the inverter input, either directly or via a simple filter. In that case, the PWM control ensures also the regulation of the output wave fundamental amplitude.

5.1.4.2 From Current-Source Inverter to PWM Rectifier

We have already noted the similarity which exists between a current-source inverter and a naturally commuted rectifier as studied in Vol. 1. Only the nature of switches may vary.

We have studied until now the case of a converter operating as a genuine current-source inverter. This converter may also operate as PWM-controlled rectifier.

In the first case, it is fed by a generator which fixes the DC current amplitude; it feeds a load on which it imposes the frequency of the AC current by the commutation frequency control.

In the second case, the converter is fed by an AC voltage supply with fixed frequency and amplitude; it feeds a DC current load across which the DC voltage value is to be regulated.

To show the similarity more clearly, the usual presentation of the current-source inverter (Fig. 5.10a) can be modified giving that which is normally used for a full-bridge rectifier (Fig. 5.10b). Moreover, if the reference direction of the

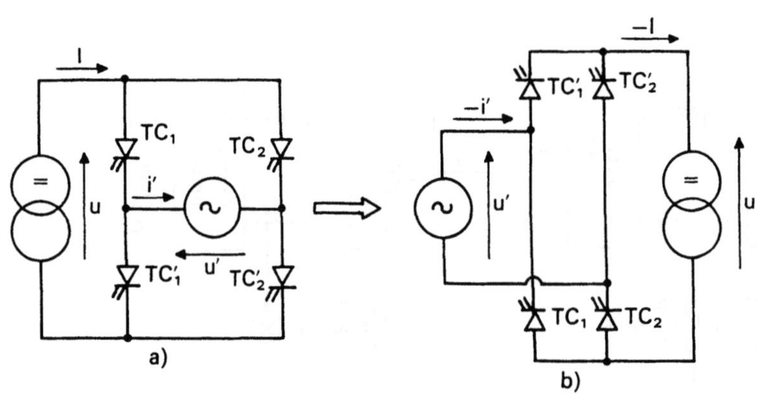

Fig. 5.10

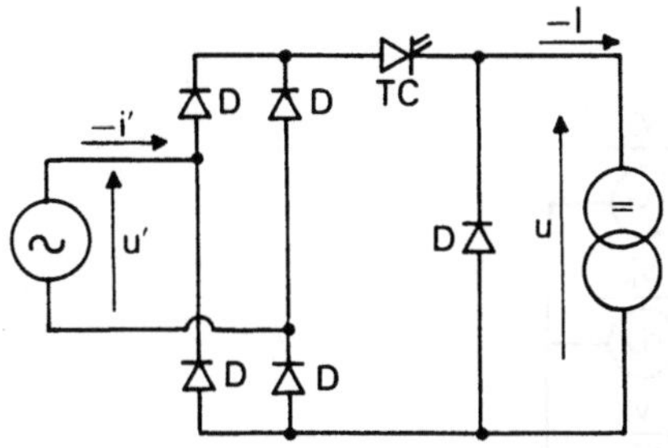

Fig. 5.11

current is reversed so as to take account of the normal direction of energy flow, the signs adopted for rectifiers can once more be found.

The rectifier with four controlled switches in Fig. 5.10b allows to vary the phase shift between the voltage u' and the current i' fundamental.

If the fundamental of $-i'$ has only to be put in phase with voltage u', the configuration in Fig. 5.10b can be replaced by that of Fig. 5.11. The latter is more economical.

Apppendix D deals with the study of these Pulse Width Modulated rectifiers.

5.2 Full-Bridge Three-Phase Inverter

The full-bridge three-phase current-source inverter comprises six controlled semiconductor switches. The load can be star-connected with a non connected neutral, or delta-connected. Figure 5.12 gives both the diagrams and indicates the notations which are used.

5.2.1 Full-Wave Control

One of the three switches connected to the same terminal of the DC current supply must always be turned on in order to prevent to put this supply in open circuit. There must only be one in order to avoid any short-circuiting across the voltage load.

If each switch in only turned on once per cycle and if a balanced three-phase current system is required at the output, this means that each "switch" must be turned on for a third of a cycle. The controls of two "switches" connected to a same input terminal are shifted by a third of a cycle; those of the two "switches" connected to a same output terminal by half a cycle.

The switches are thus turned on as follows:

TC_1 from $\omega t = 0$ to $\omega t = 2\pi/3$, TC'_1 from $\omega t = \pi$ to $\omega t = \pi + 2\pi/3$,

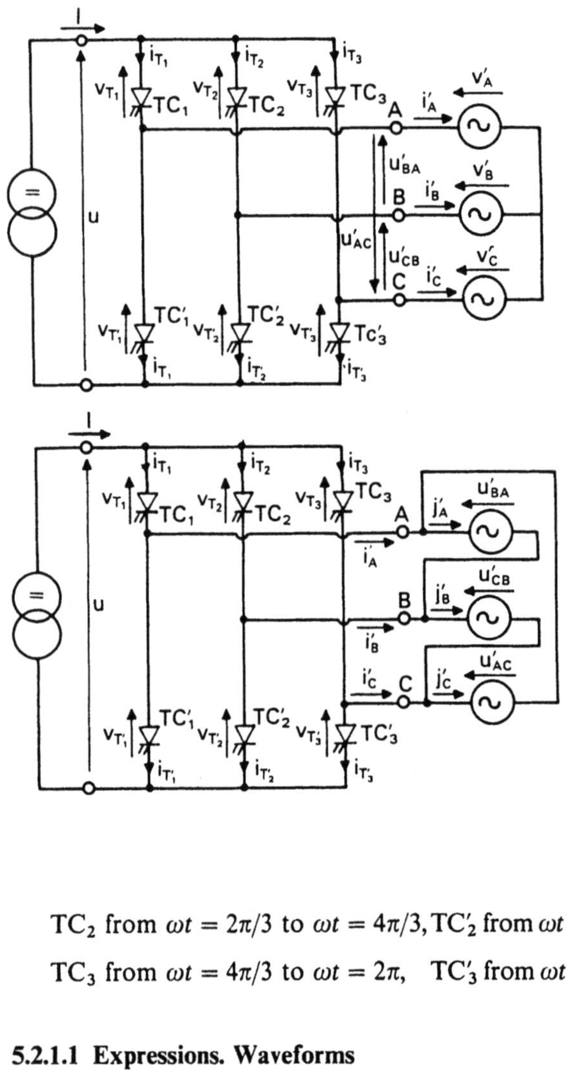

Fig 5.12

TC_2 from $\omega t = 2\pi/3$ to $\omega t = 4\pi/3$, TC'_2 from $\omega t = \pi + 2\pi/3$ to $\omega t = \pi + 4\pi/3$

TC_3 from $\omega t = 4\pi/3$ to $\omega t = 2\pi$, TC'_3 from $\omega t = \pi + 4\pi/3$ to $\omega t = \pi + 2\pi$.

5.2.1.1 Expressions. Waveforms

- The turn-on of the various switches determines *six periods* within each cycle T of the output currents. Only the first two need to be examined, since the currents and voltages in the three phases are identical to within $T/3$ or $2/T3$.
  - For $0 < \omega t < \pi/3$, TC_1 and TC'_2 are on:

$$i'_A = I, \quad i'_B = -I, \quad i'_C = 0$$

$$u = u'_{BA}$$

$$v_{T_1} = v_{T'_2} = 0, \quad v_{T'_1} = v_{T_2} = u'_{BA}, \quad v_{T_3} = -u'_{AC}, \quad v_{T'_3} = -u'_{CB}$$

$$i_{T_1} = i_{T'_2} = I, \quad i_{T'_1} = i_{T_2} = i_{T_3} = i_{T'_3} = 0.$$

– For $\pi/3 < \omega t < 2\pi/3$, TC_1 and TC'_3 are on:

$$i'_A = I, \qquad i'_B = 0, \qquad i'_C = -I$$

$$u = -u'_{AC}$$

$$v_{T_1} = v_{T'_3} = 0, \qquad v_{T'_1} = v_{T_3} = -u'_{AC}, \qquad v_{T_2} = u'_{BA}, \qquad v_{T'_2} = u'_{CB}$$

$$i_{T_1} = i_{T'_3} = I, \qquad i_{T'_1} = i_{T_2} = i_{T'_2} = i_{T_3} = 0.$$

- In plotting the waveforms, when *the load is star-connected*, it can be noted that, since current i'_A is equal to $+I$ for $0 < \omega t < 2\pi/3$, to $-I$ for $\pi < \omega t < 5\pi/3$, and zero for the remainder of the cycle, its fundamental has the following form:

$$I'_1\sqrt{2}\sin(\omega t + \pi/6).$$

Since φ denotes the phase lag of this fundamental with respect to the corresponding voltage, the phase-to-neutral voltages across the load phases can be expressed as:

$$v'_A = V'\sqrt{2}\sin(\omega t + \pi/6 + \varphi)$$

$$v'_B = V'\sqrt{2}\sin(\omega t + \pi/6 - 2\pi/3 + \varphi)$$

$$v'_C = V'\sqrt{2}\sin(\omega t + \pi/6 - 4\pi/3 + \varphi)$$

and the phase-to-phase voltages between the output terminals are written as:

$$u'_{BA} = v'_A - v'_B = U'\sqrt{2}\sin(\omega t + \pi/3 + \varphi)$$

$$u'_{CB} = v'_B - v'_C = U'\sqrt{2}\sin(\omega t + \pi/3 - 2\pi/3 + \varphi)$$

$$u'_{AC} = v'_C - v'_A = U'\sqrt{2}\sin(\omega t + \pi/3 - 4\pi/3 + \varphi)$$

with: $U' = V'\sqrt{3}$.

Figure 5.13 shows, from top to bottom, the switch turn-on periods and the waveforms of v'_A and i'_A, of v'_B and i'_B, of v'_C and i'_C, of u'_{BA}, u'_{CB}, u'_{AC}, of u, of i_{T_1} and v_{T_1},

(a) for an inductive load ($\varphi > 0$)
(b) for a capacitive load ($\varphi < 0$).

- When *the load is delta connected*, currents j'_A, j'_B, j'_C are linked to currents i'_A, i'_B, i'_C at the inverter output by

$$i'_A = j'_A - j'_C, \qquad i'_B = j'_B - j'_A, \qquad i'_C = j'_C - j'_B.$$

From this the following can be deduced:

$$j'_A = (i'_A - i'_B)/3, \qquad j'_B = (i'_B - i'_C)/3, \qquad j'_C = (i'_C - i'_A)/3$$

provided that the currents in the load phases do not have any zero-sequence

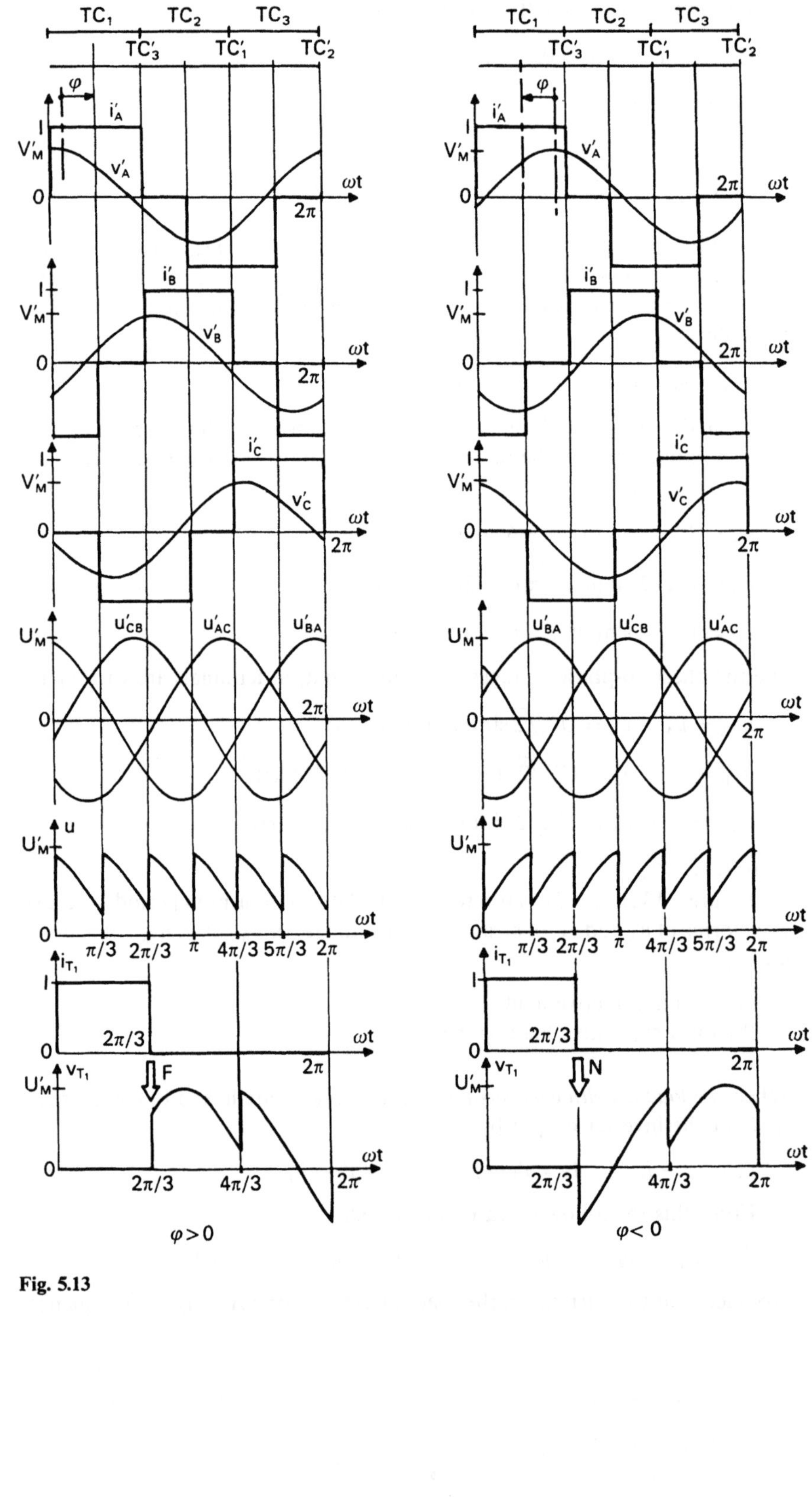

TC₁ TC₂ TC₃
TC'₃ TC'₁ TC'₂
φ
i'A
v'A
V'M
ωt
2π
i'B
v'B
i'C
v'C
u'CB
u'AC
u'BA
U'M
u
π/3 2π/3 π 4π/3 5π/3 2π
iT1
2π/3
vT1
F
4π/3
φ>0
N
u'BA
u'CB
u'AC
φ< 0

Fig. 5.13

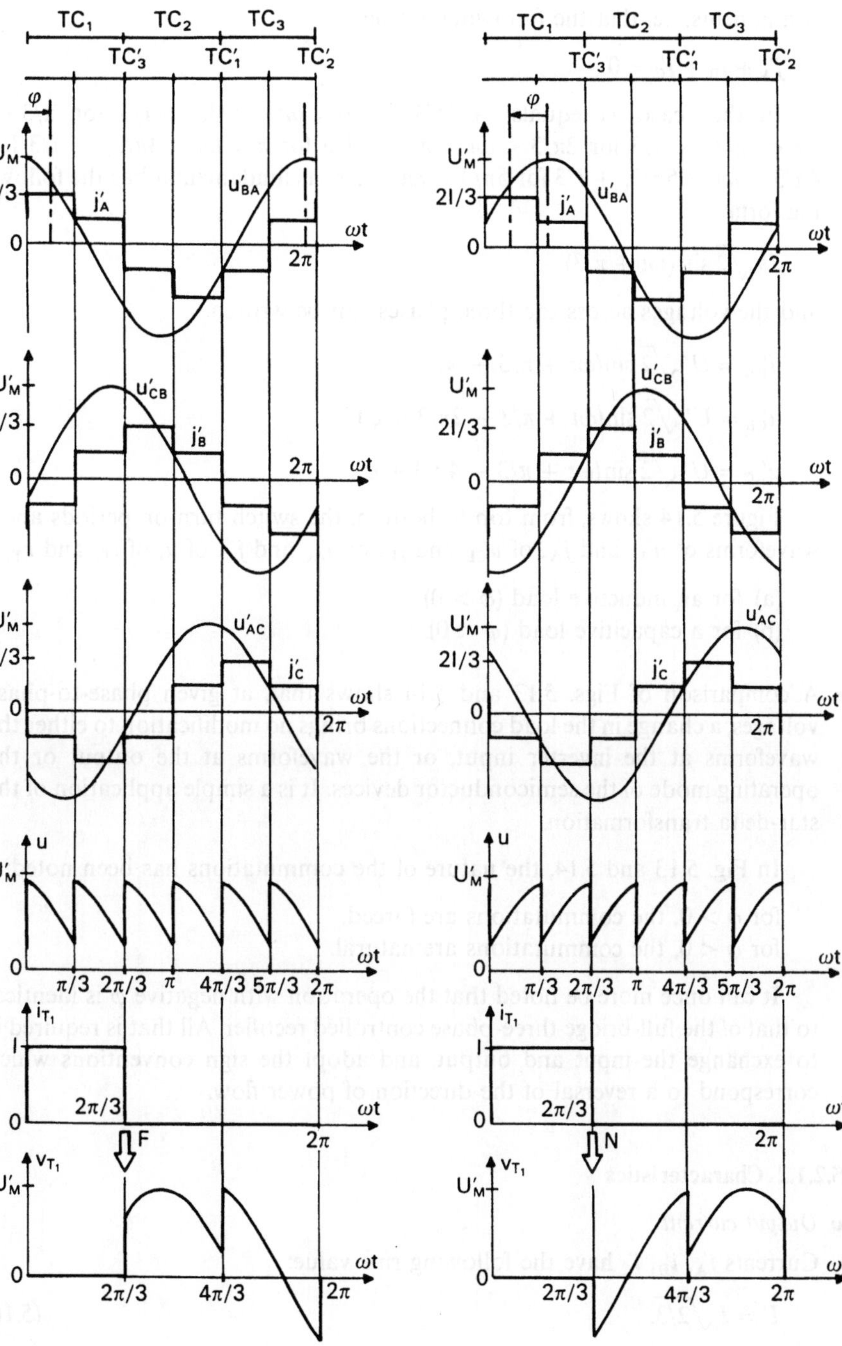

TC₁ TC₂ TC₃
TC'₃ TC'₁ TC'₂
φ
U'M
2I/3
0
j'A
u'BA
ωt
2π
u'CB
j'B
u'AC
j'C
u
π/3 2π/3 π 4π/3 5π/3 2π
iT1
I
F
N
vT1
4π/3
φ>0
φ<0

Fig. 5.14

components, i.e. that the following is true:

$$j'_A + j'_B + j'_C = 0.$$

In that case, j'_A equals $+2I/3$ for $0 < \omega t < \pi/3$, $+I/3$ for $\pi/3 < \omega t < 2\pi/3$, $-I/3$ for $2\pi/3 < \omega t < \pi$, $-2I/3$ for $\pi < \omega t < 4\pi/3$, $-I/3$ for $4\pi/3 < \omega t < 5\pi/3$, $+I/3$ for $5\pi/3 < \omega t < 2\pi$. Its fundamental has the following form:

$$J'_1\sqrt{2}\sin(\omega t + \pi/3)$$

and the voltages across the three phases can be written:

$$u'_{BA} = U'\sqrt{2}\sin(\omega t + \pi/3 + \varphi)$$

$$u'_{CB} = U'\sqrt{2}\sin(\omega t + \pi/3 - 2\pi/3 + \varphi)$$

$$u'_{AC} = U'\sqrt{2}\sin(\omega t + \pi/3 - 4\pi/3 + \varphi).$$

Figure 5.14 shows, from top to bottom, the switch turn-on periods and the waveforms of u'_{BA} and j'_A, of u'_{CB} and j'_B, of u'_{AC} and j'_C, of u, of i_{T_1} and v_{T_1}

(a) for an inductive load ($\varphi > 0$)
(b) for a capacitive load ($\varphi < 0$).

- A comparison of Figs. 5.13 and 5.14 shows that, at given phase-to-phase voltages, a change in the load connections brings no modification to either the waveforms at the inverter input, or the waveforms at the output or the operating mode of the semiconductor devices. It is a simple application of the star-delta transformation.

In Fig. 5.13 and 5.14, the nature of the commutations has been noted:

for $\varphi > 0$, the commutations are forced,
for $\varphi < 0$, the commutations are natural.

It can once more be noted that the operation with negative φ is identical to that of the full-bridge three-phase controlled rectifier. All that is required is to exchange the input and output, and adopt the sign conventions which correspond to a reversal of the direction of power flow.

5.2.1.2 Characteristics

- *Output current*

– Currents i'_A, i'_B, i'_C have the following rms value:

$$I' = I\sqrt{2/3}. \tag{5.16}$$

The rms value of the fundamental is given by:

$$I'_1 = I\sqrt{6}/\pi. \tag{5.17}$$

Their series expansion contains all the uneven harmonics except those of rank 3 or multiple of 3. The value of the existing harmonics is inversely proportional to their rank:

$$\frac{I'_{2k+1}}{I'_1} = \frac{1}{2k+1}.$$

The harmonic rate is equal to 0.311.

- If the load is delta-connected, the values relative to the polygonal currents j'_A, j'_B, j'_C can be directly deduced from those calculated for the line currents:

  rms value:

$$J' = I'/\sqrt{3} = I\sqrt{2}/3 \tag{5.18}$$

  rms value of the fundamental:

$$J'_1 = I'_1/\sqrt{3} = I\sqrt{2}/\pi \tag{5.19}$$

  rms value of the existing harmonics:

$$\frac{J'_{2k+1}}{J'_1} = \frac{I'_{2k+1}/\sqrt{3}}{I'_1/\sqrt{3}} = \frac{1}{2k+1} \tag{5.20}$$

- *Semiconductor devices*

In each switch, the current is equal to I for a third of a cycle and to zero for the rest of the cycle.

In the semiconductor devices, the current thus has

a maximum value

$$i_{T\,max} = I$$

a mean value

$$i_{T\,mean} = I/3$$

a rms value

$$i_{T\,rms} = I/\sqrt{3}$$

and the current which is to be commuted is equal to I.

The voltage across the off-switches is positive at times, negative at others; its value is comprised between $-U'\sqrt{2}$ and $+U'\sqrt{2}$

- *Input voltage*

The input voltage u has a cycle equal to a sixth of that of the output currents. When feeding an inductive load, it can be expressed as:

for $0 < \omega t < \pi/3$, $\quad u = u'_{BA} = U'\sqrt{2}\sin(\omega t + \varphi + \pi/3)$.

- It has a maximum instantaneous value of:

$$\begin{aligned} u_{max} &= U'\sqrt{2}, && \text{if } 0 < \varphi < \pi/6 \\ &= U'\sqrt{2}\sin(\pi/3 + \varphi), && \text{if } \pi/6 < \varphi < \pi \end{aligned}$$

and a minimum instantaneous value of:

$$\begin{aligned} u_{min} &= U'\sqrt{2}\sin(2\pi/3 + \varphi), && \text{if } 0 < \varphi < 5\pi/6 \\ &= -U'\sqrt{2}, && \text{if } 5\pi/6 < \varphi < \pi. \end{aligned}$$

Ripple Δu of the input voltage, equal to $u_{max} - u_{min}$, is at its minimum for $\varphi = 0$ or $\varphi = \pi$. Its maximum corresponds to $\varphi = \pi/2$ and has a value $U'\sqrt{2}$.

It should be noted that, for $0 < \varphi < \pi/3$ or for $2\pi/3 > \varphi > \pi$, voltage u is unidirectional and that voltage reversibility of the DC current supply is not required.

- The input voltage has

a mean value of:

$$U = (3/\pi)U'\sqrt{2}\cos\varphi \tag{5.21}$$

an rms value of:

$$U_{rms} = U'\sqrt{2}\sqrt{\frac{1}{2} + \frac{3\sqrt{3}}{4\pi}\cos 2\varphi} \tag{5.22}$$

Giving the quadratic sum of the harmonics

$$\sqrt{U_{rms}^2 - U^2} = U'\sqrt{2}\sqrt{0.044 - 0.0424\cos 2\varphi} \tag{5.23}$$

- The Fourier series of voltage u comprises the terms of 6ω, 12ω, 18ω, $\cdots$, angular frequency. The rms value of the term with $6k\omega$ angular frequency is equal to :

$$U_{6k} = U'\frac{6|\cos\varphi|}{\pi(36k^2 - 1)}\sqrt{1 + 36k^2\tan^2\varphi} \tag{5.24}$$

Figure 2.23 in Chap. 2, which gives I/I', I_h/I' as a function of φ for the voltage-source inverter, also gives U/U', U_h/U' for the current-source inverter. As it has been already noted, these curves can be used either for φ negative or for φ positive, since the values under study do not depend on the sign of φ.

5.2.2 PWM Control with Sinusoidal Modulation

The PWM control of the full-bridge three-phase current-source inverter allows to:

- ensure the regulation of the output current fundamental amplitude,

- improve the harmonic content of these currents by shifting the harmonic spectrum towards higher frequencies.

Such results can be achieved by using either a sinusoidal modulation or a computed modulation.

Whatever the control law adopted, one and only one of the three switches TC_1, TC_2, TC_3 must be on, and one and only one of the three switches TC'_1, TC'_2, TC'_3 must be on.

This restriction was not to be found with the voltage-source inverter where the three-phase operation could be deduced from the single-phase one. It thus imposes certain conditions on the determination of the angles. In sinusoidal modulation, the quality of a control law can only be entirely judged by its results.

In the case of sinusoidal modulation, we will present one of the most widely-used techniques, based on a principle proposed by T. Onishi and K. Okitsu in International Power Electronic Conference, Tokyo, 1983.

5.2.2.1 Determination of Conduction Periods

The wanted output currents with an angular frequency ω are denoted by:

$$i'_{Aw} = rI \sin \omega t$$

$$i'_{Bw} = rI \sin(\omega t - 2\pi/3)$$

$$i'_{Cw} = rI \sin(\omega t - 4\pi/3)$$

with r, the current regulation ratio.

The cycle is divided into six equal parts. During the first, modulation is carried out by TC_1, TC_2, TC_3, during the second by TC'_1, TC'_2, TC'_3, during the third by $TC_1, TC_2, TC_3, \ldots$

Only the first two parts need to be studied as the voltages and currents in the three phases are identical to within T/3 or 2T/3.

- *First sixth of the cycle* $(0 < \omega t < \pi/3)$

- The on-periods of TC_1 are determined by comparing the reference i'_{Aw} with a modulation or carrier wave M_1, made up of positive triangles (Fig. 5.15).

  The carrier has a frequency *mf*, *m* being an uneven multiple of 3. Its amplitude is equal to *I*.

  TC_1 is on during the periods when i'_{Aw} is greater than M_1.

- The on periods of TC_3 can be obtained by comparing i'_{Cw} with a second carrier M_2, identical to M_1 to within a shift which corresponds to a half-cycle of M_1 (or M_2). When i'_{Cw} is greater than M_2, TC_3 is on.

- Throughout the periods when TC_1 and TC_3 are simultaneously off, TC_2 remains on, in order to avoid opening the DC current supply circuit.

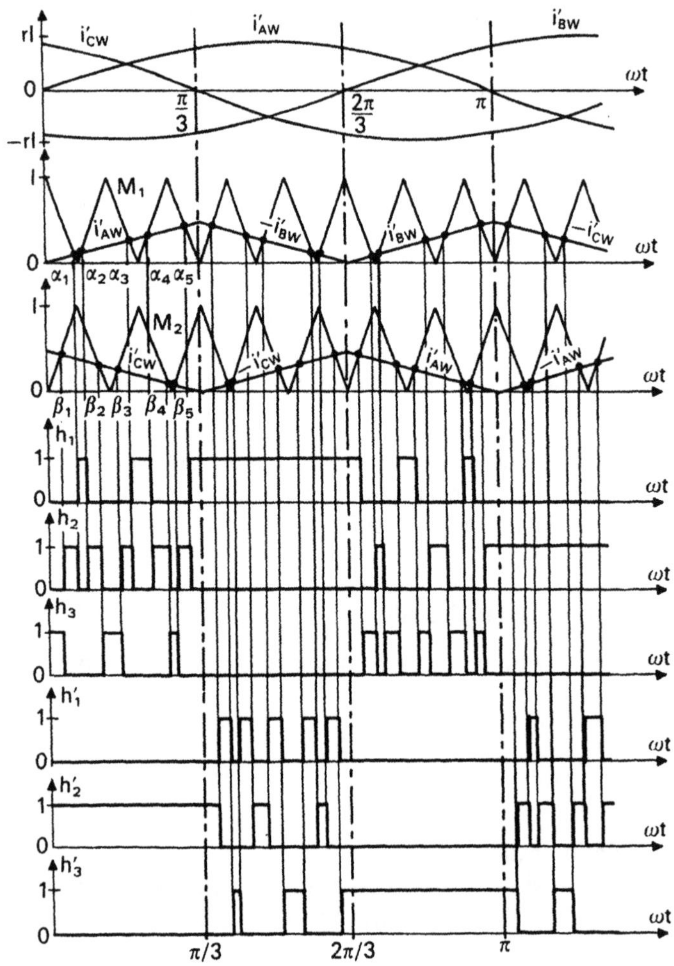

Fig. 5.15

- Throughout this first sixth of the cycle T, switch TC'_2 remains on; it is flown by $-i_B$, equal to I, when TC_1 or TC_3 are on; it short-circuits current supply I when TC_2 is on.

- *Second sixth of the cycle* ($\pi/3 < \omega t < 2\pi/3$).

The waveforms of i'_B and i'_C are imposed by fixing the conduction periods of TC'_2 and TC'_3, respectively by comparing

the absolute value of i'_{Bw} with M_1
the absolute value of i'_{Cw} with M_2.

When neither TC'_2 nor TC'_3 are turned on, TC'_1 is turned on.
Switch TC_1 remains on all the time.

- *Following sixth of the cycle*

For the third sixth of the period, the same procedure is followed as for the first, with the following substitutions being made:

i'_{Bw} for i'_{Aw}, i'_{Aw} for i'_{Cw}
TC_2 for TC_1, TC_3 for TC_2, TC_1 for TC_3, TC'_3 for TC'_2.

For the fourth part, the procedure is as for the second:

– i_{Cw} is compared with M_1 for the control of TC'_3,
– i_{Aw} is compared with M_2 for the control of TC'_1,

TC'_2 is on when TC'_1 and TC'_3 are off,
TC_2 remains permanently on, . . . ,

Figure 5.15, shows, for $m = 15$,

- the wanted output currents $i'_{Aw}, i'_{Bw}, i'_{Cw}$
- the determination of the intersections with M_1 of i'_{Aw}, of $-i'_{Bw}$, of i'_{Bw}, of $-i'_{Cw}, \ldots$
- the determination of the intersections with M_2 of i'_{Cw}, of $-i'_{Cw}$, of i'_{Aw}, of $-i'_{Aw}, \ldots$
- the on periods of the switches $TC_1, TC_2, TC_3, TC'_1, TC'_2, TC'_3$ represented by existence functions $h_1, h_2, h_3, h'_1, h'_2, h'_3$.

The existence function of a switch is 1 when it is on and zero when off.

5.2.2.2 Current Expressions. Waveforms

- The *existence function of switches* can be conveniently used in establishing the current expressions.

  Since one of the three switches TC_1 or TC_2 or TC_3 is on,

$$h_1 + h_2 + h_3 = 1 \tag{5.25}$$

  Similarly, since TC'_1 or TC'_2 or TC'_3 is on,

$$h'_1 + h'_2 + h'_3 = 1 \tag{5.25'}$$

  For the three currents i'_A, i'_B, i'_C to be identical to within $T/3$ or $2T/3$, the modulation ratio m must – as already noted – be an uneven integer which is multiple of 3.

$$\boxed{m = 3(2k - 1)}, \quad k \in [1, 2, 3, \ldots ,] \tag{5.26}$$

  For every sixth of a cycle of the output values there are thus $(2k - 1)/2$ cycles of carriers M_1 and M_2, and $(2k - 1)$ intersections between the latter and the reference currents.

– *For the first sixth of a cycle*, since the on periods of TC_1 are fixed by comparing i'_{Aw} with M_1, existence function h_1 is defined by means of $(2k - 1)$

angles $\alpha_1, \alpha_2, \ldots, \alpha_i, \ldots, \alpha_{2k-1}$ which are roots of

$$r \sin \alpha_j = m/\pi(j\pi/m - \alpha_j), \quad \text{for uneven } j \tag{5.27}$$

$$r \sin \alpha_j = m/\pi[\alpha_j - (j-1)\pi/m], \quad \text{for even } j \tag{5.27'}$$

h_1 has a value of 1 from α_1 to α_2, from α_3 to $\alpha_4, \ldots$, and from α_{2k-1} to $\pi/3$.

- Since the on periods of TC_3 are fixed by comparing i'_{Cw} with M_2, function h_3 is defined by $(2k-1)$ angles $\beta_1, \beta_2, \ldots, \beta_j, \beta_{2k-1}$ roots of

$$r \sin(\beta_j - 4\pi/3) = m/\pi[\beta_j - (j-1)\pi/m], \quad \text{for uneven } j \tag{5.28}$$

$$r \sin(\beta_j - 4\pi/3) = m/\pi(j\pi/m - \beta_j), \quad \text{for even } j \tag{5.28'}$$

h_3 has a value of 1 from 0 to β_1, from β_2 to $\beta_3, \ldots$, from β_{2k-2} to β_{2k-1}. It can be noted that

$$\beta_1 = \pi/3 - \alpha_{2k-1}, \qquad \beta_2 = \pi/3 - \alpha_{2k-2}, \quad \ldots, \quad \beta_{2k-1} = \pi/3 - \alpha_1$$

there are thus only $2k - 1$ distinct angles to be determined.

- Equation (5.25) gives h_2 by the following:

$$h_2 = 1 - h_1 - h_3.$$

- Since switch TC'_2 is on throughout this sixth of a cycle,

$$h'_1 = 0, \qquad h'_2 = 1, \qquad h'_3 = 0.$$

- *For the second sixth of a cycle*, the on periods of TC'_3 are given by comparing M_2 with $-i_{Cw}$. Figure 5.15 shows that h'_3 has a value of 1 from $\alpha_1 + \pi/3$ to $\alpha_2 + \pi/3$, from $\alpha_3 + \pi/3$ to $\alpha_4 + \pi/3, \ldots$, from $\alpha_{2k-1} + \pi/3$ to $2\pi/3$.

  TC'_2 turn-on is determined by the intersections of M_1 with $-i'_{Bw}$. Existence function h'_2 has a value of 1 from $\pi/3$ to $\beta_1 + \pi/3$, from $\beta_2 + \pi/3$ to $\beta_3 + \pi/3, \ldots$, from $\beta_{2k-2} + \pi/3$ to $\beta_{2k-1} + \pi/3$.

  The other existence functions are given by

$$h'_1 = 1 - h'_2 - h'_3$$

$$h_1 = 1; \qquad h_2 = 0; \qquad h_3 = 0.$$

- *Throughout the remaining cycle*, the existence functions can be deduced from the values which were determined during the first two sixths. Thus:

$$h_1(\omega t) = h_2(\omega t + 2\pi/3) = h_3(\omega t + 4\pi/3)$$

$$h'_1(\omega t) = h'_2(\omega t + 2\pi/3) = h'_3(\omega t + 4\pi/3).$$

- Using the existence functions, the *expressions of the different variables* can be determined immediately from the input current and the output voltages. In the case of a star-connected load, by using the notations found in the first diagram of Fig. 5.12, the following can be obtained:

- for the output currents

$$i'_A = (h_1 - h'_1)I$$
$$i'_B = (h_2 - h'_2)I$$
$$i'_C = (h_3 - h'_3)I \tag{5.29}$$

- for the input voltage

$$u = (h_1 - h'_1)v'_A + (h_2 - h'_2)v'_B + (h_3 - h'_3)v'_C \tag{5.30}$$

- for the current in TC_1 and the voltage across it

$$i_{T_1} = h_1 I$$
$$v_{T_1} = (h_1 - 1)v'_A + h_2 v'_B + h_3 v'_C \tag{5.31}$$

- for the current in TC'_1 and the voltage across it

$$i_{T'_1} = h'_1 I$$
$$v_{T'_1} = (1 - h'_1)v'_A - h'_2 v'_B - h'_3 v'_C. \tag{5.32}$$

Figure 5.16 shows, for $m = 15$ and $r = 0.6$,

- the determination of the commutation angles
- the conduction diagrams of the six semiconductor switches
- the waveforms of i'_A and v'_A
- the waveforms of i'_B and v'_B
- the waveforms of i'_C and v'_C
- the waveform of current i_{T_1}
- the waveform of voltage v_{T_1}
- the waveform of input voltage u.

Voltages v'_A, v'_B, v'_C are assumed to be sinusoidal. The case of an inductive load ($\varphi > 0$) has been chosen.

As in the case of voltage-source inverters, the PWM control requires the use of controlled turn-on/turn-off switches, whatever the sign of φ.

5.2.2.3 Characteristics

The modulation is characterised by

- *the modulation ratio m*, which can be equal to 9, 15, 21, 27, . . .
- *the current regulation ratio r*, which is the ratio of the reference current amplitude to value I of the input current.

Since two switches connected to the same input terminal cannot be turned on simultaneously – for example during the first sixth of a cycle –, the on-periods of TC_3 must be located between the on-periods of TC_1 and vice-versa. On

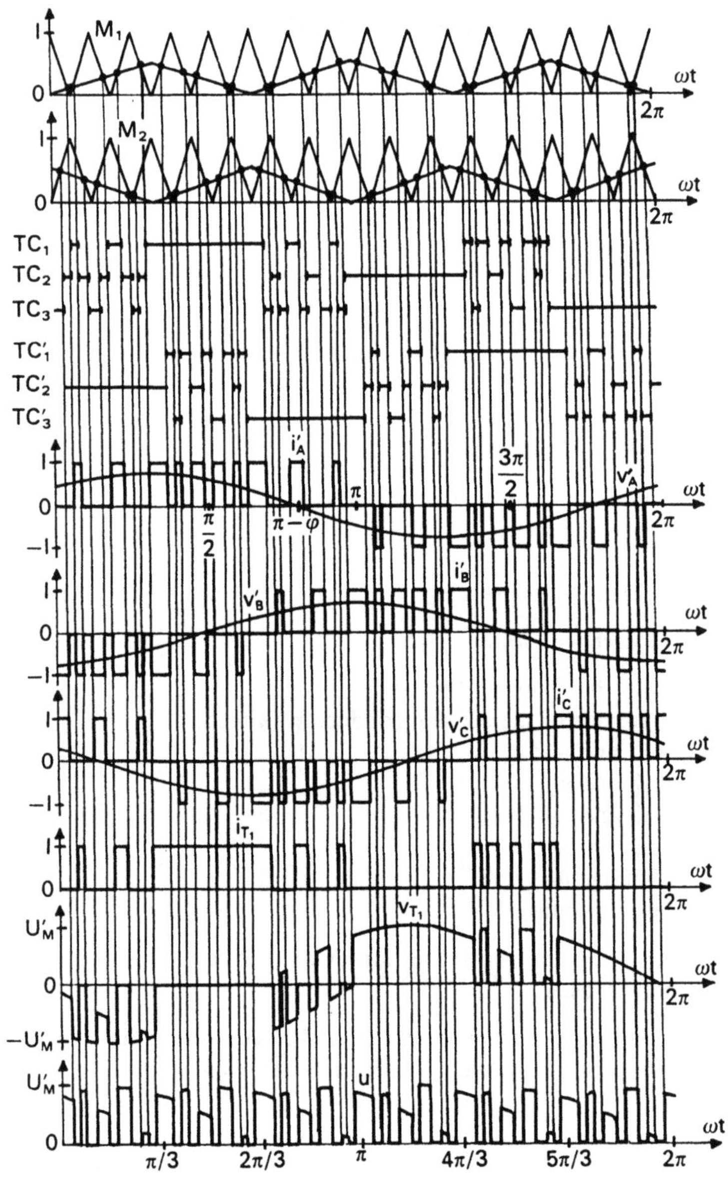

Fig. 5.16

account of the shift between the two carriers M_1 and M_2, i'_{Aw} crosses a positive slope of M_2 when i'_{Cw} crosses a negative slope of M_1 and vice-versa. As a result, the following

$$\frac{i'_{Aw}}{I} + \frac{i'_{Cw}}{I} < 1$$

is sufficient to avoid any overlapping of the on-periods of TC_1 and TC_3.

As $i'_{Aw} + i'_{Cw}$ is equal to $-i'_{Bw}$, this sum has an amplitude equal to rI, giving the maximum theoretical value of r

$$r_{max} = 1. \tag{5.33}$$

In fact, r must be kept at a value less than 1 in order to preserve a minimal duration at each on-period which takes into account the non-zero duration of commutations.

- *Output currents*

Currents i'_A, i'_B, i'_C form a balanced three-phase system. Furthermore, for each of them, the negative half-cycle reproduces the positive one to within a sign and each half-cycle is symmetrical in relation to its mid-point.

Only the development of i'_A from $\omega t = 0$ to $\omega t = \pi/2$ (Fig. 5.17) needs therefore to be followed, in order to characterise the output currents.

– The output currents have the following rms value:

$$I' = I\sqrt{\frac{4}{\pi}\left[\frac{\pi}{3} + \sum_{j=1}^{2k-1} (-1)^j \alpha_j\right]}; \tag{5.34}$$

angles $\alpha_1, \alpha_2, \ldots, \alpha_{2k-1}$, being roots of (Eq. (5.27), must be numerically computed.

Variations in the ratio I'/I as a function of r are only slightly dependent of m.

– The rms value of the fundamental is given by:

$$I'_1 = I\frac{2\sqrt{6}}{\pi}\sum_{j=1}^{2k-1} (-1)^{j-1}\cos(\alpha_j + \pi/6) \tag{5.35}$$

The difference between the fundamental and the rms value of the reference currents is very slight for $m = 9$ and can be ignored for higher values of the

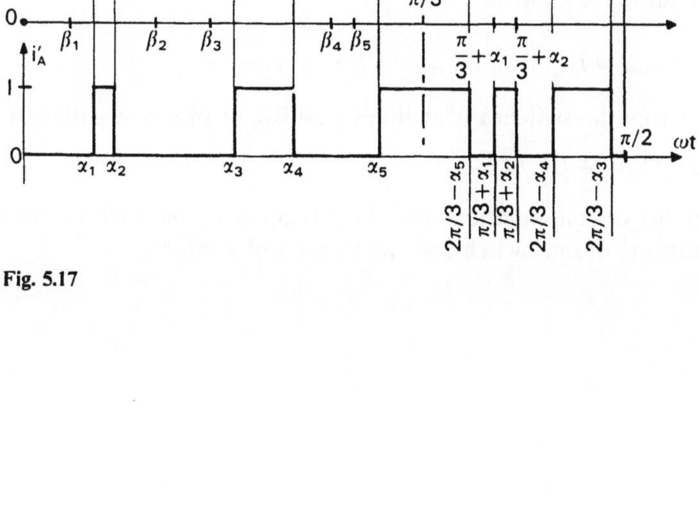

Fig. 5.17

modulation ratio. In all case, the following can be used:

$$I'_1 \simeq \frac{r}{\sqrt{2}} I \tag{5.36}$$

– The Fourier series of the output currents contains all the uneven harmonics except those or rank 3 or multiple of 3. The harmonic of rank h has an rms value of:

$$I'_h = \left| \frac{2\sqrt{6}}{h\pi} I \sum_{j=1}^{2k+1} (-1)^{(j-1)} \cos[h(\alpha_j + \pi/6)] \right| \tag{5.37}$$

As in the case of the voltage-source inverter, the harmonics can be grouped into families:

- the family centred on frequency mf which mainly comprises harmonics

  $m - 2$ and $m + 2$, $m - 4$ and $m + 4$

- the family centred on frequency $2mf$ which mainly comprises harmonics

  $2m - 1$ and and $2m + 1$ . . .

Fig. 5.18 shows, for $m = 9$ and 15, the variations as a function of r

of rms value I' of the output currents,
of rms value I'_1 of their fundamental,
of rms value I'_h of the first harmonics of any noticeable value.

All these values are referred to value I of the DC input current.

For values of m greater than 15, the curves plotted for this value of m can be used, since the variations of I'/I, I'_1/I, $I'_{m\pm2}/I$, $I'_{m\pm4}/I$, as a function of r, are virtually the same.

- *Semiconductor switches*

Each switch lets current I flow during its on-periods. For reasons of symmetry, the sum of the conduction periods of each switch is equal to the third of a cycle. The different current values in the semiconductor devices are the same as in the case of the full-wave control:

$$i_{T\,\max} = I, \qquad I_{T\,\text{rms}} = I/\sqrt{3}, \qquad I_{T\,\text{mean}} = I/3.$$

The voltage across the switches is at times positive, at others negative; it varies from $-U'\sqrt{2}$ to $+U'\sqrt{2}$.

The total number of commutations per cycle is equal to $4m$, each corresponding to the turn-off of one switch and the turn-on of another.

• *Input voltage*

The input voltage u has a cycle equal to a sixth of that of the output currents. During period $(0, T/6)$, it has the following expression:

$$u = h_1 u'_{BA} - h_3 u'_{CB}.$$

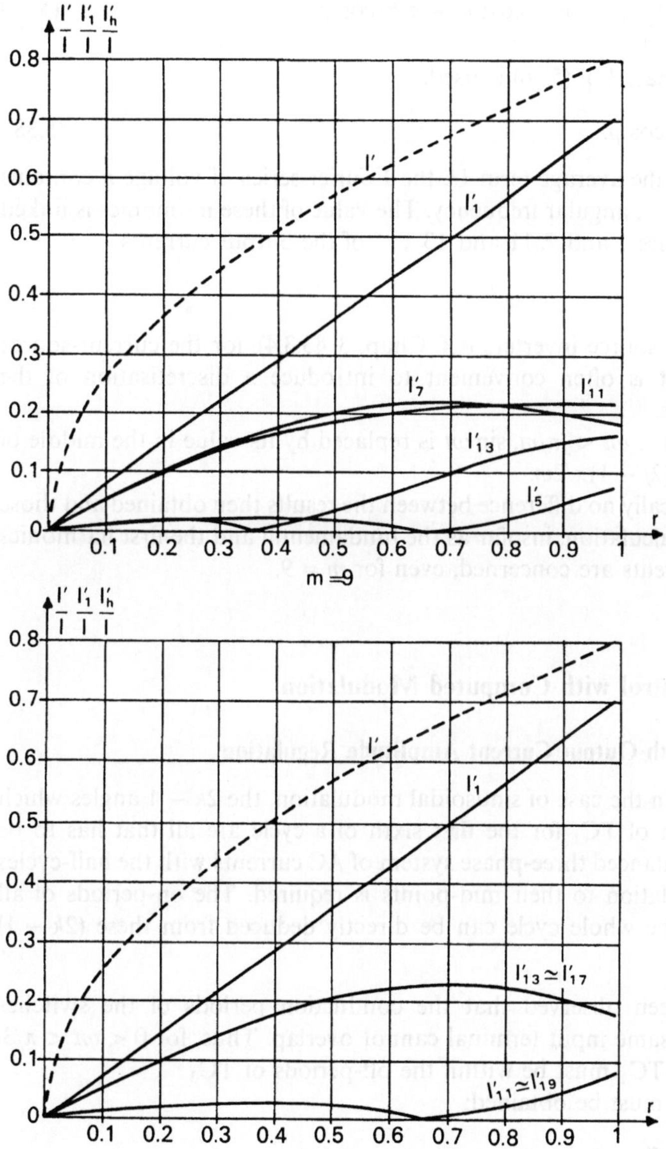

Fig. 5.18

Its mean value can be obtained by expressing that input and output power are equal:

$$UI = 3V'I'_1 \cos\varphi.$$

Equation (5.35) gives:

$$U = U' \frac{6\sqrt{2}}{\pi} \sum_{j=1}^{2k+1} (-1)^{j-1} \cos(\alpha_j + \pi/6) \cos\varphi \tag{5.38}$$

or, if the approximate Eq. (5.36) is used,

$$U \simeq U' \sqrt{3/2}\, r \cos\varphi. \tag{5.38'}$$

In addition to the average term U, the Fourier series of voltage u contains terms of 6ω, 12ω, ... angular frequency. The value of these harmonics is linked to that of harmonics 5 and 7, 11 and 13, ... of the output currents.

- *Remark*

As for the voltage-source inverters (Cf. Chap. 3, §3.3.4), for the current-source inverter control it is often convenient to introduce a discretisation of the reference sinusoids.

For $(j-1)\pi/m < \omega t < j\pi/m$, $\sin\omega t$ is replaced by its value in the middle of the period i.e. $\sin(2j-1)\pi/2m$.

There is practically no difference between the results then obtained and those given by precise calculation, insofar as the fundamental and the first harmonics of the output currents are concerned, even for $m = 9$.

5.2.3 PWM Control with Computed Modulation

5.2.3.1 Control with Output Current Amplitude Regulation

As has been seen in the case of sinusoidal modulation, the $2k-1$ angles which fix the conduction of TC_1 for the first sixth of a cycle are all that has to be determined, if a balanced three-phase system of AC currents with the half-cycles symmetrical in relation to their mid-points is required. The on-periods of all switches during the whole cycle can be directly deduced from these $(2k-1)$ angles.

It has also been observed that the conduction periods of the switches connected to the same input terminal cannot overlap. Thus, for $0 < \omega t < \pi/3$, the on-periods of TC_1 must be within the off-periods of TC_3.

The following must be obtained:

$$\pi/3 - \alpha_{2k-1} \leqslant \alpha_1$$

$$\pi/3 - \alpha_{2k-2} \geqslant \alpha_2$$

....

$\pi/3 - \alpha_2 \geqslant \alpha_{2k-2}$

$\pi/3 - \alpha_1 \leqslant \alpha_{2k-1}$

This leads to the following k conditions:

$\alpha_1 + \alpha_{2k-1} \geqslant \pi/3$

$\alpha_2 + \alpha_{2k-2} \leqslant \pi/3$

$\alpha_3 + \alpha_{2k-3} \geqslant \pi/3$

. . . .

and

$2\alpha_k \leqslant \pi/3$, if k is even

$2\alpha_k \geqslant \pi/3$, if k is uneven

The knowledge of angles α_j allows for the determination of the output currents fundamental by means of Eq. (5.35), and of their harmonics by Eq. (5.37).

In the computed modulation, the $2k - 1$ angles can be determined, by trying to find the roots of a system of $2k - 1$ equations formed by:

- equation (5.35) in which a given value is imposed on I'_1,
- $2k - 2$ equations with the form (5.37) in which a value of zero is imposed on the first $2k - 2$ uneven harmonics with a rank different of 3 or multiple of 3,

the k conditions stated above remaining satisfied.

5.2.3.2 Remark Concerning the Three-Phase PWM Rectifier

In the previous paragraphs, a study was made of the operation – in sinusoidal modulation or computed modulation – of the current-source inverter

- fed by a generator which fixes the DC current waveform,
- feeding a load on which it imposes a three-phase system of AC currents with a frequency fixed by the commutation rhythm.

As for the single-phase inverter (Cf. §5.1.4.2), the full-bridge three-phase inverter which has just been studied can act as a PWM controlled rectifier when it links

- a three-phase AC voltage generator, with fixed frequency and amplitude
- to a DC current load across which the DC voltage has to be varied.

In many cases, the nature of the generator voltage supply requires a capacitive filter to be added at the rectifier input (Cf Appendix D, §3).

5.2.3.3 Control Improving only the Output Current Waveform

As noted during the study of the single-phase inverters, the input current amplitude of a current-source inverter is often regulated by a converter placed

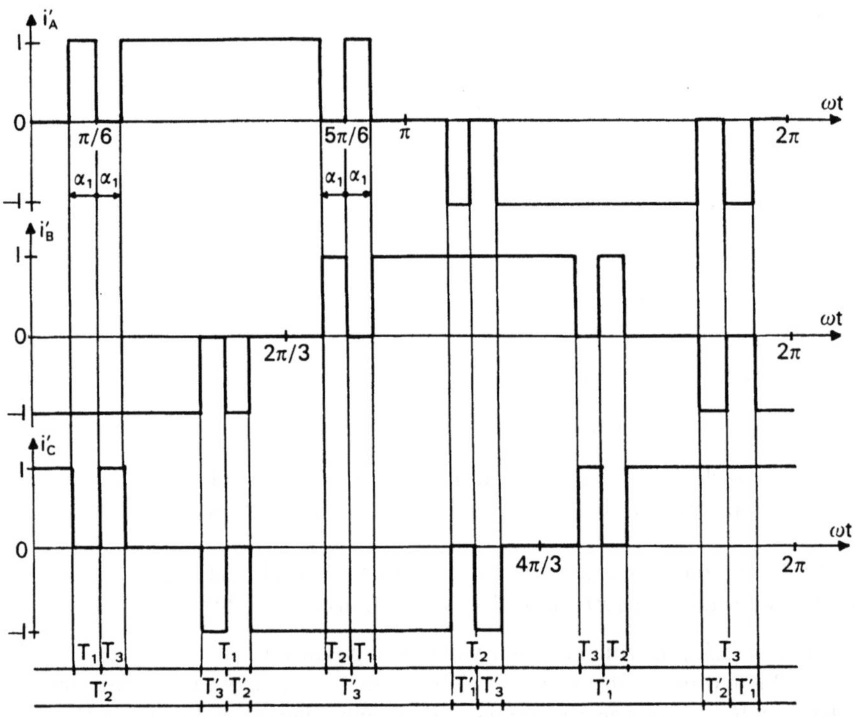

Fig. 5.19

ahead, which supplies the inverter via a smoothing inductance. In this case, the role of the PWM control can be confined to improving the output current waveform by eliminating the first harmonics.

In order to optimise the amplitude of the fundamental of these currents, waveforms are required for which the current supply is no longer cyclically short-circuited via two series-connected switches.

To achieve this, the commutations must – for each sixth of a cycle – take place not between the three switches connected to a same current supply terminal but between two switches connected to this terminal.

Figures 5.19 and 5.20 give respectively, the current output waveforms and the conduction periods of the switches which correspond:

- to the elimination of harmonic 5,
- to the elimination of harmonics 5 and 7.

In the first case, only angle α_1, equal to 12°, is needed to characterise the waveforms. In the second, the waveforms are characterised by angles α_1 and α_2 with values of 16°25 and 22°05.

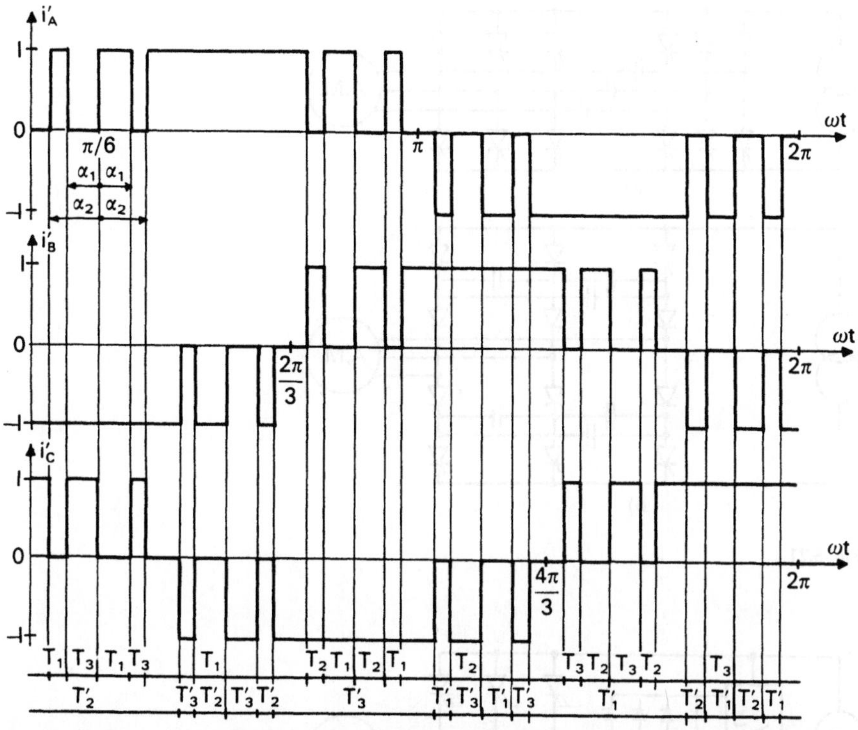

Fig. 5.20

5.3 Feeding an Inductive Load

In Chap. 1, it was noted that the autonomous current-source inverters are mainly used for supplying asynchronous motors of medium- or high-power at variable frequency.

The asynchronous motor is not a perfect voltage load. In Vol. 2 (Chap. 4, §5.1) it was shown that, in studying the operation of this machine fed by a converter, it could be modelled by a balanced three-phase circuit with each phase consisting of the series connection of a sinusoidal emf and an inductance N_1, equal to the total leakage inductance of the machine.

To prevent this inductance from producing excessive voltage spikes across the motor and the inverter, the current rate of variation in the windings must be limited during commutations.

- To achieve this, thyristor inverters are frequently used. In these, the thyristor auxiliary turn-off circuits naturally ensure this current rate of variation limitation in the windings. For these inverters, the following types are mainly found:

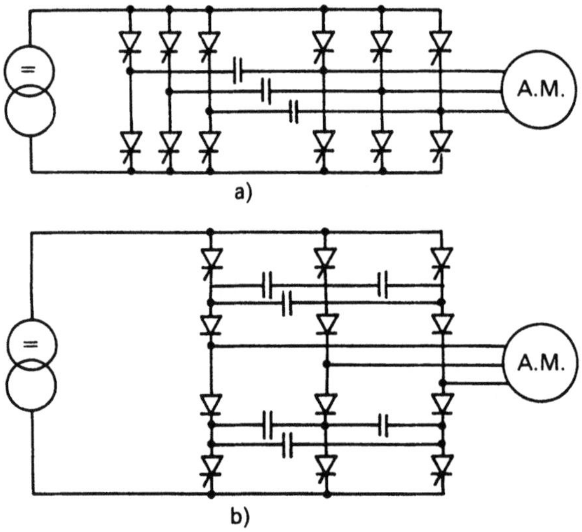

Fig. 5.21

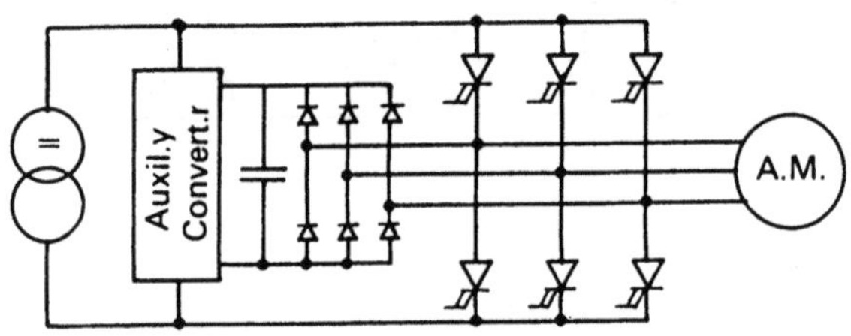

Fig. 5.22

- inverters using an auxiliary turn-off bridge (Fig. 5.21a),
- self commuting inverters (Fig. 5.21b).

– With the introduction of GTO's, the inverters develop in such a way that the controlled turn-off of the switches no longer requires the use of auxiliary turn-off circuits. In this case, the limitation of the current rate of variation in the phases is usually ensured by a voltage "clamping" circuit. Such a circuit (Fig. 5.22) stores energy at each commutation, which is recovered by an auxiliary converter and restored to the load or supply.

– In the GTO inverters, the inductive nature of the load can also be compensated for by connecting a capacitive filter in parallel (Fig. 5.23). In this case, the PWM needs to be used for the filter to remain at a reasonable size.

It should be noted that the PWM control is also used with the thyristor inverters or the GTO inverters with clamping circuit. However, in this case, this

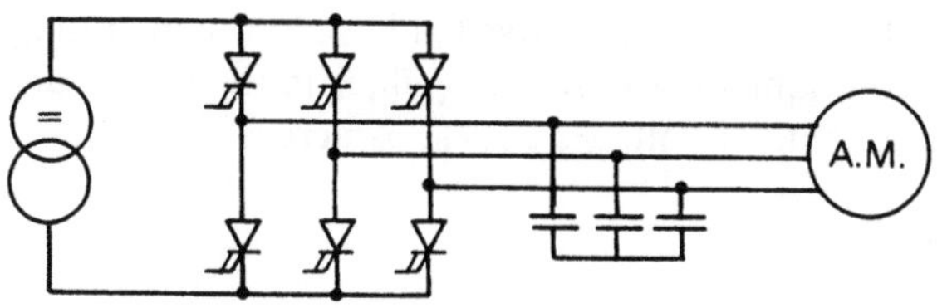

Fig. 5.23

type of control is restricted to the low operating frequencies, on account of the duration of commutation phenomena.

5.3.1 Thyristor Current-Source Inverter with Auxiliary Commutation Bridge

Figure 5.24 shows the diagram of this inverter and indicates the notations adopted in its study. Six auxiliary thyristors T_{1A}, T_{2A}, T_{3A}, T'_{1A}, T'_{2A}, T'_{3A} and three capacitors C_1, C_2, C_3 of similar capacitance C have been added to the six main thyristors T_1, T_2, T_3, T'_1, T'_2, T'_3.

The load is assumed to be star-connected; each phase includes an inductance N_1 and an emf with rms value E' and angular frequency ω:

$$e'_A = E'\sqrt{2}\sin\omega t$$

$$e'_B = E'\sqrt{2}\sin(\omega t - 2\pi/3)$$

$$e'_C = E'\sqrt{2}\sin(\omega t - 4\pi/3)$$

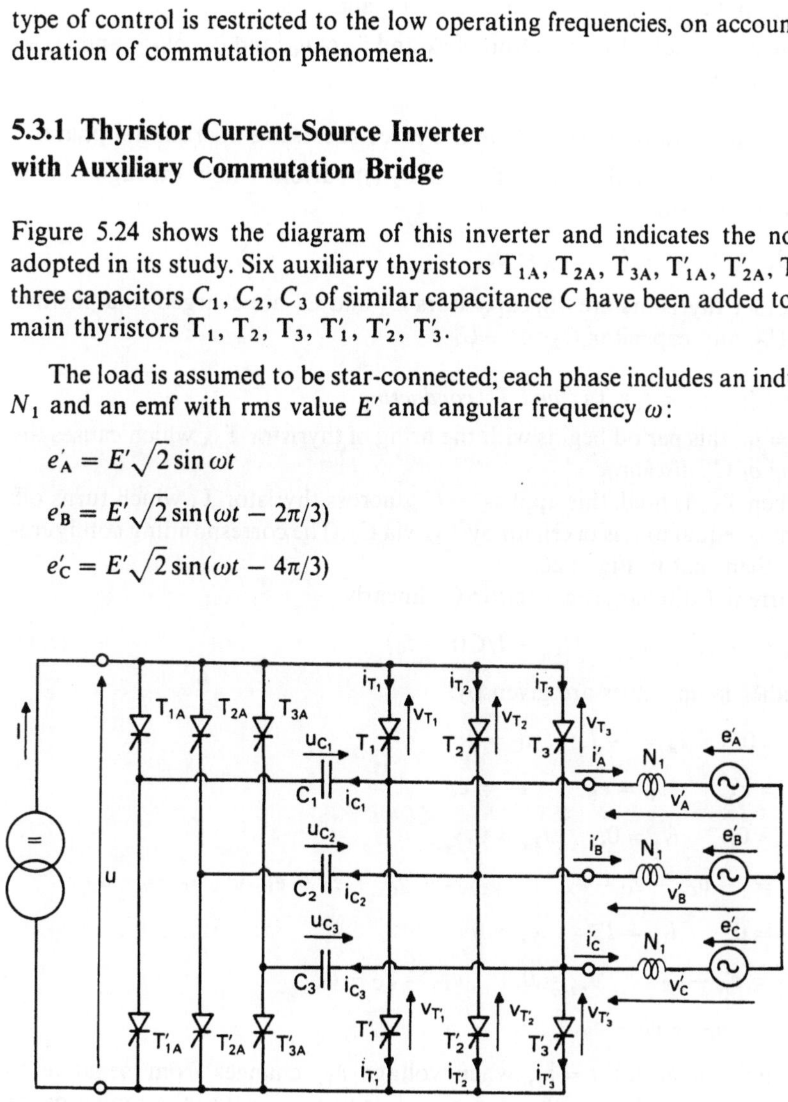

Fig. 5.24

The study will deal only with feeding an inductive load, i.e. to the operating mode where the fundamental components of currents i'_A, i'_B, i'_C in the phases are lagging the fundamental components of voltages v'_A, v'_B, v'_C across them.

5.3.1.1 Operation. Waveforms

Only the operation during a sixth of a cycle of the output currents needs to be studied. Currents i'_A, i'_B, i'_C are in fact identical to within a third of a cycle and their negative half-cycles reproduce their positive half-cycles to within a sign. The same applies to v'_A, v'_B, v'_C, to i_{C_1}, i_{C_2}, i_{C_3} and to u_{C_1}, u_{C_2}, u_{C_3}.

Figure 5.25 enables the evolution of v'_A and i'_A, of u_{C_3} and i_{C_3}, of v_{T_1} and i_{T_1}, of u to be followed.

During the sixth of a cycle, there are three successive operational phases.

It starts as follows: thyristors T_3 and T'_2 let current I flow through phases C and B of the load:

$$i'_C = I, \qquad i'_B = -I, \qquad i'_A = 0.$$

The other thyristors are off; capacitors C_3 and C_1 are charged to a positive voltage U_{C_0} and capacitor C_2 to $-U_{C_0}$.

- *Phase 1* ($t_0 < t < t_1$): T_{3A} *and* T'_2 *conducting*

- For $t = t_0$, this period begins with the firing of thyristor T_{3A} which causes the *starting of* C_3 *discharge*.

  When T_{3A} is fired, this applies $-U_{C0}$ across thyristor T_3 which turns off: current i'_C, equal to I, is taken up by T_{3A} via C_3. The corresponding configuration is then that in Fig. 5.26.

  Current I discharges capacitor C_3 linearly:

$$i_{C_3} = -I, \qquad u_{C_3} = U_{C_0} - I/C(t - t_0) \tag{5.39}$$

  The other main values are given by:

$$i'_A = 0, \qquad i'_B = -I, \qquad i'_C = I$$

$$v'_A = e'_A, \qquad v'_B = e'_B, \qquad v'_C = e'_C$$

$$i_{T_1} = 0, \qquad i_{T_2} = 0, \qquad i_{T_3} = 0,$$

$$v_{T_1} = -u_{C_3} + e'_C - e'_A, \qquad v_{T_2} = -u_{C_3} + e'_C - e'_B, \qquad v_{T_3} = -u_{C_3}$$

$$i_{T'_1} = 0, \qquad i_{T'_2} = I, \qquad i_{T'_3} = 0$$

$$v_{T'_1} = e'_A - e'_B, \qquad v_{T'_2} = 0, \qquad v_{T'_3} = e'_C - e'_B$$

$$U = -u_{C_3} + e'_C - e'_B.$$

- This phase ends, for $t = t_1$, when voltage v_{T1} changes from negative to positive; thyristor T_1, which has received its firing signal at the same time as T_{3A}, then begins conducting.

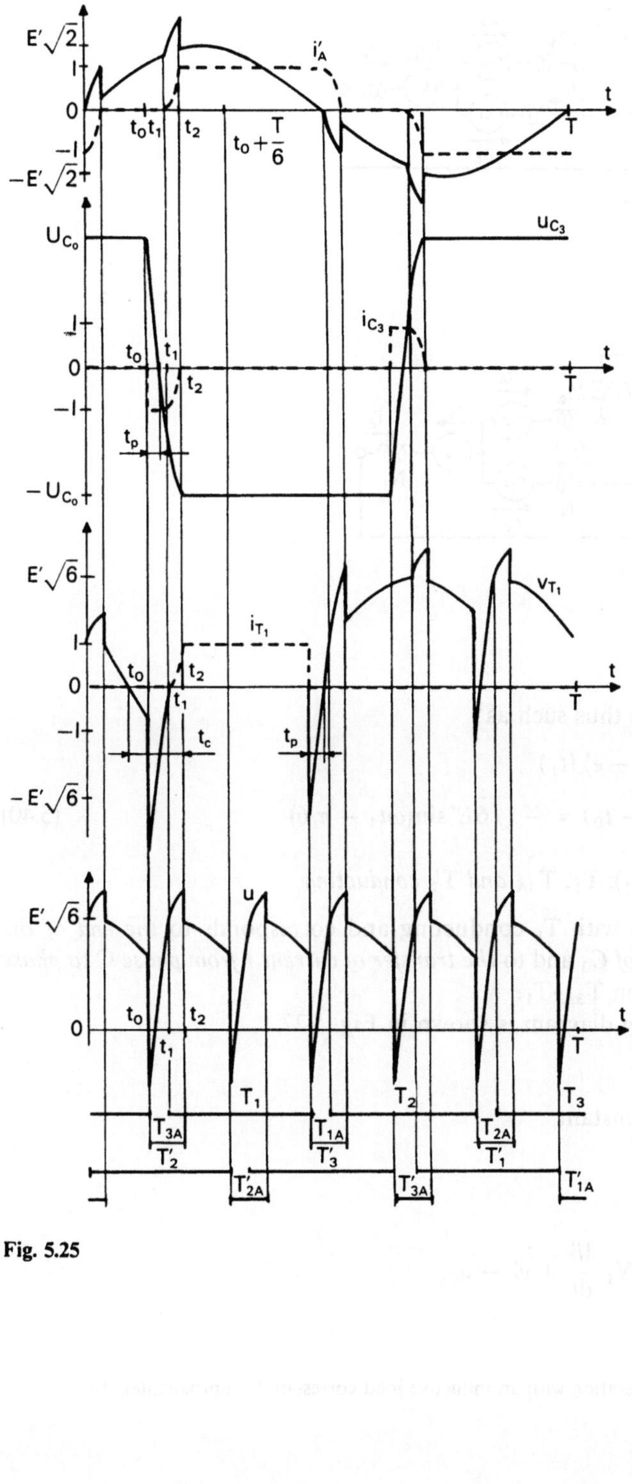

E′√2
I
0
−I
−E′√2
i′A
t
t0 t1 t2
t0 + T/6
T
UC0
uC3
iC3
t0
t1
t2
tp
−UC0
E′√6
vT1
iT1
tc
−E′√6
u
T1
T2
T3
T3A
T1A
T2A
T′2
T′3
T′1
T′2A
T′3A
T′1A

Fig. 5.25

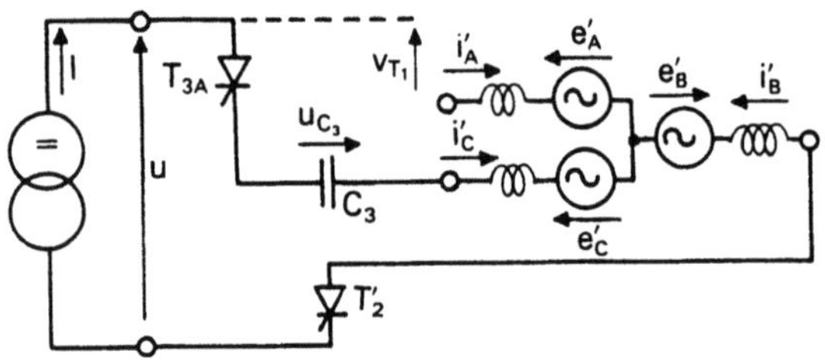

Fig. 5.26

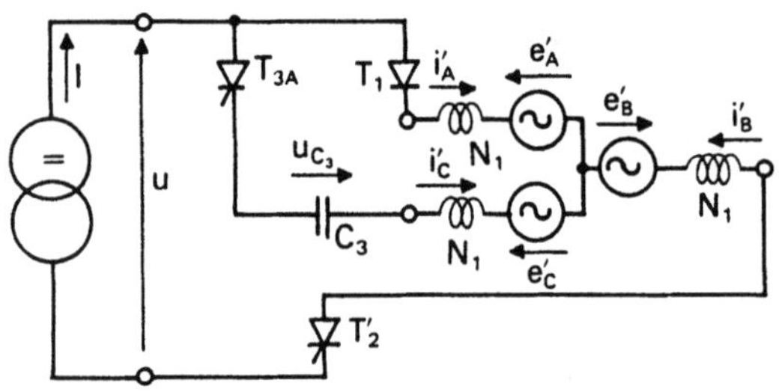

Fig. 5.27

Instant $t = t_1$ is thus such as[1]

$$u_{C_3}(t_1) = e'_C(t_1) - e'_A(t_1)$$

$$U_{C_0} - (I/C)(t_1 - t_0) = -\sqrt{6}E' \sin(\omega t_1 - \pi/6) \tag{5.40}$$

- *Phase 2* ($t_1 < t < t_2$): T_1, T_{3A} *and* T'_2 *conducting*

This period starts with T_1 conducting and corresponds to *the end of the reversal of the charge of* C_3 and to *the transfer of current I from phase* C *to phase* A – i.e. to commutation T_{3A}–T_1.

The corresponding diagram is shown in Fig. 5.27.

– From

$$i'_A + i'_C = I = \text{constant}$$

$$i'_C = -C \frac{du_{C_3}}{dt}$$

$$N_1 \frac{di'_A}{dt} + e'_A = N_1 \frac{di'_C}{dt} + e'_C - u_{C_3}$$

[1] As will be seen later, operation with an inductive load corresponds approximately to $0 < \omega t_1 - \pi/6 < \pi$

the differential equation giving i'_A can be deduced:

$$2N_1 \frac{d^2 i'_A}{dt^2} + i'_A = I + C\frac{d}{dt}(e'_C - e'_A)$$

Since:

$$i'_A(t_1) = 0$$

and that:

$$\frac{di'_A}{dt}(t_1) = \frac{1}{N_1}[e'_C(t_1) - e'_A(t_1) - u_{C_3}(t_1)] = 0,$$

current i'_A can be expressed as:

$$\begin{aligned} i'_A = {} & I[1 - \cos\omega'(t - t_1)] \\ & - \frac{\sqrt{6}E'}{2N_1\omega'}\frac{(\omega/\omega')^2}{1 - (\omega/\omega')^2}\sin(\omega t_1 - \pi/6)\sin\omega'(t - t_1) \\ & + \frac{\sqrt{6}E'}{2N_1\omega'}\frac{\omega/\omega'}{1 - (\omega/\omega')^2}[\cos(\omega t_1 - \pi/6)\cos\omega'(t - t_1) - \cos(\omega t - \pi/6)] \end{aligned} \tag{5.41}$$

by denoting:

$$\omega' = 1/\sqrt{2N_1 C}.$$

The expression of u_{C_3} can be deduced from that of i'_A by:

$$u_{C_3} = e'_C - e'_A - 2N_1\frac{di'_A}{dt}$$

which gives:

$$\begin{aligned} u_{C_3} = {} & -\sqrt{6}E'\sin(\omega t - \pi/6) - 2N_1\omega' I\sin\omega'(t - t_1) \\ & + \sqrt{6}E'\frac{(\omega/\omega')^2}{1 - (\omega/\omega')^2}\sin(\omega t_1 - \pi/6)\cos\omega'(t - t_1) \\ & + \sqrt{6}E'\frac{\omega/\omega'}{1 - (\omega/\omega')^2} \\ & \cdot[\cos(\omega t_1 - \pi/6)\sin\omega'(t - t_1) - (\omega/\omega')\sin(\omega t - \pi/6)]. \end{aligned} \tag{5.42}$$

The input voltage, calculated by

$$u = e'_A + N_1\frac{di'_A}{dt} - e'_B$$

can be expressed as:

$$u = \sqrt{6}E'\sin(\omega t + \pi/6) + N_1\omega' I \sin\omega'(t - t_1)$$
$$- \sqrt{6}\frac{E'}{2}\frac{(\omega/\omega')^2}{1-(\omega/\omega')^2}\sin(\omega t_1 - \pi/6)\cos\omega'(t - t_1)$$
$$- \sqrt{6}\frac{E'}{2}\frac{\omega/\omega'}{1-(\omega/\omega')^2}$$
$$\cdot[\cos(\omega t_1 - \pi/6)\sin\omega'(t-t_1) - (\omega/\omega')\sin(\omega t - \pi/6)] \quad (5.43)$$

The expressions of the other variables can easily be deduced from those of i'_A, u_{C_3} and u:

$$i'_B = -I, \qquad i'_C = I - i'_A$$
$$v'_A = e'_A + N_1\frac{di'_A}{dt}, \qquad v'_B = e'_B, \qquad v'_C = e'_C - N_1\frac{di'_A}{dt}$$
$$i_{T_1} = i'_A, \qquad i_{T_2} = 0, \qquad i_{T_3} = 0$$
$$v_{T_1} = 0, \qquad v_{T_2} = u, \qquad v_{T_3} = -u_{C_3}$$
$$i_{T'_1} = 0, \qquad i_{T'_2} = I, \qquad i_{T'_3} = 0$$
$$v_{T'_1} = u, \qquad v_{T'_2} = 0, \qquad v_{T'_3} = u + u_{C_3}$$

- This period ends for $t = t_2$, when i'_A reaches value I and i'_C falls to zero, leading to T_{3A} being turned off. The value of t_2 is thus the root of the equation:

$$0 = \cos\omega'(t_2 - t_1) + \frac{\sqrt{6}E'}{2N_1\omega' I}\frac{(\omega/\omega')^2}{1-(\omega/\omega')^2}\sin(\omega t_1 - \pi/6)\sin\omega'(t_2 - t_1)$$
$$- \frac{\sqrt{6}E'}{2N_1\omega' I}\frac{\omega/\omega'}{1-(\omega/\omega')^2}$$
$$\cdot[\cos(\omega t_1 - \pi/6)\cos\omega'(t_2 - t_1) - \cos(\omega t_2 - \pi/6)]. \quad (5.44)$$

At instant $t = t_2$, the voltage across C_3 is equal to $-U_{C_0}$ for reasons of symmetry. In fact, C_3 is now charged so as to ensure the turn-off of T'_3 by the firing of T'_{3A} at instant $t = t_0 + T/2$, according to a process similar to that described at the beginning of the first phase.

From $u_{C_3}(t_2) = -U_{C_0}$, the value of U_{C_0} can be deduced:

$$U_{C_0} = \sqrt{6}E'\sin(\omega t_1 - \pi/6) + 2N_1\omega' I\sin\omega'(t_2 - t_1)$$
$$- \sqrt{6}E'\frac{(\omega/\omega')^2}{1-(\omega/\omega')^2}\sin(\omega t_1 - \pi/6)\cos\omega'(t_2 - t_1)$$
$$- \sqrt{6}E'\frac{\omega/\omega'}{1-(\omega/\omega')^2}$$
$$\cdot[\cos(\omega t_1 - \pi/6)\sin\omega'(t_2 - t_1) - \omega/\omega'\sin(\omega t_2 - \pi/6)]. \quad (5.45)$$

– *Remark*: If the transfer of current I from one phase to the next occurs rapidly in relation to the rate of variation of the emf's e'_A, e'_B, e'_C, in the differential equation which gives i'_A the sinusoidal voltage

$$e'_C - e'_A = \sqrt{6}E' \sin(\omega t - \pi/6)$$

can be replaced by a constant voltage equal to the value of $e'_C - e'_A$ at instant $t = t_1$.

The expressions of i'_A, u_{C_3} and u are then greatly simplified and become:

$$i'_A = I[1 - \cos \omega'(t - t_1)] \tag{5.41'}$$

$$u_{C_3} = -\sqrt{6}E' \sin(\omega t - \pi/6) - 2N_1 \omega' I \sin \omega'(t - t_1) \tag{5.42'}$$

$$u = \sqrt{6}E' \sin(\omega t + \pi/6) + N_1 \omega' I \sin \omega'(t - t_1). \tag{5.43'}$$

In the case where the load is an asynchronous motor, these simplified equations give a sufficiently precise description of the transfer of current from one phase to the next. This explains why they will be used in the following part of this study.

• *Phase 3* ($t_2 < t < t_0 + T/6$): T_1 *and* T'_2 *conducting*

Starting from $t = t_2$, current I flows through phases A and B, via thyristors T_1 and T'_2

$$i'_A = I, \quad i'_B = -I, \quad i'_C = 0$$

$$v'_A = e'_A, \quad v'_B = e'_B, \quad v'_C = e'_C$$

$$i_{T_1} = I, \quad i_{T_2} = 0, \quad i_{T_3} = 0;$$

$$v_{T_1} = 0, \quad v_{T_2} = e'_A - e'_B, \quad v_{T_3} = e'_A - e'_C$$

$$i_{T'_1} = 0, \quad i_{T'_2} = I, \quad i_{T'_3} = 0$$

$$v_{T'_1} = e'_A - e'_B, \quad v_{T'_2} = 0, \quad v_{T'_3} = e'_C - e'_B$$

$$u = e'_A - e'_B.$$

This period ends for $t = t_0 + T/6$ with the firing of T'_{2A}.

5.3.1.2 Commutation Time. Reverse Bias Time. Choice of C

– Equation (5.41′) gives the following duration of the second phase:

$$t_2 - t_1 = \pi/2\omega' = (\pi/2)\sqrt{2N_1 C}. \tag{5.46}$$

By including this in (5.42′), one gets:

$$U_{C_0} = -u_{C_3}(t_2) = \sqrt{6}E' \sin(\omega t_1 - \pi/6) + \sqrt{(2N_1/C)}I. \tag{5.47}$$

Equations (5.39) and (5.40) then give the following expressions of u_{C_3} during the first phase and of the duration of the latter:

$$u_{C_3} = \sqrt{6}E'\sin(\omega t_1 - \pi/6) + \sqrt{(2N_1/C)}I - (I/C)(t - t_0)$$

$$\begin{aligned} t_1 - t_0 &= (C/I)[2\sqrt{6}E'\sin(\omega t_1 - \pi/6) + \sqrt{(2N_1/C)}I] \\ &= (2\sqrt{6}E'/\omega' I)\sqrt{(C/2N_1)}\sin(\omega t_1 - \pi/6) + 1/\omega'. \end{aligned} \tag{5.48}$$

- *total commutation time* t_c, equal to $t_2 - t_0$, is equal to:

$$t_c = (1/\omega')[\pi/2 + (2\sqrt{6}E'/I)\sqrt{(C/2N_1)}\sin(\omega t_1 - \pi/6) + 1]. \tag{5.49}$$

The *reverse bias time* t_p is, in the case of thyristor T_3 for example, equal to the period which separates the firing of T_{3A} from the instant when voltage u_{C_3} falls to zero.

$$t_p = (\sqrt{6}E'/\omega' I)\sqrt{(C/2N_1)}\sin(\omega t_1 - \pi/6) + 1/\omega'. \tag{5.50}$$

Equation (5.48) shows that the value of t_1, which plays a role in the expressions of U_{C_0}, t_c and t_p, depends not only on I, E', C, N_1 and ω, but also on the instant $t = t_0$ when commutation begins.

- The values of U_{C_0} and t_c are at their maximum for $\omega t_1 - \pi/6$ equal to $\pi/2$, and are then equal to:

$$U_{C_0\,\max} = \sqrt{6}E' + \sqrt{(2N_1/C)}I \tag{5.51}$$

$$t_{c\,\max} = (2\sqrt{6}/\omega' I)\sqrt{(C/2N_1)} + 1/\omega'(\pi/2 + 1) \tag{5.52}$$

The reverse bias time is minimal for ωt_1 equal to $\pi/6$, and then has the following value:

$$t_{p\,\min} = 1/\omega' = \sqrt{2N_1 C}. \tag{5.53}$$

- *The choice of capacitance* C of capacitors C_1, C_2, C_3 must take into account the fact that increasing C

  • increases the reverse bias time t_p,
  • reduces amplitude U_{C_0} of the voltage across the capacitors,
  • but increases commutation time t_c.

The minimum value of C is imposed by t_p, and its maximum by the condition

$$t_{c\,\max} \leqslant T/6 \quad \text{or} \quad \omega t_{c\,\max} \leqslant \pi/3$$

By including these in (5.52), this condition gives:

$$C \leqslant C_{\max} = \frac{(1+\pi/2)^2}{48}\frac{N_1 I^2}{E'^2}\left[\sqrt{1 + \frac{8\pi}{\sqrt{6}(1+\pi/2)^2}\frac{E'}{N_1\omega I}} - 1\right]^2 \tag{5.54}$$

The value of C_{max} enables a lower limit for $U_{C_0 max}$ to be determined.

$$U_{C_0 max} \geqslant \sqrt{6}E' \left[1 + \frac{4}{1+\pi/2} \frac{\sqrt{N_1 \omega I/E'}}{\sqrt{\dfrac{N_1 \omega I}{E'} + \dfrac{8\pi}{6(1+\pi/2)^2}} - \sqrt{\dfrac{N_1 \omega I}{E'}}} \right].$$

5.3.1.3 Operation with Partial Overlap of Commutations

In the preceding study we assumed that the total duration of the commutations was less than the sixth of a cycle of the output currents. There is thus no overlap between two successive commutations, one carried out between two switches connected to one of the input terminals and the other between two switches connected to the other input terminal.

Partial overlapping of the successive commutations modifies the waveforms of the input voltage and the voltages across the switches but does not change the *current transfer mechanism* as studied, if the following conditions are fulfilled.

- the total duration of a commutation must be less than a third of a cycle of the output currents. Indeed, it has been assumed that, at instant when T_{3A} is fired, current I was flowing through phase C and thyristor T_3. Current I must thus have been completely diverted through phase C and T_3 a *third* of a cycle after the firing of T_{2A}.
- the phenomenon of overlapping itself must last less than a *sixth* of a cycle of the output values. Indeed, the current in phase B has been assumed to be constant during the rise of i'_A and the fall of i'_C.

Figure 5.28 gives an example of waveforms of v'_A and i'_A, of u_{C_3} and i_{C_3}, of v_{T_1} and i_{T_1}, and of u. Figure 5.29 shows the equivalent circuits during the four phases of commutation $T_3 \rightarrow T_1$:

- the reversal of u_{C_3} begins while T'_{1A} is still conducting,
- when the rise of i'_A ends, the firing of T'_{2A} has already turned off T'_2.

The two conditions to be respected can be written as:

$$t_2 - t_0 = (1/\omega')\left[\frac{\pi}{2} + \frac{2\sqrt{6}E'}{I}\sqrt{\frac{C}{2N_1}}\sin(\omega t_1 - \pi/6) + 1\right] \leqslant T/3$$

$$t_2 - t_1 = (\pi/2)\sqrt{2N_1 C} \leqslant T/6.$$

The acceptable maximum value of the capacitance of the capacitors corresponds to the smallest of the following values:

$$C_{max} = \frac{(1+\pi/2)^2}{48}\frac{N_1 I^2}{E'^2}\left[\sqrt{1 + \frac{16\pi}{\sqrt{6}(1+\pi/2)^2}\frac{E'}{N_1 \omega I}} - 1\right] \tag{5.55}$$

$$C_{max} = \frac{T^2}{18\pi^2 N_1}.$$

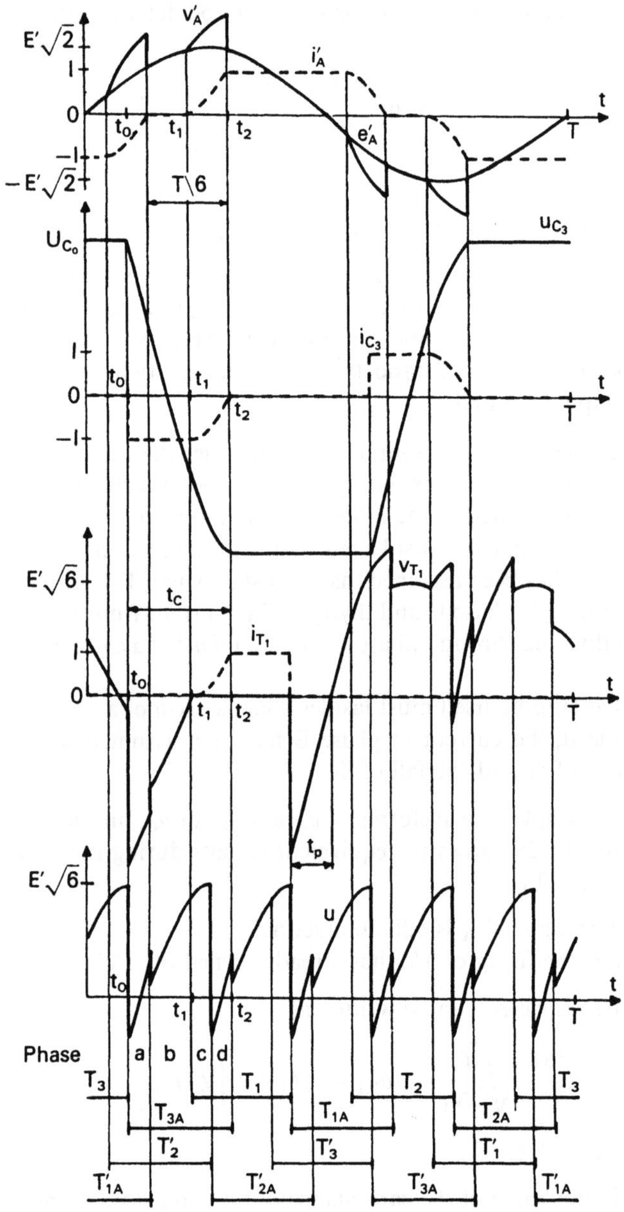

Fig. 5.28

5.3.1.4 Variation on the Basic Configuration

If the operating mode without overlap between commutations is only tolerated, the configuration in Fig. 5.24 can be replaced by the *single-turn-off capacitor* configuration shown in Fig. 5.30.

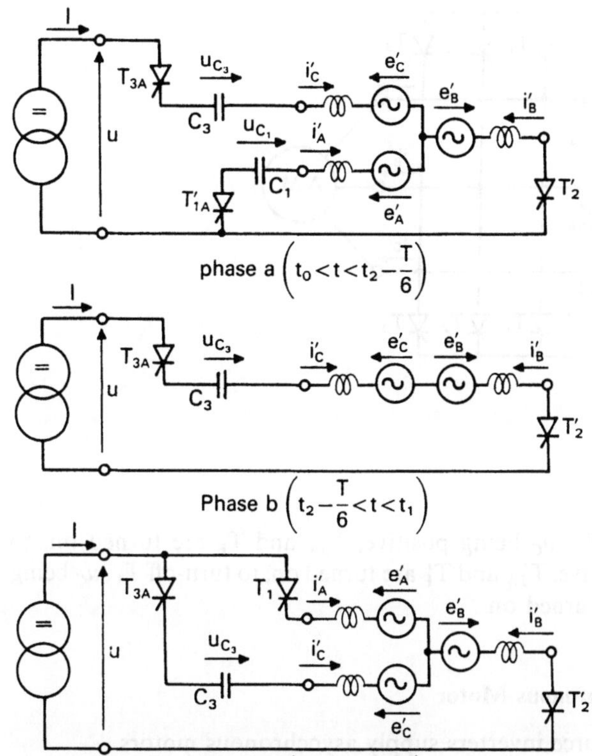

Fig. 5.29

Thyristors T_{1A} and T'_{1A}, T_{2A} and T'_{2A}, T_{3A} and T'_{3A} ensure that capacitor C is connected to the main thyristor to be turned off. Thyristors T_E and T'_E close the turn-off circuit, depending on whether a thyristor of the group T_1, T_2, T_3 or of the group T'_1, T'_2, T'_3 has to be turned off.

The capacitor is now subject to a discharge and recharge every sixth of a cycle.

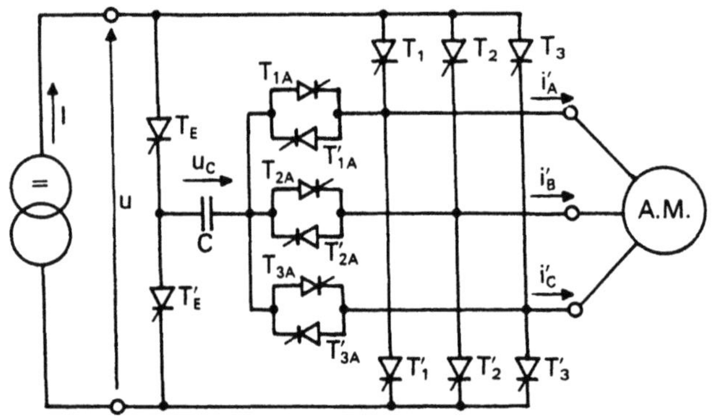

Fig. 5.30

In order to turn off T_3, u_C being positive, T_{3A} and T_E are turned on; to turn-off T'_2, u_C being negative, T'_{2A} and T'_E are turned on; to turn-off T_1, u_C being positive, T_{1A} and T_E are turned on . . .

5.3.1.5 Feeding an Asynchronous Motor

In most cases, current-source inverters supply asynchronous motors.

- *Inductance N_1 order of magnitude*

– In studying the inverter operation, each phase was assimilated to an inductance N_1, series-connected to a sinusoidal emf with the same angular frequency ω as the supply current fundamental.

In fact, this consists in replacing the classical diagram (Fig. 5.31a) in which

r_1 is the stator resistance,
l_1, the stator leakage inductance,
L_μ, the magnetizing inductance,
r_2, the rotor resistance referred to the stator,
s, the slip,
l_2, the rotor leakage inductance referred to the stator,

by the diagram of Fig. 5.31b.

To achieve this, r_1 is ignored, L_μ placed after l_2 and N_1 used to denote the $l_1 + l_2$ assembly.

The e.m.f. E' is the voltage across L_μ in parallel with r_2/s.

The current and voltage fundamental are linked to this emf by

$$\underline{V}'_1 = \underline{E}' + jN_1\omega\underline{I}'_1,$$

equation corresponding to the phasor diagram in Fig. 5.31c.

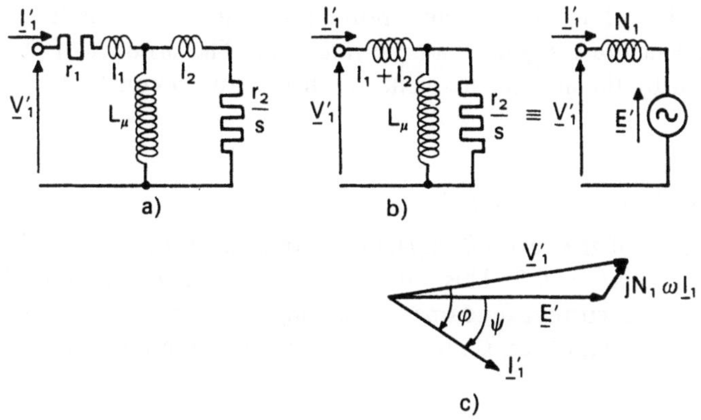

Fig. 5.31

- The value of E' and the lagging phase-shift ψ of $\underline{I}'_1$ related to $\underline{E}'$ are given by:

$$E' = \frac{(r_2/s)L_\mu\omega}{\sqrt{r_2^2/s^2 + L_\mu^2\omega^2}} I'_1, \qquad \psi = \text{arc}\tan(r_2/sL_\mu\omega).$$

- The subscript n will be added to the values of the variables when the latter correspond to the rated operating mode.

  For this type of operation, the voltage drop across the total leakage reactance N_1 is about 10% that of the rated supply voltage.

  The diagram in Fig. 5.31c shows that, under these conditions, there is little difference between E' and V'_1:

$$E'_n \simeq V'_{1n}, \qquad N_1\omega_n I'_{1n} \simeq 0.1\, V'_{1n}. \tag{5.56}$$

  At no load (very small value of s), there is little difference between current I'_1 and the magnetizing current absorbed by L_μ. This current is about 40% that of the rated current

$$I_{\mu n} = \frac{E'_n}{L_\mu\omega_n} \simeq 0.4 I'_{1n}. \tag{5.57}$$

- *Regulation of current I*

Generally, when an asynchronous motor is supplied from a current-source inverter, the value I of the input current is regulated,

- so as to maintain the *flux* (and thus the magnetizing current I_μ) at a constant level, when the supply angular frequency is less than the motor rated angular frequency. The voltage across each motor phase is therefore nearly proportional to the frequency.

$$V'_1 \simeq (\omega/\omega_n)\, V'_{1n}$$

- so as to maintain the *voltage* across each motor phase at a constant level, when angular frequency ω is higher than the rated one. The product of the magnetizing current by the angular frequency is then virtually constant.

$$I_\mu \simeq (\omega_n/\omega) I_{\mu n}.$$

- *Angular frequency ω' order of magnitude*

The maximum value of the turn-off capacitors capacitance C_{max} was calculated by writing $\omega t_1 - \pi/6 = \pi/2$. This value of ωt_1 (see the upper part of Fig. 5.25) corresponds to a current fundamental lagging by approximately $\pi/2$ the emf. or the voltage fundamental, i.e. to the no-load operation for which

$$I'_1 \simeq I_\mu$$

Taking into account that the speed variation range of an asynchronous motor fed by a current-source inverter normally goes from zero to twice the rated speed, this gives:

$$\omega_{max} = 2\omega_n$$

and thus

$$I_\mu(\omega_{max}) \simeq 0.2 I'_{1n}.$$

If the value $(\sqrt{6}/\pi)\, I$ given by Eq. (5.16) which neglects overlapping effect is taken for I', and if the order of magnitude of I' for no-load operation at maximum frequency is taken into account, Eq. (5.54) gives:

$$C_{max} = \frac{0.0075}{N_1 \omega_n^2} \tag{5.58}$$

If an operational mode with partial overlapping of commutations is accepted, Eq. (5.55) gives a value of C_{max} which is approximately twice of the previous one.

$$C_{max} = \frac{0.017}{N_1 \omega^2} \tag{5.58'}$$

Angular frequency ω' which comes into effect during commutations is thus such that:

$$\frac{\omega_n}{\omega'} = \omega_n \sqrt{2 N_1 C} \simeq 0.12 \quad \text{(or 0.18)}$$

and $E'/N_1\omega'$ which appears in the current expressions during commutations is:

$$\frac{E'}{N_1 \omega'} < \frac{V'_{1n}}{N_1 \omega'} \simeq 1.2 I'_{1n} \quad (\text{or } 1.8 I'_{1n})$$

By introducing these values into Eq. (5.41), (5.42) and (5.43) it can be seen that the approximated Eq. (5.41′), (5.42′) and (5.43′) give a sufficiently precise description of the current I transfer mechanism from one phase to the next.

- *Reverse bias time*

The reverse bias time of thyristors is at its minimum for $\omega t_1 - \pi/6 = 0$ or for $E' = 0$, and

$$t_{\mathrm{p\,min}} = 1/\omega' = \sqrt{2N_1C}$$

The order of magnitude of $t_{\mathrm{p\,min}}$ can be deduced from (5.58) or (5.58′)

$$t_{\mathrm{p\,min}} \simeq 0.12/\omega_n \text{ or } 0.18/\omega_n$$

according to whether partial overlapping is accepted or not.

For a motor with a rated frequency of 50 Hz, this gives:

$$t_{\mathrm{p\,min}} \simeq 380\ \mu\mathrm{S}\ (\text{or } 570\ \mu\mathrm{S}).$$

As can be seen, this time corresponds to the reverse recovery time of conventional thyristors. Unlike the case of choppers or voltage-source inverters with thyristors, no use needs be made of fast thyristors.

5.3.1.6 Commutation Effects

The commutations have little effects on the output current waveforms because of the short duration of the overlap periods. As in the case of the rectifiers studied in Vol. 1, the influence of the commutations on

- the rms value of currents i'_A, i'_B, i'_C
- the rms value of their fundamental

can be ignored.

On the other hand, the commutations have a considerable effect on the voltages at the inverter input and output. They cause noticeable spikes to appear here, as can be seen on the waveforms in Fig. 5.25 and 5.28.

We will examine the effects of the commutations in the case where there is no partial overlapping of successive commutations.

- *Effects on the input voltage*

The input voltage u has a cycle equal to a sixth of that T of the output currents. During one of its cycles – e.g. the one from $t = t_0$ to $t = t_0 + T/6$, it has the following successive expressions:

for $t_0 < t < t_1$,

$$u = e'_C - e'_B + u_{C_3} \simeq \sqrt{6}E' \cos \omega t - \sqrt{6}E' \sin(\omega t_1 - \pi/6)$$
$$- \sqrt{(2N_1/C)}I + \frac{I}{C}(t - t_0)$$

for $t_1 < t < t_2$,

$$u = e'_A - e'_B + N_1 \frac{di'_A}{dt} \simeq \sqrt{6}E' \sin(\omega t + \pi/6) + N_1\omega' I \sin \omega'(t - t_1)$$

for $t_2 < t < t_0 + T/6$,

$$u = e'_A - e'_B = \sqrt{6}E' \sin(\omega t + \pi/6)$$

The mean value of voltage u is equal to:

$$U = (3\sqrt{6}/\pi)E'[\cos(\omega t_1 - \pi/6) - (\omega/\omega')\sin(\omega t_1 - \pi/6)] \quad (5.59)$$

In view of the order of magnitude indicated for $N_1 \omega I_n$,

- the voltage minimum occurs in $t = t_0$ and has a value of

$$u_{min} = \sqrt{6}E' \sin(\omega t_0 + \pi/2) - U_{C_0}$$

- the maximum occurs in $t = t_2$ and has a value of

$$u_{max} = \sqrt{6}E' \sin(\omega t_2 + \pi/6) + N_1 \omega I'.$$

- *Phase shift of the output current fundamental related to the emf's*

If the losses in the inverter are ignored, the current i'_A, i'_B, i'_C, fundamental components are lagging the corresponding emf's e'_A, e'_B, e'_C by an angle ψ which is such that:

$$UI = 3E'I'_1 \cos\psi.$$

By taking I'_1 equal to $(\sqrt{6}/\pi)\ I$ and taking into account Eq. (5.59), the following applies:

$$\cos\psi = \cos(\omega t_1 - \pi/6) - (\omega/\omega')\sin(\omega t_1 - \pi/6).$$

In view of the very small value of ω/ω', this equation can be simplified and be written as:

$$\cos\psi \simeq \cos(\omega t_1 - \pi/6 + \omega/\omega')$$

that is:

$$\psi \simeq \omega t_1 - \pi/6 + \omega/\omega'. \quad (5.60)$$

In view of the very slight difference between ψ and φ, it can be seen that feeding an inductive load corresponds approximately to

$$0 < \omega t_1 - \pi/6 < \pi$$

as was assumed at the beginning of commutation analysis.

- *Stresses on semiconductor devices*

The commutations reduce the mean value and the rms value of the *current* in the main thyristors on account of:

- the diversion of the current by the auxiliary thyristors during the linear discharge phase of the turn-off capacitors,
- the overlap of the conduction periods between auxiliary and main thyristors.

On the other hand, the voltage stresses on the semiconductor devices are greatly increased by the *voltage spikes* introduced by the commutations. An upper limit of the peak voltage is given by

$$v_{T\,max} = 2\sqrt{6}E' + N_1\omega' I. \tag{5.61}$$

To determine the maximum voltage more precisely would require the relation between the value of I and the phase shift ψ between the output current fundamental and the emf's to be taken into account.

5.3.2 Self-Commuted Current-Source Inverter with Thyristors

In order to ensure the turn-off of each thyristor by the firing of the next one, to the six main thyristors $T_1, T_2, T_3, T'_1, T'_2, T'_3$ are added six diodes $D_1, D_2, D_3, D'_1, D'_2, D'_3$ and six capacitors $C_1, C_2, C_3, C'_1, C'_2, C'_3$ of a same capacitance C as shown in Fig. 5.32. In this figure the notations used are also indicated.

We will show that the operating of the self-commuted inverter is similar to that of the inverter with an auxiliary turn-off bridge.

To achieve this, we adopt the same assumptions concerning the supply and the load, and represent these by the same equivalent circuit. The expressions of the emf's e'_A, e'_B, e'_C are the same.

Similarly, we will limit our study to the supply of an inductive load.

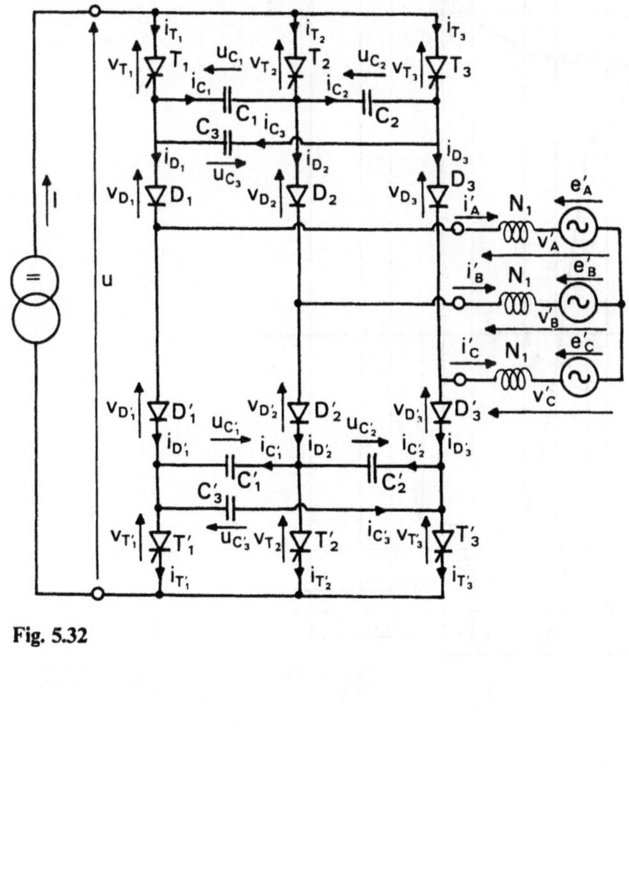

Fig. 5.32

5.3.2.1 Operation. Waveforms

The waveforms shown in Fig. 5.33 enable the evolution of the following voltages and currents to be followed:

v'_A and i'_A, v'_B and i'_B, v'_C and i'_C, u_{C_1} and i_{C_1}, v_{T_1} and i_{T_1}, v_{D_1} and i_{D_1}, u.

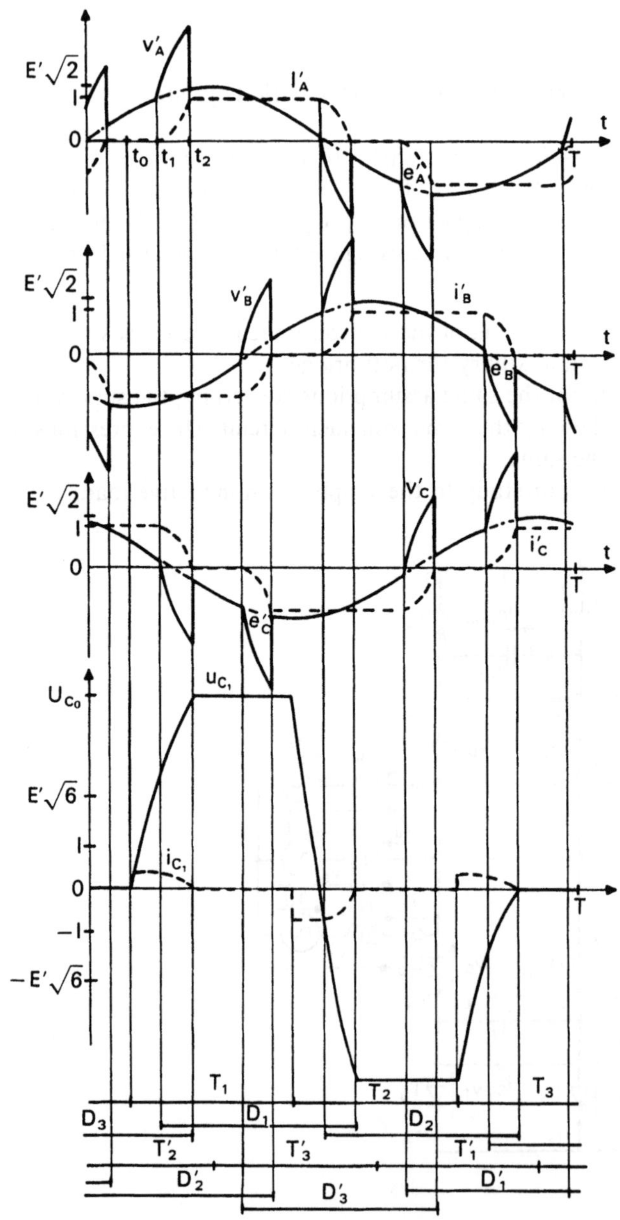

Fig. 5.33

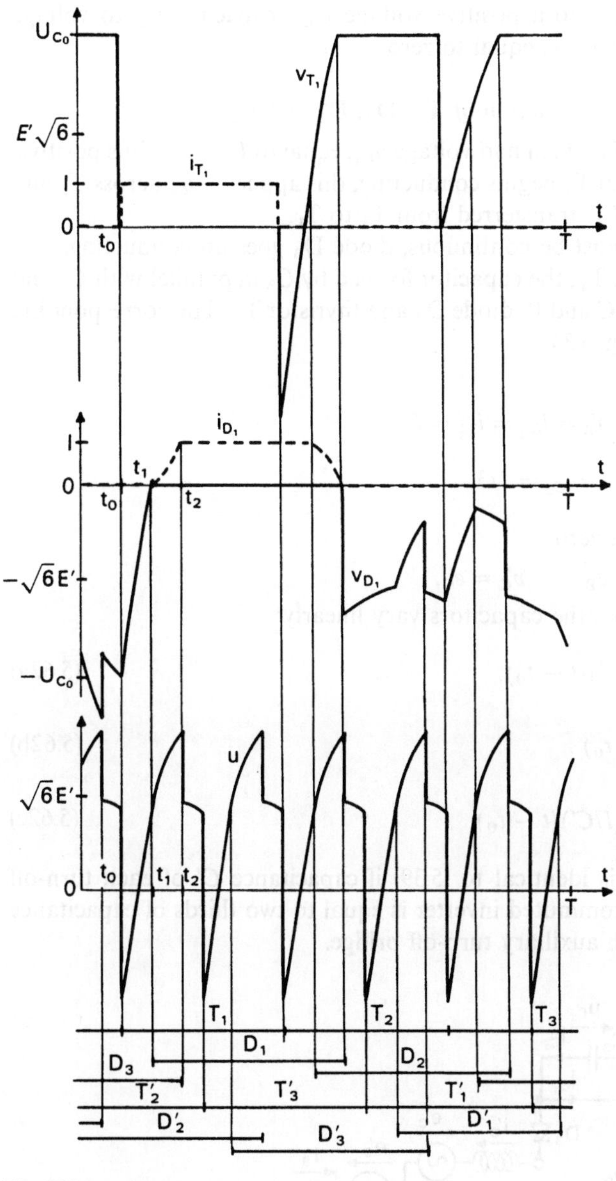

Fig. 5.33 (continued)

Once again, the operation needs only be studied for a sixth of the cycle T of the output currents. We follow the sixth one which begins by the transfer of current I from phase C to phase A, i.e. by commutation $T_3 \rightarrow T_1$.

The following state is taken as starting point:

- thyristors T_3 and T'_2 are conducting current I which flows via diodes D_3 and D'_2 and phases B and C of the load ($i'_C = I$; $i'_B = -I$; $i'_A = 0$). The other thyristors are blocked.

- capacitor C_3 is charged to a positive voltage U_{C_0}, capacitor C_2 to voltage $-U_{C_0}$; voltage across C_1 is equal to zero.

- *Phase 1* ($t_0 < t < t_1$). *Conduction of* T_1, D_3, D'_2 *and* T'_2.

– For $t = t_0$, thyristor T_1 which had voltage u_{C_3}, equal to U_{C_0} and thus positive, across it, is fired. When T_1 begins conducting, this applies $-u_{C_3}$ across T_3 and turns it off. Current I is transferred from T_3 to T_1.

Since current i'_C must be continuous, diode D_3 goes on conducting.

Current I goes via T_1, the capacitor formed by C_3 in parallel with C_1 and C_2, diode D_3, phases C and B, diode D'_2 and thyristor T'_2. The corresponding diagram is that of Fig. 5.34.

This thus gives:

$$i_{T_1} = i_{D_3} = i'_C = -i'_B = i_{D'_2} = i_{T'_2} = I$$

$$i_{C_3} = -\frac{2}{3}I; \qquad i_{C_1} = i_{C_2} = I/3$$

the other currents are zero.

$$v'_A = e'_A, \qquad v'_B = e'_B, \qquad v'_C = e'_C.$$

The voltages across the capacitors vary linearly:

$$u_{C_3} = U_{C_0} - \frac{2}{3}(I/C')(t - t_0) \tag{5.62a}$$

$$u_{C_1} = \frac{1}{3}(I/C')(t - t_0) \tag{5.62b}$$

$$u_{C_2} = -U_{C_0} + \frac{1}{3}(I/C')(t - t_0). \tag{5.62c}$$

Equation (5.62a) is identical to (5.39) if capacitance C' of each turn-off capacitor of the self-commuted inverter is equal to two thirds of capacitance C of the inverter with auxiliary turn-off bridge.

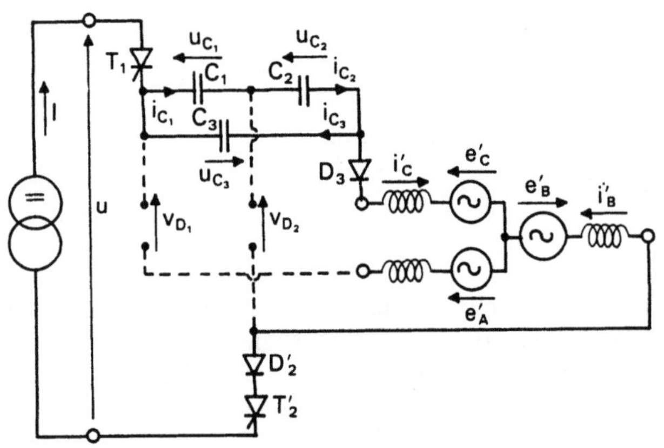

Fig. 5.34

The input voltage is given by:

$$u = -u_{C_3} + e'_C - e'_B.$$

- During this phase, the voltages across diodes D_1 and D'_2 can be expressed as:

$$v_{D_1} = -u_{C_3} + e'_C - e'_A$$

$$v_{D_2} = u_{C_2} + e'_C - e'_B.$$

Voltage v_{D1} is assumed to be negative at instant $t = t_0$ and voltage v_{D2} to be negative throughout period $(t_0, t_0 + T/6)$. The conditions which must be fulfilled for this to occur will be seen at the end of the analysis.

- The first phase ends for $t = t_1$ when voltage v_{D_1} is no longer negative and diode D_1 begins conducting. The value of t_1 is given by:

$$u_{C_3}(t_1) = e'_C(t_1) - e'_A(t_1)$$

$$U_{C_0} - \frac{2}{3}(I/C')(t_1 - t_0) = -\sqrt{6}\,E' \sin(\omega t_1 - \pi/6) \tag{5.63}$$

For $C' = 2C/3$, Eq. (5.63) is identical to Eq. (5.40).

- During this period and throughout the remainder of the sixth of a cycle under study, the values concerning the semiconductor devices and the capacitors in the lower part of Fig. 5.32 diagram remain unchanged.

$$i_{T'_1} = i_{D'_1} = i_{T'_3} = i_{D'_3} = i_{C'_1} = i_{C'_2} = i_{C'_3} = 0$$

$$i_{T'_2} = i_{D'_2} = I$$

$$u_{C'_1} = -U_{C_0}, \qquad u_{C'_2} = +U_{C_0}, \qquad u_{C'_3} = 0$$

$$v_{T'_1} = U_{C_0}, \qquad v_{T'_2} = 0, \qquad v_{T'_3} = -U_{C_0}.$$

Only the voltages across diodes D'_1 and D'_3 are linked to the variations in voltages across the load:

$$v_{D'_1} = -e'_B + v'_A, \qquad v_{D'_2} = 0, \qquad v_{D'_3} = -e_B + v'_C.$$

- *Phase 2* $(t_1 < t < t_2)$: T_1 *and* D_1, D_3, D'_2 *and* T'_2 *conducting*

Starting from $t = t_1$, diode D_1 is conducting, current i'_C decreases and current i'_A increases. The corresponding configuration is that of Fig. 5.35 as long as voltage v_{D_2}, equal to $-e'_B + e'_C - N\,\mathrm{d}i'_C/\mathrm{d}t - u_{C_2}$, remains negative.

- The equations which enable the differential equations giving i'_A to be laid down are:

$$i'_A + i'_C = I = Cte$$

$$i'_C = -\frac{3}{2} i_{C_3} = -\frac{3}{2} C' \frac{\mathrm{d}u_{C_3}}{\mathrm{d}t}$$

$$N_1 \frac{\mathrm{d}i'_A}{\mathrm{d}t} + e'_A = N_1 \frac{\mathrm{d}i'_C}{\mathrm{d}t} + e'_C - u_{C_3}.$$

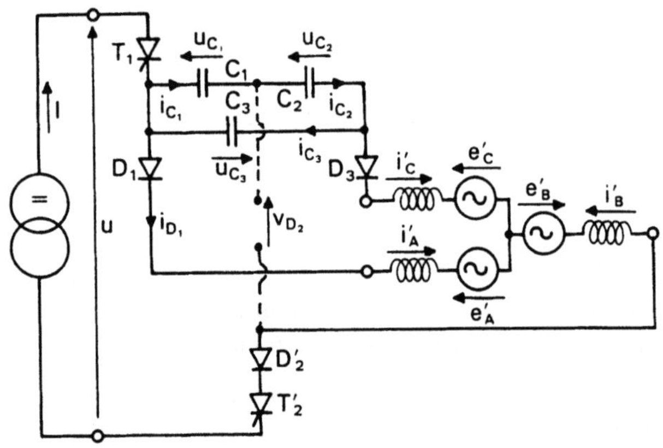

Fig. 5.35

These are the same as those which were found during the study of phase 2 of the inverter with auxiliary turn-off bridge, if C' is replaced by $2C/3$.

Denoting:

$$\omega' = \frac{1}{\sqrt{3N_1C'}}.$$

Eqs. (5.41), (5.42) or (5.43) can be used.

It has been shown that the variations of e'_A and e'_C during this period can be ignored. Expressions of i'_A, u_{C3} and u can thus once more be written:

$$i'_A = I[1 - \cos\omega'(t - t_1)] \tag{5.64}$$

$$u_{C_3} = -\sqrt{6}E'\sin(\omega t_1 - \pi/6) - 2N_1\omega' I \sin\omega'(t - t_1) \tag{5.65}$$

$$u = \sqrt{6}\,E'\sin(\omega t + \pi/6) + N_1\omega' I \sin\omega'(t - t_1). \tag{5.66}$$

- The other variables can be deduced from i'_A, u_{C_3} and u by:

$$i'_B = I, \qquad i'_C = I - i'_A$$

$$v'_A = e'_A + N_1\frac{di'_A}{dt}, \qquad v'_B = e'_B, \qquad v'_C = e'_C - N_1\frac{di'_A}{dt}$$

$$i_{T_1} = I, \qquad i_{D_1} = i'_A, \qquad i_{D_3} = i'_C$$

$$i_{C_3} = -\frac{2}{3}i'_C, \qquad i_{C_1} = i_{C_2} = \frac{1}{3}i'_C$$

$$u_{C_1} = -\frac{1}{2}(u_{C_3} - U_{C_0}), \qquad u_{C_2} = -U_{C_0} - \frac{1}{2}(u_{C_3} - U_{C_0}).$$

- The end of this period, corresponding to the end of the $T_3 \to T_1$ commutation, occurs for $t = t_2$ when current i'_C falls to zero.

This then gives:

$$u_{C_3}(t_2) = -U_{C_0} = -[\sqrt{6}\,E'\sin(\omega t_1 - \pi/6) + \sqrt{(4N_1/3C')}\,I]; \qquad (5.67)$$

With $C' = 2C/3$, Eqs. (5.47) and (5.67) are identical.

At instant $t = t_2$,

$$u_{C_1} = U_{C_0}; \qquad u_{C_2} = 0.$$

- *Phase 3* ($t_2 < t < t_0 = T/6$): T_1 *and* D_1, D'_2 *and* T_2 *conducting*

From the end of commutation $T_3 \rightarrow T_1$, until the beginning of the $T'_2 \rightarrow T'_3$ commutation, current I flows via T_1 and D_1, phases A and B of the load, D'_2 and T'_2.

$$v'_A = e'_A, \qquad v'_B = e'_B, \qquad v'_C = e'_C$$

$$u_{C_1} = U_{C_0}, \qquad u_{C_2} = 0, \qquad u_{C_3} = -U_{C_0}.$$

5.3.2.2 Conditions Required for Ensuring the Foreseen Operation

So that the charge and discharge of the turn-off capacitors occurs as foreseen, the diodes must be effectively blocked during certain phases. Thus, during the period $(t_0, t_0 + T/6)$ under study,

v_{D_1} must be negative at the beginning of the first phase,
v_{D_2} must be negative throughout all the three phases,
v_{D_3} must be negative during the third phase.

– The voltage across diode D_1 can be expressed, during the first phase, as:

$$v_{D_1} = -u_{C_3} + e'_C - e'_A = -u_{C_3} - \sqrt{6}\,E'\sin(\omega t - \pi/6)$$

For $t = t_0$, u_{C_3} equals U_{C0}; diode D_1 remains blocked if

$$U_{C_0} > -\sqrt{6}E'\sin(\omega t_0 - \pi/6).$$

Since, when feeding an inductive load, $\omega t_1 - \pi/6$ varies between 0 and π and there is little difference between ωt_0 and ωt_1, this condition is fulfilled if U_{C_0} is positive.

– The voltage across D_2 during period (t_0, t_1) is given by:

$$v_{D_2} = u_{C_2} + e'_C - e'_B = u_{C_2} + E'\sqrt{6}\cos\omega t.$$

Voltage u_{C_2}, equal to $-U_{C_0}$ for $t = t_0$, increases linearly half as fast as u_{C_3} decreases; it can thus be expressed as:

$$u_{C_2} = -U_{C_0} + \frac{1}{2}[U_{C_0} - u_{C_3}(t_1)](t - t_0)/(t_1 - t_0)$$

If $u_{C_3}(t_1)$ is replaced by $e'_C(t_1) - e'_A(t_1)$ and U_{C_0} by the value given by equation (5.67), the following results:

$$v_{D_2} = -[\sqrt{6}E'\sin(\omega t_1 - \pi/6) + 2\sqrt{(N_1/3C')}I]$$
$$+ [\sqrt{6}E'\sin(\omega t_1 - \pi/6) + \sqrt{(N_1/3C')}I](t - t_0)/(t_1 - t_0)$$
$$+ \sqrt{6}E'\cos\omega t$$

Since $\omega t_1 - \pi/6$ is to be found between 0 and π when feeding an inductive load, the sum of two terms in $\sin(\omega t - \pi/6)$ of the expression of v_{D_2} is negative at the start of the second phase and zero at the end of the latter. Voltage v_{D_2} remains negative throughout this phase, if the following occurs:

$$\sqrt{(N_1/3C')}I[-2 + (t - t_0)/(t_1 - t_0)] + \sqrt{6}E'\cos\omega t \leqslant 0$$

For this condition to be fulfilled at instant $t = t_1$, even in the case where ωt_1 equals $\pi/6$, it is sufficient that

$$\sqrt{(N_1/3C')}I \geqslant (\sqrt{3}/2)\sqrt{6}E'.$$

- During period (t_1, t_2), the voltage across D_2 is given by

$$v_{D_2} = u_{C_2} + N_1\frac{di'_C}{dt} + e'_C - e'_B.$$

If u_{C_2} is replaced by $-u_{C_3}/2 - U_{C_0}/2$, and di'_C/dt by $C\,du_{C_2}/dt$, and if the expression of u_{C_3} given by (5.65) is taken into account, this gives, after simplifications

$$v_{D_2} = -\sqrt{(N_1/3C')}I + \sqrt{6}E'\cos\omega t.$$

It will be seen that condition (5.68) which ensures the blocking of D_2 during the third phase also guarantees the blocking of D_2 during the second.

- Finally, during the third phase, the voltages across diodes D_2 and D_3 are given by

$$v_{D_2} = -u_{C_1} + e'_A - v'_B$$
$$v_{D_3} = -u_{C_1} + e'_A - v'_C.$$

Whatever the operating conditions, voltages v_{D2} and v_{D3} will remain negative over this period, if the following applies[1]:

$$\sqrt{(N_1/3C')}I > \sqrt{6}E' \tag{5.68}$$

---

[1] A more precise calculation of the minimal value of $I\sqrt{N_1/3C'}$, ensuring the blocking of D_2 would require the phase shift ψ between the fundamental of current i'_A, i'_B, i'_C and the electromotive forces e'_A, e'_B, e'_C to be taken into account.

5.3.2.3 Remarks

- For C' equal to $2C/3$, the values of period $t_1 - t_0$ and $t_2 - t_1$ are identical to those obtained with the inverter with auxiliary turn-off bridge. The same consequently applies to the reverse bias time of the thyristors and to the maximum value of C. However, the latter must be kept below the maximum value indicated by Eq. (5.68).
- The diodes D_1, D_2, D_3 and D'_1, D'_2, D'_3 in the configuration shown in Fig. 5.32 could naturally be replaced by suitably controlled thyristors, in order to avoid any risk of untimely conduction caused by an insufficient initial charge of the turn-off capacitors.

  In that case, the advantage of the classic self-commuted inverter which needs only six thyristors disappears, while four semiconductor devices are still found in the path of current I instead of two in the case of the inverter with auxiliary turn-off bridge.

5.3.3 GTO Current Inverter with Clamping Circuit

Figure 5.36 shows the basic diagram of a GTO inverter with a clamping circuit and a commutation energy recovery converter.

On this figure, the following can be found:

- the current supply comprising, in this case, a rectifier feeding the inverter via a smoothing inductance L,
- the bridge with six GTO thyristors, T_1, T_2, T_3 and T'_1, T'_2, T'_3, which forms the inverter as such,

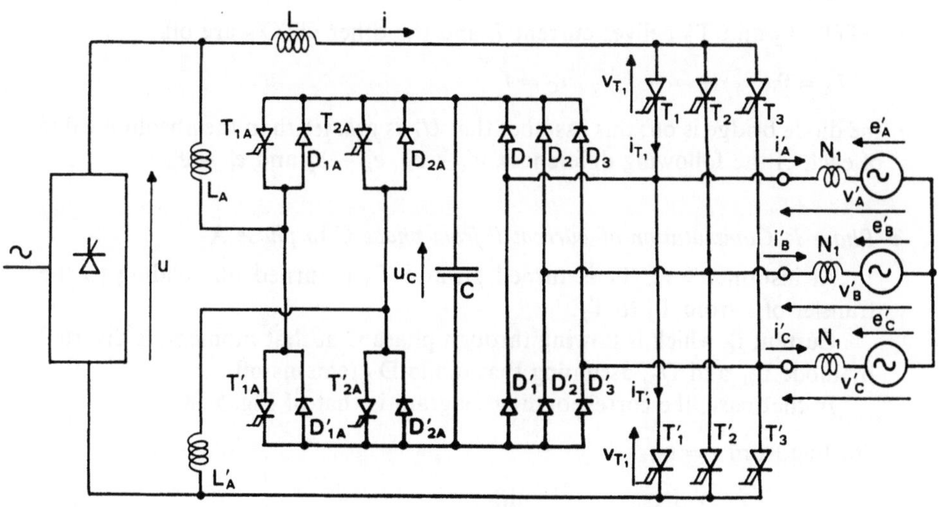

Fig. 5.36

- the clamping circuit formed by the bridge with diodes, D_1, D_2, D_3 and D'_1, D'_2, D'_3 and capacitor C.
- the circuit which recovers the energy stored in C at each commutation. In this case, it is a four-quadrant chopper made up of four current-reversible switches with controlled turn-off/turn-on, T_{1A} and D_{1A}, T_{2A} and D_{2A}, T'_{1A} and D'_{1A}, T'_{2A} and D'_{2A}. Via inductances L_A and L'_A, it delivers in parallel with the input converter.

5.3.3.1 Operation

The operation will be studied over the sixth of an output values cycle T. During this period, two phases can be distinguished:

- the first corresponds to the commutation of current i from one phase to the next;
- the second to normal conduction, with current i flowing via two phases series-connected.

It will be assumed that the recovery of energy stored in capacitor C during commutation, will occur during the normal conduction phase.

Inductance L and capacitance C will be assumed to have sufficiently high values to enable the ripples of current i and voltage u_C to be neglected. It is thus possible to equate these quantities with their mean values I and U_C.

Once again the study will be confined to the feeding of an inductive load. Figure 5.37 enables the evolution of the main variables to be followed.

The following state, which occurs cyclically in steady state, is taken as the starting point:

- GTO's T_3 and T'_2 deliver current I, and the other GTO's are off

$$i'_A = 0, \qquad i'_B = -I, \qquad i'_C = I$$

- the diode bridge is off; this assumes that U_C is greater than the absolute value of each of the following differences, $e'_A - e'_B$, $e'_B - e'_C$ and $e'_C - e'_A$.

- *Phase 1: Commutation of current I from phase* C *to phase* A

  At instant $t = t_0$, T_1 is turned on and T_3 is turned off, leading to the transfer of I from T_3 to T_1.

  Current i'_C which is flowing through phase C at that moment, is diverted by diode D_1 and D'_3, assuming that diode D_2 remains off.

  In that case, the corresponding diagram is that of Fig. 5.38.

– Starting from $t = t_0$,

$$U = e'_A + N_1 \frac{di'_A}{dt} - e'_C - N_1 \frac{di'_C}{dt}$$

with

$$i'_A + i'_C = i_{T_1} = I; \qquad i'_A(t_0) = 0; \qquad i'_C(t_0) = I$$

The following can be deduced:

$$\frac{di'_A}{dt} = \frac{U_C - e'_A + e'_C}{2N_1} = \frac{U_C - \sqrt{6}E'\sin(\omega t - \pi/6)}{2N_1} \tag{5.69}$$

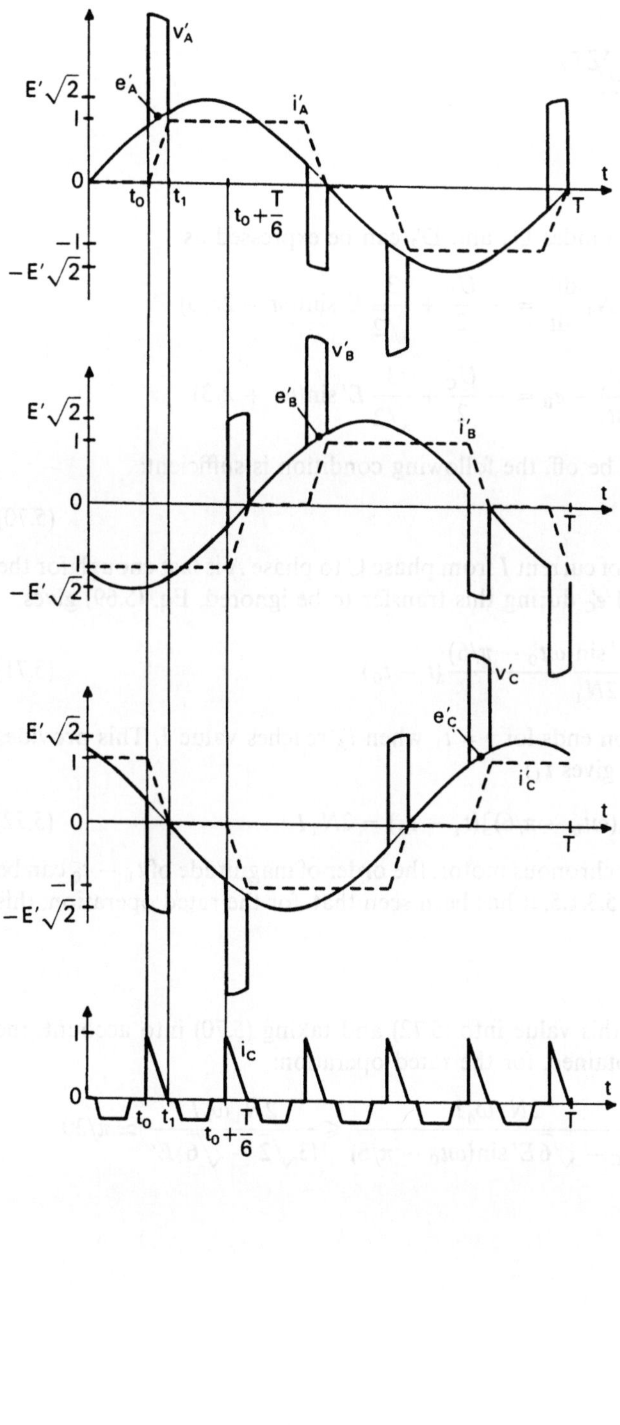

Fig. 5.37

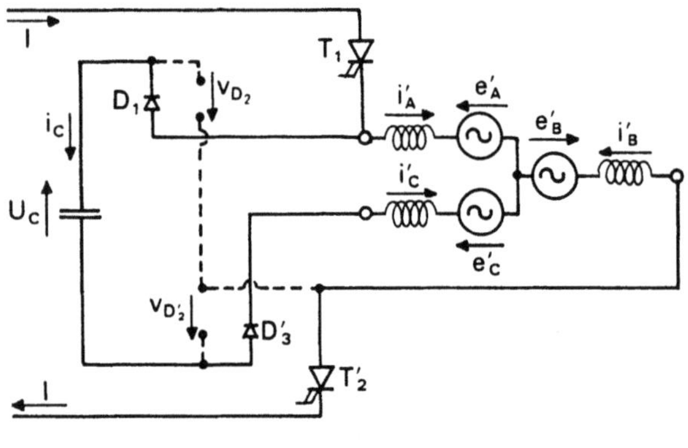

Fig. 5.38

– The voltages across diodes D_2 and D'_2 can be expressed as:

$$v_{D_2} = e'_B - e'_A - N_1 \frac{di'_A}{dt} = -\frac{U_C}{2} + \frac{3}{\sqrt{2}} E' \sin(\omega t - 2\pi/3)$$

$$v_{D'_2} = e'_C + N_1 \frac{di'_A}{dt} - e'_B = -\frac{U_C}{2} + \frac{3}{\sqrt{2}} E' \sin(\omega t + \pi/3)$$

For these diodes to be off, the following condition is sufficient:

$$U_C \geqslant 3\sqrt{2}\,E' \tag{5.70}$$

– If the commutation of current I from phase C to phase A is fast enough for the variations of e'_A and e'_C during this transfer to be ignored, Eq. (5.69) gives

$$i'_A = \frac{U_C - \sqrt{6}E' \sin(\omega t_0 - \pi/6)}{2N_1}(t - t_0) \tag{5.71}$$

The commutation ends for $t = t_1$ when i'_A reaches value I. This provides the equation which gives t_1:

$$[U_C - \sqrt{6}E' \sin(\omega t_0 - \pi/6)](t_1 - t_0) = 2N_1 I \tag{5.72}$$

– If the load is an asynchronous motor, the order of magnitude of $t_1 - t_0$ can be fixed. In paragraph 5.3.1.5, it has been seen that, for the rated operation, this gives:

$$N_1 \omega_n I \simeq 0.1E'.$$

By transferring this value into (5.72) and taking (5.70) into account, the following can be obtained, for the rated operation:

$$\omega_n(t_1 - t_0) = \frac{2N_1\omega_n I}{U_C - \sqrt{6}E' \sin(\omega t_0 - \pi/6)} \leqslant \frac{2N_1\omega_n I}{(3\sqrt{2} - \sqrt{6})E'} \simeq \pi/20$$

- *Phase 2: Normal conduction; recovery of commutation energy*

- Starting from $t = t_1$ and up until instant $t = t_0 + T/6$ when T'_3 is controlled at turn-on and T'_2 at turn-off, current I flows via T_1, phase A, phase B and T'_2.

$$i'_A = -i'_B = I, \qquad i'_C = 0, \qquad v'_A = e'_A, \qquad v'_B = e'_B, \qquad v'_C = e'_C.$$

- During this period, the commutation energy recovery circuit is brought into operation. It is a four-quadrant chopper of the type studied in volume 3.

  The switch duty ratio is regulated so that the mean voltage at the chopper output is equal to the mean voltage at the rectifier output. Capacitor C thus discharges in parallel with the input rectifier, via the chopper and inductances L_A and L'_A.

  The current provided by the chopper must remain less than the current absorbed by the motor. Since the charge stored in C during a commutation is equal to $(1/2)\, I(t_1 - t_0)$, the discharge time must be equal to at least half the commutation time. As a result, the latter cannot exceed 2/3 of the period of time between two successive commutations.

5.3.3.2 Variations on the Recovery Method

The energy stored in capacitor C during each commutation can be recovered by sending it directly back to the power supply of the assembly via *a transformer* ensuring the galvanic insulation needed. In this case, the recovery can be permanent.

- *If the supply is* AC (Fig. 5.39), a thyristor bridge operating as a non-autonomous inverter can be used. The inductance L_A ensures the smoothing of the current delivered by the capacitor.
- *If the supply is* DC (Fig. 5.40), the energy recovery can be carried out by using an autonomous voltage-source inverter and a diode rectifier. The diagram shows, between the supply and the current-source inverter, the inverter which enables the sign of voltage u to be reversed.

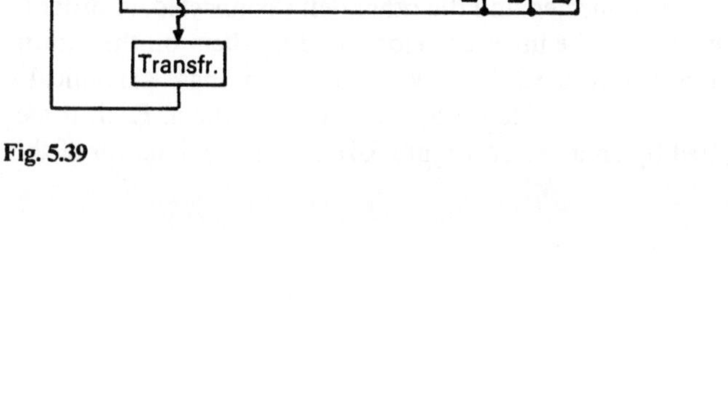

Fig. 5.39

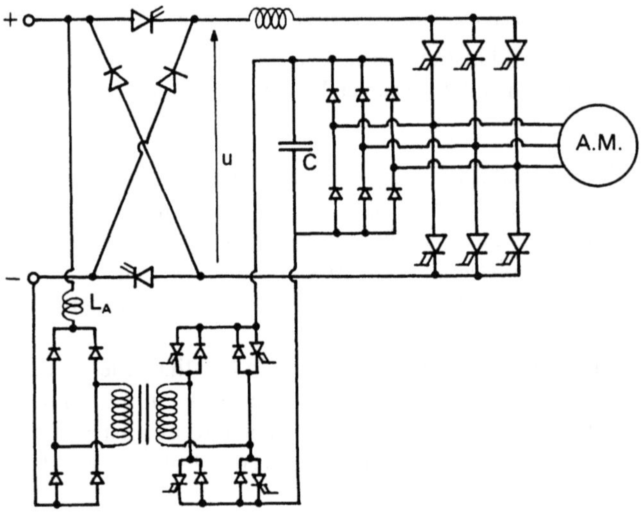

Fig. 5.40

5.4 Notes on the Configuration with a Pulse Number Greater than 6

Feeding an asynchronous motor by a current-source inverter produces torque pulsations.

The current harmonics of rank 5 and 7 give rise to a torque component with a frequency equal to 6 times the current fundamental frequency. Harmonics 11 and 13 produce a torque at a frequency equal to 12 times that of the supply frequency.

The pulsating torque can be reduced by increasing the pulse number of the power supplied to the machine.

5.4.1 Configuration with Pulse Number Equal to 12

One frequently adopted solution is to supply the motor by means of two series-connected inverters, the controls of which being shifted by a twelfth of a cycle.

As the operating frequency of the inverters must vary widely, it is not possible to use, as in the case of rectifiers used in Vol. 1, a transformer with two primary windings, one star-connected and the other delta-connected, in order to combine the output currents of the inverter before sending them on the motor.

For this reason, the motor used has two stator windings, galvanically independent and with a shift of 30 electric degrees between them. Each three-phase winding is supplied by an inverter. Figure 5.41 shows the diagram of this

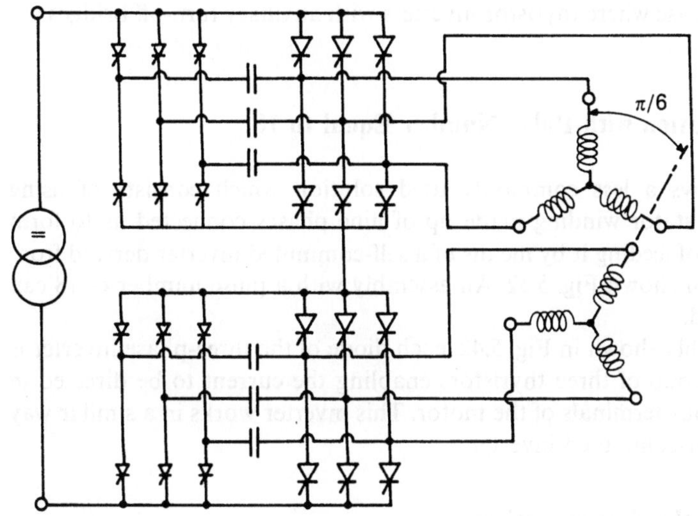

Fig. 5.41

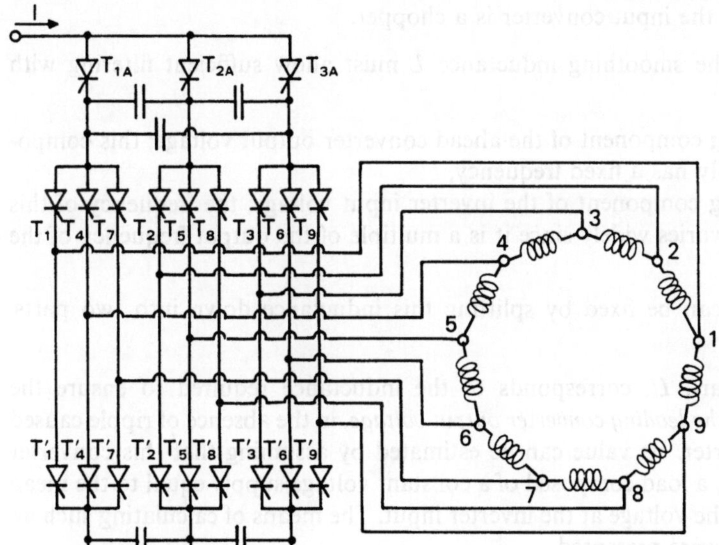

Fig. 5.42

assembly in the case where thyristor inverters with auxiliary turn-off bridges are used.

5.4.2 Configuration with Pulse Number Equal to 18

Figure 5.42 shows a less commonly used solution which consists of using a motor with a stator winding made up of nine phases connected as to form a polygon[1] and of feeding it by means of a self-commuted inverter derived from the configuration shown Fig. 5.32. An assembly with a pulse number of 18 can thus be obtained.

In the assembly shown in Fig. 5.42, each diode of the three-phase inverter is replaced by a group of three thyristors enabling the current to be directed in turn to the various terminals of the motor. This inverter works in a similar way to the classic self-commuted inverter.

5.5 Notes on the Computation of the Current-Source Inverter Input Filter

The DC current supply I is usually obtained by supplying the inverter, via a smoothing inductance, by means of a converter connected to a voltage supply, and by regulating the current it delivers.

If the supply is AC, the intermediate converter is a controlled rectifier. If the supply is DC, the input converter is a chopper.

In all cases, the smoothing inductance L must allow sufficient filtering with regard to

- the pulsating component of the ahead converter output voltage; this component normally has a fixed frequency,
- the pulsating component of the inverter input voltage; the frequency of this component varies widely since it is a multiple of the output frequency of the inverter.

A size for L can be fixed by splitting this inductance down into two parts, L' and L''.

- The first part, L', corresponds to the inductance required to ensure the *filtering of the leading converter output voltage*, in the absence of ripple caused by the inverter. Its value can be estimated by assuming that this converter feeds, via L', a load composed of a constant voltage supply equal to the mean value U of the voltage at the inverter input. The means of calculating such an inductance were presented
  - in volume 1, for rectifiers,
  - in volume 3, for choppers.

[1] As shown for the rectifiers in Vol. 1, the polygon grouping gives an use of the windings which is better than the star grouping as soon as the number of phases exceeds 3.

- The second part L'', corresponds to the inductance required to ensure the *filtering of the inverter input voltage*, in the absence of the ripple caused by the leading converter. Its value can be estimated by assuming that this converter delivers a constant voltage equal to U.

  The inverter input voltage is similar to the output voltage of a rectifier operating with a firing delay angle α equal to $\omega t_1 - \pi/6$, with voltage spikes introduced by the commutations.

  Apart from cases close to the maximum operating frequency of the inverter, these spikes correspond to very fast variations as compared with the cycle of the inverter input voltage ripple. These spikes usually have little influence on current i absorbed by the latter.

  Except for the commutation spikes and until the rated frequency of the motor is reached, the voltage at the inverter input has an amplitude proportional to its operating frequency, as shown in paragraph 5.3.1.5. As a result, the current i ripple, caused by fluctuations in voltage u, is virtually constant and independent of the operating frequency. Inductance L'' can thus be calculated so as to reduce the input current ripple in the inverter to its required value when voltage u corresponds to the rated voltage and rated frequency of the motor. L'' can be calculated in a similar way as the smoothing inductance of a rectifier.

  In the case of operating modes at a frequency above the rated frequency, the input voltage amplitude remains constant; current i ripple thus tends to decrease when the operating frequency of the inverter increases.

Bibliography

Almost the only use made of Current-Source Inverters at the present time is in the production of variable-speed drives with asynchronous induction motor.

Before noting publications on the latter type of drive, we considered it is of interest to indicate where bibliographical reference to the various types of variable-speed drives are to be found, in the four volumes which make up this series:

- *DC motor supplied by a rectifier* (Vol. 1, Chap. 9);
- *DC motor supplied by a chopper* (Vol. 3, Chap. 4),
- *AC self-commuted synchronous motor supplied by a voltage-source inverter* (Vol. 1, Chap. 7),
- *AC synchronous or asynchronous motor supplied by a cycloconverter* (Vol. 2, Chap. 5),
- *AC asynchronous motor supplied by AC regulator* (Vol. 2, Chap. 4),
- *AC asynchronous motor with chopper-controlled slip* (Vol. 3, Chap. 4),
- *AC synchronous or asynchronous motor supplied by a voltage-source inverter* (Vol. 4, Chap. 3),
- *AC asynchronous motor supplied by a current-source inverter* (Vol. 4, Chap. 5).

If the current-source inverter is always studied in association with a motor, certain articles put greater emphasis on the inverter and others on the inverter-motor assembly, particularly concerning its control.

- *Current-source inverters without PWM*, i.e. in three-phase operation with six steps per cycle, are still studied, owing to the difficulties involved in following commutations.

– Certain publications [1–11] deal with the *classical structure* (with six thyristors, six diodes and six capacitors) which we have studied or with other configurations which we have presented. Two inverters are occasionally associated in order to obtain twelve steps per cycle.

– Other publications [12–22] study *variations*, notably the chopper-inverter assembly where the former ensures the commutations and the latter merely directs the currents as required.

- In order to reduce the output AC current and input DC current harmonics, *current-source inverters with PWM* are produced [23–35].

Since the output current frequency varies, the PWM frequency-modulation ratio is variable and there is a change to six steps per cycle beyond a certain output frequency value

Initially, such publications dealt with assembly using ordinary thyristors: now they study converters using GTO thyristors with clamping circuit.

- We have divided the publications on the *current source inverter – asynchronous motor assembly into three groups*:

- The first [36–47] deals with *steady-state characteristics* of an induction motor supplied by the inverter and notably with the torque pulsation. Several articles examine the reduction in this pulsation brought about by use of machines with twin stator windings (see also Refs. 87 to 107 at the end of this bibliography).

- Many articles study at length the *stability and control* aspects [48–70] of the CSI-powered motor, the inverter-motor assembly being unstable in open-loop operation. This gives rise to publications concerning closed-loop controls enabling a good level of stability to be obtained and the dynamic characteristics to be improved.

  Recent publications deal with vector control which requires knowledge of the machine parameters and especially of the time constant of the rotor winding and its variations. Several studies concern ways of eliminating the rotor-position detector.

- We have classified separately articles dealing with a CSI supplying an induction motor *shunted by capacitors* [71–74]. The machine–capacitor configuration is equivalent to a capacitive impedance and enables the inverter to operate in natural commutation. Any problems are related to the risks of capacitor resonance with the magnetizing inductance of the machine.

- The use of an *induction motor in electric traction* [75–86] has been the subject of numerous studies. Some of the latter have been followed by actual productions.

  Certain publications are especially concerned with a specific problem of electric traction: injecting harmonics into the catenary and risking interference with the signalisation signals.

  The other articles indicated deal with motors powered either by voltage-source inverter or by current-source inverter.

  For these two reasons, we have treated these articles separately from those dealing with other applications of the induction motor in AC drives.

- The articles in the final group [87–107] deal with the *effects of feeding an asynchronous motor by means of an inverter*, thus non-sinusoidal voltages or currents. This particularly concerns torque pulsations, noise and extra losses.

These studies examine the quantitative evaluation of these effects and the means by which they can be reduced – either by acting on the signals supplied by the inverter, or by the construction of the motor.

Current-Source Inverters without PWM

– Conventional Diagrams

1. Philips K (1972) Current source converter for AC motor drives, *IEEE Trans. Ind. Appl.*, **8**(6), 679–683.
2. Revankar GN, Bashir A (1977) Effect of circuit and induction motor parameters on current source inverter operation, *IEEE Trans. Ind. Electron. Control Instrum.*, **24**(1), 126–132.
3. Palaniappan R, Vithayathil J, Datta S (1979) Principle of a dual source converter for AC motor drive, *IEEE Trans. Ind. Appl.*, **15**(4), 445–452.
4. Steigerwald RL (1979) Characteristics of a current-fed inverter with commutation applied through load neutral point, *IEEE Trans. Ind. Appl.*, **15**(5), 538–553.
5. Subrahmanyam V, Yuvarajan S, Ramaswami B (1980) Analysis of commutation of current inverter feeding an induction motor load, *IEEE Trans. Ind. Appl.*, **16**(3), 332–340.
6. Subrahmanyam V (1983) Analysis of commutation of a current-source inverter feeding an induction motor, using 2-axis variables, *Proc. Inst. Electr. Eng., Part B*, **130**(5), 355–359.
7. Saito S, Nakagawa T, Hirata A, Miyazaki M (1984) New application of current-type inverter, *IEEE Trans. Ind. Appl.*, **20**(1), 226–235.
8. Min BJ, Jarc DA (1984) Considerations in testing an stabilizing frequency beating of a current source inverter, *IEEE Trans. Ind. Appl.*, **20**(5), 1290–1295.
9. Deng D, Lipo TA (1986) A modified control method for fast-response current source inverter drives, *IEEE Trans. Ind. Appl.*, **22**(4), 653–665.
10. Osman RH (1987) A method for compensation of variable commutation delay in current source inverters, *IEEE Trans. Ind. Appl.*, **23**(6), 1055–1060.
11. Chattopadhyay AK, Meher N (1988) A generalized approach to steady-state analysis of a current-source inverter with induction motor load including commutation overlap, *IEEE Trans. Ind. Electron.*, **35**(3), 434–441.

– Improvements. Variations

12. Ho HH (1974) A new and simple three-phase thyristor inverter for variable frequency operation, *IEEE Trans. Ind. Appl.*, **10**(6), 787–796.
13. Martinez S, Aldana F (1978) Current-source double DC-side forced commutation inverter, *IEEE Trans. Ind. Appl.*, **14**(6), 581–593.
14. Chin TH, Nakano M (1980) A coupled-reactor-commutating sinusoidal-wave inverter, *IEEE Trans. Ind. Appl.*, **16**(1), 96–102.
15. Palaniappan RG, Vithayathil J (1980) High-frequency current source inverter, *IEEE Trans. Ind. Appl.*, **16**(3), 431–438.
16. Cho GH, Park SB (1981) Novel six-step and twelve-step current source inverters with DC-side commutation and energy rebound, *IEEE Trans. Ind. Appl.*, **17**(5), 524–532.
17. Drury W, Jones BL, Brown JE (1982) Performance of current-fed inverter system with controlled flywheeling applied to supply convertor, *Proc. Inst. Electr. Eng., Part B*, **129**(5), 262–270.
18. Cho GH, Jeon SJ, Park SB (1983) Optimum design of a new DC–SC circuit, *Proc. Inst. Electr. Eng., Part B*, **130**(3), 171–180.
19. Osman RH (1984) A simple energy absorbing circuit for current source inverters, *IEEE Trans. Ind. Appl.*, **20**(6), 1448–1452.
20. Palaniappan RG (1985) Voltage clamping circuits for CSI/IM drives, *IEEE Trans. Ind. Appl.*, **21**(2), 429–447.
21. Wu B, Dewan SB, Sen PC (1988) A modified current-source inverter for a multiple induction drive system, *IEEE Trans. Power Electron.*, **3**(1), 10–16.
22. Cho GH, Park SB (1989) A new current source inverter with simultaneous recovery and commutation, *IEEE Trans. Ind. Appl.*, **25**(1), 162–171.

Current-Source Inverters with PWM

23. Kaimoto M, Hashii M, Yanase T, Nakano T (1982) Performance improvement of current source inverter-fed induction motor drives, *IEEE Trans. Ind. Appl.*, **18**(6), 703–711.
24. Bowes SR, Bullough RI (1984) PWM switching strategies for current-fed inverter drives, *Proc. Inst. Electr. Eng., Part B*, **131**(5), 195–202.
25. Nonaka S, Neba Y (1986) New GTO current source inverters with pulsewidth modulation control techniques, *IEEE Trans. Ind. Appl.*, **22**(4), 666–672.
26. Betz RE, Evans RJ, Cook BJ (1986) Optimal pulsewidth modulation for current source inverters, *IEEE Trans. Ind. Electron.*, **33**(3), 318–324.
27. Namuduri C, Sen PC (1986) Optimal pulsewidth modulation for current source inverters, *IEEE Trans. Ind. Appl.*, **22**(6), 1052–1072.
28. Nonaka S, Naba Y (1987) Analysis of PWM GTO current source inverter-fed induction motor drive system, *IEEE Trans. Ind. Appl.*, **23**(2), 256–262.
29. Biswas SK, Iyengar BSR, Vithayathil J (1987) A new PWM control method for three-phase autosequentially commutated current source inverters, *IEEE Trans. Ind. Appl.*, **23**(3), 554–560.
30. Viriya P, Kubota H, Matsuse K (1987) New PWM-controlled GTO converter, *IEEE Trans. Power Electron.*, **2**(4), 373–381.
31. Biswas SK, Basak B., Sathiakumar S, Vithayathil J (1987) A new three-phase current source inverter with flexible PWM capability, *IEEE Trans. Ind. Appl.*, **23**(5), 921–926.
32. Itoh R (1988) Simplified configuration of GTO current source inverter for induction motor drives, *Proc. Inst. Electr. Eng., Part B*, **135**(5), 218–223.
33. Nonaka S, Naba Y (1989) A PWM GTO current-source converter–inverter system with sinusoidal inputs and outputs, *IEEE Trans. Ind. Appl.*, **25**(1), 76–85.
34. Bowes SR, Bullough RI (1989) Novel PWM controlled series-connected current-source inverter drive, *Proc. Inst. Electr. Eng., Part B*, **136**(2), 69–82.
35. Itoh R (1989) Method of reducing computation in the analysis of PWM current-source GTO convertor, *Proc. Inst. Electr. Eng., Part B*, **136**(2), 105–112.

Current Source Inverted-Fed Induction Motor

– Steady-State Analysis

36. Lipo TA, Cornell EP (1975) State variable steady-state analysis of a controlled current induction motor drive, *IEEE Trans. Ind. Appl.*, **11**(6), 704–712.
37. Lesenne J, Ravalitera G, Seguier G (1978) Les composantes relatives de YH Ku. Application à l'étude analytique du moteur asynchrone alimenté par un onduleur autonome fonctionnant en commutateur de courant, *Rev. Gén. Electr.*, **87**(3), 216–224.
38. Nacabal M (1978) Alimentation en courant des machines asynchrones à fréquence variable, *Rev. Gén. Electr.*, **87**(4), 277–282.
39. Lataire Ph (1978) Problèmes recontrés dans l'alimentation de moteurs asynchrones à partir du 380V, *Rev. Gén. Electr.*, **87**(4), 287–290.
40. Rojat G, Mertz JL Foggia A (1979) Theoretical and experimental analysis of a two-phase inverter-fed induction motor, *IEEE Trans. Ind. Appl.*, **15**(6), 601–605.
41. Joshi A, Dewan SB (1981) Modified steady-state analysis of the current source-inverter and squirrel cage motor drive, *IEEE Trans. Ind. Appl.*, **17**(1), 50–57.
42. Mc Murray W (1983) Hybrid computer simulations of a controlled current inverter under normal and fault conditions, *IEEE Trans. Ind. Appl.*, **19**(2), 268–280.
43. Fuchs FW, Mueller-Hellmann A (1983) Control methods for reducing the inductance in the DC link of current source inverters, *IEEE Trans. Ind. Appl.*, **19**(5), 699–707.
44. Lipo TA (1983) Design and control techniques for extending high frequency operation of a CSI induction motor drive, *IEEE Trans. Ind. Appl.*, **19**(5), 744–753.
45. Bowes SR, Bullough R (1984) Steady-state performance of current-fed pulse-width-modulated inverter drives, *Proc. Inst. Electr. Eng., Part B*, **131**(3), 113–132.
46. Gopakumar S, Sathiakumar S, Biswas SK and Vithayathil J (1984) Modified current source inverter fed induction motor drive with reduced torque pulsations, *Proc. Inst. Electr. Eng., Part B*, **131**(4), 159–164.
47. Kubota H, Matsuse K, Ree JH (1986) Analysis of new current source GTO inverter-fed induction motor drive, *IEEE Trans. Power Electron.*, **1**(4), 210–214.

– Control

48. Sawaki N, Sato N (1977) Steady-state and stability analysis of induction motor driven by current source inverter, *IEEE Trans. Ind. Appl.*, **13**(3), 244–253.
49. Cornell EP, Lipo TA (1977) Modeling and design of controlled current induction motor drive systems, *IEEE Trans. Ind. Appl.*, **13**(4), 321–330.
50. de Fornel B, Noyes D (1977) Transient characteristics of an asynchronous machine current fed from a static convertor, *Proc. Inst. Electr. Eng.*, **124**(10), 884–886.
51. Plunkett AB, D'atre JD, Lipo TA (1979) Synchronous control of static AC induction motor drive, *IEEE Trans. Ind. Appl.*, **15**(4), 430–437.
52. Garces LJ (1980) Parameter adaptation for the speed controlled static AC drive with squirrel-cage induction motor, *IEEE Trans. Ind. Appl.*, **16**(2), 173–178.
53. Walker LH, Espalage PM (1980) A high performance controlled-current inverter drive, *IEEE Trans. Ind. Appl.*, **16**(2), 193–202.
54. Dote Y (1980) Stabilization of controlled current induction motor drive system via new nonlinear state observer, *IEEE Trans. Ind. Electron. Control Instrum.*, **27**(2), 77–81.
55. Abbas MA, Novotny DW (1982) A stator voltage-controlled current source inverter induction motor drive, *IEEE Trans. Ind. Appl.*, **18**(3), 219–229.
56. de Fornel B, Pietrzak-David M, Gauvrit M (1982) State-space compensation applied to the control of a current fed induction motor, *Proc. Inst. Electr. Eng., Part B*, **129**(4), 221–226.
57. Ito T, Yamaguchi T, Ueda R, Mochizuki T, Takata S (1983) Analysis of field orientation control of current source inverter drive induction systems, *IEEE Trans. Ind. Appl.*, **19**(2), 206–210.
58. Joetten R, Maeder G (1983) Control method for good dynamic performance induction motor drives based on current and voltage as measured quantities, *IEEE Trans. Ind. Appl.*, **19**(3), 356–363.
59. Krishnan R, Lindsay JF, Stefanovic VR (1983) Design of angle-controlled current source inverter-fed induction motor drive, *IEEE Trans. Ind. Appl.*, **19**(3), 370–378.
60. de Fornel B, Bach JL, Hapiot JC, Pietrzak-David M (1984) Numerical speed control of current-fed asynchronous machine by measurement of supply voltages, *Proc. Inst. Electr. Eng., Part B*, **131**(4), 165–169.
61. Kim HG, Sul SK, Park MH (1984) Optimal efficiency drive of a current source inverter fed induction motor by flux control, *IEEE Trans. Ind. Appl.*, **20**(6), 1453–1459.
62. Matsuo T, Lipo TA (1985) A rotor parameter identification scheme for vector-controlled induction motor drives, *IEEE Trans. Ind. Appl.*, **21**(3), 624–632.
63. Nordin KB, Novotny DW, Zinger DS (1985) The influence of motor parameter deviations in feedforward field orientation drive systems, *IEEE Trans. Ind. Appl.*, **21**(4), 1009–1015.
64. Bolognani S, Buja GS (1985) Control system design of a current inverter induction motor drive, *IEEE Trans. Ind. Appl.*, **21**(5), 1145–1153.
65. Koyama M, Yano M, Kamiyama I, Yanos S (1986) Microprocessor-based vector control system for induction motor drives with rotor time constant identification function, *IEEE Trans. Ind. Appl.*, **22**(3), 453–459.
66. Sul K, Park MH (1988) A novel technique for optimal efficiency control of a current-source inverter-fed induction motor, *IEEE Trans. Power Electron.*, **3**(2), 192–199.
67. Liaw CM, Pan CT, Chen YC (1988) An adaptive controller for current-fed induction motor, *IEEE Trans. Aerosp. Electron. Syst.*, **24**(3), 250–261.
68. Zhang J, Barton TH (1989) Microprocessor-based primary current control for a cage induction motor drive, *IEEE Trans. Power Electron.*, **4**(1), 73–82.
69. Chattopadhyay AK, Meher N (1989) Microprocessor implementation for a state feedback control strategy for a current source inverter-fed induction motor drive, *IEEE Trans. Power Electron.*, **4**(2), 279–288.
70. Liu CH, Hwu CC, Feng YF (1989) Modeling and implementation of a microprocessor-based CSI-fed induction motor-drive using field-oriented control, *IEEE Trans. Ind. Appl.*, **25**(4), 588–597.

– Use of Capacitors Parallel-Connected Across the Motor

71. Watson DB (1982) Performance of induction motor driven from self-commutating inverter, *Proc. Inst. Electr. Eng., Part B*, **129**(5), 248–250.

72. Hombu M, Ueda S, Ueda A, Matsuda Y (1985) A new current source GTO inverter with sinusoidal output voltage and current, *IEEE Trans. Ind. Appl.*, **21**(4), 1192–1198.
73. Hombu M, Ueda S, Ueda A (1987) A current source GTO inverter with sinusoidal inputs and outputs, *IEEE Trans. Ind. Appl.*, **23**(2), 247–255.
74. Itoh R (1989) Stability of induction motor drive controlled by current-source inverter, *Proc. Inst. Electr. Eng., Part B*, **136**(2), 83–88.

Inverter-Induction Motor Drives for Electric Traction

75. Plunkett AB, Plette DL (1977) Inverter-induction motor drive for transit cars, *IEEE Trans. Ind. Appl.*, **13**(1), 26–37.
76. Lieanu W, Muller-Hellmann A, Skudelny H (1980) Power converters for feeding asynchronous traction motors of single-phase AC vehicles, *IEEE Trans. Ind. Appl.*, **16**(1), 103–110.
77. Kielgas H, Nill R (1980) Converter propulsion systems with three-phase induction motors for electric traction vehicles, *IEEE Trans. Ind. Appl.*, **16**(2), 222–233.
78. Taufiq JA, Goodman CJ, Mellitt B (1986) Railway signalling compatibility of inverter fed induction motor drives for rapid transit, *Proc. Inst. Electr. Eng., Part B*, **133**(2), 71–84.
79. Taufiq JA, Mellitt B, Goodman CJ (1986) Novel algorithm for generating near optimal PWM waveforms for AC traction drives, *Proc. Inst. Electr. Eng., Part B*, **133**(2)?, 85–94.
80. Loth WK, Renfrew AC (1986) Optimised microelectronic power control of an drive for an electric vehicle, *Proc. Inst. Electr. Eng., Part B*, **133**(4), 291–298.
81. Palais G (1986) Automotrice à deux niveaux: un bilan positif pour la transmission asynchrone, *Rev. Gén. Electr.*, **95**(7), 46–51.
82. Abraham L (1988) Power electronics in German railway propulsion, *Proc. IEEE*, **76**(4), 472–480.
83. Mouneime ZS, Mellitt B (1988) Modelling of inverter-controlled asynchronous drives for system simulation in DC railways, *Proc. Inst. Electr. Eng., Part B*, **135**(55), 210–217.
84. Holtz J, Klein HJ (1989) The propagation of harmonic currents generated by inverter-fed locomotives in the distributed overhead supply system, *IEEE Trans. Power Electron.*, **4**(2), 168–174.
85. Taufiq JA, Xiaoping J (1989) Fast accurate computation of the DC-side harmonics in a traction VSI drive, *Proc. Inst. Electr. Eng., Part B*, **136**(4), 176–187.
86. Taufiq JA, Mellitt B, Xiaoping J (1989) Input impedance of inverter equipment and its significance in low-frequency track circuits, *Proc. Inst. Electr. Eng., Part B*, **136**(5), 243–248.

Effects on Induction Motor of Supplying by Inverter

87. Robertson SDT, Hebbar KM (1970) Torque pulsation in induction motors with inverter drives, *IEEE Trans. Ind. Gen. Appl.*, **7**(2), 318–323.
88. Jacovides LJ (1973) Analysis of induction motor drives with a nonsinusoidal voltage using Fourier analysis, *IEEE Trans. Ind. Appl.*, **9**(6), 741–747.
89. Lipo TA, Turnbull FG (1975) Analysis and comparison of two types of squarewave inverters, *IEEE Trans. Ind. Appl.*, **11**(2), 137–147.
90. Wiart A (1975) Association d'onduleurs et de machines asynchrones, *Rev. Gén. Electr.*, **84**(11), 809–820.
91. Finidori Ch, Tholozan G (1976) Contribution à l'"analyse des associations d'onduleurs et de machines tournantes, *Rev. Gén. Electr.*, **85**(1), 1–17.
92. Avadhanlu TV, Saxena RB (1979) Torque pulsation minimization in a variable-speed inverter-fed induction motor drive system, *IEEE Trans. Power Appar. Syst.*, **98**(1), 13–18.
93. De Buck FCG (1979) Losses and parasitic torques in electric motors subjected to PWM waveforms, *IEEE Trans. Ind. Appl.*, **15**(1), 47–53.
94. Yang SJ, Timar PL (1980) The effect of harmonic currents on the noise of the three-phase induction motor, *IEEE Trans. Power Appar. Syst.*, **99**(1), 307–310.
95. Green RM, Boys JT (1982) Inverter AC-drive efficiency, *Proc. Inst. Electr. Eng., Part B*, **129**(2), 75–81.
96. Venkatesan K, Lindsay JF (1982) Comparative study of the losses in voltage and current source inverter fed induction motors, *IEEE Trans. Ind. Appl.*, **18** (3), 240–246.

97. Murphy JMD, Egan MG (1983) A comparison of PWM strategies for inverter-fed induction motors, *IEEE Trans. Ind. Appl.*, **19**(3), 363–369.
98. Bowes SR, Clare J (1983) Steady-state performance of PWM inverter drives, *Proc. Inst. Electr. Eng., Part B*, **130**(4), 229–244.
99. De Buck FCG, Gistelinck P, De Backer D (1984) A simple but reliable loss model for inverter-supplied induction motors, *IEEE Trans. Ind. Appl.*, **20**(1), 190–202.
100. Andresen EC, Bieniek K (1984) On the torques and losses of voltage- and current-source inverter drives, *IEEE Trans. Ind. Appl.*, **20**(2), 321–327.
101. Avadhanlu TV (1984) Evaluation and comparison of two types of induction-motor drives, *Proc. Inst. Electr. Eng., Part B*, **131**(3), 105–112.
102. Kirschen DS, Novotny DW, Suwanwisoot W (1984) Minimizing induction motor losses by excitation control in variable frequency, *IEEE Trans. Ind. Appl.*, **20**(5), 1244–1250.
103. Williamson S, Cann RG (1984) A comparison of PWM switching strategies on the basis of drive system efficiency, *IEEE Trans. Ind. Appl.*, **20**(6), 1460–1472.
104. Peak SC, Oldenkamp JL (1985) A study of system losses in a transistorized inverter-induction motor drive system, *IEEE Trans. Ind. Appl.*, **21**(1), 248–258.
105. Pavithran KN, Parimelalagan R, Sridhara Rao S, Holtz J (1987) Optimum design of an induction motor for operation with current source inverters, *Proc. Inst. Elecdtr. Eng., Part B*, **134**(1), 1–8.
106. Bowes SR, Bullough RI (1987) Harmonic minimisation in microprocessor controlled current fed PWM inverter drives, *Proc. Inst. Electr. Eng., Part B*, **134**(1), 25–41.
107. Belmans RJM, D'hont L, Vandeput AJ, Geysen W (1987) Analysis of the audible noise of three-phase squirrel-cage induction motors supplied by inverters, *IEEE Trans. Ind. Appl.*, **23**(5), 842–847.

Chapter 6

Resonant Inverters

A slightly damped resonant circuit shows a pseudo-cyclical free response, i.e. it tends to oscillate. If it is supplied at a frequency close to its pseudo-oscillating frequency, stable forced oscillations are produced, and the current through it or the voltage across it has a waveform close to a sine wave, depending on whether it is a series or parallel resonant circuit. This is the principle used in resonant inverters.

In Chap. 1, Sect. 1.3, we presented the major differences between these inverters and the usual type of autonomous inverters:

- the load must be a slightly damped oscillating circuit. This frequently requires an inductance or a capacitance to be added to the load itself;
- the operating frequency is imposed by the load.

These inverters offer two main advantages:

- the output variable which is not imposed by the supply is virtually sinusoidal and thus does not need to be filtered;
- the controlled commutations may always be of the same type, thus making the semiconductor devices easier to control.

We will present the operation and characteristics of the series inverter and then of the parallel inverter. We will end this chapter with notes on resonant supplies.

6.1 Series Resonant Inverter

The series inverter is a *voltage-source inverter*, comprising, for example, a four-switch bridge (Fig. 6.1) delivering at a cycle T an output voltage u'

- equal to $+U$ for $0 < t < T/2$
- equal to $-U$ for $T/2 < t < T$.

The angular frequency of this voltage fundamental component is denoted:

$$\omega = 2\pi/T. \tag{6.1}$$

The load comprises a resistance R, an inductance L and a capacitance C series-connected. R and L are normally the parameters of the load itself; C is the capacitance added to create the oscillating circuit.

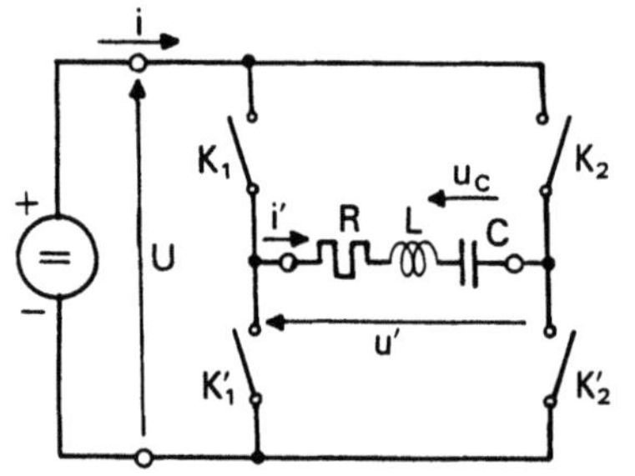

Fig. 6.1

First of all, we will describe the response, in steady state, of the RLC circuit supplied by rectangular pulses, before showing the conditions under which this response can be approximated by using the first harmonic method. This will enable the control and the characteristics of the series inverter to be described.

6.1.1 Response of the Series Resonant Circuit to a Square-Wave Voltage

Voltage u_C across C and current i' in the circuit are linked to voltage u' by

$$Ri' + L\frac{di'}{dt} + u_C = u'$$

with

$$i' = C\,du_C/dt.$$

This gives

$$\begin{cases} LC\dfrac{d^2 u_C}{dt^2} + RC\dfrac{du_C}{dt} + u_C = u & (6.2) \\ i' = C\,du_C/dt. & (6.3) \end{cases}$$

The circuit is often characterised by

- its natural angular frequency, in the absence of damping ($R = 0$), ω_0
- its damping coefficient α
- or the ratio ζ of the latter to ω_0.

$$\boxed{\omega_0 = \frac{1}{\sqrt{LC}} \qquad \alpha = \frac{RC}{2LC} = \frac{R}{2L} \qquad \zeta = \frac{\alpha}{\omega_0} = \frac{R}{2}\sqrt{\frac{C}{L}}} \qquad (6.4)$$

6.1.1.1 High Damping ($\zeta > 1$)

- If ζ is greater than 1, u_C has the following form:

$$u_C = u_{Cp} + A_1 \exp(r_1 t) + A_2 \exp(r_2 t) \qquad (6.5)$$

using u_{Cp} to denote the particular solution of the differential Eq. (6.2) with a second member, and r_1 and r_2 the roots of the characteristic equation:

$$r_1 = -\alpha + \sqrt{\alpha^2 - \omega_0^2} = -\zeta\omega_0\left(1 - \sqrt{1 - \frac{1}{\zeta^2}}\right)$$
$$r_2 = -\alpha - \sqrt{\alpha^2 - \omega_0^2} = -\zeta\omega_0\left(1 + \sqrt{1 - \frac{1}{\zeta^2}}\right). \tag{6.6}$$

Neither u_C nor its derivative can be subject to any discontinuity and thus the integration constants A_1 and A_2 are given by

$$A_1 = \frac{1}{r_2 - r_1}\left[r_2(u_{C_0} - u_{Cp_0}) - \left(\frac{du_C}{dt}\right)_0 + \left(\frac{du_{Cp}}{dt}\right)_0\right]$$
$$A_2 = \frac{1}{r_1 - r_2}\left[r_1(u_{C_0} - u_{Cp_0}) - \left(\frac{du_C}{dt}\right)_0 + \left(\frac{du_{Cp}}{dt}\right)_0\right] \tag{6.7}$$

with u_{C_0}, $(du_C/dt)_0$. u_{Cp_0}, $(du_{Cp}/dt)_0$, values of u_C and of its derivative, of u_{Cp} and of its derivative at instant $t = 0$.

- In the case under study, for period $[0, T/2]$,

$$LC\frac{d^2u_C}{dt^2} + RC\frac{du_C}{dt} + u_C = U$$

$u_{Cp} = U$ thus $u_{Cp_0} = U$; $(du_{Cp}/dt) = 0$ thus $(du_{Cp}/dt)_0 = 0$

u_{C_0} and $(du_C/dt)_0$, equal to i'_0/C, are different from zero.

Voltage u_C and current i' can thus be expressed, during the first half-cycle of u', by

$$u_c = U + \frac{u_{C_0} - U}{r_2 - r_1}(r_2 \exp r_1 t - r_1 \exp r_2 t) - \frac{i'_0}{C(r_2 - r_1)}(\exp r_1 t - \exp r_2 t) \tag{6.8}$$

$$i' = \frac{C(u_{C_0} - U)}{r_2 - r_1} r_1 r_2 (\exp r_1 t - \exp r_2 t) - \frac{i'_0}{r_2 - r_1}(r_1 \exp r_1 t - r_2 \exp r_2 t). \tag{6.9}$$

- *For steady-state operation,*

$$u_C\left(t + \frac{T}{2}\right) = -u_C(t)$$
$$i'\left(t + \frac{T}{2}\right) = -i'(t) \tag{6.10}$$

and, in particular,

$$u_C(T/2) = -u_{C_0}, \qquad i'(T/2) = -i'_0$$

u_{C_0} and i'_0 can be deduced by

$$-u_{C_0} = U + \frac{u_{C_0} - U}{r_2 - r_1}\left(r_2 \exp r_1 \frac{T}{2} - r_1 \exp r_2 \frac{T}{2}\right) - \frac{i'_0}{C(r_2 - r_1)}\left(\exp r_1 \frac{T}{2} - \exp r_2 \frac{T}{2}\right)$$

$$-i'_0 = \frac{C(u_{C_0} - U)}{r_2 - r_1} r_1 r_2 \left(\exp r_1 \frac{T}{2} - \exp r_2 \frac{T}{2}\right) - \frac{i'_0}{r_2 - r_1}\left(r_1 \exp r_1 \frac{T}{2} - r_2 \exp r_2 \frac{T}{2}\right)$$

which gives

$$u_{C_0} = U \frac{(r_2 - r_1)\left[-1 + \exp(r_1 + r_2)\frac{T}{2}\right] + (r_2 + r_1)\left[\exp r_1 \frac{T}{2} - \exp r_2 \frac{T}{2}\right]}{(r_2 - r_1)\left[1 + \exp r_1 \frac{T}{2} + \exp r_2 \frac{T}{2} + \exp(r_1 + r_2)\frac{T}{2}\right]} \tag{6.11}$$

$$i'_0 = 2CU \frac{r_1 r_2 \left(\exp r_1 \frac{T}{2} - \exp r_2 \frac{T}{2}\right)}{(r_2 - r_1)\left[1 + \exp r_1 \frac{T}{2} + \exp r_2 \frac{T}{2} + \exp(r_1 + r_2)\frac{T}{2}\right]}. \tag{6.12}$$

- The load is capacitive if $1/C\omega$ is greater than $L\omega$, i.e. if ω is less than ω_0. If ω is greater than ω_0, the load is inductive.

  Figure 6.2 gives the waveforms of voltage u_C and current i'

  (a) for $\zeta = 4$ and $\omega = \omega_0/4$

  $$\left(\text{or } \frac{1}{C\omega} = \frac{R}{2} = 16 L\omega\right)$$

  (b) for $\zeta = 4$ and $\omega = \omega_0$

  $$\left(\text{or } \frac{1}{C\omega} = \frac{R}{2} = L\omega\right)$$

  (c) for $\zeta = 4$ and $\omega = 4\,\omega_0$

  $$\left(\text{or } L\omega = \frac{R}{2} = \frac{16}{C\omega}\right).$$

  Neither voltage u_C nor current i' can be subject to any discontinuity. Sharp variations of voltage u' are found in $L(\mathrm{d}i'/\mathrm{d}t)$.

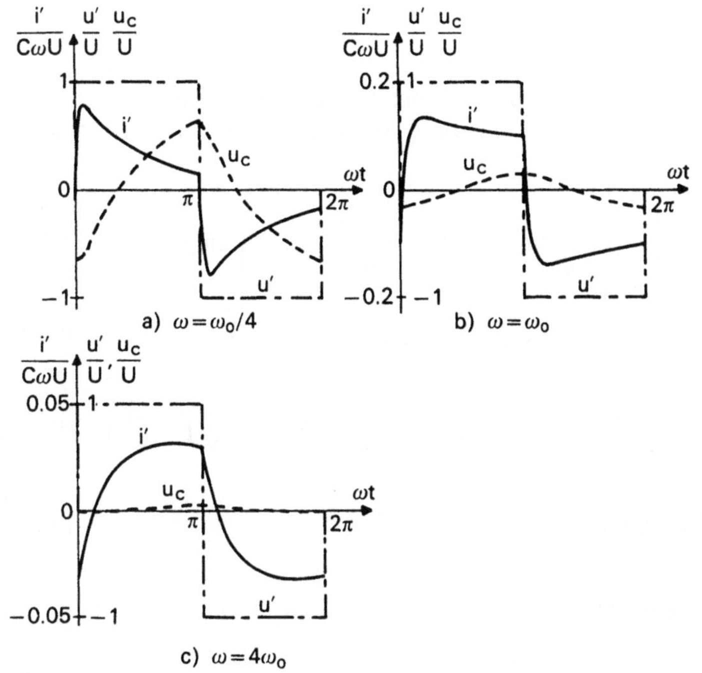

Fig. 6.2

It can be seen that, for high values of ζ, whatever the value of ω/ω_0, the waveform of i' cannot be assimilated to a sinewave.

6.1.1.2 Low Damping ($\zeta < 1$)

- When ζ is less than 1, the roots of the characteristic equation are complex conjugate quantities. Voltage u_C can conveniently be expressed in the following form:

$$u_C = u_{Cp} + [A'_1 \cos(\omega_0 \sqrt{1-\zeta^2}\, t) + A'_2 \sin(\omega_0 \sqrt{1-\zeta^2}\, t)] \exp(-\zeta \omega_0 t). \tag{6.13}$$

The integration constants can be deduced from the initial values of u_C and u_{Cp} and their derivatives by means of

$$A'_1 = u_{C0} - u_{Cp0}$$

$$A'_2 = \frac{1}{\omega_0\sqrt{1-\zeta^2}} \left[\left(\frac{du_C}{dt}\right)_0 - \left(\frac{du_{Cp}}{dt}\right)_0 + \zeta\omega_0 (u_{C0} - u_{Cp0}) \right]. \tag{6.14}$$

– In this case, for period $[0, T/2]$,

$$LC \frac{d^2 u_C}{dt^2} + RC \frac{du_C}{dt} + u_C = U$$

thus

$u_{C_p} = U = u_{C_{p0}}$; $(\mathrm{d}u_{C_p}/\mathrm{d}t)_0 = 0$

u_{C_0} and $(\mathrm{d}u_C/\mathrm{d}t)_0$, equal to i'_0/C, are different from zero.

This provides the expressions of u_C and i' during the first half-cycle of voltage u':

$$u_C = U + \left\{\left[\frac{\zeta}{\sqrt{1-\zeta^2}}(u_{C_0} - U) + \frac{i'_0}{C\omega_0\sqrt{1-\zeta^2}}\right]\sin(\omega_0\sqrt{1-\zeta^2}\,t) + (u_{C_0} - U)\cos(\omega_0\sqrt{1-\zeta^2}\,t)\right\}\exp(-\zeta\omega_0 t) \tag{6.15}$$

$$i' = \left\{-\left[\frac{C\omega_0}{\sqrt{1-\zeta^2}}(u_{C_0} - U) + \frac{\zeta}{\sqrt{1-\zeta^2}}i'_0\right]\sin(\omega_0\sqrt{1-\zeta^2}\,t) + i'_0\cos(\omega_0\sqrt{1-\zeta^2}\,t)\right\}\exp(-\zeta\omega_0 t). \tag{6.16}$$

– *In steady-state operation,*

$$u_C\left(t + \frac{T}{2}\right) = -u_C(t); \quad i'\left(t + \frac{T}{2}\right) = -i'(t)$$

u_{C_0} and i'_0 can be obtained by writing:

$$u_C(T/2) = -u_{C_0}, \quad i'(T/2) = -i'_0,$$

thus giving:

$$u_{C_0} = -U\frac{1 - M\exp\left(-\zeta\omega_0\frac{T}{2}\right) - \dfrac{\sin^2\left(\omega_0\sqrt{1-\zeta^2}\frac{T}{2}\right)}{N(1-\zeta^2)}\exp(-\zeta\omega_0 T)}{1 + M\exp\left(-\zeta\omega_0\frac{T}{2}\right) + \dfrac{\sin^2\left(\omega_0\sqrt{1-\zeta^2}\frac{T}{2}\right)}{N(1-\zeta^2)}\exp(-\zeta\omega_0 T)} \tag{6.17}$$

$$i'_0 = -\frac{\dfrac{2C\omega_0 U}{N\sqrt{1-\zeta^2}}\sin\left(\omega_0\sqrt{1-\zeta^2}\frac{T}{2}\right)\exp\left(-\zeta\omega_0\frac{T}{2}\right)}{1 + M\exp\left(-\zeta\omega_0\frac{T}{2}\right) + \dfrac{\sin^2\left(\omega_0\sqrt{1-\zeta^2}\frac{T}{2}\right)}{N(1-\zeta^2)}\exp(-\zeta\omega_0 T)} \tag{6.18}$$

by writing

$$M = \cos\left(\omega_0\sqrt{1-\zeta^2}\,\frac{T}{2}\right) + \frac{\zeta}{\sqrt{1-\zeta^2}}\sin\left(\omega_0\sqrt{1-\zeta^2}\,\frac{T}{2}\right)$$

$$N = 1 + \left[\cos\left(\omega_0\sqrt{1-\zeta^2}\,\frac{T}{2}\right) - \frac{\zeta}{\sqrt{1-\zeta^2}}\sin\left(\omega_0\sqrt{1-\zeta^2}\,\frac{T}{2}\right)\right]$$
$$\times \exp\left(-\zeta\omega_0\frac{T}{2}\right). \tag{6.19}$$

- The initial value i'_0 of current i' is zero, if

$$\omega = \omega_0\sqrt{1-\zeta^2}. \tag{6.20}$$

If this condition is fulfilled, current i' has the following expression, during the positive half-cycle of voltage u',

$$i' = \frac{C\omega}{\sqrt{1-\zeta^2}}(U - u_{C_0})\sin\omega t \exp\left(-\frac{\zeta}{\sqrt{1-\zeta^2}}\,\omega t\right) \tag{6.21}$$

with

$$u_{C_0} = -U\coth\left(\frac{\zeta}{\sqrt{1-\zeta^2}}\,\frac{\pi}{2}\right)$$

and its waveform is all the closer to a sine wave as ζ is smaller.

If ω is less than $\omega_0\sqrt{1-\zeta^2}$, i'_0 is positive.

If ω is greater than $\omega_0\sqrt{1-\zeta^2}$, i'_0 is negative.

- Figure 6.3 shows the waveforms of voltage u_C and current i', plotted, for ζ equal to 1/4, thus $\omega_0\sqrt{1-\zeta^2}$ equal to $0.96825\,\omega_0$, and successively

(a) for $\omega = \omega_0/1.5$
(b) for $\omega = \omega_0/1.1$
(c) for $\omega = 0.96825\omega_0$
(d) for $\omega = \omega_0$
(e) for $\omega = 1.1\omega_0$
(f) for $\omega = 1.5\omega_0$.

These waveforms show that:

- current i'_0 is positive or negative according to whether ω is greater or less than $\omega_0\sqrt{1-\zeta^2}$. The sign of i'_0 gives the nature of the commutations to be carried out;
- voltage u_C and current i' have a sinusoidal form if ω is close to $\omega_0\sqrt{1-\zeta^2}$. This is clearer when ζ is smaller.
- the amplitudes of u_C and i' decrease when ω moves away from $\omega_0\sqrt{1-\zeta^2}$.

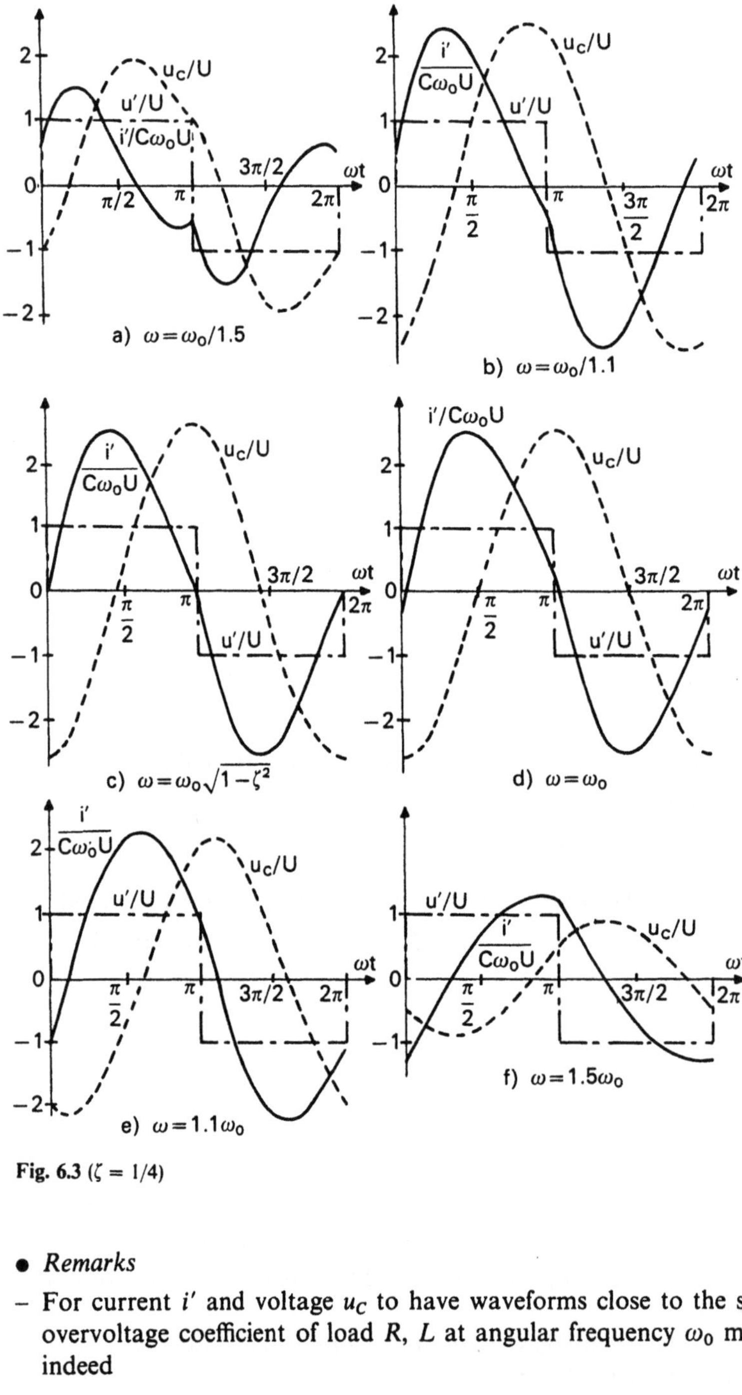

Fig. 6.3 ($\zeta = 1/4$)

- *Remarks*

– For current i' and voltage u_C to have waveforms close to the sinusoid, the overvoltage coefficient of load R, L at angular frequency ω_0 must be high; indeed

$$\frac{L\omega_0}{R} = \frac{1}{2\zeta}. \tag{6.22}$$

- Condition (6.20) corresponds to the equality between the pseudo-cycle of i' and the cycle of u'.

  If $\omega_0\sqrt{1-\zeta^2}$ equals $k\omega$, with k equal to 2, 3, . . . , i'_0 is still zero. Equation (6.21) can still be used, but i' describes k pseudo-half-cycle during each half-cycle of u':

  - the resulting waveform bears little resemblance to the quasi-sinusoidal waveform of same frequency as u', which was wanted.
  - current i' falls to zero k times per half-cycle of voltage u'. If the switches are controlled at the same frequency as u', conduction becomes discontinuous.

Usually, $(\omega_0\sqrt{1-\zeta^2})/2$ is considered to be the *minimal value* of ω giving a normal operating mode.

6.1.2 "First Harmonic" Method

The inverter imposes voltage u' across the oscillating circuit. From the series expansion of u', the value of the fundamental and the harmonics of the various currents and other voltages is determined. It is then possible to deduce

- which currents and voltages can be reduced to their fundamentals,
- under which conditions this approximation can be acceptable.

- The series expansion of voltage u' is

$$u' = U'_1\sqrt{2}\sin\omega t + U'_3\sqrt{2}\sin 3\omega t + \ldots + U'_n\sqrt{2}\sin n\omega t + \ldots$$

with

$$U'_1 = \frac{2\sqrt{2}}{\pi}U \quad \text{and} \quad U'_n = \frac{U'_1}{n}.$$

The current i' fundamental has an rms value of

$$I'_1 = \frac{U'_1}{\sqrt{R^2 + \left(L\omega - \dfrac{1}{C\omega}\right)^2}} \tag{6.23}$$

it is lagging the fundamental of u' by an angle φ_1 such that

$$\tan\varphi_1 = \frac{1}{R}\left(L\omega - \frac{1}{C\omega}\right). \tag{6.24}$$

The fundamental components u_{C_1} and u_{RL_1} of voltages across C and across the R, L assembly have the following rms values:

$$U_{C_1} = \frac{U'_1}{C\omega\sqrt{R^2 + \left(L\omega - \dfrac{1}{C\omega}\right)^2}}, \qquad U_{RL_1} = U'_1\sqrt{\frac{R^2 + L^2\omega^2}{R^2 + \left(L\omega - \dfrac{1}{C\omega}\right)^2}}. \tag{6.25}$$

If ω/ω_0 and ζ are introduced into the expressions of I'_1, φ_1, U_{C1} and U_{RL1}, they become

$$I'_1 = \frac{C\omega_0 U'_1}{\sqrt{4\zeta^2 + \left(\frac{\omega}{\omega_0} - \frac{\omega_0}{\omega}\right)^2}} \tag{6.26}$$

$$\tan\varphi_1 = \frac{1}{2\zeta}\left(\frac{\omega}{\omega_0} - \frac{\omega_0}{\omega}\right) \tag{6.27}$$

$$U_{C_1} = \frac{\omega_0}{\omega}\frac{U'_1}{\sqrt{4\zeta^2 + \left(\frac{\omega}{\omega_0} - \frac{\omega_0}{\omega}\right)^2}} \tag{6.28}$$

$$U_{RL_1} = U'_1\sqrt{\frac{4\zeta^2 + (\omega/\omega_0)^2}{4\zeta^2 + \left(\frac{\omega}{\omega_0} - \frac{\omega_0}{\omega}\right)^2}}. \tag{6.29}$$

- Figure 6.4 shows, for several values of ζ, the variations of $I'_1/C\omega_0 U'_1$, of φ_1, of U_{C_1}/U'_1 and of U_{RL_1}/U'_1 as a function of ω/ω_0. These curves also give $I'_n/C\omega_0 U'_1$, φ_n, U_{C_n}/U'_1 and U_{RL_n}/U'_1 providing that these ratios or this angle are read for $n\omega/\omega_0$.

- *Current.* If ω/ω_0 is close to unity, the ratio $I'_1/C\omega_0 U'_1$ is close to $1/2\zeta$; ratio I'_n/U'_n for the various harmonics is virtually equal to $1/nL\omega$ and thus is only slightly dependent on ζ. Reducing i' to its fundamental thus becomes all the more acceptable as $I'_1/C\omega_0 U'_1$ is high, i.e. as ζ smaller and ω/ω_0 closer to unity.
- *Phase.* The R, L, C circuit is capacitive for ω/ω_0 less than unity, and inductive for ω/ω_0 greater than 1. The phase inversion is sharper when ζ decreases.
- *Voltage across C.* If ω/ω_0 is close to unity, the ratio U_{C_n}/U'_n for the harmonics is little different from $(\omega_0/n\omega)^2$. If ω/ω_0 is close to unity and ζ small, the first harmonic hypothesis is even more valid for voltage u_C than for current i', especially if ω/ω_0 is greater than unity.
- *Voltage across R, L.* The ratio U_{RL_n}/U'_n giving the value of the harmonics across the R, L load tends towards unity when $n\omega/\omega_0$ increases. This voltage cannot be reduced to its fundamental since the discontinuities of voltage u' are also found in u_{RL}.

6.1.3 Operation and Control of the Inverter

As with all voltage-source inverters, the switches of the series resonant inverters consist of controlled semiconductor devices with diodes in antiparallel. Figure 6.5 shows the diagram of the full-bridge voltage-source inverter.

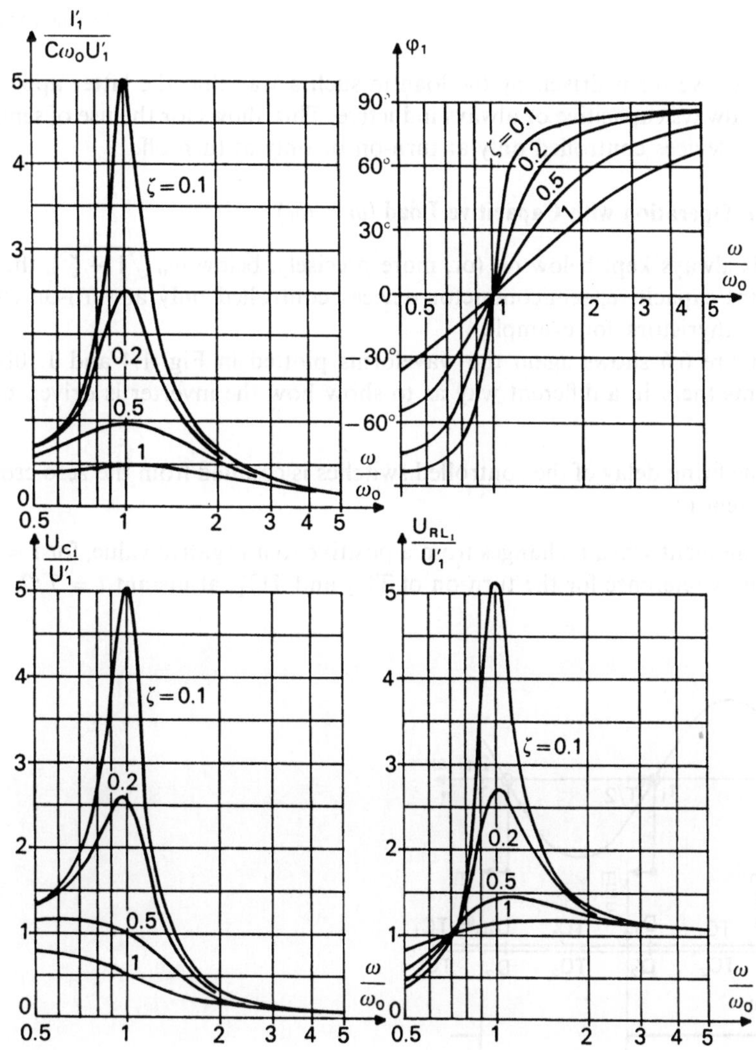

$\frac{I'_1}{C\omega_0 U'_1}$
$\zeta=0.1$
0.2
0.5
1
$\frac{\omega}{\omega_0}$
φ_1
90°
60°
30°
0
−30°
−60°
$\zeta=0.1$
0.2
0.5
1
$\frac{U_{C_1}}{U'_1}$
$\frac{U_{RL_1}}{U'_1}$
0.5
1
2
3
4
5

Fig. 6.4

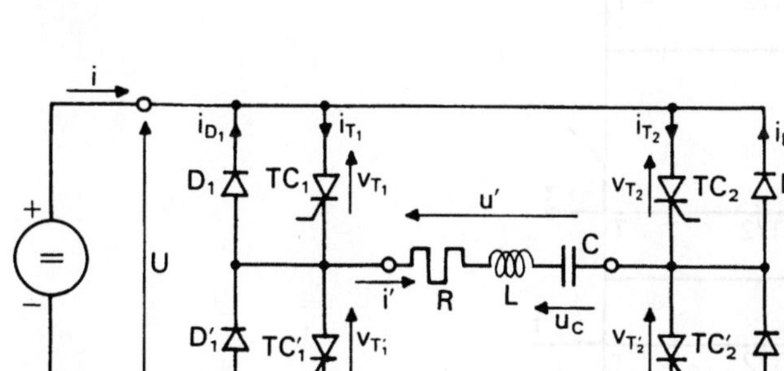

i
U
i_{D_1}
D_1
TC_1
i_{T_1}
v_{T_1}
u'
i_{T_2}
v_{T_2}
TC_2
D_2
i_{D_2}
i'
R
L
C
u_C
D'_1
TC'_1
$v_{T'_1}$
$i_{D'_1}$
$i_{T'_1}$
$v_{T'_2}$
TC'_2
D'_2
$i_{T'_2}$
$i_{D'_2}$

Fig. 6.5

The inverter is driven by the load in such a way that the latter appears as either always capacitive or always inductive. This allows for the use of semiconductor devices controlled only at turn-on or only at turn-off.

6.1.3.1 Operation with Capacitive Load ($\omega < \omega_0$)

If ω is always kept below ω_0 (or, more precisely, below $\omega_0\sqrt{1-\zeta^2}$), the load remains capacitive; semiconductor devices controlled only at turn-on can be used – thyristors, for example.

Figure 6.6 shows again the waveforms plotted in Fig. 1.4 and 1.10b, but presents them in a different way as to show how the inverter is driven by the load.

The firing delay of the controlled switches is counted from the zero crossing of current i':

- the moment when i' changes from a positive to a negative value, for $t = t_1$, is used as reference for the turn-on of TC_2 and TC'_1, at instant $t = T/2$.

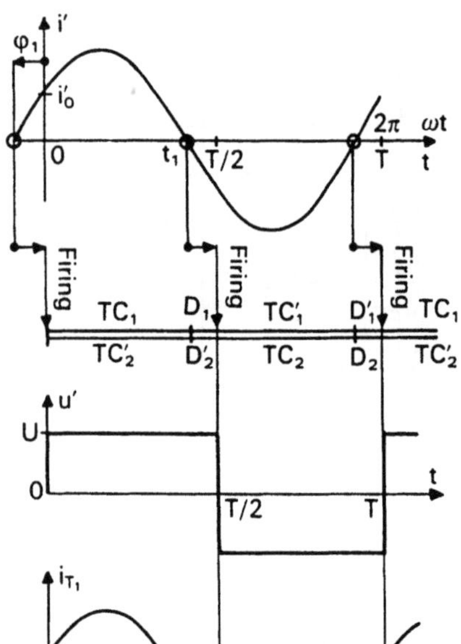

Fig. 6.6

- the other zero crossing of i', for $t = t_1 + T/2$, is used as reference for the turn-on of TC_1 and TC'_2, at instant $t = T$.

When the *delay* $T/2 - t_1$, between the moments when i' falls to zero and the turn-on signals, *increases*, the frequency decreases and the greater is the distance from the resonance: for given values of R, L, C, the various values relative to the output (current i', voltages u_C and u_{RL}) *decrease*.

6.1.3.2 Operation with Inductive Load ($\omega > \omega_0$)

If ω is maintained at a value higher than $\omega_0\sqrt{1-\zeta^2}$, the load is always inductive: semiconductor devices only controlled at turn-off can be used, as *dual thyristors*.

Figure 6.7 shows the waveforms of Fig. 1.3 and 1.10b, with a suitably adapted presentation.

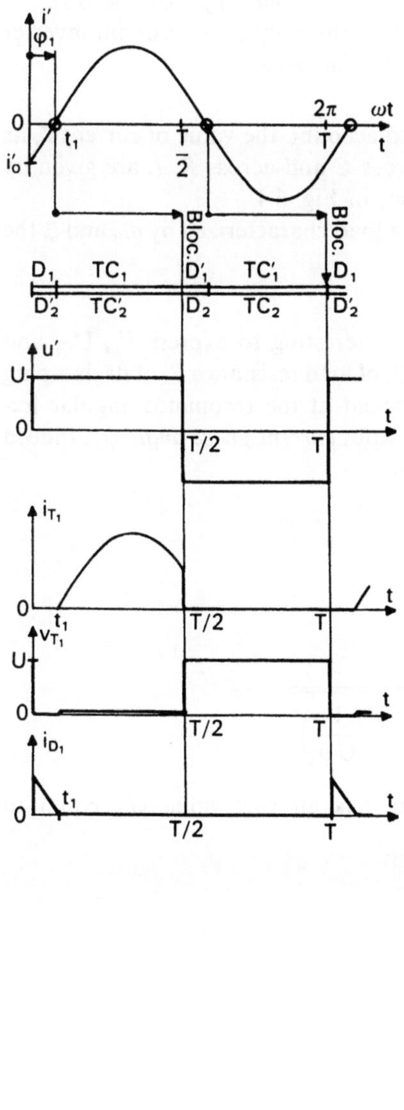

Fig. 6.7

The control of turn-ons is fixed from the moments when current i' crosses zero:

- when i' changes from a negative to a positive value, for $t = t_1$, i.e. the spontaneous turn-on of TC_1 and TC'_2 can be used as a reference for the controlled turn-off of these devices at instant $t = T/2$.
- the start of TC'_1 and TC_2 conducting, for $t = t_1 + T/2$, can be used as a reference for controlling their turn-off at $t = T$.

The *shorter* the *conduction period*, $T/2 - t_1$, of the controlled semiconductor devices, the higher the operating frequency of the inverter and the difference from the resonance: current i', voltages u_C and u_{RL} *decrease*.

6.1.4 Characteristics

6.1.4.1 "First Harmonic" Approximation

If the output current i' is reduced to its fundamental component, the phase shift φ_1 of the latter in relation to the voltage u' fundamental component is given by the angle φ_1 between the zero crossings of i' and u' (see Figs. 6.6 and 6.7).

Since the inverter is controlled by acting on this angle, the resonant inverter is said to be *controlled by the frequency* or *by the phase*.

- If only the fundamental of u' is taken into account, the value of current i', its phase angle, the value of the voltage across C and across R, L are given by Eqs. (6.26)–(6.29) or by the characteristics in Fig. 6.4.

  These relations and graphs show, for a load characterized by ω_0 and ζ, the effects of *frequency* $\omega/2\pi$.

- According to the control law used, it is interesting to express I'_1, U_{C_1} and U_{RL_1} as a function of the supply voltage U, of load resistance R, of its damping coefficient ζ (or of its overvoltage coefficient at the resonance angular frequency $L\omega_0/R$, equal to 2ζ) and of the output *current phase angle* φ_1. Indeed the control is carried out by acting on φ_1.

– Since

$$\tan \varphi_1 = \frac{1}{R}\left(L\omega - \frac{1}{C\omega}\right)$$

$$\cos \varphi_1 = \frac{1}{\sqrt{1 + \mathrm{tg}^2 \varphi_1}} = \frac{R}{\sqrt{R^2 + \left(L\omega - \dfrac{1}{C\omega}\right)^2}}.$$

Furthermore, the fundamental of u' has an rms value U'_1 equal to $4U/\pi\sqrt{2}$.

Incorporating this into (6.23) gives

$$I'_1 = \frac{4}{\pi\sqrt{2}}\frac{U}{R}\cos\varphi_1. \tag{6.30}$$

If i' is reduced to its fundamental component, the power P supplied to the load is thus equal to

$$P = RI'^2_1 = \frac{8}{\pi^2}\frac{U^2}{R}\cos^2\varphi_1. \tag{6.31}$$

- Equations (6.25) can be written:

$$U_{C_1} = \frac{1}{C\omega}I'_1 = \frac{1}{LC\omega_0^2}\frac{L\omega_0}{R}\cdot\frac{\omega_0}{\omega}\frac{4}{\pi\sqrt{2}}U\cos\varphi_1$$

$$U_{RL_1} = \sqrt{R^2 + L^2\omega^2}\,I'_1 = R\sqrt{1 + \left(\frac{L\omega_0}{R}\frac{\omega}{\omega_0}\right)^2}\frac{4}{\pi\sqrt{2}}\frac{U}{R}\cos\varphi_1$$

which gives

$$\left.\begin{aligned} U_{C_1} &= \frac{1}{2\zeta}\frac{\omega_0}{\omega}\frac{4}{\pi\sqrt{2}}U\cos\varphi_1 \\ U_{RL_1} &= \frac{1}{2\zeta}\sqrt{4\zeta^2 + \left(\frac{\omega}{\omega_0}\right)^2}\frac{4}{\pi\sqrt{2}}U\cos\varphi_1. \end{aligned}\right\} \tag{6.32}$$

To express ω/ω_0 as a function of φ_1, we start with

$$\tan\varphi_1 = \frac{1}{2\zeta}\left(\frac{\omega}{\omega_0} - \frac{\omega_0}{\omega}\right)$$

giving

$$\left(\frac{\omega}{\omega_0}\right)^2 - 2\zeta\left(\frac{\omega}{\omega_0}\right)\tan\varphi_1 - 1 = 0.$$

Only the solution of this equation which gives a positive value of ω/ω_0 should be retained:

$$\frac{\omega}{\omega_0} = \zeta\tan\varphi_1 + \sqrt{1 + \zeta^2\tan^2\varphi_1}. \tag{6.33}$$

- Figure 6.8 shows the variations, as functions of φ_1, of $I'_1/(U/R)$ and $P/(U^2/R)$ and then, for different values of ζ, of U_{C_1}/U and U_{RL_1}/U.

  At the end of this paragraph, we will give our reasons for not having plotted the various curves up to $\varphi_1 = -\pi/2$.

- If current i' is assumed to be sinusoidal, the results of the general study of the voltage-source inverters (Chap. 2), made according this hypothesis, can be used.

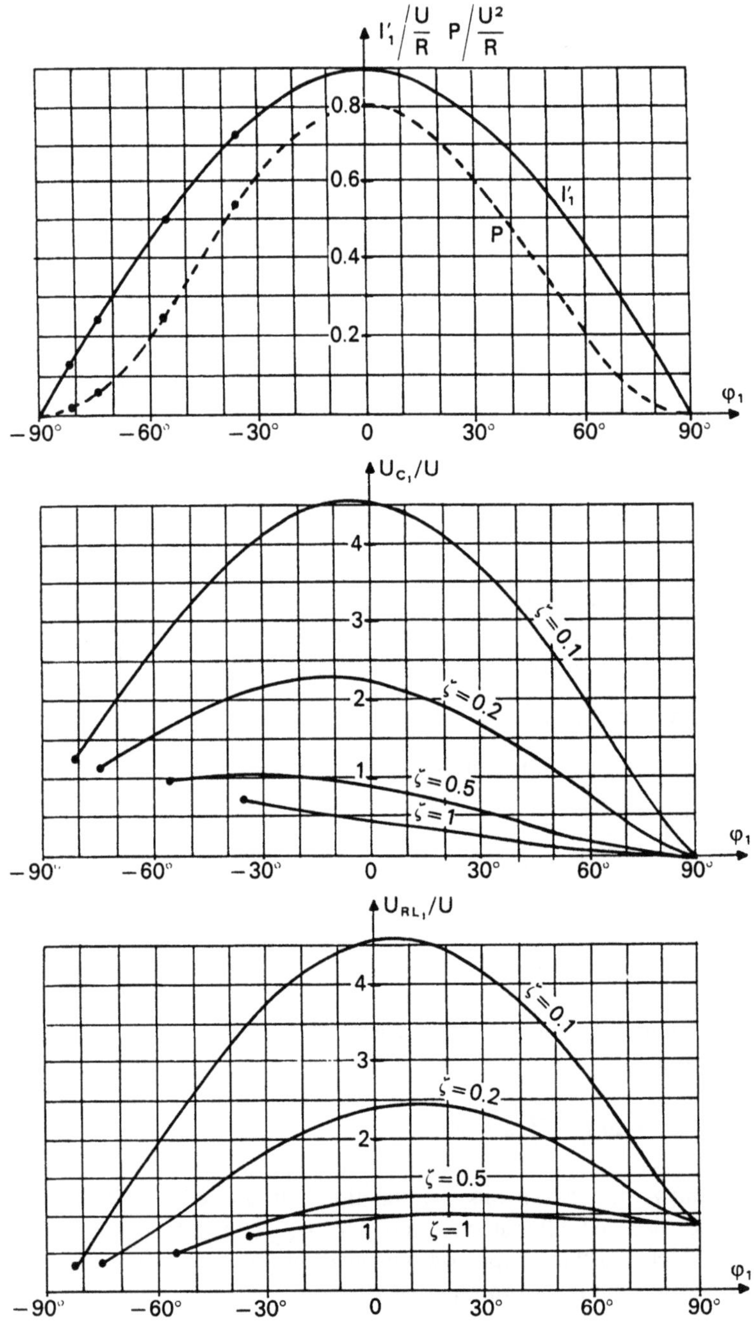

$I'_1 / \frac{U}{R}$ $P / \frac{U^2}{R}$
0.8
0.6
0.4
0.2
I'_1
P
φ_1
−90° −60° −30° 0 30° 60° 90°
U_{C_1}/U
4
3
2
1
ζ=0.1
ζ=0.2
ζ=0.5
ζ=1
φ_1
−90° −60° −30° 0 30° 60° 90°
U_{RL_1}/U
4
3
2
1
ζ=0.1
ζ=0.2
ζ=0.5
ζ=1
φ_1
−90° −60° −30° 0 30° 60° 90°

Fig. 6.8

More particularly, the equations which give the currents in the semiconductor devices as a function of the output current rms value I'_1 and phase angle φ_1 can be used.

The current in the controlled semiconductor devices has

- an rms value of

$$I_{T\mathrm{rms}} = I'_1 \sqrt{2} \sqrt{\frac{\pi - |\varphi_1|}{4\pi} + \frac{\sin 2|\varphi_1|}{8\pi}} \tag{6.34}$$

- a mean value of

$$I_{T\mathrm{mean}} = I'_1 \sqrt{2} \frac{1 + \cos\varphi_1}{2\pi}. \tag{6.35}$$

The current in the diodes has

- an rms value of

$$I_{D\mathrm{rms}} = I'_1 \sqrt{2} \sqrt{\frac{|\varphi_1|}{4\pi} - \frac{\sin 2|\varphi_1|}{8\pi}} \tag{6.36}$$

- a mean value of

$$I_{D\mathrm{mean}} = I'_1 \sqrt{2} \frac{1 - \cos\varphi_1}{2\pi}. \tag{6.37}$$

The input current has an rms value of I'_1 and the following mean value:

$$I = \frac{2}{\pi} I'_1 \sqrt{2} \cos\varphi_1. \tag{6.38}$$

If (6.30) is taken into account, Eqs. (6.34)–(6.38) show how, for a given U/R, the currents in the semiconductor devices and the current delivered by the source vary as a function of φ_1.

Currents I'_1, $I_{T\mathrm{rms}}$, $I_{T\mathrm{mean}}$ and I reach their highest values for the minimal value of $|\varphi_1|$; it is for these values that the current through the controlled devices must be rated.

For a given U/R, the current in the diodes, which is zero when φ_1 is zero, increases when $|\varphi_1|$ increases, reaches a maximum value and then decreases;

$I_{D\mathrm{rms}}$ is at its maximum for $|\varphi_1| = 54°3$ and then has a value of $0.14U/R$.
$I_{D\mathrm{mean}}$ is at its maximum for $|\varphi_1| = 60°$ and then has a value of $0.0507U/R$.

- The study carried out in Sects. 1.1 and 1.2 shows the extent to which the different values can be reduced to their fundamental.

  The "first harmonic" approximation is all the more acceptable as ζ is smaller and ω/ω_0 becomes closer to unity. When ω/ω_0 increases in relation to unity, the error made by only taking the fundamental into account is less important than when ω/ω_0 moves away from 1 by decreasing.

In Fig. 6.8, we have thus stopped plotting the characteristics giving U_{C_1}/U and U_{RL_1}/U at the value of φ_1 which, for each value of ζ, gives $\omega/\omega_0 = 1/2$. We have indicated the corresponding points on the curves which show the variations in the relative values of I'_1 and P.

6.1.4.2 Limitations

- *Inverter with thyristors* ($\omega/\omega_0 < 1$; negative φ_1)

– The minimal value $|\varphi_{1\,\mathrm{min}}|$ of angle φ_1, which corresponds to the minimal conduction time of the diodes $|\varphi_{1\,\mathrm{min}}|/\omega$, is imposed by the thyristor reverse recovery time t_q

$$|\varphi_1| > \omega t_q$$

or

$$|\varphi_1| > 2\pi \frac{t_q}{T}. \tag{6.39}$$

An increase in the relative value t_q/T of the thyristor reverse recovery time induces a reduction in the maximum power which can be supplied to the load:

$$P_{\mathrm{max}} = \frac{8}{\pi^2} \frac{U^2}{R} \cos^2 \varphi_{1\,\mathrm{min}}.$$

But, more particularly, it leads to working even further below the resonant frequency. The condition

$$\tan|\varphi_1| > \tan \omega t_q$$

gives, if (6.27) is taken into account,

$$\frac{1}{2\zeta}\left(\frac{\omega_0}{\omega} - \frac{\omega}{\omega_0}\right) > \tan \omega t_q$$

Since ζ equals $R/2L\omega_0$, by approximating $\tan \omega t_q$ by ωt_q, this condition can be written:

$$\frac{\omega}{\omega_0} < \frac{1}{\sqrt{1 + 2\pi \dfrac{t_q/T}{L\omega/R}}} \tag{6.40}$$

The higher the value of t_q/T and the lower that of $L\omega/R$ at the maximum operating frequency, the more the latter must be below the resonant frequency.

– Furthermore, as seen at the end of the previous paragraph, decreasing P or I' by reducing ω/ω_0 means moving towards the area of ω/ω_0 where the "first harmonic" approximation rapidly becomes unacceptable.

If ω/ω_0 fell below the value $\sqrt{1 - \zeta^2/2}$, current i' would cross the zero axis twice during each half-cycle of u'. Periods at zero i' would appear, unless the controlled semiconductor devices were turned on twice per half-cycle of u'. Whatever the solution adopted, the waveform of i' would be very different from a sinusoid with angular frequency ω.

- It can thus be seen that the range of variations of ω/ω_0 which can be used is quite limited:
  - because of t_q, ω/ω_0 must be noticeably less than 1;
  - ω/ω_0 must be greater than 1/2 in order to have an acceptable waveform of current i'.

- *Inverter with dual thyristors* ($\omega/\omega_0 > 1$; positive φ_1)

- The problem of the reverse recovery time disappears. The only requirement is to remain in the foreseen operational mode by giving the conduction period of the dual thyristors a value below that of the half-cycle.

- When, in order to decrease P or I', the frequency is increased, there is a move towards the zone of ω/ω_0 where the "first harmonic" approximation remains acceptable. The upper limit of the range of the operational frequencies is only determined by technological reasons.

- The series resonant inverter with dual thyristors operates in the most advantageous part of the resonance curves ($\omega/\omega_0 > 1$).

6.1.5 Starting. Protection

6.1.5.1 Initiating the Load Oscillation

Starting the series resonant inverter from the quiescent state (zero current i', zero voltage u_C) poses no particular problem.

If, in Eqs. (6.16) and (6.15), i'_0 equals 0 and u_{C_0} equals 0, the following can be obtained, from the application of voltage U in $t = 0$:

$$i' = \frac{C\omega_0 U}{\sqrt{1-\zeta^2}} \sin(\omega_0 \sqrt{1-\zeta^2}\, t) \exp(-\zeta\omega_0 t)$$

$$u_C = U\left\{1 - \left[\frac{\zeta}{\sqrt{1-\zeta^2}} \sin(\omega_0\sqrt{1-\zeta^2}\, t) + \cos(\omega_0\sqrt{1-\zeta^2}\, t\right]\right.$$
$$\left.\times \exp(-\zeta\omega_0 t)\right\}$$

- *In the case of the thyristor inverter* (Fig. 6.9), current i', which has begun to flow from $t = 0$ by the firing of TC_1 and TC'_2, falls to zero for $t = t_1$. From $t = t_1$ to $t = t_2$, diodes D_1 and D'_2 conduct. Firing TC_2 and TC'_1 in $t = t_1 + \Delta t$ enables the negative half-cycle of i' to continue and causes the reversal of u'. The reversal of i' at instant $t = t_3$ induces, in $t = t_3 + \Delta t$, the firing of TC_1 and TC'_2.

- *In the case of the dual thyristor inverter* (Fig. 6.10), thyristors TC_1 and TC'_2 which have begun conducting for $t = 0$ receive the turn-off command Δt later; current i' is then switched from TC_1 to D'_1, from TC'_2 to D_2; voltage u' is reversed. During the period t_1, t_2, diodes D'_1 and D_2 conduct. For $t = t_2$, TC'_1 and TC_2 begin conducting when i' falls to zero. Δt after the detection of this moment, the blocking of TC'_1 and TC_2 is controlled ...

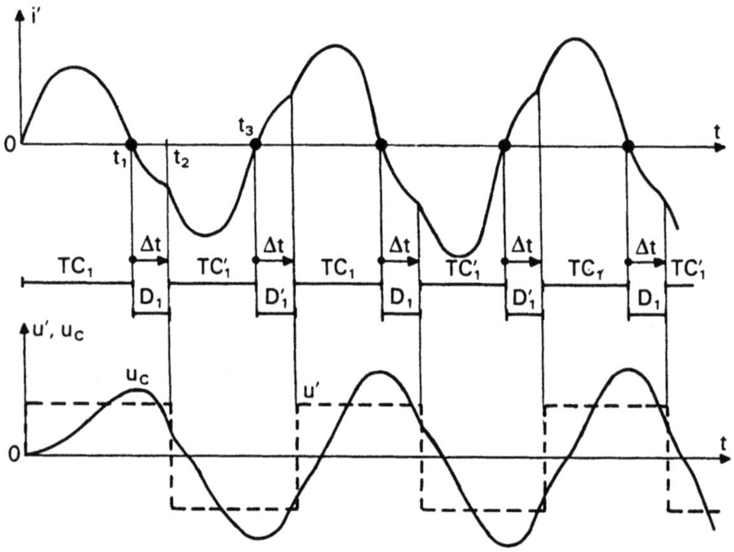

Fig. 6.9

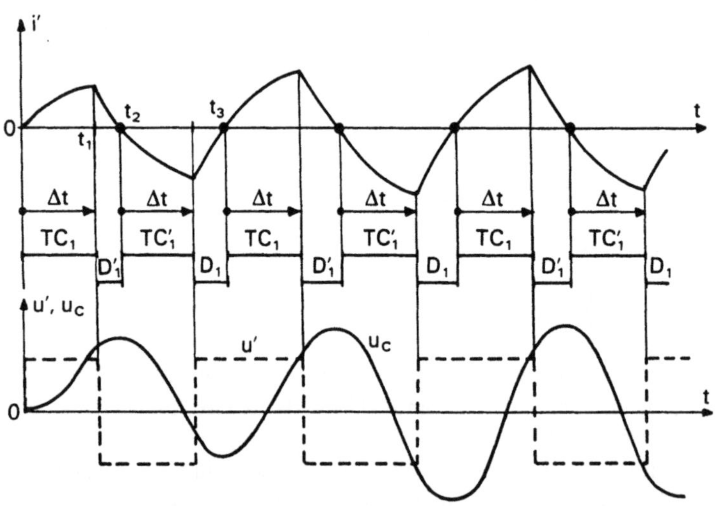

Fig. 6.10

6.1.5.2 Limiting the Stresses on the Semiconductor Devices. Faults

- As with all resonant inverters, the semiconductor devices of the series inverter have only one type of commutation to ensure. This reduces the commutation losses and enables the frequency to be increased. Moreover, it makes it easier to limit the stresses which the semiconductor devices have to sustain.

In the thyristor inverter, the thyristors are turned off when the current i' falls to zero. In this case, all that needs to be added are inductances limiting $\mathrm{d}i/\mathrm{d}t$ at firing. Capacitors across the thyristors prevent them from being turned on accidentally when the positive voltage reappears across them.

In the dual thyristor inverter, the thyristors begin conducting when current i' naturally changes polarity. Capacitors needs only to be placed across the switches to limit $\mathrm{d}v/\mathrm{d}t$ when the voltage is reapplied at their terminals when they are turned off.

- Since the charging circuit of capacitor C is series-connected with the R, L load, this ensures self protection of the inverter to a certain extent.

  A fault in the load (open circuit, short-circuit of the R, L load) causes the oscillations to stop. To interrupt the inverter operation, the control signals have merely to be interrupted.

  In case of faults in the generation of these signals, on account of C. there can be no DC component in the current supplied to the load.

  However, as in all the voltage-source inverters, there is a risk of short-circuiting the DC voltage supply by simultaneously firing the two thyristors series-connected across this source.

6.1.6 Variants

6.1.6.1 Two-Switch Configuration

- The full-bridge configuration with full-wave control is the most widely used. Normally, the inverter is supplied by the mains through a diode rectifier and an input filter composed of elements L_e, C_e (Fig. 6.11a).

- At low power, the two-switch voltage-source inverter with a capacitive divider is often used (Fig. 6.11b), the latter acting as the input filter capacitance (see Chap. 2, Sect. 2.5.3).

  Capacitors C_{e1} and C_{e2}, which have a capacitance C_e much higher than C in the resonant circuit, have little effect on the value to be considered in studying the latter; indeed

  $$C\frac{C_e}{2}\bigg/\left(C+\frac{C_e}{2}\right) \text{ is only slightly different from } C.$$

- The great difference between the value to be given to the filtering capacitor and that to be given to the divider capacitors (particularly if the inverter is operating at high frequency) often leads to the voltage division being carried out with the he!p of the oscillating circuit capacitance (Fig. 6.11c). The circuit has R, L, C as constants if two capacitors with a capicatance equal to $2C$, as indicated, are series-connected.

- An asymmetrical divider can be used. At its limit, the configuration to be found is that of Fig. 6.11d where load R, L, C is supplied by undirectional voltage square-waves.

- In many cases, a transformer must be placed between the inverter itself and the load (Fig. 6.12a). This transformer provides voltage adaptation and galvanic insulation (it is thus possible to suppress the transformer feeding the rectifier supplying the inverter).

  The capacitor of the oscillating circuit can be placed either on the secondary or the primary side of the transformer. The leakage inductance of

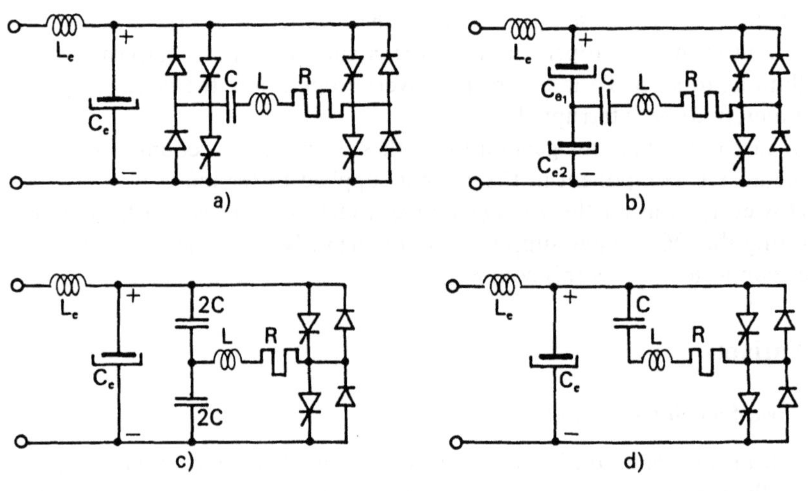

Fig. 6.11

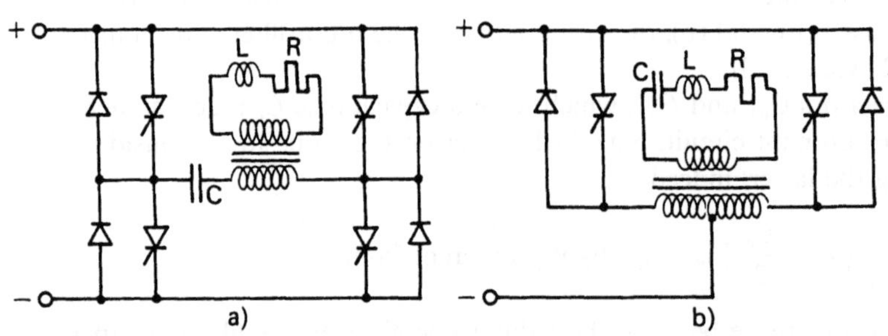

Fig. 6.12

the latter is added to the inductance of the load itself in order to make up the inductance of the oscillating circuit.

- The use of an output transformer often leads, at low power levels, to use a centre-tapped transformer configuration for the inverter (Fig. 6.12b).

6.1.6.2 Use of the Shifted Control for the Full-Bridge Inverter

The full-bridge series resonant inverter (Fig. 6.5) was studied with a full-wave control, i.e. the simultaneous control of TC_1 and TC'_2, of TC_2 and TC'_1. The results obtained can be easily transposed to the two-switch inverters.

By using again (Fig. 6.13) the diagram of Fig. 2.13, the full-bridge inverter with a full-wave control operates on the axis corresponding to zero shift angle β:

- with the configuration using semiconductor devices controlled only at turn-on, operation takes place in the lower part of zone 4. In order to reduce the current, the frequency is decreased and the operating point is moved from the left of A towards C;
- with the configuration where switches are controlled only at turn-off, operation takes place in the lower part of zone 2. In order to reduce the current, the frequency is increased, thus moving the operating point from the right of A towards C.

- If the output variable value has to be varied by acting on the shift β between the two half-bridge controls, while keeping the same types of semiconductor device, φ_1 has once more to be varied by acting on the frequency in order to remain within the same operating zone:
  - to the left of segment AD for the thyristor configuration,
  - to the right of segment AB for the dual thyristor configuration.

  Any action on β should be accompanied by an action on φ_1; this obviously should lead to a delicate and unsatisfactory form of control.

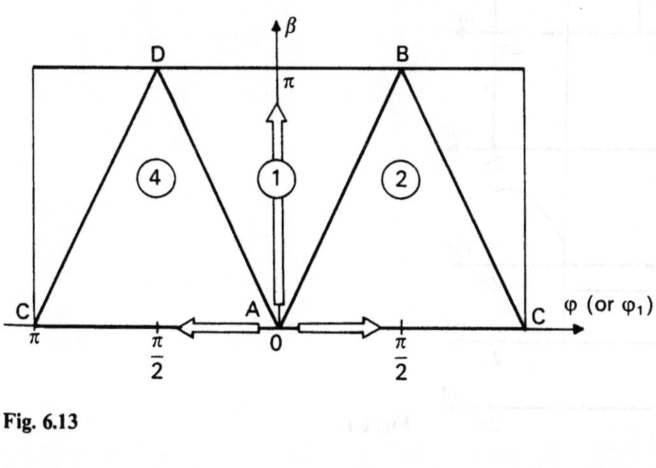

Fig. 6.13

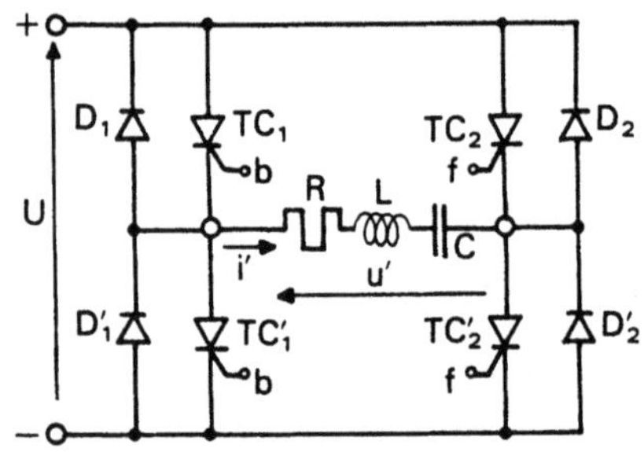

Fig. 6.14

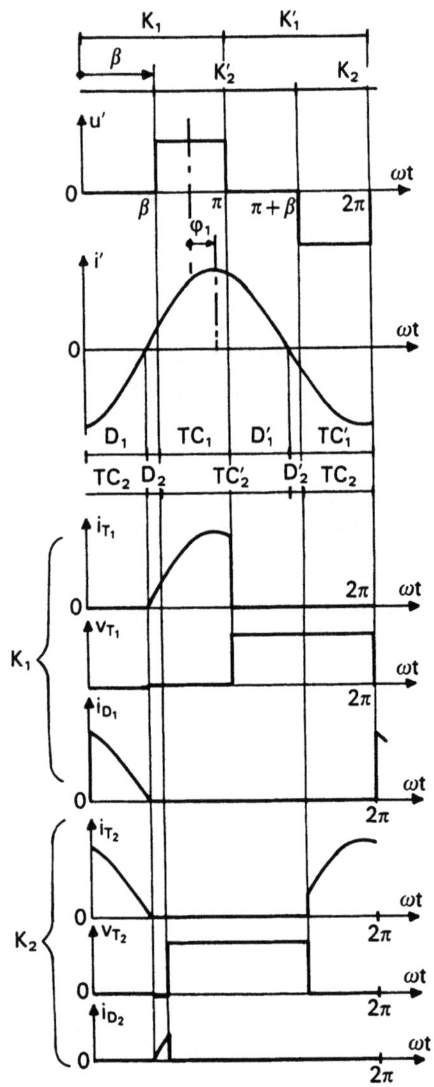

Fig. 6.15

– The output variable value can be varied, while opeating *at constant frequency*, by using only β shift. In order to do this, an *asymmetrical bridge* must be used, i.e. one composed of two half-bridges using different types of controlled devices. The operating point must remain in zone 1 of Fig. 6.13.

In such an asymmetrical bridge (Fig. 6.14), semiconductor devices TC_1 and TC'_1 have a controlled turn-off (the letter b next to their control terminal stands for blocking). Semiconductor devices TC_2 and TC'_2 have a controlled turn-on (the letter f next to their control terminal stands for firing).

Figure 6.15 shows an example of the resulting waveforms.

For this type of operation to be possible, the following must apply

$$\beta > 2|\varphi_1|$$

The conduction time of diodes D_1 and D'_1 is $(\beta/2 + \varphi_1)/\omega$ and that of diodes D_2 and D'_2 is $(\beta/2 - \varphi_1)/\omega$. Even if the load is supplied at its resonant frequency ($\varphi_1 = 0$), β cannot be reduced to zero. The duration of commutation and of recovery time must be taken into consideration.

Furthermore when β is increased in order to decrease the output variable value, the hypothesis of a sinusoidal current is difficult to accept when β approaches π. The voltage u' harmonics take on very high relative values (see Fig. 2.15).

Operation at fixed frequency with an asymmetrical bridge seldom replaces operation at variable frequency for the traditional applications of the series resonant inverter, since this assumes a clearly defined resonant frequency. This is rarely the case. However, it makes it possible to conceive new applications.

6.2 Parallel Resonant Inverter

Duality enables the majority of the characteristics of the parallel resonant inverter to be deduced from those of the series resonant inverter. The study of the former can thus be made shorter.

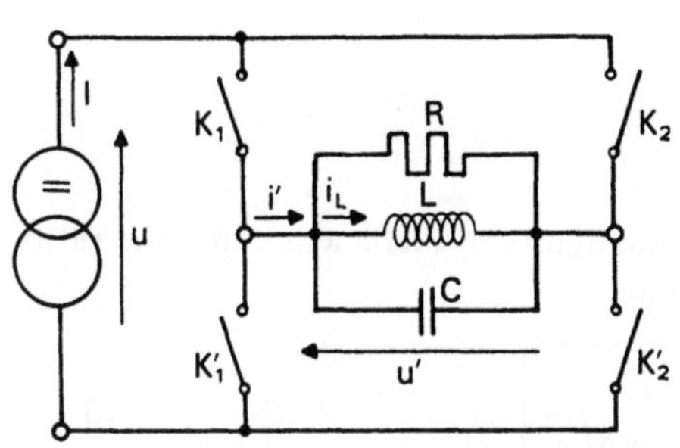

Fig. 6.16

The parallel inverter is a current-source inverter comprising, for example, a four-switch full-bridge (Fig. 6.16) which supplies the load with a current i'

equal to $+I$ for one half-cycle ($0 < t < T/2$),
to $-I$ for the other ($T/2 < t < T$).

The load consists of the parallel assembly of a capacitor C, an inductor L and a resistance R. The latter stands in for the whole of the active power supplied to the load.

6.2.1 Response of the Parallel Resonant Circuit to a Square-Wave Current

6.2.1.1 Analytical Expressions of Variables

Current i_L in the inductance L as well as output voltage u' are linked to current i' which the inverter imposes by

$$C\frac{\mathrm{d}u'}{\mathrm{d}t} + i_L + \frac{u'}{R} = i'$$

with

$$L\,\mathrm{d}i_L/\mathrm{d}t = u'$$

This thus gives:

$$\begin{cases} C\,L\dfrac{\mathrm{d}^2 i_L}{\mathrm{d}t^2} + \dfrac{L}{R}\dfrac{\mathrm{d}i_L}{\mathrm{d}t} + i_L = i' & (6.41) \\[2ex] u' = L\dfrac{\mathrm{d}i_L}{\mathrm{d}t} & (6.42) \end{cases}$$

Because of L and C, neither current i_L nor its derivative, equal to u'/L, can be subject to any discontinuity.

A comparison between Eqs. (6.41) and (6.42), on the one hand, and (6.2) and (6.3) on the other, shows that all the expressions worked out for the series inverter can be used for the parallel inverter, providing that we replace

U by I, i by u,
u' by i', i' by u',
u_C by i_L,
L by C, C by L, R by $1/R$.

The resonant angular frequency ω_0, the damping coefficient α and the ratio ζ of α to ω_0 now have the following values:

$$\boxed{\omega_0 = \frac{1}{\sqrt{LC}} \qquad \alpha = \frac{1}{2RC} \qquad \zeta = \frac{1}{2R}\sqrt{\frac{L}{C}}} \;. \tag{6.43}$$

For low values of ζ, Fig. 6.3 gives the waveforms of i'/I (instead of u'/U), of i_L/I (instead of u_C/U) and of $u'/L\,\omega_0 I$ (instead of $i'/C\,\omega_0 U$).

The nature of the commutations to be carried out depends on the polarity of u'_0:

- for ω less than $\omega_0\sqrt{1-\zeta^2}$, u'_0 is positive,
- for ω greater than $\omega_0\sqrt{1-\zeta^2}$, u'_0 is negative.

6.2.1.2 "First Harmonic" Method

Current i' has the following series expansion

$$i' = I'_1\sqrt{2}\sin\omega t + I'_3\sqrt{2}\sin 3\omega t + \ldots + I'_n\sqrt{2}\sin n\omega t + \ldots$$

with

$$I'_1 = \frac{2\sqrt{2}}{\pi} I \quad \text{and} \quad I'_n = \frac{I'_1}{n}.$$

The voltage u' fundamental component has the following rms value:

$$U'_1 = \frac{I'_1}{\sqrt{\dfrac{1}{R^2} + \left(C\omega - \dfrac{1}{L\omega}\right)^2}}; \tag{6.44}$$

it lags the fundamental of i' with an angle whose tangent is equal to $R(C\omega - 1/L\omega)$. Since the phase lag of the current on the voltage is counted positively, the following must be used:

$$\tan\varphi_1 = -R\left(C\omega - \frac{1}{L\omega}\right) \tag{6.45}$$

The fundamental components of current i_L in inductance L and current i_{RC}, in the assembly composed of R and C parallel-connected, have the following rms values:

$$I_{L_1} = \frac{I'_1}{L\omega\sqrt{\dfrac{1}{R^2} + \left(C\omega - \dfrac{1}{L\omega}\right)^2}}, \quad I_{RC_1} = I'_1\sqrt{\frac{\dfrac{1}{R^2} + C^2\omega^2}{\dfrac{1}{R^2} + \left(C\omega - \dfrac{1}{L\omega}\right)^2}} \tag{6.46}$$

If, in the expressions of U'_1, φ_1, I_{L_1} and I_{RC_1}, respectively, ω_0 and ζ as defined by (6.43) are introduced, the following can be obtained:

$$U'_1 = \frac{L\omega_0 I'_1}{\sqrt{4\zeta^2 + \left(\dfrac{\omega}{\omega_0} - \dfrac{\omega_0}{\omega}\right)^2}} \tag{6.47}$$

$$\tan\varphi_1 = -\frac{1}{2\zeta}\left(\frac{\omega}{\omega_0} - \frac{\omega_0}{\omega}\right) \tag{6.48}$$

$$I_{L_1} = \frac{\omega_0}{\omega}\frac{I'_1}{\sqrt{4\zeta^2 + \left(\frac{\omega}{\omega_0} - \frac{\omega_0}{\omega}\right)^2}} \tag{6.49}$$

$$I_{RC_1} = I'_1\sqrt{\frac{4\zeta^2 + (\omega/\omega_0)^2}{4\zeta^2 + \left(\frac{\omega}{\omega_0} - \frac{\omega_0}{\omega}\right)^2}}\,. \tag{6.50}$$

A comparison between Eqs. (6.47), (6.49), (6.50) and Eqs. (6.26), (6.28), (6.29), as worked out for the series resonant inverter, shows that the curves in Fig. 6.4 can be used to follow the variations of

$$U'_1/L\omega_0 I'_1,\ I_{L_1}/I'_1,\ I_{RC_1}/I'_1$$

(instead of $I'_1/C\omega_0 U'_1$, U_{C_1}/U'_1, U_{RL_1}/U'_1) as functions of ω/ω_0.

The conclusions to be deduced from a study of these curves are similar to those given for the series resonant inverter:

- Voltage u' can be reduced to its fundamental, if ζ is low and ω/ω_0 close to 1. The approximation becomes more valid if ω/ω_0 is close to 1 from a higher value than from a lower value.
- If u' can be reduced to its fundamental, this can also be done for current i_L; the approximation is even better.
- The current in the capacitor cannot be reduced to its fundamental since the discontinuities of current i' are to be found in this current.

The characteristics which give φ_1 as a function of ω/ω_0, for various values of ζ, in Fig. 6.4 can be used, providing that the sign of φ_1 is reversed:

- for ω/ω_0 less than 1, the load is inductive,
- for ω/ω_0 greater than 1, the load is capacitive.

6.2.2 Inverter Operation and Control

As in the case with all current-source inverters, the parallel resonant inverter is composed of controlled semiconductor devices which must be capable of withstanding reverse voltages. Figure 6.17 shows the full-bridge current-source inverter configuration.

By using an appropriate control strategy, ω/ω_0 can always be kept above 1 or always below 1, so that the semiconductor devices have only one type of commutation to carry out.

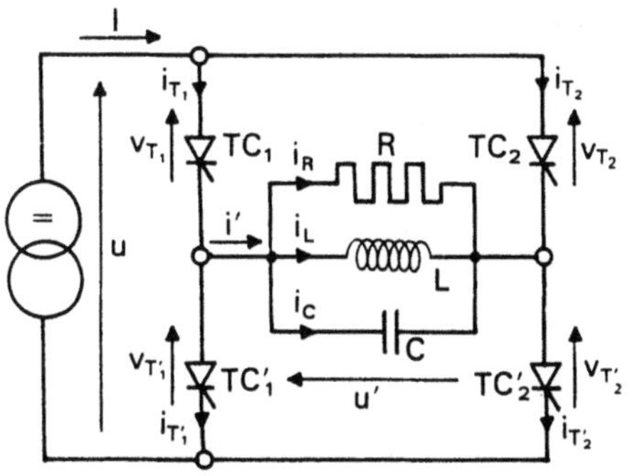

Fig. 6.17

- If the operating angular frequency ω is kept below the resonant angular frequency ω_0, the load will always appear to be *inductive*. The waveforms are as in Figs. 1.6 and 1.11b.

  This operation is rarely used for two main reasons.

  - The semiconductor devices have to ensure forced turn-off. However, in the current-source inverters, the inductances limiting di/dt bring about high-voltage spikes across the semiconductor devices at the end of turn-off (see Fig. 1.18b).
  - Operation occurs in the unfavourable side of the characteristics of Fig. 6.4, i.e. where ω/ω_0 is less than 1. It has been seen (Sect. 6.1.4.2) that the variable which was assumed to be sinusoidal and was used as a reference for the control – in this case, voltage u' – was no longer sinusoidal when ω/ω_0 was close to 1/2.

  We will thus limit our study to the operation with ω/ω_0 greater than 1.

- If ω is always maintained greater than ω_0 by the control, the load appears to be *capacitive*. The waveforms are those shown in Fig. 1.7 and 1.11b.

  The semiconductor devices have only to ensure turn-ons; the turn-offs are ensured by the turn-ons of the following semiconductor devices. Simple symmetrical thyristors can be used (or GTO thyristors or series assemblies of transistors and diodes used as thyristors).

In Fig. 6.18 the various waveforms have once again been plotted in order to show driving by the load:

- the change of u' from a negative to a positive value is taken as a reference for the firing control of TC_2 and TC'_1;
- u' zero crossing – in the opposite direction – is taken as a reference for the firing (Δt later) of TC_1 and TC'_2.

The shorter the delay Δt at turn-on, the higher the frequency and the distance from the resonance.

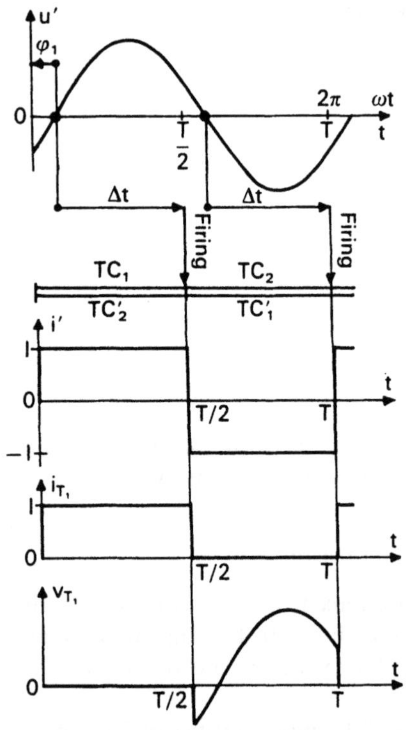

Fig. 6.18

6.2.3 Characteristics

6.2.3.1 "First Harmonic" Approximation

- If the output voltage u' is assumed to be sinusoidal, the angular distance between the moments when u' and i' fall to zero gives phase shift φ_1.

If only the fundamental component I'_1 of current i' is taken into account the values of voltage u' and current i_L are linked to that of i' by Eqs. (6.47) and (6.49). Equation (6.48) gives phase shift φ_1.

In Fig. 6.4, it is possible to read the variations of

$$\frac{U'_1}{L\omega_0 I'_1}, \quad \frac{I_{L_1}}{I'_1}, \quad \varphi_1$$

as a function of ω/ω_0 and for different values of ζ.

If voltage u' is assumed to be sinusoidal, all the results worked out in Chap. 5 for the full-wave operation ($\beta = 0$) of the full-bridge current-source inverter can be used, since this study was made on the assumption that u' was sinusoidal. More especially,

- the voltage across the semiconductor devices has a peak value of

$$V_{T\max} = U'_1\sqrt{2} \tag{6.51}$$

- the input voltage of the inverter has a mean value of

$$U = \frac{2}{\pi} U'_1 \sqrt{2} \cos \varphi_1. \tag{6.52}$$

- In most cases, the parallel resonant inverter is supplied by the mains via a rectifier and an inductor. The inverter, in such cases, is fed by a supply which
  - is a genuine current source, from the point of view of the commutation,
  - but which has the static characteristic of a voltage source.

  The mains *imposes the mean voltage U at the inverter input*; the load and the control determine the value of I'_1.

  It can thus be useful to express the various values as functions of U, of φ_1 and of resistance R.

Equation (6.52) gives U'_1 *by*

$$U'_1 = \frac{\pi}{2\sqrt{2}} \frac{U}{\cos \varphi_1}. \tag{6.53}$$

The power supplied to the load thus has the following value, if u' is reduced to its fundamental,

$$P = \frac{U_1'^2}{R} = \frac{\pi^2}{8} \frac{U^2}{R} \frac{1}{\cos^2 \varphi_1}. \tag{6.54}$$

Since the power is also equal to $U'_1 I'_1 \cos \varphi_1$, I'_1 can be deduced

$$I'_1 = \frac{P}{U'_1 \cos \varphi_1} = \frac{\pi}{2\sqrt{2}} \frac{U}{R} \frac{1}{\cos^2 \varphi_1} \tag{6.55}$$

as well as the current supplied to the inverter

$$I = \frac{\pi}{2\sqrt{2}} I'_1 = \frac{\pi^2}{8} \frac{U}{R} \frac{1}{\cos^2 \varphi_1}. \tag{6.56}$$

The current in the inductor is obtained by writing

$$I_{L_1} = \frac{U'_1}{L\omega} = \frac{U'_1}{R} \frac{R}{L\omega} = \frac{U'_1}{R} \frac{RC\omega_0}{LC\omega_0^2} \frac{\omega_0}{\omega} = \frac{U'_1}{R} \frac{1}{2\zeta} \frac{\omega_0}{\omega}.$$

Thus

$$I_{L_1} = \frac{\pi}{4\sqrt{2}} \frac{1}{\zeta} \frac{\omega_0}{\omega} \frac{U}{R} \frac{1}{\cos \varphi_1}. \tag{6.57}$$

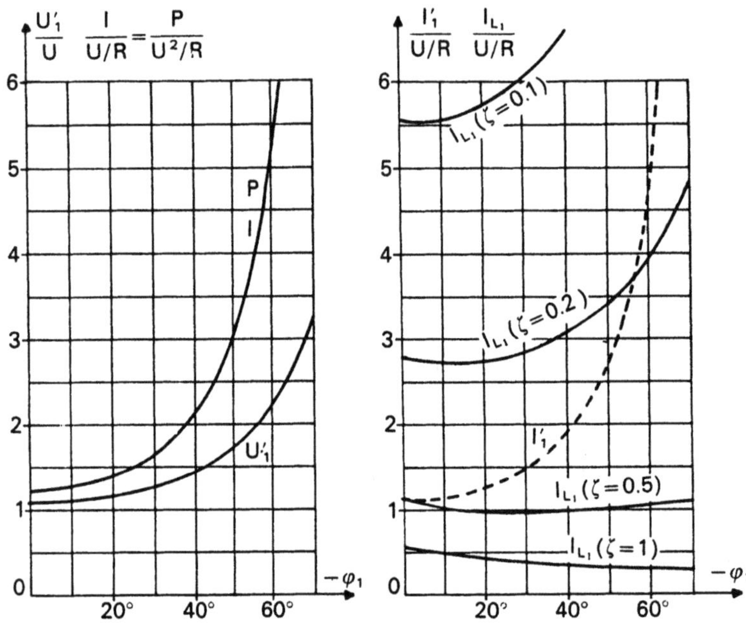

Fig. 6.19

Since $\tan \varphi_1$ equals $-(1/2\zeta)(\omega/\omega_0 - \omega_0/\omega)$, the ratio ω/ω_0 is linked to φ_1 by

$$\frac{\omega_0}{\omega} = \zeta \tan \varphi_1 + \sqrt{1 + \zeta^2 \tan^2 \varphi_1} \tag{6.58}$$

Figure 6.19 shows the variations, as functions of $-\varphi_1$,

- of U'_1/U and of $I/(U/R)$ or $P/(U^2/R)$
- of $I'_1/(U/R)$ and $I_{L_1}/(U/R)$, for several value of ζ.

Power P, supplied to the load, is increased by moving ω away from the resonant angular frequency. The control characteristic which gives P (or U'_1 or I) as a function of φ_1 is independent of ζ.

This does not apply to the characteristics giving I_{L_1} as a function of φ_1.

The stresses on the semiconductor devices, which are functions of U'_1 and I, depend only on φ_1.

6.2.3.2 Limitations

The minimal value of $-\varphi_1$ is imposed by the recovery time t_q of the semiconductor devices. The following condition has to be respected:

$$|\varphi_1| > 2\pi \frac{t_q}{T}. \tag{6.59}$$

With the help of Eq. (6.48), this gives the minimal value of the ratio ω/ω_0.

If ω/ω_0 or $-\varphi_1$ are increased, in order to increase the power supplied to the load, there is also an increase in the current delivered by the supply, the current in the semiconductor devices, the voltage across them, the current in the capacitor and, if ζ is small, the current in the inductor.

The ratings of all the elements of the converter limit the maximum value of ω/ω_0.

6.2.4 Starting. Protection

6.2.4.1 Initiating the Load Oscillation

If the nature of the supply – equivalent to a constant voltage source U series-connected to an inductance with a very high value L_S – is taken into account, the control of the inverter switches does not enable the oscillations of the parallel resonant circuit to start.

As was emphasized several times in Chap. 5, the current-source inverter operating in natural commutation is the equivalent of a rectifier operating as a line-commuted inverter. The load acts as an AC voltage source and ensures the commutations.

The load must be made to oscillate by providing it with the necessary energy, $(1/2)CU'^2_{max}$, and then by maintaining the oscillations by means of the inverter.

To supply the necessary energy to the oscillating circuit the following operations can be carried out:

- either discharge an auxiliary capacitor C' into this circuit,
- or establish a current in the smoothing inductance before starting the inverter.

– In the first solution (Fig. 6.20a), a capacitor C' is charged, using a DC voltage source of value E greater than U.

C' is then discharged into the oscillating circuit by means of the auxiliary thyristor T′. Inductance L' limits the current peak in T′ and makes the current flowing through T′ pseudo-cyclical. This ensures that T′ is turned off at the end of a half-pseudo-cycle.

The resonant circuit then begins to oscillate. The oscillations are maintained by firing thyristors TC_1 and TC'_2 and then TC_2 and TC'_1 by taking the moments when voltage u' crosses zero as references.

– The second solution (Fig. 6.20b) consists in charging the inductance L_S by deriving, at the inverter input, an auxiliary thyristor T′ series-connected with a large capacitance C' shunted by a resistance.

When i has reached a sufficiently high value, the turn-on of TC_1 and TC'_2 is controlled. Since U'_C is positive and u' zero, this turn-on applies a reverse voltage across thyristor T′ and blocks it. The resonant circuit is fed by a current step equal to the current flowing through T′.

The resonant circuit oscillates. The inverter thyristors can be controlled by taking the moments when u' falls to zero as references.

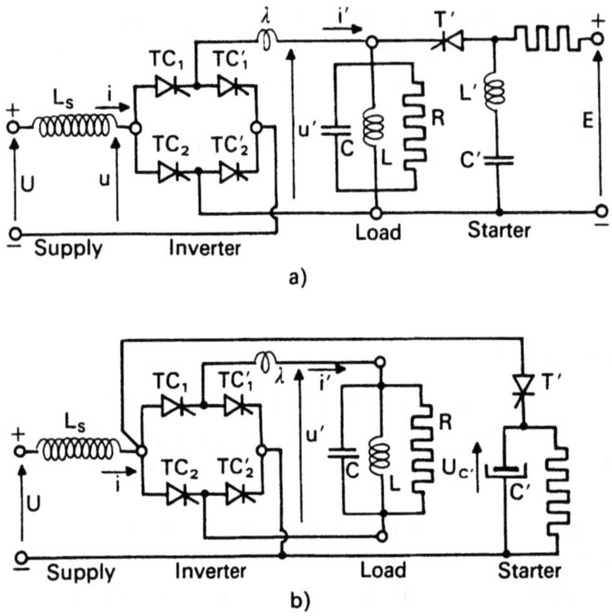

Fig. 6.20

6.2.4.2 Protection of the Semiconductor Devices. Faults

- The di/dt of the current must be limited when the semiconductor devices are fired. Since commutations occur between TC_1 and TC_2, and between TC'_1 and TC'_2, a small inductance λ series-connected with the load enables this limitation to be ensured. Inductance λ has been represented in both diagrams of Fig. 6.20.

  Throughout the commutations, the four thyristors conduct together:

  $$\lambda(di'/dt) + u' = 0; \qquad di'/dt = -u'/\lambda.$$

  Assuming that u' remains constant and equal to u'_0 during the commutations, the duration t_c of the latter is equal to

  $$t_c = 2\lambda I/|u'_0|.$$

- The parallel resonant inverter protects itself less effectively that the series resonant inverter.

  In the case of any error in the generation of control signals, current i' in the load and voltage u' across it show a DC component.

  If the load is short-circuited, voltage u' which ensured the commutations, falls to zero; the conducting thyristor remains on: the inverter is short-circuited.

If the load inductance is put in open circuit, the operating frequency of the converter moves towards excessive values. If the capacitive path is removed, the inverter will short-circuit, since the commutations are no longer ensured.

Finally, the operation of the inverter cannot be stopped by interrupting the firing signals. The conducting thyristor continues to conduct and operation in short-circuit will result. The supply of the inverter must be cut off to stop it.

This explains why the parallel resonant inverter is often supplied by a thyristor rectifier. If there is any problem, the input current of the inverter can be cut off by increasing the delay angle of this rectifier.

6.2.5 Variants

- In the case of the parallel resonant inverter, the same types of variants can be found as for the series resonant inverter (Sect. 6.1.6)

- When an adaptation transformer is needed, it should be placed, if possible, at the inverter output, rather than at the rectifier input: it is then lighter and less costly, since it is operating at a frequency which is normally much higher.

  It is then possible to connect either the whole or a part of the oscillating circuit to the secondary winding of the transformer. Depending on the case, the leakage inductance of the transformer acts as a di/dt limiting inductance or acts as a part of the inductance L of the oscillating circuit.

- When a transformer is needed, a centre-tapped transformer can be used and the two-switch configuration adopted (Chap. 5, Sect. 5.1.1). In certain cases, the capacitor of the oscillating circuit can be derived between the input terminals of the transformer.

- With the full-bridge configuration, the power supplied to a given load can be varied by acting on angle β (Chap. 5, Sect. 5.1.2) and no longer on angle φ_1. In this case, the bridge is asymmetrical, TC_1 and TC_2, for example, having controlled turn-on, and TC'_1 and TC'_2 controlled turn-off. Operation then takes place at zero or nearly zero φ_1, i.e. at the resonant frequency or in its immediate proximity.

- If the parallel resonant inverter is powered by *a true DC current source* (see Chap. 5, Sect. 5.1.4.1), the power supplied to the load can be varied by varying value I of the inverter input current.

  Use of this control mode is made considerably easier, in high-power applications, by using GTO thyristors.

  When u' goes from a negative to a positive value, the turn-on of TC_1 and TC'_2 and the turn-off of TC_2 and TC'_1 are controlled (Fig. 6.21). When u' goes from a positive to a negative value, current I is transferred from TC_1 towards TC_2 and from TC'_2 towards TC'_1. The commutations thus occur under a very low voltage u' and commutation losses are greatly reduced.

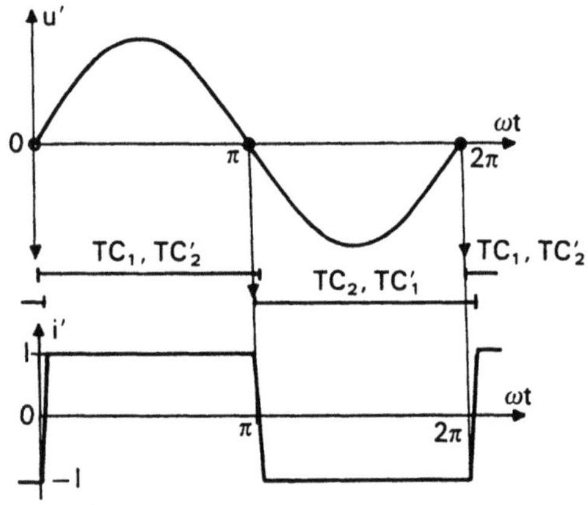

Fig. 6.21

The inverter operates nearly at the oscillating circuit resonance frequency. If ζ is low, the waveform of voltage u' is virtually sinusoidal. The equations worked out in Chap. 5, Sect. 5.1.2.3 can be used directly, by making $\varphi = 0$ and $\beta = 0$.

6.3 Resonant Supplies

In the study of switch-mode power supplies at the end of Vol. 3 we showed the advantages of the use of a higher intermediate frequency in producing the AC–DC transformation at low or medium power.

In the symmetrical configurations (Vol. 3, Chap. 6, Sect. 6), the transformer operating at this intermediate frequency is AC fed. In order to do this, the chopper is replaced by an inverter.

There is thus, in cascade, the diode rectifier connected to the mains the medium-frequency inverter, the transformer and the output rectifier.

In resonant supplies, the inverter feeds a resonant circuit. To achieve this, an inductor and a capacitor are added; the DC load, placed at the assembly output and seen through the output rectifier and the transformer, forms the resistive element of the oscillating circuit.

Resonant supplies are used increasingly frequently to replace symmetrical switch-mode power supplies.

- The disadvantage of the former is that they require an increase in the power rating of the switches and that regulation becomes more difficult.
- However, they have the advantage of enabling the choice to be made of the types of commutation to be carried out, simplifying the snubbers, reducing the electromagnetic interferences caused by commutations, reducing the losses in the semiconductor devices and thus increasing the possible operation frequency of the inverter.

The development of resonant inverters is closely linked to that of the use of dual thyristor

Many different configurations are possible but we will confine our study to the supplies which use voltage-source inverters since we have seen that their operation is easier to control.

We will thus only deal with

- the series resonant inverter
- the series-parallel inverter.

We will also limit our study to the first harmonic approximation to determine the characteristics.

6.3.1 Supply with a Series Resonant Inverter

6.3.1.1 Supply Configuration

Figure 6.22 gives the diagram of the cascade which forms the series resonant inverter supply.

The mains voltage is rectified by a diode bridge and then filtered. The load circuit of the voltage-source inverter comprises series-connection of inductor L, capacitor C and the primary winding of the n_1/n_2 ratio transformer. The secondary current $n_1 i'/n_2$ is rectified by a diode bridge. Capacitor C_d filters the output voltage.

The voltage U is assumed to be constant at the inverter input; the output voltage ripple U_d and output current ripple I_d are ignored.

The input filter must include a capacitor C_e with a capacitance sufficiently high to supply the inverter from a voltage-source; the frequency of the first harmonic of current i is twice that of the operating frequency f of the inverter. The output current of the second rectifier also has a frequency equal to $2f$. The higher the value of f, the greater the reduction that can be made in the capacitance C_e and C_d of the two capacitors.

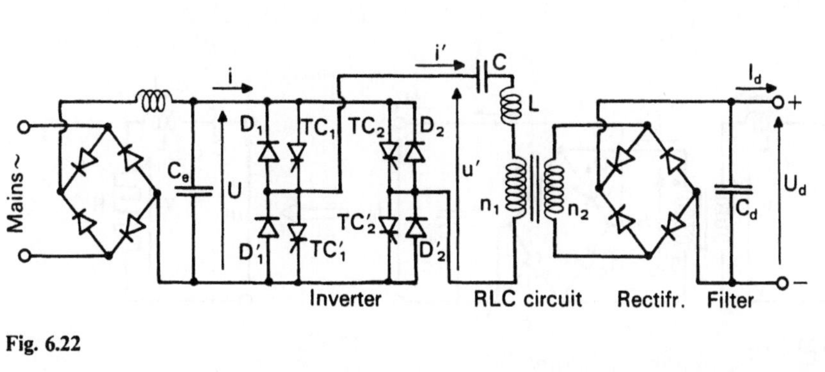

Fig. 6.22

The assembly can be varied in many ways:

- the input rectifier is a three-phase six-diode bridge if the converter has a three-phase supply;
- a two-switch inverter can be used if a capacitive divider is placed at the input;
- L or C or both of these can be placed at the secondary winding of the transformer;
- finally, a transformer with a centre-tapped secondary winding and a two-diode output rectifier can be used.

In order to vary U_d at a given current I_d or to keep U_d constant despite the variations in I_d, the inverter frequency must be varied.

In order to operate in the right side of the resonance curves and to be able to use switches controlled only at turn-off, ω/ω_0 must be kept above 1.

6.3.1.2 Inverter Output

Figure 6.23 shows the part of the diagram corresponding to the inverter load as well as the equivalent RLC circuit to which this load must be reduced in order to use the equations worked out for the study of the series resonant inverter (Sect. 6.1.4.1).

- *Currents*

If the magnetizing current is ignored, the secondary current i_S of the transformer is equal to $n_1 i'/n_2$. It is sinusoidal, according to the first harmonic approximation, and has an amplitude of $(n_1/n_2)I'_1\sqrt{2}$.

Current i_S flows through diodes D_A and D_D when positive, and through diodes D_C and D_B when negative. The mean value of the rectified current i'' is equal to $(2/\pi)\,(n_1/n_2)I'_1\sqrt{2}$.

Capacitor C_d absorbs the variations of i''. The mean value of i'' is thus equal to the output current I_d.

$$I_d = \frac{2}{\pi}\frac{n_1}{n_2} I'_1\sqrt{2}.$$

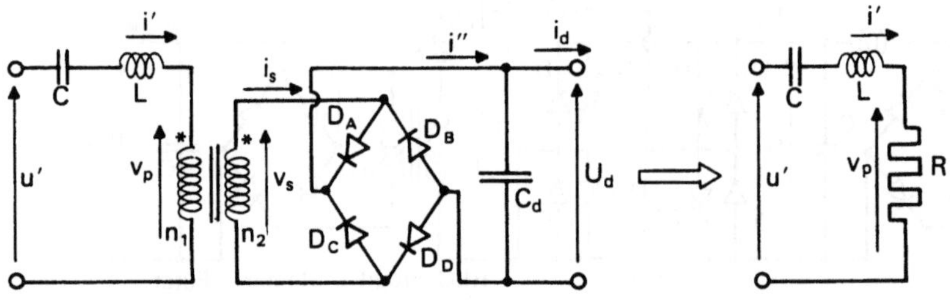

Fig. 6.23

Hence I'_1 as a function of I_d:

$$I'_1 = \frac{\pi}{2\sqrt{2}} \frac{n_2}{n_1} I_d. \tag{6.60}$$

- *Voltages*

The output voltage v_S of the transformer is equal to $+U_d$ when D_A and D_D are conducting, and to $-U_d$ when D_C and D_B are conducting. The voltage at the transformer input v_P, equal to $(n_1/n_2)v_S$, is made up of alternating square-waves. Its fundamental has an rms value of

$$V_{P_1} = \frac{2\sqrt{2}}{\pi} \frac{n_1}{n_2} U_d. \tag{6.61}$$

The value of U_d can thus be deduced from that of V_{P_1} by

$$U_d = \frac{\pi}{2\sqrt{2}} \frac{n_2}{n_1} V_{P_1}. \tag{6.62}$$

- *Equivalent resistance*

Figure 6.24 shows the waveform of i'' deduced from that of i', and then waveform of v_P deduced from that of U_d. It can be seen that the fundamental v_{P1} of the transformer input voltage is in phase with the primary current i'.

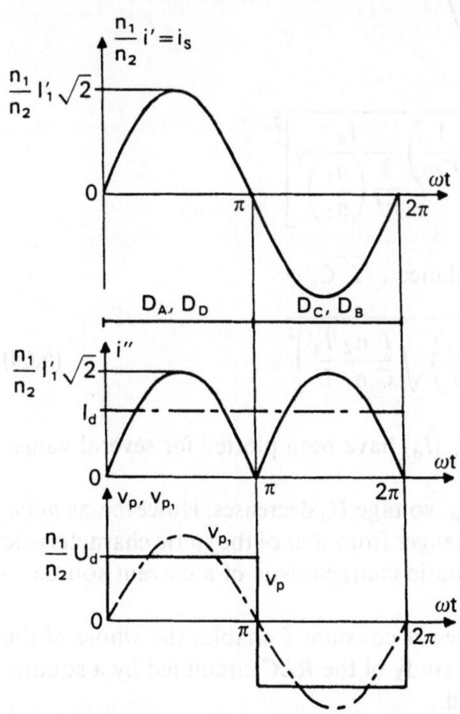

Fig. 6.24

In the case of the fundamental, referred to the current i' circuit, the DC load behaves as a resistance R with a value of

$$R = \frac{V_{P_1}}{I'_1} = \frac{8}{\pi^2}\left(\frac{n_1}{n_2}\right)^2 \frac{U_d}{I_d} \tag{6.63}$$

6.3.1.3 Characteristics

- In order to work out the equation showing how U_d varies as a function of I_d and ω/ω_0, we start from

$$V_{P_1} = RI'_1 = \frac{RU'_1}{\sqrt{R^2 + \left(L\omega - \dfrac{1}{C\omega}\right)^2}}.$$

V_{P_1} gives U_d by (6.62); fundamental component U'_1 of voltage u' equals $(2\sqrt{2}/\pi)U$; R is given by (6.63). By substitution, the following can be found:

$$U_d = \frac{\dfrac{\pi}{2\sqrt{2}}\dfrac{n_2}{n_1}\dfrac{2\sqrt{2}}{\pi}U}{\sqrt{1 + \left(\dfrac{L\omega - \dfrac{1}{C\omega}}{\dfrac{8}{\pi^2}\left(\dfrac{n_1}{n_2}\right)^2\dfrac{U_d}{I_d}}\right)^2}}$$

giving

$$U_d = \sqrt{\left(\frac{n_2}{n_1}\right)^2 U^2 - \left[\left(L\omega - \frac{1}{C\omega}\right)\frac{I_d}{\dfrac{8}{\pi^2}\left(\dfrac{n_1}{n_2}\right)^2}\right]^2}$$

or, by introducing ω/ω_0 and impedance $\sqrt{L/C}$,

$$U_d = \frac{n_2}{n_1}U\sqrt{1 - \left[\frac{\pi^2}{8}\left(\frac{\omega}{\omega_0} - \frac{\omega_0}{\omega}\right)\sqrt{\frac{L}{C}}\frac{n_2}{n_1}\frac{I_d}{U}\right]^2}. \tag{6.64}$$

In Fig. 6.25 the curves giving $U_d\,(I_d)$ have been plotted for several values of ω/ω_0 using this equation.

When ω/ω_0 increases at given I_d, voltage U_d decreases. However, as ω/ω_0 increases, the shape of the curves changes from that of the static characteristic of a voltage source to that of the static characteristic of a current source.

- Plotting in the same graph of curves at constant ζ enables the whole of the results obtained during the general study of the *RLC* circuit fed by a square-wave voltage (Sect. 6.1.2) to be used.

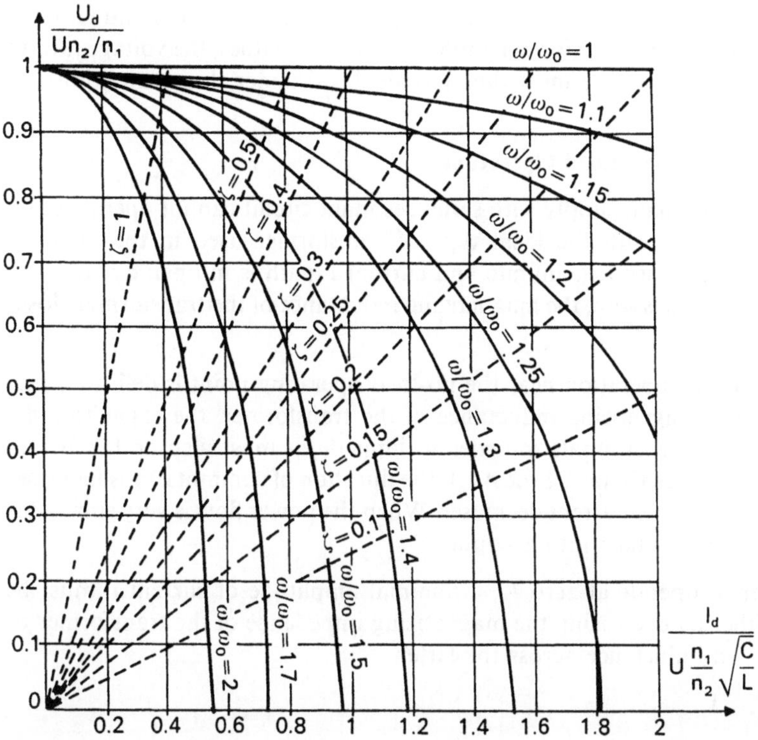

Fig. 6.25

As ζ equals $(R/2)\sqrt{C/L}$, the following can be deduced from Eq. (6.63):

$$\frac{4}{\pi^2}\left(\frac{n_1}{n_2}\right)^2\frac{U_d}{I_d}\sqrt{\frac{C}{L}}=\zeta.$$

This equality can be written in the form

$$\frac{\dfrac{U_d}{U n_2/n_1}}{\dfrac{I_d}{U\dfrac{n_1}{n_2}\sqrt{\dfrac{C}{L}}}}=\frac{\pi^2}{4}\zeta. \tag{6.65}$$

The curves at constant ζ are straight lines: some have been plotted in dashed lines in Fig. 6.25.

- For the rated value of current I_d, current I'_1 is at its maximum [Eq. (6.60)]. The current rating of the inverter semiconductor devices and the different elements of the assembly must therefore be determined for that value.

For the rated values of I_d and U_d, the voltages across inductor L, capacitor C and the transformer show their highest values; the voltage ratings of the different elements must thus be calculated under these conditions.

6.3.1.4 Remark on No-Load Operation

In theory, the resonant supply with series resonant circuit cannot operate at no load. If I_d is zero, with the ideal type of transformer used in the previous calculations, resistance R is infinite and current i' path is in open circuit.

In practice, if I_d is zero, the magnetizing inductance of the transformer closes the circuit.

A study of the operation near the no-load operating point is delicate since, apart from the magnetizing inductance of the transformer, the capacitors derived at the switch terminals in order to limit dv/dt must also be taken into account. As noted in Chap. 4, Sect. 4.1.1, the duration of commutations increases when the commutated current decreases. When the power flow is very small, this duration has to be taken into account.

In order to operate at zero I_d, a minimal amplitude of current i' must be ensured, either by decreasing the magnetizing impedance of the transformer or by deriving an inductance across the latter.

6.3.2 Supply with a Series–Parallel Resonant Inverter

6.3.2.1 The Series–Parallel Inverter

The series–parallel inverter is a *voltage-source inverter* supplying a series resonant circuit with the load connected across the capacitor. Fig. 6.26 shows the configuration with a full-bridge inverter in the case where the load is a pure R resistance.

This connection of the load across the capacitor enables the series resonant inverter to be used when a highly variable load has to be fed. In particular, it enables no-load operation (infinite value of the load resistance R).

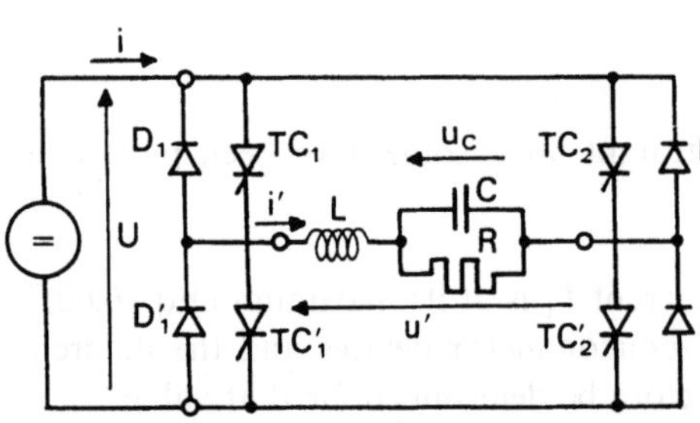

Fig. 6.26

- Voltage u_C across C or R and current i' in L are linked to the square-wave voltage u' supplied by the voltage-source inverter by

$$L\frac{di'}{dt} + u_C = u'$$

with

$$i' = C\frac{du_C}{dt} + \frac{u_C}{R}$$

giving

$$\begin{cases} LC\dfrac{d^2u_C}{dt^2} + \dfrac{L}{R}\dfrac{du_C}{dt} + u_C = u' & (6.66) \\[2ex] i' = C\dfrac{du_C}{dt} + \dfrac{u_C}{R} & (6.67) \end{cases}$$

Neither u_C nor its derivative can be subject to any discontinuity.

The resonant angular frequency ω_0, the damping coefficient α and the ratio ζ of α to ω_0 now have the following values:

$$\boxed{\omega_0 = \frac{1}{\sqrt{LC}} \qquad \alpha = \frac{1}{2RC} \qquad \zeta = \frac{1}{2R}\sqrt{\frac{L}{C}}}\,. \tag{6.68}$$

At given values of ω/ω_0 and of α or ζ, the general expression of u_C is the same as in the case of the series resonant inverter. However, the equations which give the integration constants as functions of u_{C_0} and i'_0 are different on account of the difference between Eqs. (6.3) and (6.67).

- The rms value I'_1 of fundamental of the current delivered by the inverter is linked to that U'_1 of voltage u' by

$$I'_1 = U'_1\sqrt{\frac{1 + R^2C^2\omega^2}{L^2\omega^2 + R^2(1 - LC\omega^2)^2}}\,. \tag{6.69}$$

The fundamental of i' lags that of u' by an angle φ_1 such that

$$\tan\varphi_1 = \frac{L\omega}{R} - RC\omega + RLC^2\omega^3. \tag{6.70}$$

The fundamental of voltage u_C across C or R has the following rms value:

$$U_{C_1} = \frac{RU'_1}{\sqrt{L^2\omega^2 + R^2(1 - LC\omega^2)^2}}\,. \tag{6.71}$$

The fundamental of voltage u_L across inductance L has the following rms value:

$$U_{L_1} = L\omega I'_1 = U'_1 \sqrt{\frac{L^2\omega^2 + R^2L^2C^2\omega^4}{L^2\omega^2 + R^2(1 - LC\omega^2)^2}}. \tag{6.72}$$

By introducing ω/ω_0 and ζ in the preceding four equations, they become

$$I'_1 = C\omega_0 U'_1 \sqrt{\frac{4\zeta^2 + \left(\frac{\omega}{\omega_0}\right)^2}{4\zeta^2 + \left[1 - \left(\frac{\omega}{\omega_0}\right)^2\right]^2}} \tag{6.73}$$

$$\tan\varphi_1 = -\frac{1}{2\zeta}\frac{\omega}{\omega_0}\left[1 - \left(\frac{\omega}{\omega_0}\right)^2\right] + 2\zeta\frac{\omega}{\omega_0} \tag{6.74}$$

$$U_{C_1} = \frac{U'_1}{\sqrt{4\zeta^2 + \left[1 - \left(\frac{\omega}{\omega_0}\right)^2\right]^2}} \tag{6.75}$$

$$U_{L_1} = U'_1 \sqrt{\frac{\left(2\zeta\frac{\omega}{\omega_0}\right)^2 + \left(\frac{\omega}{\omega_0}\right)^4}{4\zeta^2 + \left[1 - \left(\frac{\omega}{\omega_0}\right)^2\right]^2}}. \tag{6.76}$$

Figure 6.27 gives the variations of $I'_1/C\omega_0 U'_1$, of φ_1, of U_{C_1}/U'_1 and of U_{L_1}/U'_1 as functions of ω/ω_0, for some values of ζ.

A study of these curves shows that current i' can be reduced to its fundamental component if ζ is low and if ω/ω_0 is close to 1, more especially if it is greater than 1. Under the same conditions, voltage u_C can be reduced to its fundamental; the approximation is, in fact, even more valid. On the contrary, reducing u_L to its fundamental component gives a rather poor approximation, since the sharp variations in voltage u' are found in u_L.

The curves which give φ_1 show notable differences from those which were found in the case of the series resonant converter. For ω/ω_0 less than unity, the load is only capacitve if ζ is sufficiently small. For ω/ω_0 greater than 1, the load is always inductive. If the same type of commutation is always required, the operation must thus take place at ω/ω_0 greater than 1, and switches with only turn-off control must be used.

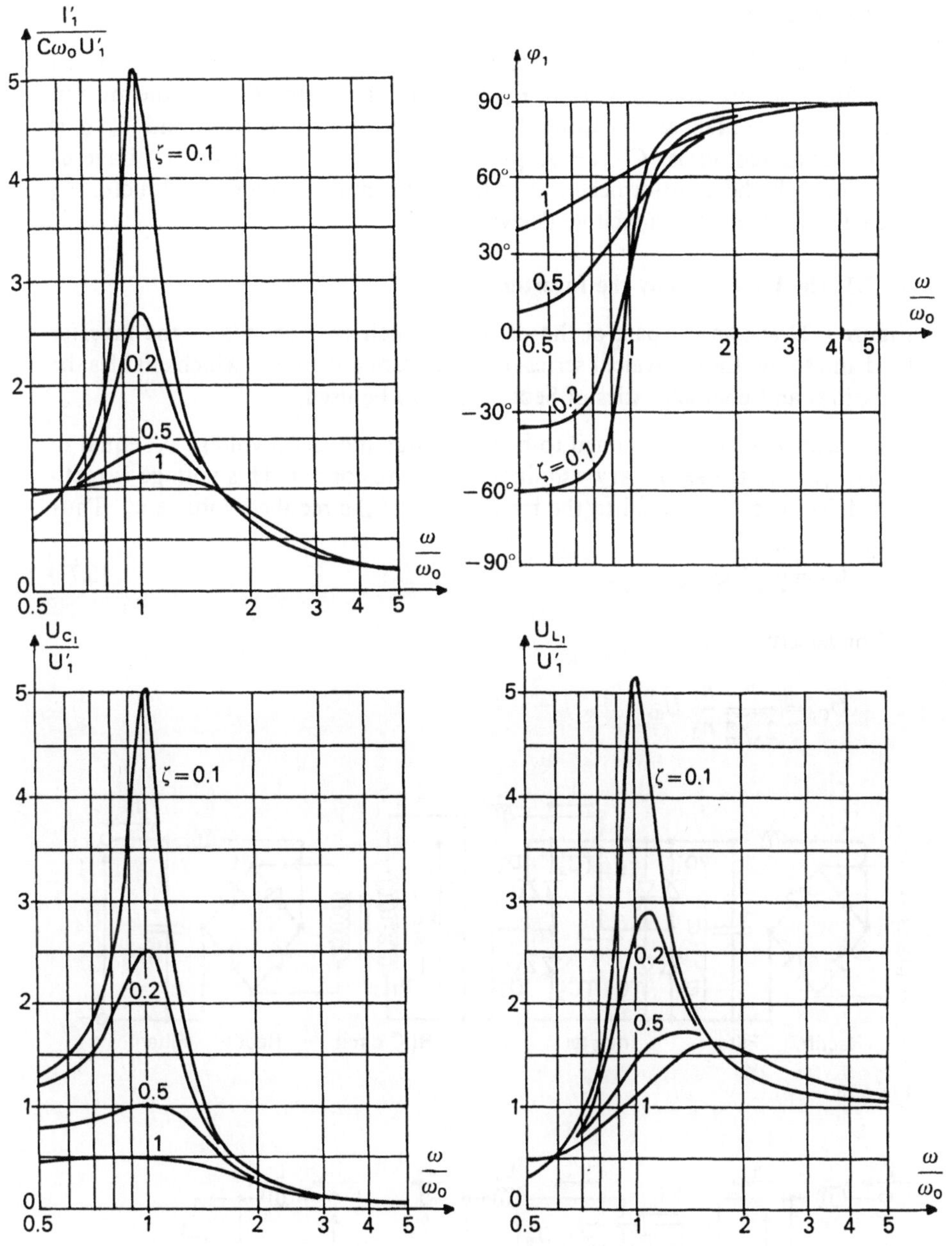
$\frac{I'_1}{C\omega_0 U'_1}$
$\zeta=0.1$
0.2
0.5
1
$\frac{\omega}{\omega_0}$
0.5
1
2
3
4
5
φ_1
90°
60°
30°
0
−30°
−60°
−90°
$\frac{U_{C_1}}{U'_1}$
$\frac{U_{L_1}}{U'_1}$

Fig. 6.27

6.3.2.2 Supply Configuration

Figure 6.28 gives the resonant supply configuration with a series–parallel oscillating circuit. It differs from Fig. 6.22 in the connection of the output transformer–rectifier set: instead of series-connecting it with L and C, it is shunt-connected across C. The output rectifier is now voltage-fed (voltage u_C seen through the transformer); the output filter must include an inductor L_d placed in front of the capacitive derivation.

6.3.2.3 The Load Seen by the Inverter

Figure 6.29 shows the part of the diagram which corresponds to the inverter load and shows the equivalent series–parallel resonant circuit, which enables the equations and characteristics of Sect. 6.3.2.1 to be used.

- *Voltage* u_C, which is assumed to be sinusoidal, induces a voltage v_S, equal to $(n_2/n_1)u_C$, in the secondary winding of the transformer. This voltage is rectified. Voltage U_d is equal to the mean value of the rectified voltage u'_d. Thus

$$U_d = \frac{2}{\pi}\frac{n_2}{n_1} U_{C_1}\sqrt{2}. \qquad (6.77)$$

Conversely,

$$U_{C_1} = \frac{\pi}{2\sqrt{2}}\frac{n_1}{n_2} U_d.$$

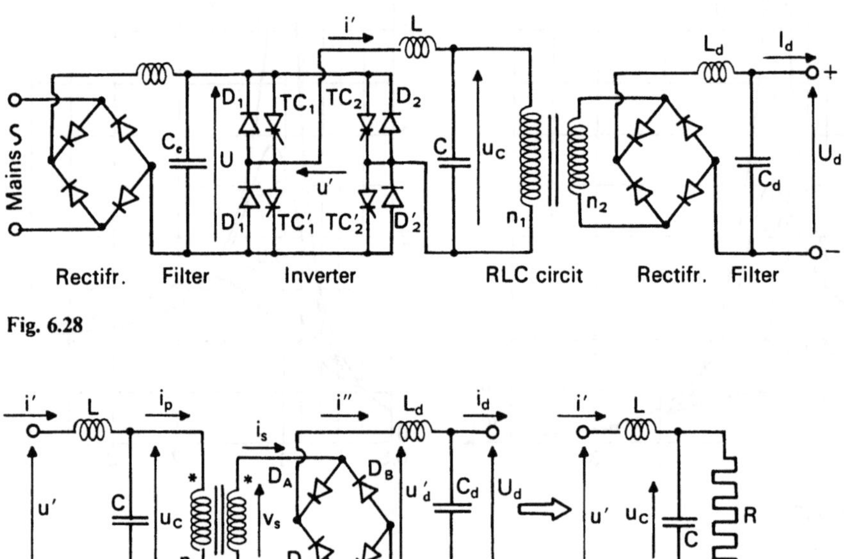

Fig. 6.28

Fig. 6.29

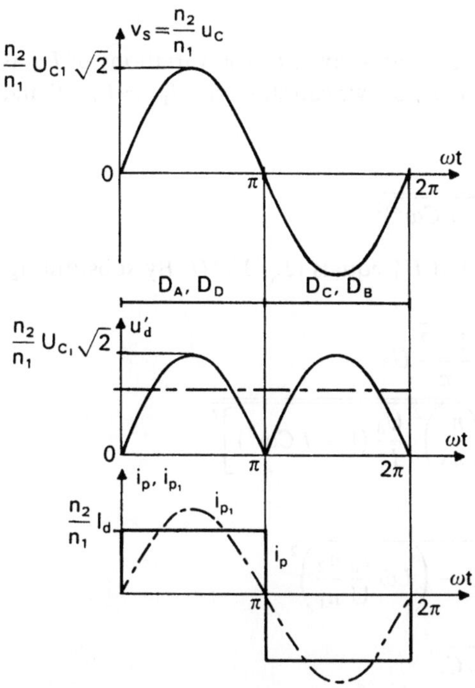

Fig. 6.30

- If inductance L_d is sufficient, the ripple of *current i''* can be ignored; this current remains constantly equal to I_d. The secondary current of the transformer is equal, at times to $+I_d$ and, at others, to $-I_d$. If the magnetizing current is ignored, the primary current i_P is equal to $n_2 i_S/n_1$. Its fundamental has an rms value of:

$$I_{P_1} = \frac{1}{\sqrt{2}} \frac{4}{\pi} \frac{n_2 I_d}{n_1} \tag{6.78}$$

Conversely,

$$I_d = \frac{\pi}{2\sqrt{2}} \frac{n_1}{n_2} I_{P_1}.$$

- Figure 6.30 gives a plot of the rectified voltage u'_d waveform deduced from that of u_C. Diodes D_A and D_D are conducting during the positive half-cycle of u_C, and diodes D_C and D_B during the negative half-cycle. The conduction diagrams enable the waveform of currents i_S and i_P to be plotted.

The fundamental of i_P is in phase with u_C. As far as concerns the fundamental, the assembly which is shunt-connected across C is equivalent to a *resistance* with the following value:

$$R = \frac{U_{C_1}}{I_{P_1}} = \frac{\pi^2}{8} \left(\frac{n_1}{n_2}\right)^2 \frac{U_d}{I_d} \tag{6.79}$$

6.3.2.4 Characteristics

- In order to plot the curves which give, at a given U, the variations of U_d as a function of I_d for different values of ω/ω_0, we can start with $I_{P_1} = U_{C_1}/R$ and use Eq. (6.71).

$$I_{P_1} = \frac{U_{C_1}}{R} = \frac{U_1'}{\sqrt{L^2\omega^2 + R^2(1 - LC\omega^2)^2}}\,.$$

I_{P_1} is given by (6.78), R by (6.79) and U_1' equals $(2\sqrt{2}/\pi)U$. By substituting, this gives

$$\frac{2\sqrt{2}}{\pi}\frac{n_2}{n_1} I_d = \frac{\dfrac{2\sqrt{2}}{\pi} U}{\sqrt{L^2\omega^2 + \left[\dfrac{\pi^2}{8}\left(\dfrac{n_1}{n_2}\right)^2 \dfrac{U_d}{I_d}(1 - LC\omega^2)\right]^2}}\,.$$

This gives

$$U_d = \frac{8}{\pi^2}\frac{1}{|1 - LC\omega^2|} U \frac{n_2}{n_1}\sqrt{1 - \left(L\omega \frac{I_d}{U}\frac{n_2}{n_1}\right)^2}$$

or, by introducing ω/ω_0 and $\sqrt{L/C}$.

$$U_d = \frac{n_2}{n_1} U \frac{8}{\pi^2}\frac{1}{\left|1 - \left(\dfrac{\omega}{\omega_0}\right)^2\right|}\sqrt{1 - \left(\sqrt{\frac{L}{C}}\frac{\omega}{\omega_0}\frac{n_2}{n_1}\frac{I_d}{U}\right)^2}\,. \tag{6.80}$$

This equation has been used to plot (Fig. 6.31) the curves $U_d(I_d)$ for some values of ω/ω_0. U_d has been referred to $(n_2/n_1)U$ and I_d to $U(n_2/n_1)/(n_2/n_1)^2 \sqrt{L/C}$.

- In order to use Eqs. (6.73)–(6.76) as well as the characteristics of Fig. 6.27, the curves corresponding to constant value of ζ have been added on Fig. 6.31.

  Since ζ equals $\dfrac{1}{2R}\sqrt{\dfrac{L}{C}}$, Eq. (6.79) gives

$$\frac{4}{\pi^2}\left(\frac{n_2}{n_1}\right)^2 \frac{I_d}{U_d}\sqrt{\frac{L}{C}} = \zeta$$

or

$$\frac{\sqrt{\dfrac{L}{C}}\dfrac{n_2}{n_1}\dfrac{I_d}{U}}{\dfrac{U_d}{\dfrac{n_2}{n_1}U}} = \frac{\pi^2}{4}\zeta.$$

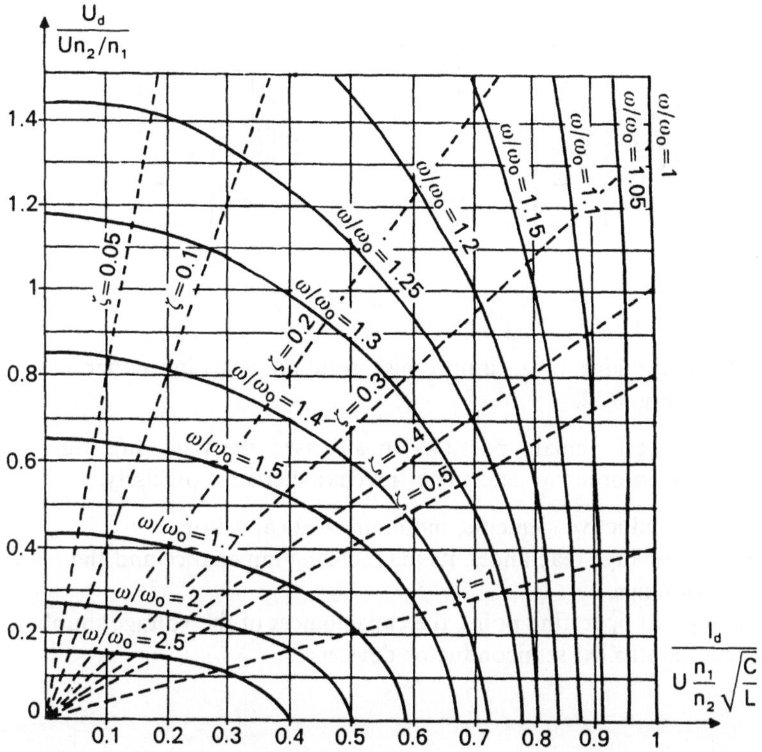

Fig. 6.31

The curves at constant ζ are straight lines with a slope which increases when ζ becomes lower.

- No-load operation is possible. It is possible to change from no-load operation to short-circuit operation. In the latter case, ζ is greater than unity, but the inverter feeds a purely inductive circuit and the semiconductor devices which were intended for such a flow can carry out the commutations.

6.3.3 Double Resonance Supplies

A study of the characteristics $U_d(I_d)$ of Fig. 6.24 and 6.31 shows that:

- the series resonant converter acts as a voltage supply with convenient regulation by means of ω/ω_0 in the case of high relative values of I_d;
- the series–parallel resonant converter acts in a similar way for the low relative values of I_d.

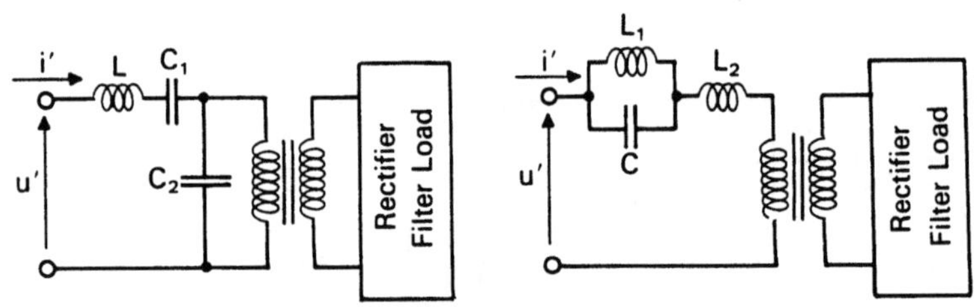

Fig. 6.32

This leads to the idea of a compromise solution by using either of the configurations in Fig. 6.32.

Other combinations remain possible. In all cases, the following must be taken into account in order to determine the characteristics precisely:

- the losses in the inductive elements, inductors and capacitors,
- the transformer leakage reactances, its magnetizing impedance and the resistances of its windings,
- around the no-load operating point, the capacitances of the voltage snubbers of the inverter controlled semiconductor devices.

Bibliography

- The first publications dealing with *resonant inverters* [1–16] provide a classical study of series or parallel or series–parallel inverters. Those following examine control techniques, possibilities offered by new semiconductor devices and series or parallel assemblies of elementary inverters.
- *Induction heating* [17–25] is the traditional application of resonant inverters. The publications mainly deal with the starting of oscillations as well as with control and protection procedures. The development of equipment is directly linked to that of the inverters themselves.
- *The DC–DC resonant converters* [26–59] are resonant inverters, feeding – via a transformer – a diode rectifier which supplies the output DC voltage. The inverter can be supplied by a battery or by a diode rectifier connected to the mains; we have presented these converters – under the title of *resonant supplies* – in the latter way.

  There are numerous recent studies of resonant converters. The various elements of the assembly have been the subject of increasingly detailed studies into their operation; this has improved the determination of the characteristics. A great many articles deal with control.

  The various possible configurations are dealt with and lead to comparisons which enable the configuration and control mode best adapted to each area of use to be determined.
- DC–DC resonant converters form the most basic cascades converters which make use of resonant inverters. Under the heading of "*converters with high-frequency link*", we have brought together the articles dealing with the cascades made up of a resonant inverter and one or more other elementary converters with controlled semiconductor devices.

  The symmetrical diagram of a reversible DC–DC converter can be obtained with a resonant inverter and a fully-controlled rectifier.

  With a resonant inverter followed by a rectifier followed by another inverter, a DC–AC converter can be obtained; the second inverter often operates in natural commutation if it feeds the mains. The same DC–AC transformation can be achieved without using the DC intermediary: each half-cycle of the output voltage must be made up of a suitable number of high-frequency half-cycles supplied by the resonant inverter.

To change from one AC system to another, two controlled switch full-bridge converters connected by an oscillating circuit can be used. The latter acts as an intermediate energy tank.

Resonant Inverter Analysis

1. Revankar GN, Gadag SA (1973) Analysis of high frequency inverter circuit, *IEEE Trans. Ind. Electron. Control Instrum.*, **20**(3), 178–182.
2. Revankar GN, Gadag SA (1974) A high frequency bridge inverter with series parallel compensated load, *IEEE Trans. Ind. Electron. Control. Instrum.*, **21**(1), 18–21.
3. Revankar GN, Karande DS (1974) Voltage-fed high-frequency bridge inverter circuit, *IEEE Trans. Ind. Electron. Control Instrum.*, **21**(4), 226–229.
4. Kasturi R (1975) An analysis of series inverter circuits, *IEEE Trans. Ind. Electron. Control Instrum.*, **22**(4), 515–519.
5. Ludu ND (1978) Basic operating characteristics of a high-frequency inverter with capacitive voltage multiplier, *IEEE Trans. Ind. Appl.*, **14**(3), 264–269.
6. Roda MR, Revankar GN (1978) Voltage-fed discontinuous current mode high frequency inverter for induction heating, *IEEE Trans. Ind. Electron. Control Instrum.*, **25**(3), 226–232.
7. Kassakian JG (1982) A new current mode sine wave inverter, *IEEE Trans. Ind. Appl.*, **18**(3), 273–278.
8. Slonim MA, Biringer PP (1982) Analysis of the transient and steady-state processes in the parallel inverter, *IEEE Trans, Ind. Electron. Control Instrum.*, **29**(4), 329–336.
9. Bredenkamp GL, Van Rhyn P (1985) The design of welder power supplies incorporating self-oscillating high-frequency transformer inverters, *Proc. Inst. Electr. Eng., Part B*, **132**(4), 202–208.
10. Oruganti R, Lee FC (1985) Resonant power processors. Part 1 State-plane analysis, *IEEE Trans. Ind. Appl.*, **21**(6), 1453–1460. Part 2 Methods of control, *ibid.*, 1461–1471.
11. Savary P, Nakaoka M, Maruhashi T (1987) High-frequency resonant inverter using current-vector control scheme and its performance evaluation, *IEEE Trans. Ind. Electron.*, **34**(2), 247–256.
12. Jain PK, Dewan SB (1989) Analysis of a clamped series inverter with a transformer coupled load, *IEEE Trans. Ind. Appl.*, **25**(1), 153–161.
13. Bhat AKS, Dewan SB (1989) A generalized approach for the steady-state analysis of resonant inverters, *IEEE Trans. Ind. Appl.*, **25**(2), 326–338.
14. Tymerski R, Vorperian V, Lee FCY (1989) DC-to-AC inversion using quasi-resonant techniques, *IEEE Trans. Power Electron.*, **4**(4), 381–390.
15. Ray B, Stuart TA (1989) An analysis of N Schwartz inverters driving a common AC bus, *IEEE Trans. Power Electron.*, **4**(4), 470–477.
16. Ray B, Stuart TA (1989) A cascaded Schwartz converter for high frequency power distribution, *IEEE Trans. Power Electron.*, **4**(4), 478–485.

Induction Heating

17. Pelly BR (1970) Latest developments in static high frequency power sources for induction heating, *IEEE Trans. Ind. Electron. Control Instrum.*, **47**(4), 297–312.
18. Landis JP (1970) A static power supply for induction heating, *IEEE Trans. Ind. Electron.*, **17**(4), 313–320.
19. de Miramon B (1976) Application des onduleurs à moyenne fréquence au chauffage et à la fusion par induction, *Rev. Gén. Electr.*, **85**(2), 165–174.
20. Zao KB, Sen PC, Premchandran G (1984) A thyristor inverter for medium-frequency induction heating, *IEEE Trans. Ind. Electron.*, **31**(1), 33–36.
21. Ferrieux JP, Keradec JP, Baudon Y (1988) A series-resonant inverter for induction heating, *Rev. Gén. Electr.*, **97**(2), 54–62.
22. Akagi H, Sawae T, Nabae A (1988) 130 KHz, 7.5 KW current source inverters using Static Induction Transistors for induction heating applications, *IEEE Trans. Power Electron.*, **3**(3), 303–309.

23. Jain P, Dewan SB (1988) Starting problems associated with a transformer coupled load in a series inverter, *IEEE Trans. Magn.*, **24**(6), 2895–2897.
24. Peng FZ, Akagi H, Nabae A, Sugarawa S (1989) High-frequency current-source inverters using SI Thyristors for induction heating applications, *IEEE Trans. Ind. Appl.*, **25**(1), 172–180.
25. Jain P, Dewan SB (1989) A starting inverter for a voltage-source series inverter with a transformer-coupled high-Q induction heating load (type A inverter), *IEEE Trans. Ind. Appl.*, **25**(4), 627–633.

DC–DC Resonant Converters

26. Schwarz FC (1970) A method for resonant current pulse modulation for power converters, *IEEE Trans. Ind. Electron Control Instrum.*, **17**(3), 209–221.
27. Schwarz FC (1976) An improved method of resonant current pulse modulation for power converters, *IEEE Trans. Ind. Electron. Control Instrum.*, **23**(2), 133–141.
28. Schwarz FC, Klaassens JB (1976) A controllable secondary multikilowatt DC current source with constant maximum power factor in its three-phase supply line, *IEEE Trans. Ind. Electron. Control Instrum.*, **23**(2), 142–150.
29. Schwarz FC, Klaassens JB (1979) A controllable 45 KW current source for DC machines, *IEEE Trans. Ind. Appl.*, **15**(4), 437–444.
30. King RJ, Stuart TA (1981) A normalized model for the half-bridge series resonant converter, *IEEE Trans. Aerosp. Electron. Syst.*, **17**(2), 190–198.
31. King RJ, Stuart TA (1982) Modeling the full-bridge series–resonant power converter, *IEEE Trans. Aerosp. Electron. Syst.*, **18**(4), 449–459.
32. King RJ, Stuart TA (1983) Transformer induced instability of the series resonant converter, *IEEE Trans. Aerosp. Electron. Syst.*, **19**(4), 474–482.
33. King RJ, Stuart TA (1983) Inherent overload protection for the series resonant converter, *IEEE Trans. Aerosp. Electron. Syst.*, **19**(6), 820–830.
34. Steigerwald RL (1984) High-frequency resonant transistor DC–DC converters, *IEEE Trans. Ind. Electron.*, **31**(2), 181–191.
35. De Haan SWH (1984) A new integral pulse module for the series resonant converter, *IEEE Trans. Ind. Electron.*, **31**(3), 255–262.
36. De Haan SWH, Huisman H (1985) Novel operation and control modes for series-resonant converters, *IEEE Trans. Ind. Electron.*, **32**(2), 150–157.
37. Kislovski AS (1986) A contribution to steady-state modeling of half-bridge series-resonant power cells, *IEEE Trans. Power Electron.*, **1**(3), 161–166.
38. Tilgenkamp NV, De Haan SWH, Huisman H (1987) A novel series-resonant converter topology, *IEEE Trans. Ind. Electron.*, **34**(2), 240–246.
39. Bhat AKS, Dewan SB (1987) Analysis and design of a high-frequency resonant converter using LCC-type commutation, *IEEE Trans. Power Electron.*, **2**(4), 291–301.
40. NGO KDT (1988) Analysis of a series resonant converter pulse-width modulated or current-controlled for low switching loss, *IEEE Trans. Power Electron.*, **3**(1), 55–63.
41. Johnson SD, Erickson RW (1988) Steady-state analysis and design of the parallel resonant converter, *IEEE Trans. Power Electron.*, **3**(1), 93–104.
42. Hennevin B, Volut S, Dupaquier A (1988) For high efficiency DC/DC power converter the double resonant circuit, *Rev. Gén. Electr.* **97**(2), 48–63.
43. Lee YS, Cheng YC (1988) Calculation of resonance current in high-frequency series-resonant DC–DC converters, *IEEE Trans. Aerosp. Electron. Syst.*, **24**(2), 124–132.
44. Klaassens JB, Moize de Chateleux WL, Van Wesenbeeck MP (1988) Phase-staggering control of a series-resonant DC–DC converter with paralleled power modules, *IEEE Trans. Power Electron.*, **3**(2), 164–173.
45. Steigerwald RT (1988) A comparison of high-bridge resonant converter topologies, *IEEE Trans. Power Electron.*, **3**(2), 174–182.
46. Deb S, Joshi A, Doradla SR (1988) A novel frequency-domain model for a parallel-resonant converter, *IEEE Trans. Power Electron.*, **3**(2), 208–215.
47. Johnson SD, Witulski AF, Erickson RW (1988) Comparison of resonant topologies in high-voltage DC applications, *IEEE Trans. Aerosp. Electron. Syst.*, **24**(3), 263–273.
48. Kang YG, Upadhyay AK (1988) Analysis and design of a half-bridge parallel resonant converter, *IEEE Trans. Power Electron.*, **3**(3), 254–265.

49. Oruganti R, Yang JJ, Lee FC (1988) Implementation of optimal trajectory control of series resonant converter, *IEEE Trans. Power Electron.*, **3**(3), 318–327.
50. Elbuluk ME, Verghese GC, Kassakian JG (1988) Sampled data modeling and digital control of resonant converters, *IEEE Trans. Power Electron.*, **3**(3), 344–354.
51. Tsai FS, Chin Y, FCY Lee (1988) State-plane analysis of a constant clamped-mode parallel-resonant converter, *IEEE Trans. Power Electron.*, **3**(3), 364–378.
52. Vandelac JP, Ziogas PD (1988) A DC to DC PWM series resonant converter operated at resonant frequency, *IEEE Trans. Ind. Electron.*, **35**(3), 451–460.
53. Gu WJ, Harada K (1988) A new method to regulate resonant converters, *IEEE Trans Power Electron.*, **3**(4), 430–439.
54. Tsai FS, Materu P, Lee FCY (1988) Constant-frequency clamped resonant converter, *IEEE Trans. Power Electron.*, **3**(4), 460–473.
55. Bhat AKS (1988) Analysis and design of a DC–DC converter using square-wave output resonant inverter, *IEEE Trans. Ind. Electron.*, **35**(4), 494–501.
56. Vorperian V (1989) Approximate small-signal analysis of the series and the parallel resonant converters, *IEEE Trans. Power Electron.*, **4**(1), 15–24.
57. Joung GB, Rim CT, Cho GH (1989) Integral cycle mode control of the series resonant converter, *IEEE Trans. Power Electron.*, **4**(1), 83–91.
58. Bhat AKS, Swamy MM (1989) Loss calculations in transistorized parallel resonant converters operating above resonance, *IEEE Trans. Power Electron.*, **4**(4), 391–401.
59 Batarseh I, Lee CQ (1989) High-frequency high-order parallel resonant converter, *IEEE Trans. Ind. Electron.*, **36**(4), 485–498.

Converters with High Frequency Link

60. Gyugyi L, and Cibulka F (1979) The high-frequency base converter – A new approach to static high-power conversion, *IEEE Trans. Ind. Appl.*, **15**(4), 420–429.
61. Schwarz FC, Moise de Chateleux WL (1981) A multi-kilowatt polyphase AC/DC converter with reversible power flow and without passive low frequency filters, *IEEE Trans. Ind. Control Instrum.*, **28**(4), 273–281.
62. Ziogas PD, Ranganathan VT, Stefanovic VR (1982) A four-quadrant current regulated converter with a high frequency link, *IEEF Trans. Ind. Appl.*, **18**(5), 499–505.
63. Steigerwald RL, Ferraro A, Turnbull FG (1983): Application of power transistors to residential and intermediate rating photovoltaic array power conditioners, *IEEE Trans. Ind. Appl.*, **19**(2), 254–267.
64. Ranganathan VT, Ziogas PD, Stefanovic VR (1983) A DC/AC power conversion technique using twin resonant HF links, *IEEE Trans. Ind. Appl.*, **19**(3), 393–400.
65. Huisman H, De Haan SWH (1985) A DC to 3-phase series-resonant converter with low harmonic distortion, *IEEE Trans. Ind. Electron.*, **32**(2), 1426149.
66. Klaassens JB (1986) DC-to-AC series-resonant converter system with high internal frequency generating synthetized waveforms for multikilowatt power levels, *IEEE Trans. Power Electron.*, **1**(1), 9–20.
67. Savary P, Nakaoka M, Maruhashi T (1986) Novel type of high-frequency link inverter photovoltaic residential applications, *Proc. Inst. Electr. Eng., Part B*, **133**(4), 279–284.
68. Pitel IJ (1986) Phase-modulated power conversion techniques for high-frequency link inverters, *IEEE Trans. Ind. Appl.*, **22**(6), 1044–1051.
69. Manias S, Ziogas PD, Olivier G (1987) Bilateral DC to DC converter using a high frequency link, *Proc. Inst. Electr. Eng., Part B*, **134**(1), 15–23.
70. Tsai FS, Oruganti R, Lee FC (1987) A novel control for bidirectional power flow of a parallel resonant converter, *IEEE Trans. Ind. Appl.* **23**(5), 937–943.
71. Klaassens JB (1987) DC–AC series-resonant system with high internal frequency generating multiphase AC waveforms for multikilowatt power levels, *IEEE Trans. Power Electron.*, **2**(3), 247–256.
72. Huisman H (1988) A three-phase to three-phase series-resonant power converter with optimal input current waveforms. Part 1: Control strategy, *IEEE Trans. Ind. Electron.*, **35**(2), 263–268. Part 2: Application and results, *ibid.*, 269–277.
73. Bhat AKS, Dewan SB (1988) DC-to-utility interface using sinewave resonant inverter, *Proc. Inst. Electr. Eng., Part B*, **135**(5), 193–201.

74. Lauw HK, Klaassens JB, Butler NG, Seely DB (1988) Variable-speed generation with the series-resonant converter, *IEEE Trans. Energy Conv.*, **3**(4), 754–765.
75. Klaassens JB (1989) Steady-state analysis of a series-resonant DC–DC converter with a bipolar flow, *IEEE Trans. Ind. Electron.*, **36**(1), 48–55.
76. Divan DM (1989) The resonant DC link converter. A new concept in static power conversion, *IEEE Trans. Ind. Appl.*, **25**(2), 317–325.
77. Bhat AKS, Dewan SB (1989) Input-output harmonics of line-current-modulated high-frequency link DC to utility interface power converters, *IEEE Trans. Power Electron.*, **4**(2), 253–264.
78. Divan DM, Skibinski G (1989) Zero-switching-loss inverters for high-power applications, *IEEE Trans. Ind. Appl.*, **25**(4), 634–643.

Appendix A: Input Filter in Voltage-Source Inverters

Since the study of voltage-source inverters with one pulse per alternance performed in Chap. 2 was already sufficiently long, we confined ourselves to the frequential study of the input filter. It was assumed that the ripple of the values to be filtered could be reduced to the peak-to-peak amplitude of their first harmonic.

The analytical study of the LC filter was carried out for choppers in Vol. 3 (Chap. 4, Sect. 2). A comparison with the results given by the first harmonic method showed considerable differences concerning, in particular, the voltage ripple. The chopper input current is composed of square-waves, whereas the voltage-source inverter input current is formed of sinusoidal arcs. It is thus to be expected that, in the latter case, the first harmonic approximation should lead to even larger differences.

Figure A.1 shows the filter, comprising a series inductor L and a shunt capacitor C, connected between the DC voltage supply U_S and the inverter. The latter imposes the filter output current i.

The purpose of the filter is to reduce the ripple Δi_S of the current i_S delivered by the supply, and the ripple Δu of the voltage u at the inverter input[1].

- In single-phase inverters, the cycle T_i of current i is equal to half the cycle T of the inverter output current.

  In three-phase inverters feeding a balanced load, the cycle T_i of the input current equals a sixth of the cycle T of the output currents.

  $$\frac{T_i}{T} = \frac{1}{2}, \quad \text{in single-phase inverters,}$$

  $$\frac{T_i}{T} = \frac{1}{6}, \quad \text{in three-phase inverters.}$$

- If the inverter has two "switches" per phase (single-phase half-bridge inverter or with centre-tapped transformer, three-phase full-bridge inverter), current

[1] For this study, we have used the calculations carried out by T. Fricheteau and X. Materne.

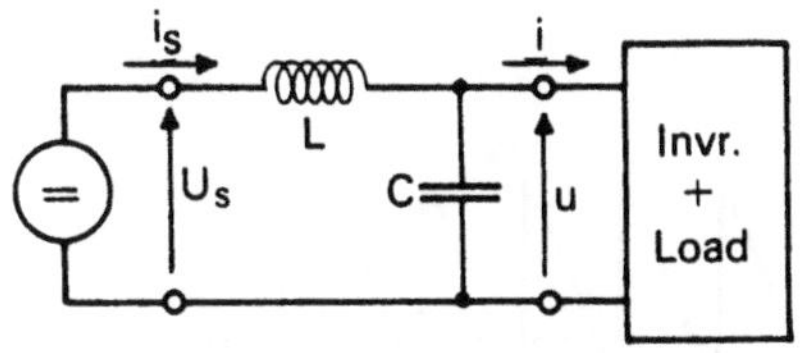

Fig. A.1

i is composed of a sinusoidal arc of amplitude I_M and of angular frequency ω during each of its cycles:

for $0 < t < T_i$, $i = I_M \sin(\omega t - \varphi_i)$

with $\omega = 2\pi/T$.

The value of I_M can be directly deduced from the amplitude I'_M of the current at the inverter output. For the push-pull inverter, for example, $I_M = (2n_2/n_1)I'_M$.

If the inverter has four "switches" per phase (single-phase full-bridge inverter, for example), during each of its cycles, current i is made of a sinusoidal arc of amplitude I_M, of angular frequency ω, followed by a zero current interval.

In both cases, the filter can be characterized by *two parameters*, k_f and k_i.

The first is defined as the ratio of the resonant angular frequency of the filter itself ω_f equal to $1/\sqrt{LC}$, to the angular frequency ω of the output current of the inverter.

The second is identified as the ratio of the characteristic impedance of the filter $\sqrt{L/C}$ to the rated impedance U_S/I_M.

$$\boxed{k_f = \frac{\omega_f}{\omega} = \frac{1}{\omega\sqrt{LC}}; \quad k_i = \sqrt{\frac{L}{C}}\frac{I_M}{U_S}} \tag{A.1}$$

A.1 Inverters with Two "Switches" per Phase

Figure A.2 gives the waveform of current i at the filter ouput in the case of a two-switch single-phase inverter and then in the case of the full-bridge three-phase inverter.

For single-phase inverters,

for $0 < \omega t < \pi$, $i = I_M \sin(\omega t - \varphi_i)$,

with $\varphi_i = \varphi$

For three-phase inverters,

for $0 < \omega t < \pi/3$, $i = I_M \sin(\omega t - \varphi_i)$,

with $\varphi_i = \varphi - \pi/3$,

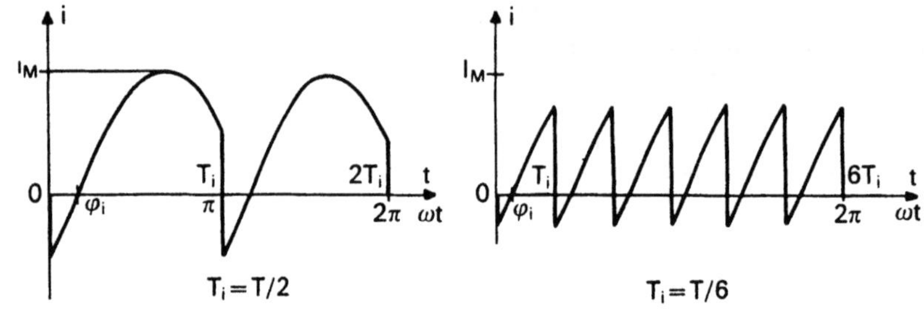

Fig. A.2

using φ to denote the current phase shift related to the fundamental of the inverter output voltage.

A.1.1 Expression of i_S and u

- Current i_S and voltage u are linked to current i and to voltage U_S by

$$i_S - C\frac{du}{dt} = i$$

$$L\frac{di_S}{dt} + u = U_S$$

where $i = I_M \sin(\omega t - \varphi_i)$ and $U_S = \text{constant}$.

From this the differential equation giving i_S can be deduced:

$$\frac{d^2 i_S}{dt^2} + \frac{i_S}{LC} = \frac{I_M}{LC}\sin(\omega t - \varphi_i)$$

Giving:

$$i_S = A\cos\omega_f t + B\sin\omega_f t + I_M \frac{k_f^2}{k_f^2 - 1}\sin(\omega t - \varphi_i)$$

and, since $u = U_S - L\,di_S/dt$

$$u = U_S - L\omega_f\left[-A\sin\omega_f t + B\cos\omega_f t + I_M\frac{k_f}{k_f^2 - 1}\cos(\omega t - \varphi_i)\right]$$

- In order to find the integration constants, it should be noted that neither i_S nor u can be subject to any discontinuity and that their cycle is equal to T_i.

$i_S(0) = i_S(T_i)$ gives:

$$A = B\frac{\sin\omega_f T_i}{1 - \cos\omega_f T_i} + I_M\frac{k_f^2}{k_f^2 - 1}\cdot\frac{\sin(\omega T_i - \varphi_i) + \sin\varphi_i}{1 - \cos\omega_f T_i}$$

$u(0) = u(T_i)$ gives:

$$A = B\frac{\cos\omega_f T_i - 1.}{\sin\omega_f T_i} + I_M\frac{k_f}{k_f^2 - 1}\cdot\frac{\cos(\omega T_i - \varphi_i) - \cos\varphi_i}{\sin\omega_f T_i}$$

B can be found by equalling the two expressions of A. By transferring B thus obtained to one or the other of the previous equations, this gives A:

$$B = I_M\frac{k_f}{2(k_f^2 - 1)}$$

$$\times\left\{\cos(\omega T_i - \varphi_i) - \cos\varphi_i - \frac{k_f}{\tan\omega_f\frac{T_i}{2}}[\sin(\omega T_i - \varphi_i) + \sin\varphi_i]\right\}$$

$$A = I_M\frac{k_f}{2(k_f^2 - 1)}$$

$$\times\left\{\frac{1}{\tan\omega_f\frac{T_i}{2}}[\cos(\omega T_i - \varphi_i) - \cos\varphi_i] + k_f[\sin(\omega T_i - \varphi_i) + \sin\varphi_i]\right\}$$

– Finally, i_S and u will be written in the following way:

$$i_S = I_M\frac{k_f}{2(k_f^2 - 1)}[A'\cos k_f\omega t + B'\sin k_f\omega t + 2k_f\sin(\omega t - \varphi_i)] \quad \text{(A.2)}$$

$$u = U_S\left\{1 - \frac{k_i k_f}{2(k_f^2 - 1)}[-A'\sin k_f\omega t + B'\cos k_f\omega t + 2\cos(\omega t - \varphi_i)]\right\} \quad \text{(A.3)}$$

with

$$A' = \frac{1}{\tan k_f\omega\frac{T_i}{2}}[\cos(\omega T_i - \varphi_i) - \cos\varphi_i]$$

$$+ k_f[\sin(\omega T_i - \varphi_i) + \sin\varphi_i]$$

$$B' = [\cos(\omega T_i - \varphi_i) - \cos\varphi_i] - \frac{k_f}{\tan k_f\omega\frac{T_i}{2}}[\sin(\omega T_i - \varphi_i) + \sin\varphi_i] \quad \text{(A.4)}$$

A.1.2 Waveforms

Equations (A.2), (A.3) and (A.4) show that, at a given ωT_i,

- the waveform of i_S/I_M only depends on k_f and φ_i;
- the waveform of u/U_S also depends on k_i; the variations of u on either side of its mean value U are proportional to I_M.

– Neither current i_S (on account of L) nor voltage u (on account of C) can be subject to discontinuities.

Since i_S is equal to $i - C\mathrm{d}u/\mathrm{d}t$, when $\mathrm{d}u/\mathrm{d}t$ is zero, current i_S is equal to i. The derivative of current i_S, being equal to $(U_S - u)/L$, cannot be subject to any discontinuity: the current i_S wave shows no angular point.

Voltage u is given by the difference $U_S - L(\mathrm{d}i_S/\mathrm{d}t)$: when $\mathrm{d}i_S/\mathrm{d}t$ is zero, u is equal to U_S. The derivative of voltage u is equal to $(i_S - i)/C$; if current i shows any discontinuity, the waveform of u shows an angular point; this is the case for all values of φ, except for zero φ.

– *For single-phase inverters,*

$$T_i = \frac{T}{2}, \qquad \omega T_i = \pi, \qquad \varphi = \varphi_i$$

The mean value I_S of current i_S is equal to that of current i:

$$I_S = \frac{2}{\pi} I_M \cos\varphi.$$

Figure A.3 gives the waveforms, for one of their cycles T_i, of voltage u and current i_S, for the case where $k_f = 0.5$ and $k_i = 1$, successively:

- for $\varphi = 0$
- for $\varphi = \pi/4$
- for $\varphi = \pi/2$.

– *For the three-phase inverter,*

$$T_i = \frac{T}{6}, \qquad \omega T_i = \frac{\pi}{3}, \qquad \varphi = \varphi_i + \frac{\pi}{3}.$$

The mean value I_S of current i_S is equal to

$$I_S = \frac{3}{\pi} I_M \cos\varphi$$

Figure A.4 shows the waveforms of voltage u and current i_S, calculated for $k_f = 2.5$ and $k_i = 1$, with $\varphi = 0$, $\varphi = \pi/4$ and $\varphi = \pi/2$, respectively.

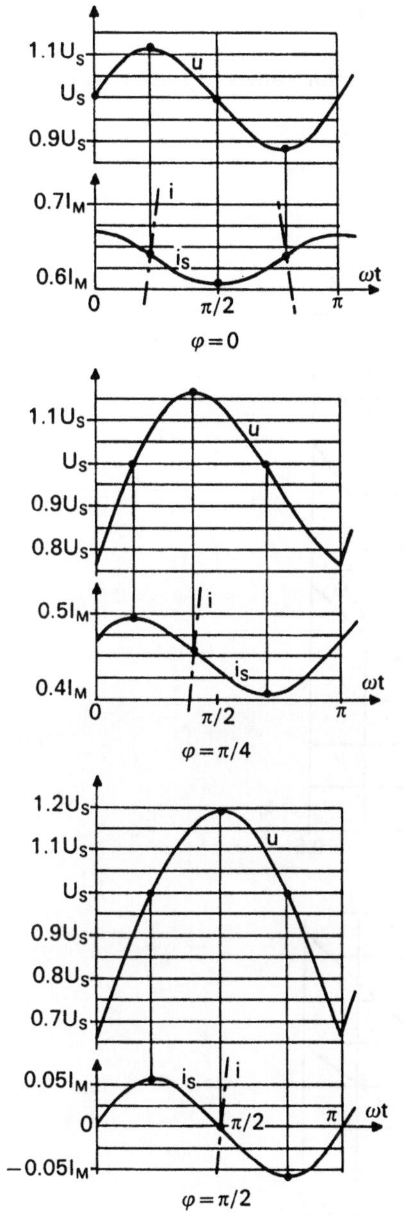

Fig. A.3 ($T_i = T/2$)

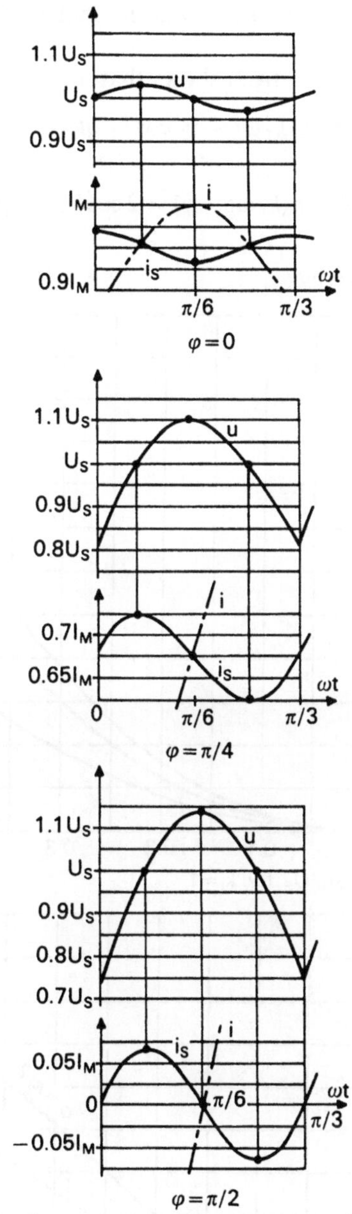

Fig. A.4 ($T_i = T/6$)

A.1.3 Ripple of i_S and u. Choice of L and C

- Equation (A.2) shows that di_S/dt falls to zero for values of t such that:

$$A' \sin k_f \omega t - B' \cos k_f \omega t = 2 \cos(\omega t - \varphi_i)$$

or

$$C' \sin(k_f \omega t - \gamma) = 2 \cos(\omega t - \varphi_i)$$

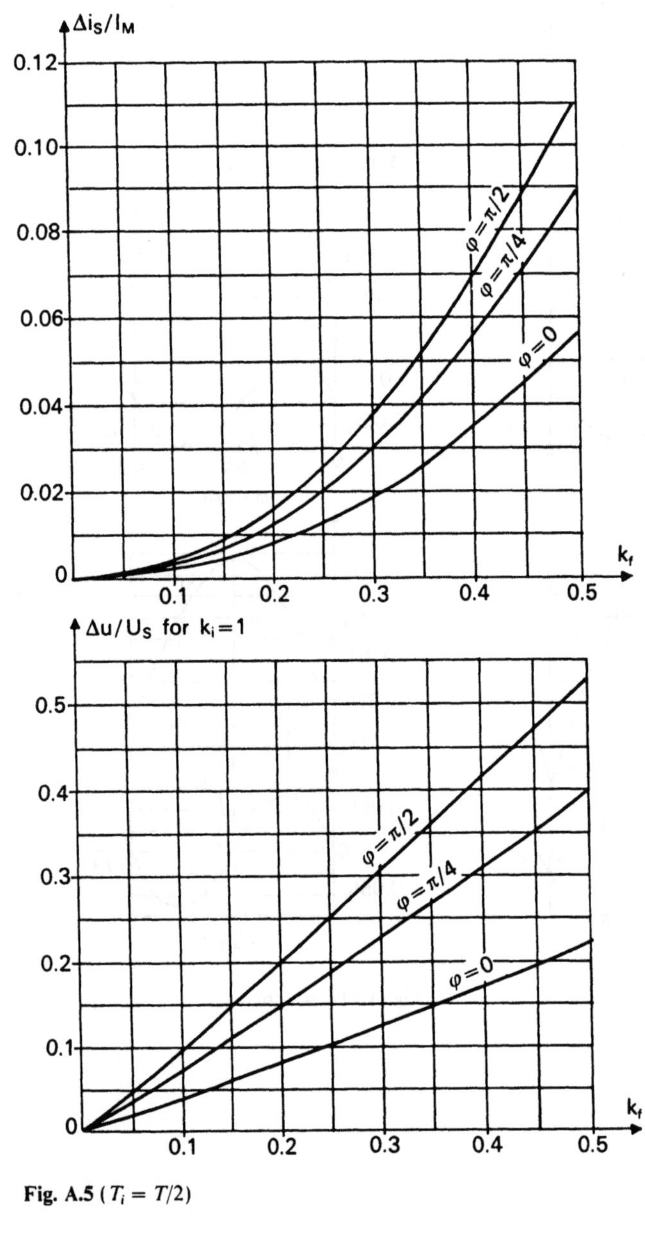

Fig. A.5 ($T_i = T/2$)

with

$$C' = A'\sqrt{1 + (B'/A')^2} \quad \text{and} \quad \gamma = \text{arc tan}\frac{B'}{A'}.$$

For φ taken between 0 and $\pi/2$, the smaller value of t solution of the above equation gives the instant at which current i_S is at its maximum, the greater one gives the instant when i_S is at its minimum. The normalised ripple of i_S

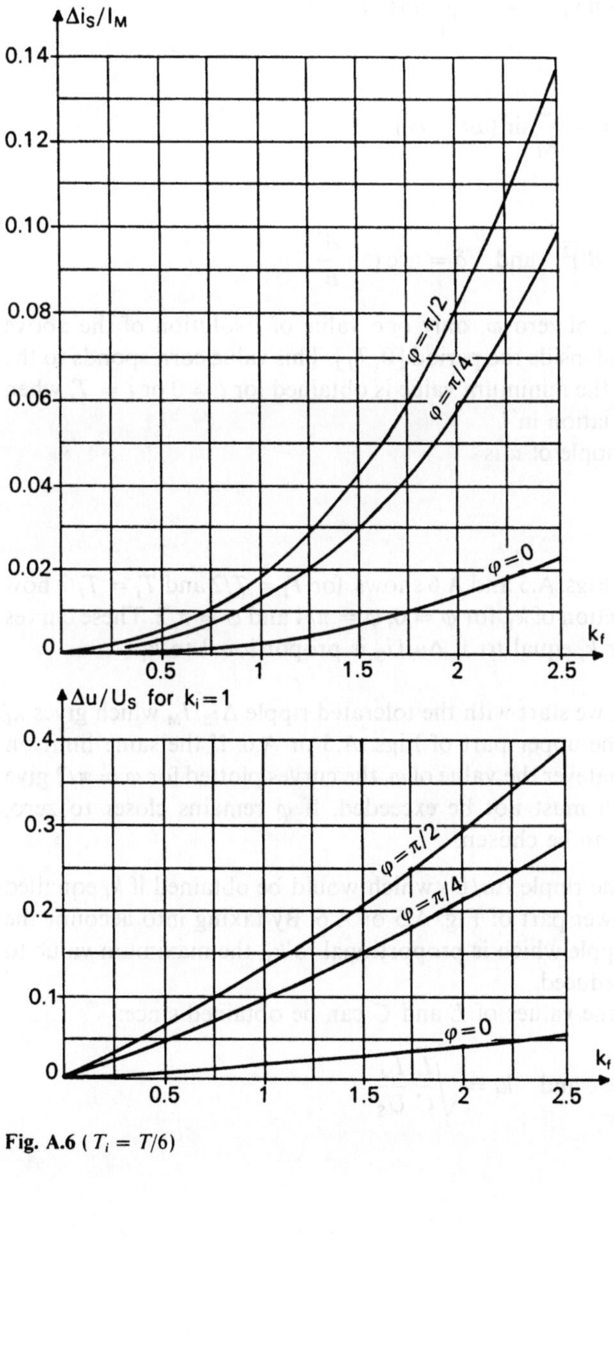

Fig. A.6 ($T_i = T/6$)

can be deduced from this:

$$\frac{\Delta i_S}{I_M} = \frac{i_{S\,max} - i_{S\,min}}{I_M}$$

The upper part of Figs. A.5 and A.6 shows, for $T_i = T/2$ and $T_i = T/6$, the variations of $\Delta i_S/I_M$ as a function of k_f for $\varphi = 0$, $\varphi = \pi/4$ and $\varphi = \pi/2$.

- Equation (A.3) indicates that the derivative of u falls to zero for t such that:

$$A' \cos k_f \omega t + B' \sin k_f \omega t = -\frac{2}{k_f} \sin(\omega t - \varphi_i)$$

or

$$C' \sin(k_f \omega t + \delta) = -\frac{2}{k_f} \sin(\omega t - \varphi_i)$$

with

$$C' = B'\sqrt{1 + (A'/B')^2} \quad \text{and} \quad \delta = \arctan\frac{A'}{B'}.$$

Except in the case of zero φ, only one value of t solution of the above equation is to be found inside the period $[0, T_i]$. This value corresponds to the maximum value of u; the minimum value is obtained for $t = 0$ or $t = T_i$ when there is a sudden variation in i.

The normalised ripple of u is

$$\frac{\Delta u}{U_S} = \frac{u_{max} - u_{min}}{U_S}$$

The lower part of Figs. A.5 and A.6 shows, for $T_i = T/2$ and $T_i = T/6$, how $\Delta u/U_S$ varies as a function of k_f, for $\varphi = 0$, $\varphi = \pi/4$ and $\varphi = \pi/2$. These curves have been plotted for k_i equal to 1. $\Delta u/U_S$ is proportional to k_i.

- In choosing L and C, we start with the tolerated ripple $\Delta i_S/I_M$ which gives k_f using the curves in the upper part of Figs. A.5 or A.6. If the same limit on $\Delta i_S/I_M$ is required, whatever the value of φ, the curves plotted for $\varphi = \pi/2$ give the value of k_f which must not be exceeded. If φ remains closer to zero, a higher value of k_f can be chosen.

Having chosen k_f, the ripple $\Delta u/U_S$ which would be obtained if k_i equalled 1 can be read on the lower part of Fig. A.5 or A.6. By taking into account the limit imposed on this ripple which is proportional to k_i, the maximum value to be given to k_i can be deduced.

From k_f and k_i, the values of L and C can be obtained since:

$$k_f = \frac{\omega_f}{\omega} = \frac{1}{\omega\sqrt{LC}} \quad \text{and} \quad k_i = \sqrt{\frac{L}{C}}\frac{I_M}{U_S},$$

Thus:

$$C = \frac{I_M}{U_S \omega k_f k_i}; \quad L = \frac{k_i}{k_f} \frac{U_S}{I_M \omega} \tag{A.5}$$

- *Remarks*

- Ripples $\Delta i_S/I_M$ and $\Delta u/U_S$ are the same for φ, for $-\varphi$, for $\pi + \varphi$ and for $\pi - \varphi$.
- A comparison between the results obtained and those calculated by the first harmonic method shows that the latter gives satisfactory results for $\Delta i/I_M$ as soon as k_f is small enough, but that it considerably underestimates the value of $\Delta u/U_S$.

  For example, for $T_i = T/2$, $k_f = 0.5$, $\varphi = 90°$, the first harmonic method gives

  $\Delta u/U_S = 0.45\,k_i$, instead of $0.53\,k_i$.

  For $T_i = T/6$, $k_f = 2$, $\varphi = 90°$, it gives

  $\Delta u/U_S = 0.245\,k_i$, instead of $0.295\,k_i$

A.2 Full-Bridge Single-Phase Inverter

With the full-bridge single-phase inverter, if a shift is introduced between the controls of the two half-bridges, periods at zero current i can be made to appear. The current at the inverter input (i.e. at the filter output) has the waveform shown in Fig. A.7.

Current i can be expressed, during one of its cycles, as:

for $0 < t < \alpha T_i$, $\quad i = I_M \sin(\omega t - \varphi_i)$

for $\alpha T_i < t < T_i$, $\quad i = 0$

The cycle T_i of i is here equal to half of that T of the inverter output current.

However, we will first of all work out expressions of i_S and u which are valid for any integer value of T/T_i, since these equations can be used in other applications.

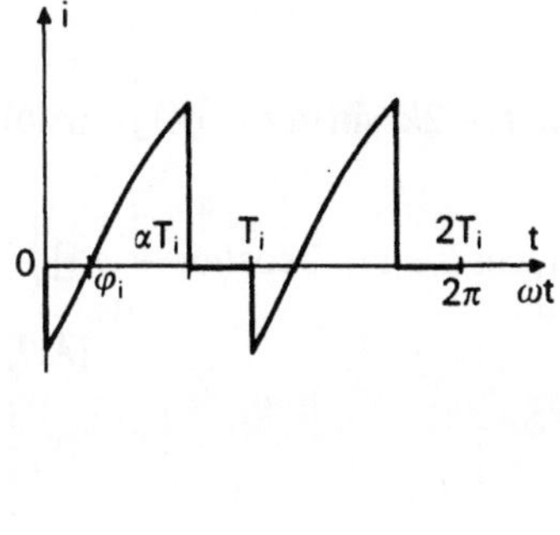

Fig. A.7

A.2.1 Expressions of i_S and u

The following equations are still valid:

$$i_S - C\frac{du}{dt} = i, \qquad L\frac{di_S}{dt} + u = U_S$$

$$\frac{d^2 i_S}{dt^2} + \frac{i_S}{LC} = \frac{i}{LC}$$

Giving the expressions of i_S and u:

- over the period $[0, \alpha T_i]$,

$$i_S = A_1 \cos \omega_f t + B_1 \sin \omega_f t + I_M \frac{k_f^2}{k_f^2 - 1} \sin(\omega t - \varphi_i)$$

$$u = U_S - L\omega_f \left[- A_1 \sin \omega_f t + B_1 \cos \omega_f t + I_M \frac{k_f}{k_f^2 - 1} \cos(\omega t - \varphi_i) \right]$$

- over the period $[\alpha T_i, T_i]$,

$$i_S = A_2 \cos \omega_f (t - \alpha T_i) + B_2 \sin \omega_f (t - \alpha T_i)$$

$$u = U_S - L\omega_f [- A_2 \sin \omega_f (t - \alpha T_i) + B_2 \cos \omega_f (t - \alpha T_i)]$$

- In order to determinate A_1, B_1, A_2, B_2, the continuous nature of current i_S and voltage u at instants $t = T_i$ and $t = \alpha T_i$ are used:

$$A_1 - I_M \frac{k_f^2}{k_f^2 - 1} \sin \varphi_i = A_2 \cos(1 - \alpha)\omega_f T_i + B_2 \sin(1 - \alpha)\omega_f T_i$$

$$B_1 + I_M \frac{k_f}{k_f^2 - 1} \cos \varphi_i = - A_2 \sin(1 - \alpha)\omega_f T_i + B_2 \cos(1 - \alpha)\omega_f T_i$$

$$A_1 \cos \alpha\omega_f T_i + B_1 \sin \alpha\omega_f T_i + I_M \frac{k_f^2}{k_f^2 - 1} \sin(\alpha\omega T_i - \varphi_i) = A_2$$

$$- A_1 \sin \alpha\omega_f T_i + B_1 \cos \alpha\omega_f T_i + I_M \frac{k_f}{k_f^2 - 1} \cos(\alpha\omega T_i - \varphi_i) = B_2$$

- As a result of these calculations, the following is obtained:
  - for $0 < t < \alpha T_i$,

$$i_S = I_M \frac{k_f}{2(k_f^2 - 1)} [A'_1 \cos k_f \omega t + B'_1 \sin k_f \omega t + 2k_f \sin(\omega t - \varphi_i)] \quad (A.6)$$

$$u = U_S \left\{ 1 - \frac{k_i k_f}{2(k_f^2 - 1)} [- A'_1 \sin k_f \omega t + B'_1 \cos k_f \omega t + 2 \cos(\omega t - \varphi_i)] \right\} \quad (A.7)$$

with

$$A'_1 = k_f \sin(\alpha\omega T_i - \varphi_i)\cos(1-\alpha)k_f\omega T_i + k_f \sin\varphi_i$$
$$+ \cos(\alpha\omega T_i - \varphi_i)\sin(1-\alpha)k_f\omega T_i$$
$$+ \frac{1}{\tan k_f\omega\frac{T_i}{2}}[\cos(\alpha\omega T_i - \varphi_i)\cos(1-\alpha)k_f\omega T_i - \cos\varphi_i$$
$$- k_f \sin(\alpha\omega T_i - \varphi_i)\sin(1-\alpha)k_f\omega T_i]$$
$$B'_1 = \cos(\alpha\omega T_i - \varphi_i)\cos(1-\alpha)k_f\omega T_i - \cos\varphi_i$$
$$- k_f \sin(\alpha\omega T_i - \varphi_i)\sin(1-\alpha)k_f\omega T_i$$
$$- \frac{1}{\tan k_f\omega\frac{T_i}{2}}[k_f \sin(\alpha\omega T_i - \varphi_i)\cos(1-\alpha)k_f\omega T_i - k_f\sin\varphi_i$$
$$+ \cos(\alpha\omega T_i - \varphi_i)\sin(1-\alpha)k_f\omega T_i]$$

- for $\alpha T_i < t < T_i$,

$$i_S = I_M \frac{k_f}{2(k_f^2-1)}[A'_2 \cos k_f\omega(t-\alpha T_i) + B'_2 \sin k_f\omega(t-\alpha T_i)] \tag{A.8}$$

$$u = U_S\left\{1 - \frac{k_i k_f}{2(k_f^2-1)}[-A'_2 \sin k_f\omega(t-\alpha T_i) + B'_2 \cos k_f\omega(t-\alpha T_i)]\right\} \tag{A.9}$$

with:

$$A'_2 = A'_1 \cos\alpha k_f\omega T_i + B'_1 \sin\alpha k_f\omega T_i + 2k_f\sin(\alpha\omega T_i - \varphi_i)$$
$$B'_2 = -A'_1 \sin\alpha k_f\omega T_i + B'_1 \cos\alpha k_f\omega T_i + 2\cos(\alpha\omega T_i - \varphi_i).$$

A.2.2 Waveforms

The waveforms of i_S/I_M is dependent on T_i/T, φ_i, α and k_f. That of u/U_S is also dependent on k_i, since the ripple of u on either side of U is proportional to I_M.

- Neither current i_S nor voltage u can be subject to any discontinuity. The equations

$$i_S = i + C\frac{du}{dt}, \qquad \frac{di_S}{dt} = \frac{U_S - u}{L}$$

show that:

- i_S is equal to i when $du/dt = 0$; i_S increases when u is lower than U_S and decreases in the opposite case; di_S/dt is not subject to any discontinuity.

- u is equal to U_S when $di_S/dt = 0$; u increases when i is lower than i_S and decreases in the opposite case; du/dt is subject to discontinuity at each sharp variation in i.

– In the case of the full-bridge single-phase inverter (see Chap. 2, Sect. 2.3.2.3):

$$T_i = \frac{T}{2}, \qquad \omega T_i = \pi, \qquad \alpha = 1 - \frac{\beta}{\pi}, \qquad \varphi_i = \varphi - \frac{\beta}{2},$$

using β to denote the phase shift between the controls of the two half-bridges, and φ to denote the inverter output current phase lag on the output voltage fundamental of the latter.

Current i_S has a mean value of

$$I_S = \frac{2}{\pi} I_M \cos\varphi \cos\frac{\beta}{2} = \frac{2}{\pi} I_M \cos\varphi \sin\frac{\alpha\pi}{2}$$

Figure A.8 shows the waveforms of voltage u and currents i_S, plotted for

$$k_f = 0.5, \qquad k_i = 1, \qquad \alpha = 0.5$$

and φ equal to 0, $\pi/4$ and $\pi/2$ respectively.

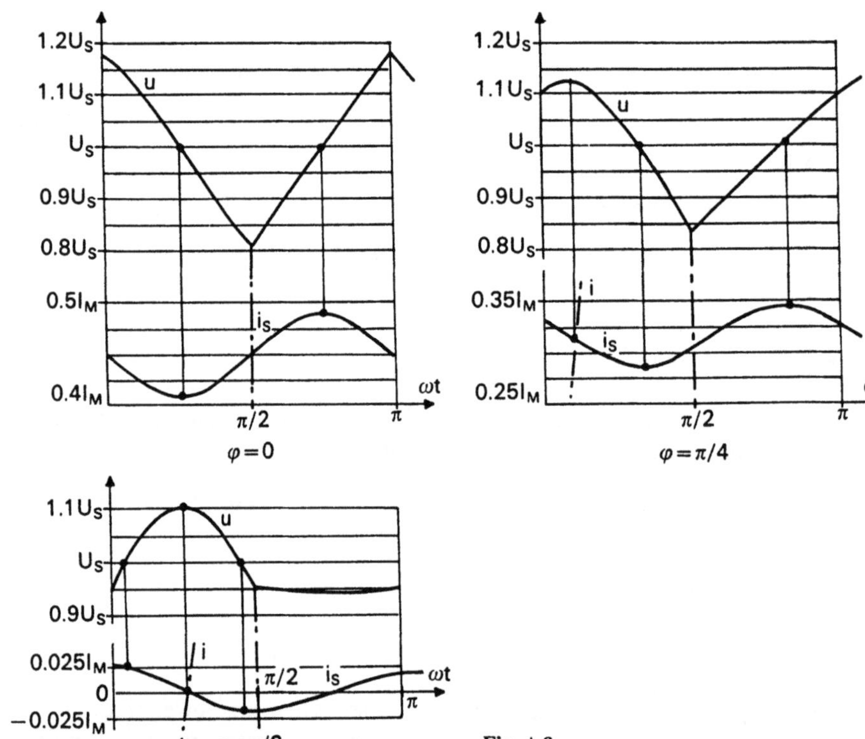

Fig. A.8

A comparison with the waveforms in Fig. A.3, drawn for the same values of k_f, k_i and φ but with α equal to 1, shows that changing from α equal to 1 to α equal to 0.5 increases $\Delta u/U_S$ and $\Delta i_S/I_M$ for $\varphi = 0$, but decreases these ripples for $\varphi = \pi/4$ and, in particular, for $\varphi = \pi/2$.

Remark. Equations (A.6), (A.7), (A.8) and (A.9) enable the following to be calculated, at a given T_i/T:

$\Delta i_S/I_M$ as a function of k_f, φ and α,
$\Delta u/U_S$ as a function of k_f, k_i, φ and α.

If ripples have to be limited to the same value, whatever, the values of φ and α, the limit values to be given to k_f and k_i must be determined for $\varphi = \pi/2$ and $\alpha = 0$. The values of L and C are then chosen as indicated in paragraph A.1.3.

Appendix B: "Sinusoidal" Voltage-Source Inverters

A sinusoidal voltage can be approximated by a wave made up of steps of equal or unequal amplitude, and of equal or unequal duration[1].

To achieve this, several voltage-source inverters are used, fed by the same DC supply (Fig. B.1); they deliver voltages u'_1, u'_2, ... each made of one rectangular pulse per half-cycle. These voltage are applied at the input of transformers with suitable transformation ratios m_1, m_2, ... By adding on the secondary voltages, a "quasi-sinusoidal" wave can be obtained.

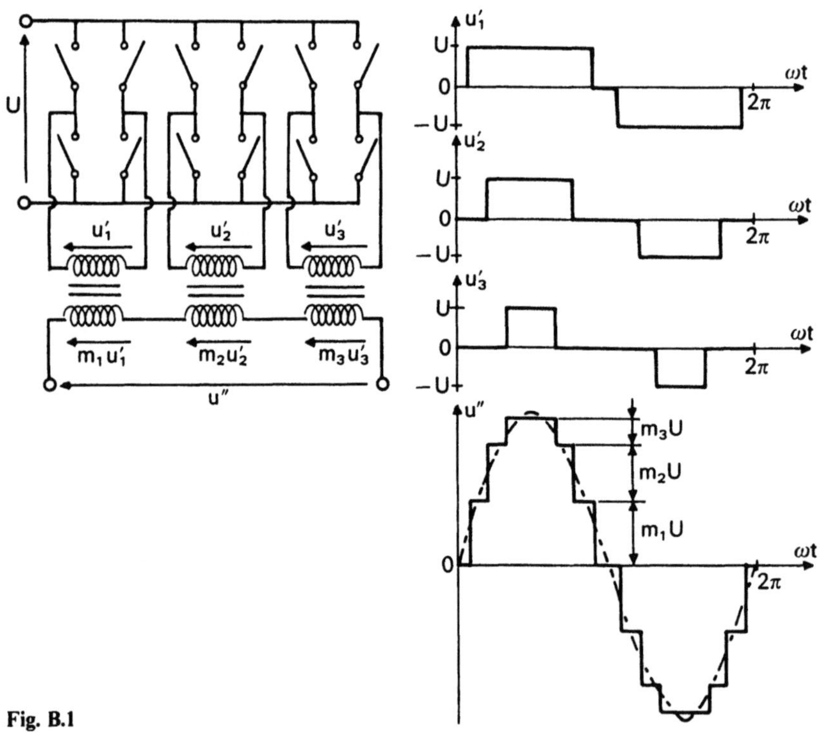

Fig. B.1

[1] In this Appendix, we have used the results of the study carried out with the aid of J. Lesenne.

Figure B.1 gives the diagram of a set comprising 3 inverters and 3 transformers and shows how the output voltage waveform is built up.

It is easy to see that this configuration is very costly. Therefore, except in very specific cases, it is only used to generate a three-phase system of quasi-sinusoidal voltages. Each transformer then comprises several windings and contributes to generating the three output voltages. For these three voltages to be identical to within $2\pi/3$ or $4\pi/3$, *steps of the same duration* must be used.

We will therefore confine our study to voltages formed of steps of equal duration.

B.1 Voltage Waveform

Let us examine (Fig. B.2) a voltage made up of $2p$ steps of equal width and with amplitude denoted $y_1, y_2, \ldots$.

In order that the series expansion of this voltage should contain only sine terms of uneven rank, the following is necessary:

$$u''(\pi + \omega t) = -u''(\omega t)$$

$$u''\left(\frac{\pi}{2} + \omega t\right) = u''\left(\frac{\pi}{2} - \omega t\right)$$

If p is even, there will be two steps of the same amplitude in the middle of each half-cycle. If p is uneven, the middle of each half-cycle will correspond to the middle of a step.

Amplitude A_n of the $n\omega$ angular frequency component is calculated by

$$A_n = \frac{2}{\pi}\int_0^{\pi} u''(\omega t)\sin n\omega t\, d\omega t$$

$$= \frac{2}{\pi}\sum_{k=1}^{p} y_k \int_{(k-1)\frac{\pi}{p}}^{k\frac{\pi}{p}} \sin n\omega t\, d\omega t$$

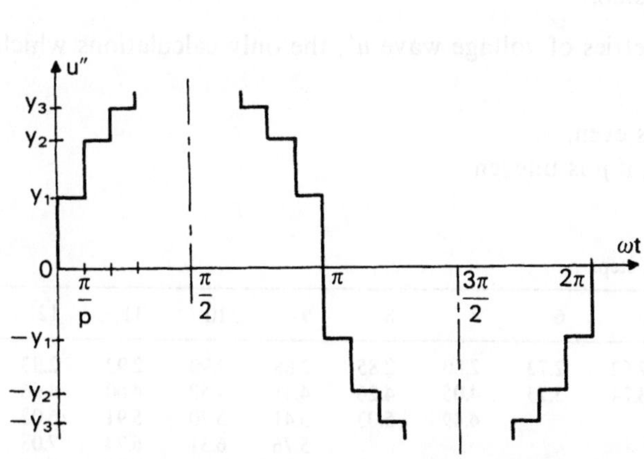

Fig. B.2

giving:

$$A_n = \frac{4}{n\pi} \sin \frac{n\pi}{2p} \sum_{k=1}^{p} y_k \sin (2k-1) \frac{n\pi}{2p}. \tag{B.1}$$

B.1.1 Calculating the Step Amplitudes

Equations (B.1) in matrix form give the terms of the series expansion of voltage u'' as functions of the amplitude of the steps.

In order to calculate y_k as a function of A_n, the matrix must be inverted. This is relatively simple, if limited to the first p values of n.

The value $U''_1\sqrt{2}$ required for the fundamental of u'' can then be given and normally the $p-1$ following terms of the series expansion are cancelled out.

From $A_1 = U''_1\sqrt{2}$, $A_2 = A_3 = \ldots = A_p = 0$ the following can be deduced:

$$\boxed{y_1 = \frac{1}{p}\frac{\pi\sqrt{2}}{2} U''_1} \tag{B.2}$$

$$y_2 = y_1 \frac{\sin(3\pi/2p)}{\sin \pi/2p}$$

$$y_3 = y_1 \frac{\sin(5\pi/2p)}{\sin \pi/2p} \ldots$$

$$\boxed{y_k = y_1 \frac{\sin[(2k-1)\pi/2p]}{\sin \pi/2p}} \ldots \tag{B.3}$$

It can be seen that the amplitude of a step is proportional to the sine of angle ωt in the middle of this step.

In view of the symmetries of voltage wave u'', the only calculations which need to be made are

$p/2$ amplitudes, if p is even,
$(p+1)/2$ amplitudes, if p is uneven.

Table B.1. Value of successive steps

p	3	4	5	6	7	8	9	10	11	12
y_2/y_1	2	2.41	2.62	2.73	2.80	2.85	2.88	2.90	2.92	2.93
y_3/y_1			3.24	3.73	4.05	4.26	4.41	4.52	4.60	4.66
y_4/y_1					4.49	5.03	5.41	5.70	5.91	6.08
y_5/y_1							5.76	6.31	6.74	7.08
y_6/y_1									7.03	7.60

Table B.1 gives the relative amplitudes of these $p/2$ or $(p + 1)/2$ steps for the first values of p.

B.1.2 Value of the Remaining Harmonics

In view of the value of y_k, the calculation of A_n by Eq. (B.1) shows that only the harmonics of rank $2k'p \pm 1$ (where k' is a positive integer) have a value other than zero, whether p be even or uneven.

The remaining harmonics have the following amplitude:

$$A_{2k'p\pm 1} = \frac{2p\,y_1}{\pi(2k'p \pm 1)} = \frac{1}{2k'p \pm 1} U''_1\sqrt{2} \tag{B.4}$$

Their normalised value is inversely proportional to their frequency.

B.1.3 Three-Phase Output System

If a three-phase system of quasi-sinusoidal voltages v''_A, v''_B, v''_C is generated using the same inverters, the step changes occur simultaneously for the three output voltages.

For these three waves to be identical, to within $2\pi/3$ or $4\pi/3$, an interval with a width of $2\pi/3$ must correspond to a whole number of steps.

$$2\pi/3 = k''2\pi/p$$

$$p = 3k''. \tag{B.5}$$

If this condition is achieved, the instantaneous sum of the three voltages v''_A, v''_B, v''_C is zero throughout the cycle. Indeed:

$$y_k + y_{k-\frac{2p}{3}} + y_{k-\frac{4p}{3}} = \frac{y_1}{\sin\dfrac{\pi}{2p}}\left\{\sin(2k-1)\frac{\pi}{2p} + \sin\left[(2k-1)\frac{\pi}{2p} - \frac{2\pi}{3}\right]\right.$$
$$\left. + \sin\left[(2k-1)\frac{\pi}{2p} - \frac{4\pi}{3}\right]\right\} = 0$$

B.2 Production of Optimized Waveforms

The system v''_A, v''_B, v''_C which fulfils conditions (B.3) and (B.5) is obtained by using p single-phase inverters delivering square-wave voltages $u'_1, u'_2, \ldots u'_i, \ldots u'_p$ of the same amplitude U, shifted two by two by π/p.

When applied (as in Fig. B.3) to the primary windings of p transformers with three secondary windings each, they give 3 groups of p secondary voltages which are added in order to generate v''_A, v''_B, v''_C.

$$v''_A = \sum_{i=1}^{p} m_i^1 u'_i, \qquad v''_B = \sum_{i=1}^{p} m_i^2 u'_1, \qquad v''_C = \sum_{i=1}^{p} m_i^2 u'_i \tag{B.6}$$

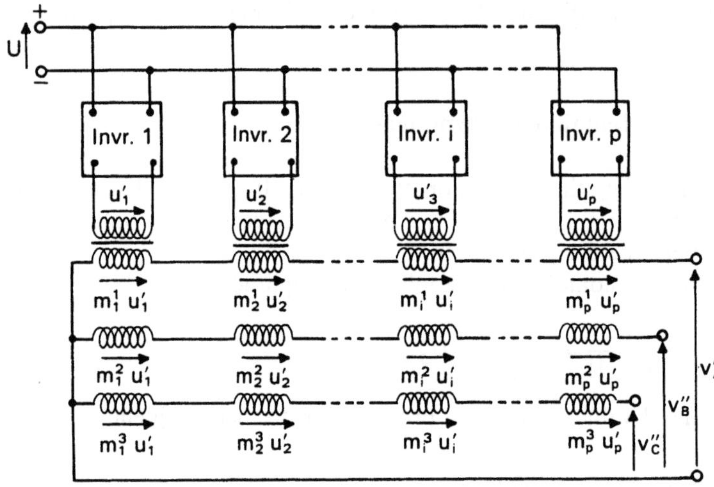

Fig. B.3

B.2.1 Calculating the Transformation Ratios

For $\omega t = 0$, when u'_1 changes from value $-U$ to $+U$, this causes v''_A to change from y_{2p} to y_1;

$$2m_1^1 U = y_1 - y_{2p} = 2y_1$$

thus:

$$m_1^1 = y_1/U \tag{B.7}$$

For $\omega t = 2\pi/2p$, the reversal of u'_2 makes v''_A change from y_1 to y_2:

$$2m_2^1 U = y_2 - y_1 = y_1 \left[\frac{\sin(3\pi/2p)}{\sin(\pi/2p)} - 1\right] = 2y_1 \cos\frac{\pi}{p}$$

For $\omega t = (i-1)2\pi/2p$, the reversal of u'_i makes v''_A vary by

$$2m_i^1 U = y_i - y_{i-1} = y_1 \frac{\sin[(2i-1)\pi/2p] - \sin[(2i-3)\pi/2p]}{\sin(\pi/2p)}$$

$$= 2y_1 \cos(i-1)\frac{\pi}{p}.$$

For $\omega t = (p-1)2\pi/2p$, the reversal of u'_p makes v''_A vary by

$$2m_p^1 U = y_p - y_{p-1} = 2y_1 \cos(p-1)\frac{\pi}{p}.$$

This gives the values of the transformation ratios related to m_1^1

$$\frac{m_2^1}{m_1^1} = \cos\frac{\pi}{p} \qquad \frac{m_3^1}{m_1^1} = \cos\frac{2\pi}{p} \qquad \dots \qquad \frac{m_i^1}{m_1^1} = \cos\,(i-1)\frac{\pi}{p}$$

Since voltages v_B'' and v_C'' are lagging v_A'' by $(2p/3)(\pi/p)$ and $(4p/3)(\pi/p)$, ratios m_i^2 and m_i^3 are given by

$$\frac{m_i^2}{m_1^1} = \cos\left(i - 1 - \frac{2p}{3}\right)\frac{\pi}{p} \qquad \frac{m_i^3}{m_1^1} = \cos\left(i - 1 - \frac{4p}{3}\right)\frac{\pi}{p}$$

This gives the general expression of the transformation ratios

$$m_i^r = \frac{y_1}{U}\cos\left[i - 1 - (r-1)\frac{2p}{3}\right]\frac{\pi}{p} \tag{B.8}$$

with r equal to 1, 2 or 3, depending on whether it concerns phases A, B or C.

B.2.2 Number of Transformer Types Required

The p transformers do not all have different ratios.

Provided that the secondary windings are connected in the suitable way, Eqs. (B.5) and (B.8) show that:

for $2p = 6$, there is only one type of transformer,
for $2p = 12$ and $2p = 18$, there are only two,
for $2p = 24$ and $2p = 30$, there are only three, ...

Table B.2 gives the three turn number ratios of each of the p transformers required, for $2p$ equal to 6, 12 and 18. All these ratios are referred to m_1^1 which is itself equal to y_1/U.

Table B.2. Values of m_i^r referred to m_1^1

Transf.r	$2p = 6$			$2p = 12$			$2p = 18$		
i	m_i^1	m_i^2	m_i^3	m_i^1	m_i^2	m_i^3	m_i^1	m_i^2	m_i^3
1	1	− 0.50	− 0.50	1	− 0.50	− 0.50	1	− 0.50	− 0.50
2	0.50	0.50	− 1	0.866	0	− 0.866	0.940	− 0.174	− 0.766
3	− 0.50	1	− 0.50	0.50	0.50	− 1	0.766	0.174	− 0.940
4				0	0.866	− 0.866	0.50	0.50	− 1
5				− 0.50	1	− 0.50	0.174	0.766	− 0.940
6				− 0.866	0.866	0	− 0.174	0.940	− 0.766
7							− 0.50	1	− 0.50
8							− 0.766	0.940	− 0.174
9							− 0.940	0.766	0.174

B.3 Variation in the Output Voltage Value

In calculating the transformation ratios as functions of the step amplitude, the voltages $u'_1, u'_2, \ldots, u'_p$ delivered by the inverters are assumed to be composed of square pulses with a width equal to π. This corresponds to operating at the maximum output voltages.

If full-bridge inverters are used, the width of the square pulses may be varied in order to make the output voltage value vary (see Chap. 2, Sect. 2.3).

If β is used to denote the width of periods at zero voltage (Fig. B.4), each voltage u' can be considered as the sum of two voltages u'_a and u'_b

consisting of square pulses with a width equal to π,
with an amplitude equal to $U/2$,
shifted between themselves by β.

The p voltages $u'_1, u'_2 \ldots, u'_p$ applied to the primaries of the p transformers are the sum of two systems of p voltages consisting of square pulses with a width equal to π shifted between themselves by $2\pi/2p$;

the first system comprises $u'_{1a}, u'_{2a}, \ldots, u'_{pa}$
the second $u'_{1b}, u'_{2b}, \ldots, u'_{pb}$.

The output voltages are the result of the superposition of these two systems

- their fundamental varies proportionately to

$$\frac{1}{2} + \frac{1}{2}\cos\beta, \quad \text{i.e. to } \cos\beta/2$$

- their harmonics of order $2k'p \pm 1$ vary proportionately to

$$\frac{1}{2} + \frac{1}{2}\cos(2k'p \pm 1)\beta, \quad \text{i.e. to } \cos(2k'p \pm 1)\beta/2.$$

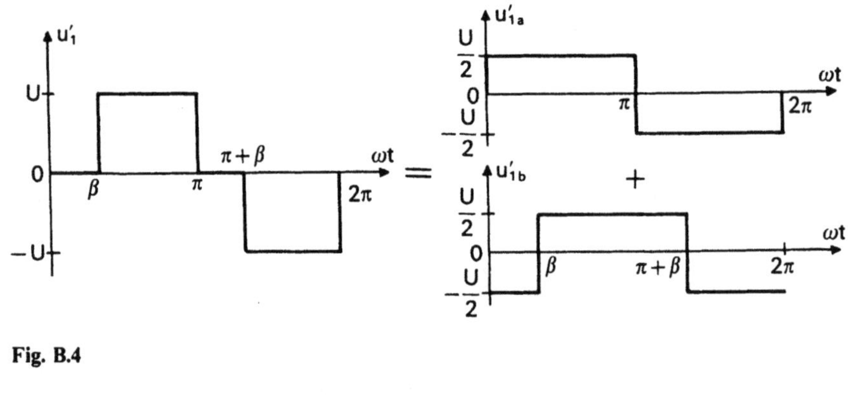

Fig. B.4

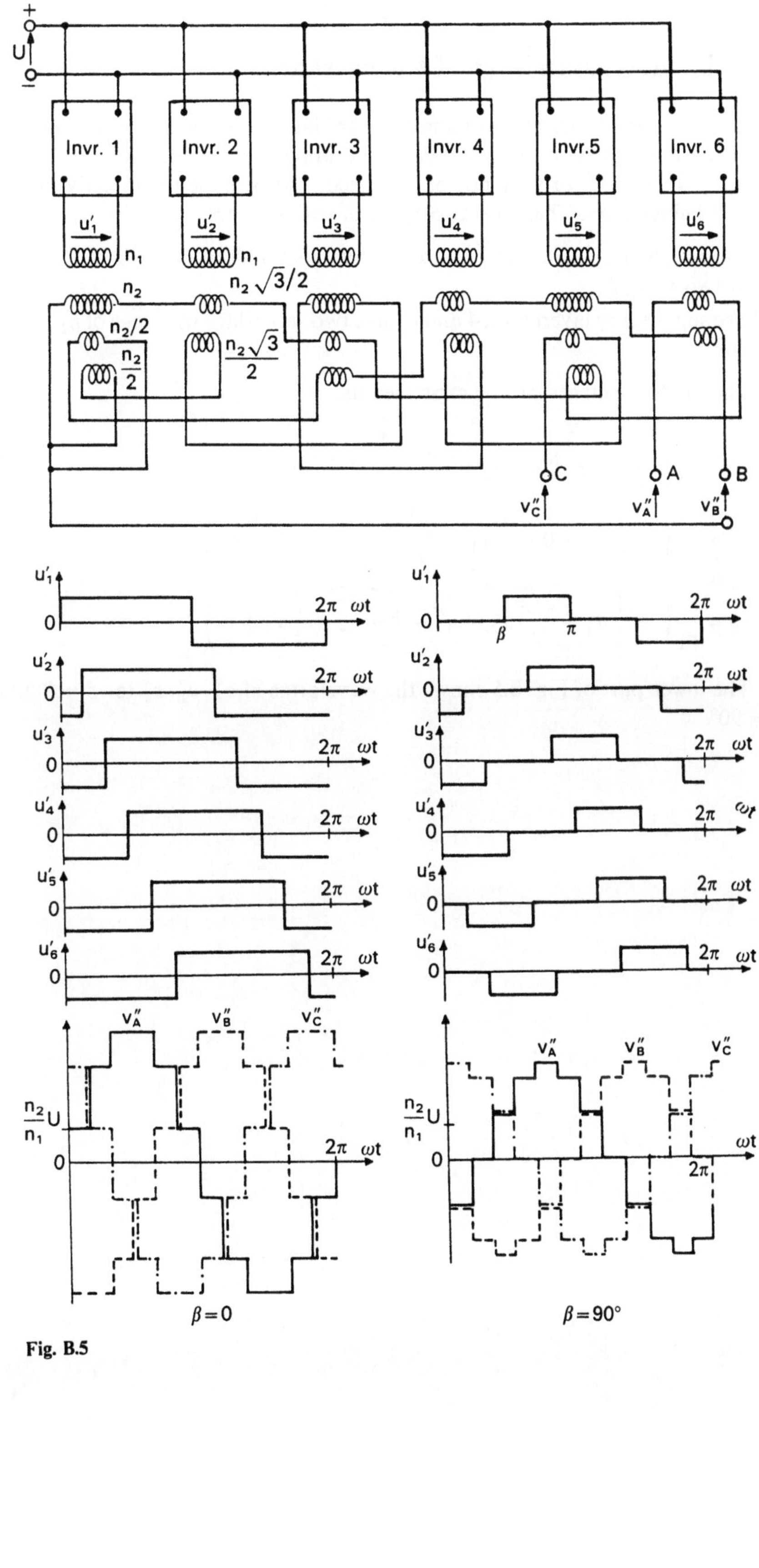

Invr. 1
Invr. 2
Invr. 3
Invr. 4
Invr.5
Invr. 6
U
u'_1
u'_2
u'_3
u'_4
u'_5
u'_6
n_1
n_2
$n_2/2$
$\frac{n_2}{2}$
$n_2\sqrt{3}/2$
$\frac{n_2\sqrt{3}}{2}$
C
A
B
v''_C
v''_A
v''_B
2π ωt
β
π
v''_A v''_B v''_C
$\frac{n_2}{n_1}U$
0
$\beta=0$
$\beta=90°$

Fig. B.5

B.4 The Most Usual Type of Configuration

The most frequently used type of quasi-sinusoidal inverters is that of $2p$ equal to 12 with a three-phase output, shown in the upper part of Fig. B.5. It enables a load – which may or may not be balanced – to be supplied directly or via a much lighter filter. There are two types of transformers:

- those supplied by inverters 1, 3 and 5 have a secondary winding of n_2 turns and two of $n_2/2$.
- those supplied by inverters 2, 4 and 6 have two secondary windings of $n_2\sqrt{3}/2$ turns.

The output voltages can be expressed as:

$$v''_A = \frac{n_2}{n_1}\left(u'_1 + \frac{\sqrt{3}}{2}u'_2 + \frac{1}{2}u'_3 + 0 - \frac{1}{2}u'_5 - \frac{\sqrt{3}}{2}u'_6\right)$$

$$v''_B = \frac{n_2}{n_1}\left(-\frac{1}{2}u'_1 + 0 + \frac{1}{2}u'_3 + \frac{\sqrt{3}}{2}u'_4 + u'_5 + \frac{\sqrt{3}}{2}u'_6\right)$$

$$v''_C = \frac{n_2}{n_1}\left(-\frac{1}{2}u'_1 - \frac{\sqrt{3}}{2}u'_2 - u'_3 - \frac{\sqrt{3}}{2}u'_4 - \frac{1}{2}u'_5 + 0\right)$$

The lower part of Fig. B.5 shows the waveforms of v''_A, v''_B, v''_C for $\beta = 0$ and $\beta = 90°$.

Appendix C: Forced Turn-Off of Thyristors in Voltage-Source Inverters

In Chap. 5 of Vol. 3, we gave an in-depth analysis of the turn-off circuits of thyristors in the directly linked choppers, classifying these circuits according to the commutation mode used.

In this Appendix, we will complete the study of the forced turn-off of thyristors by examining the main types of turn-off circuits used in the autonomous voltage-source inverters.

We base our study on the *classification* established previously and will adopt the same simplifying *hypotheses*:

- The variations in voltage across the generator and in current through the load phases will be ignored during the commutations. The latter hypothesis is justified, even though the current is AC, on account of the short duration of the commutations compared with the cycle of AC variables.
- The rise- and fall-times of the current in the semiconductor devices, as well as the reverse current spikes at turn-off will be ignored.
- The phenomena linked to the presence of snubbers limiting the $\mathrm{d}i/\mathrm{d}t$, the $\mathrm{d}v/\mathrm{d}t$ and the overvoltages will be ignored. Although their presence is necessary for a correct use of the semiconductor devices, we will not show them on the circuit diagrams under study.

C.1 Parallel Commutation by Oscillating Circuit

As noted in Vol. 3, a joint turn-off circuit is used for the two thyristors which belong to the same inverter leg.

Figure C.1 shows a half-bridge single-phase inverter fitted with a widely used type of turn-off circuit – the clamped Mc Murray commutating circuit – which comprises two auxiliary thyristors, two auxiliary diodes, the oscillating circuit made up of capacitor C and inductor L, and a damping resistance R.

Using this circuit, we will recall the main characteristics of parallel commutation by oscillating circuit. We will then study the advantages and disadvantages of various possible modifications.

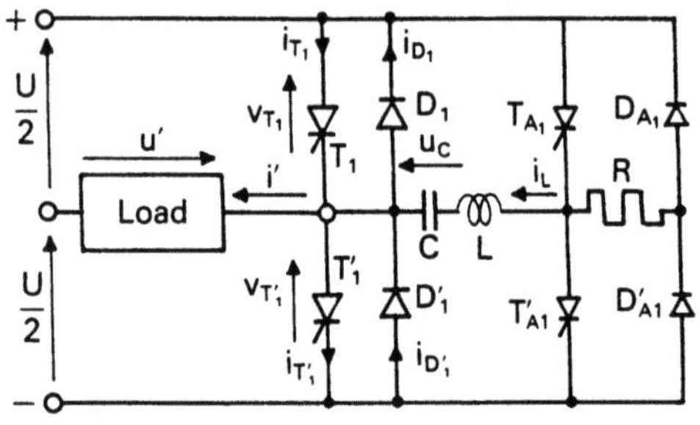

Fig. C.1

C.1.1 Successive Phases of a Commutation

As an example, we will study the forced turn-off of thyristor T_1, i.e. the transfer of a positive current i' from T_1 to diode D'_1. Figure C.2 shows the variations of u_C and i_L, of v_{T_1} and i_{T_1}, of $v_{D'_1}$ and $i_{D'_1}$, and u' during this commutation. Current i' is assumed to be constant and will be denoted by I_{OFF}.

The initial state is the following:

- as the other thyristors are turned off, thyristor T_1 conducts the current I_{OFF},
- current i_L is zero, voltage u_C equal to $+U$ and output voltage u' equal to $+U/2$.

The commutation itself can be divided into 4 phases:

- *Phase* 1 ($t_0 < t < t_1$). T_1 and T_{A_1} *conducting*

– At instant $t = t_0$, the commutation is started by the firing of the auxiliary thyristor T_{A_1}, which sustained a forward voltage equal to $+U$.

The simultaneous conduction of T_1 and T_{A_1} initiates an oscillating discharge of C:

$$U_C = U \cos \omega(t - t_0), \qquad i_L = \frac{U}{L\omega} \sin \omega(t - t_0)$$

with

$$\omega = \frac{1}{\sqrt{LC}}.$$

Current i_{T_1}, equal to $I_{OFF} - i_L$, decreases.

– For $t = t_1$, this first phase ends when i_L reaches value I_{OFF} and current i_{T_1} falls to zero. This provides the equation giving t_1:

$$\frac{U}{L\omega} \sin \omega(t_1 - t_0) = I_{OFF}, \quad \text{with } \omega(t_1 - t_0) < \frac{\pi}{2}.$$

Turning off T_1 requires an amplitude $U\sqrt{C/L}$ of current i_L greater than I_{OFF}.

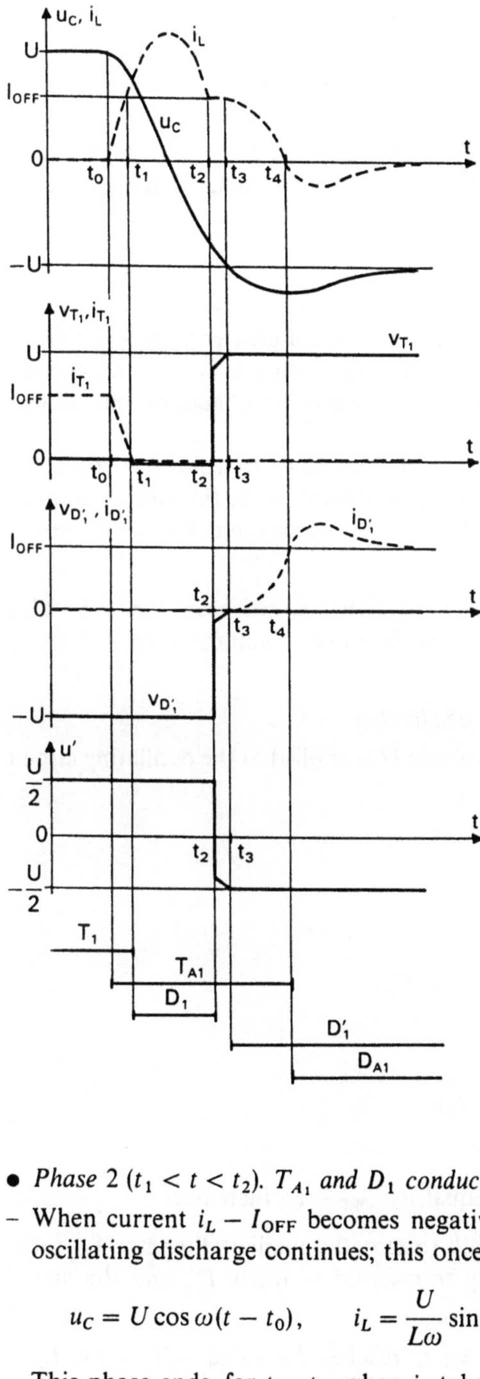

Fig. C.2

- *Phase 2* ($t_1 < t < t_2$). T_{A_1} *and* D_1 *conducting*

- When current $i_L - I_{OFF}$ becomes negative, it flows through diode D_1. The oscillating discharge continues; this once more gives:

$$u_C = U \cos \omega(t - t_0), \qquad i_L = \frac{U}{L\omega} \sin \omega(t - t_0)$$

- This phase ends, for $t = t_2$, when i_L takes again the value I_{OFF} after having reached its maximum value. Current i_{D_1}, equal to $i_L - I_{OFF}$, tends to become negative and diode D_1 turns off. The value of t_2 is given by

$$\frac{U}{L\omega}\sin\omega(t_2 - t_0) = I_{OFF}, \quad \text{with } \frac{\pi}{2} < \omega(t - t_2) < \pi$$

- When D_1 is conducting, thyristor T_1 is reverse biased. If $t_2 - t_1$ is higher than the turn-off time t_q of the thyristor, at instant $t = t_2$, the latter has recovered its ability to block a forward voltage.

- *Phase* 3 ($t_2 < t < t_3$). T_{A_1} *conducting*

At instant $t = t_2$, voltage u_C is negative but is still greater than $-U$; diode D'_1 is reverse biased and cannot begin conducting. From $t = t_2$ onwards, current I_{OFF} flows via T_{A_1}, L, C and the load. The discharge of C goes on but now at constant current.

As di/dt in the inductor L falls to zero, this leads to voltage $-u_C$ being applied across T_1 (in fact, this voltage step is applied to the thyristor at a speed lower than its critical dv/dt, owing to the snubber which is ignored in the present analysis).

The linear discharge of C ends, for $t = t_3$, when u_C reaches value $-U$. This allows for diode D'_1 to conduct and fixes the value of voltage u' at $-U/2$.

- *Phase* 4 ($t_3 < t < t_4$). T_{A_1} *and* D'_1 *conducting*

- From $t = t_3$ onwards, the supply voltage U is applied to the oscillating circuit through T_{A_1} and D'_1:

$$LC\frac{d^2u_C}{dt^2} + u_C = -U$$

with

$$u_C(t_3) = -U \quad \text{and} \quad \frac{du_C}{dt}(t_3) = -\frac{I_{OFF}}{C}$$

A new oscillation begins:

$$u_C = -U - L\omega I_{OFF}\sin\omega(t - t_3)$$

$$i_L = I_{OFF}\cos\omega(t - t_3);$$

current i_L decreases; current $i_{D'_1}$, equal to $I_{OFF} - i_L$, increases.

- This phase ends at $t = t_4$ when i_L falls to zero; the auxiliary thyristor T_{A_1} turns off. Current I_{OFF} has been totally transferred to diode D'_1 and the actual commutation is over.

- At instant $t = t_4$, the voltage across C reaches the value $-U - L\omega I_{OFF}$. Owing to the auxiliary diode D'_{A_1}, voltage u_C reverts to the value $-U$ and energy $LI_{OFF}^2/2$ is dissipated in resistance R.

- From $t = t_3$ onwards, a control signal can be applied to thyristor T'_1. This signal does not make the latter start conducting since it is reverse biased by the conduction of D'_1. In order to operate correctly, the control must be kept on T'_1 until the load current i' reversal leads to the turn-off of D'_1 and the turn-on of T'_1.

  In practice, the control of T'_1 (conversely: T_1) is maintained during the whole period when the pair $T'_1 D'_1$ (conversely: $T_1 D_1$) must be conducting.

- At the end of T_1 turn-off, voltage u_C has the value $-U$. The capacitor is thus ready to ensure the turn-off of T'_1 by firing T'_{A_1}, following a similar process to that just described for the turn-off of T_1. The turn-off of T'_1 brings voltage u_C back to value $+U$.

- *Remark*

In the preceding analysis, it was assumed that thyristor T_1 was conducting at instant $t = t_0$ when the voltage u' was required to change from $+U/2$ to $-U/2$. In fact, this is the case for full-wave operation on an inductive load.

On the other hand, in the case of Pulse Width Modulation, commutation from the pair T_1D_1 to the pair $T'_1 D'_1$ may occur when current i' goes through diode D_1 and thyristor T_1 is turned off.

In this case, firing T_{A_1} simply leads to an half-cycle of oscillation of circuit LC. This oscillating current is added to current i' going through D_1. At the end of this half-oscillation, voltage u_C is theoretically reversed.

In practice, on account of the dampings which are inevitably present in the circuit, the final voltage across the capacitor is slightly less than the theoretial voltage. In order to avoid a decrease in the amplitude of current i_L oscillation during the following commutation, the controls of T_{A_1} and T'_1 (as well as T'_{A_1} and T_1) will slightly overlap, thus applying the supply voltage to the turn-off circuit and leading to the complete charging of the capacitor.

C.1.2 Calculating the Turn-Off Circuit Elements

The value of L and C can be calculated using the necessary reverse bias time and the ratio of the current I_{OFF} to be commutated to the amplitude $I_{L\max}$ of the current in the oscillating circuit. This amplitude is equal to $U\sqrt{C/L}$.

By writing

$$\cos k\frac{\pi}{2} = \frac{I_{OFF}}{U\sqrt{C/L}}$$

$$t_p = at_q$$

with

t_q, turn-of time of the thyristors
a, a coefficient greater than 1, normally taken close to 2,

this gives:

$$C = \frac{I_{OFF} a t_q}{U k\pi \cos\frac{k\pi}{2}}$$

$$L = \frac{U a t_q}{I_{OFF}} \frac{\cos(k\pi/2)}{k\pi} .$$

C.1.3 Modifying the Turn-Off Circuit

- *First modification*

The first modification consists in replacing inductance L by two inductances L_1 and L'_1, both with a value equal to L and series-connected with the main "switches" T_1–D_1 and T'_1–D'_1 (Fig. C.3a).

Such a modification has no effect on the actual commutation mechanism.

It has the advantage of putting the inductances L_1 and L'_1 in places where they can be used to limit the di/dt applied to the thyristors and to make it easier to protect the thyristor in case of short-circuit resulting from the simultaneous conduction of the two main thyristors[1].

However, this modification has two disadvantages:
the load current flows through each inductance during a whole half-cycle of this current,

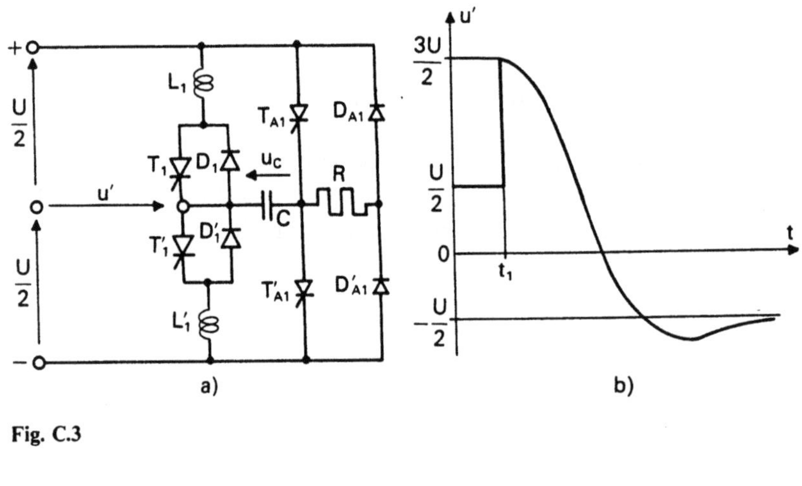

Fig. C.3

[1] There is no risk of short-circuiting the supply in the case of the series chopper and so this modification was not envisaged in Vol. 3.

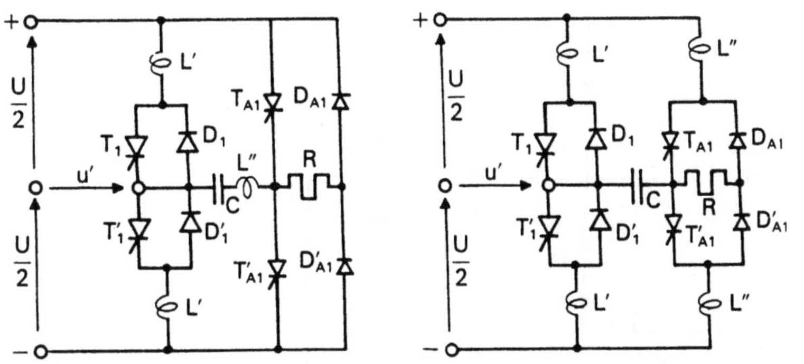

Fig. C.4

- at instant $t = t_1$, an overvoltage up to $3U/2$ appears across the load, since, throughout the conduction of T_{A_1}, there is:

$$u' = \frac{U}{2} + u_C$$

Various configurations, which have properties between those of the configuration in figure C.1 and those of its variant in Fig. C.3, can be obtained by dividing the commutation inductance into two (Fig. C.4).

For the two configurations in Fig. C.4, inductance L is equal to $L' + L''$. The second is more interesting since part of the inductance is series-connected with T_{A_1} and T'_{A_1}.

- *Second modification*

The second modification, which leads to the diagram in Fig. C.5, consists in moving

- the anode of the auxiliary thyristor T_{A_1} from the + terminal to the midpoint of the DC voltage supply,
- the cathode of the auxiliary thyristor T'_{A_1} from the − terminal to the midpoint of the DC voltage supply.

– This modification is similar to that which consists – in the case of the series chopper with parallel commutation – of moving the terminal of the turn-off circuit capacitor connected to the generator from the + terminal of the latter to its − terminal (Fig. C.6).

In Chap. 5 of Vol. 3, it was seen that this modification shifted voltage u_C by a level equal to U; it oscillates between $+U$ and $-U$ in the case of the configuration in Fig. C.6a, between 0 and $2U$ in the case of the configuration in Fig. C.6b.

In the case of Fig. C.5, circuit LC can only be made to oscillate under a voltage supply equal to $U/2$: for equal values of L and C, the oscillation of

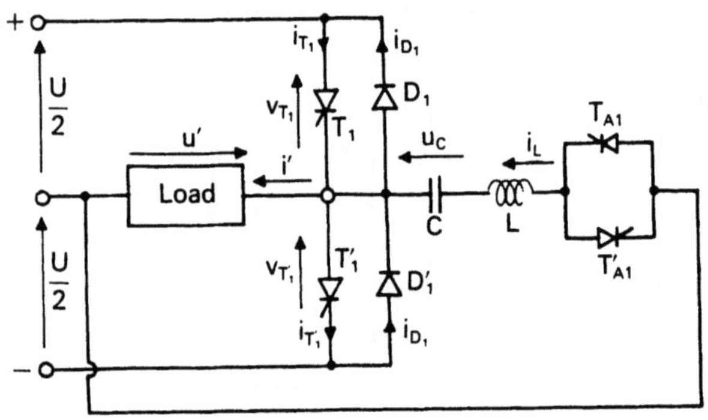

Fig. C.5

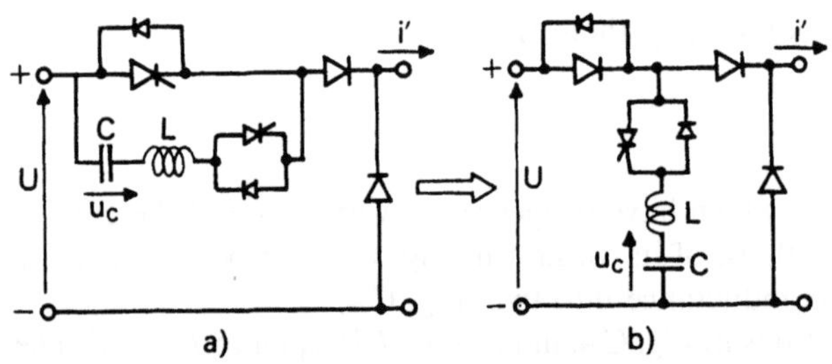

Fig. C.6

current i_L is halved in comparison with that given by the circuit in Fig. C.1. In order to maintain the same oscillation frequency and peak current, the capacitance must be doubled and the inductance halved, when changing from the circuit in Fig. C.1 to that in Fig. C.5.

– In order to describe the operation of the circuit in Fig. C.5, the *thyristor* T_1 *turn-off* will be followed. The waveforms of u_C and i_L, of v_{T_1} and i_{T_1}, of $v_{T'_1}$ and $i_{T'_1}$, of $v_{D'_1}$ and $i_{D'_1}$, and of u' are given in Fig. C.7.

- The initial state is as follows:
  only thyristor T_1 is conducting, it lets current I_{OFF} flow;
  current i_L is zero, voltage u_C equals U.

- The turn-off of T_1 is initiated by firing T_{A_1} at instant $t = t_0$. The simultaneous conduction of T_1 and T_{A_1} closes the oscillating circuit on the upper half-part of the DC supply.
  During this first phase,

$$u_C = \frac{U}{2}[1 + \cos\omega(t - t_0)]$$

$$i_L = \frac{U}{2}\sqrt{\frac{C}{L}}\sin\omega(t - t_0)$$

$$i_{T_1} = I_{OFF} - i_L$$

with $\omega = 1/\sqrt{LC}$.

- For $t = t_1$, i_L equals I_{OFF} and thyristor T_1 turns off. Oscillation goes on via diode D_1 which takes up the difference between i_L and I_{OFF}.
- For $t = t_2$, i_L once more crosses the value I_{OFF}, diode D_1 turns off, and only the auxiliary thyristor T_{A_1} remains conducting. From $t = t_2$ onwards, current i_L remains equal to I_{OFF} and causes a linear discharge of the capacitor starting from voltage

$$u_C(t_2) = \frac{U}{2}[1 + \cos\omega(t_2 - t_1)]$$

Since

$$I_{OFF} = \frac{U}{2}\sqrt{\frac{C}{L}}\sin\omega(t_1 - t_0) = \frac{U}{2}\sqrt{\frac{C}{L}}\sin\omega(t_2 - t_0),$$

with

$$\omega(t_1 - t_0) < \pi/2 \quad \text{and} \quad \omega(t_2 - t_0) > \pi/2.$$

by writing

$$x = \arcsin\left(\frac{2}{U}\sqrt{\frac{L}{C}}I_{OFF}\right) < \frac{\pi}{2}.$$

this leads to:

$$u_C(t_2) = \frac{U}{2}(1 - \cos x)$$

In order to obtain, at the end of the commutation process, a capacitor charged at voltage $-U$ and without having to add a damping circuit, thyristor T'_1 must be fired at instant $t = t_3$, such that:

$$u_C(t_3) = -u_C(t_2) = -\frac{U}{2}(1 - \cos x).$$

- From $t = t_3$ onwards, an oscillation starts in the circuit composed of the lower half-part of the DC supply, thyristor T_{A_1}, inductor L, capacitor C and the pair T'_1–D'_1.

$$u_C = -\frac{U}{2}(1 - \cos[\omega(t - t_3) + x])$$

$$i_L = \frac{U}{2}\sqrt{\frac{C}{L}}\sin[\omega(t - t_3) + x]$$

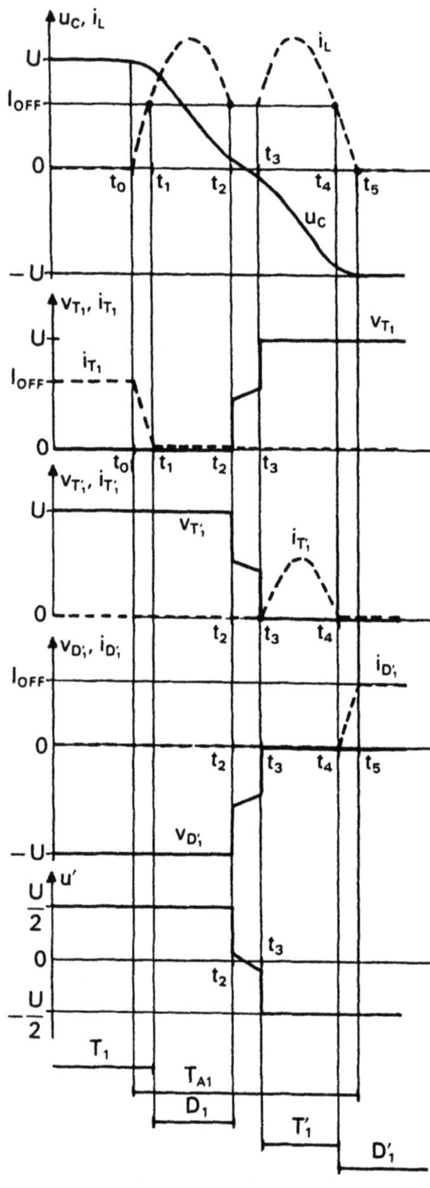

Fig. C.7

The difference between i_L and I_{OFF} flows via T'_1 for $t_3 < t < t_4$, as long as i_L is greater than I_{OFF}, and then, via D'_1, when this difference is negative.

- With the choice indicated above for t_3, when i_L falls to zero, for $t = t_5$, voltage u_C is equal to $-U$ and is simply reversed. Contrary to the configuration in Fig. C.1, there is thus no overvoltage to be eliminated by a damping circuit.

Normally, the value of t_3 is dependent on the value of current I_{OFF} in the load. The following equation is easily established.

$$t_3 - t_2 = \frac{2}{\omega}\frac{1 - \cos x}{\sin x} = \frac{2}{\omega}\tan\frac{x}{2};$$

where x is dependent on I_{OFF}. In practice, however, the peak value of current i_L is always high enough, in relation to I_{OFF}, for the following to be written:

$$\sin x = \frac{2}{U}\sqrt{\frac{L}{C}}I_{OFF} \simeq x \quad \text{and} \quad \tan\frac{x}{2} \simeq \frac{x}{2}$$

This results in

$$\omega(t_3 - t_2) \simeq x, \qquad \omega(t_2 - t_0) \simeq \pi - x$$

and thus:

$$\omega(t_3 - t_0) \simeq \pi.$$

The required effect can, therefore, be virtually obtained by firing thyristor T'_1 with a delay, in relation to the firing of T_{A_1}, equal to the half-cycle of the oscillation of circuit LC.

The reverse bias time of the thyristors is calculated in the same way as for the initial configuration. The values to be given to L and C can be deduced in the same way from the reverse bias time and the normalised value of the maximum of i_L.

- The mid-point of the voltage supply is given by the capacitve divider at the input for the single-phase half-bridge inverters.

  For the single-or three-phase full-bridge inverters, the mid-point, to which the commutation capacitors are connected, can be created by dividing the input filter capacitor into two equal parts connected in series. The commutation capacitor can even be used as a divider for each half-bridge.

Finally, as in the initial configuration, all or a part of the commutation inductance can be replaced by two identical inductances, one of which is connected in series switch T_1–D_1 and the other in series switch T'_1–D'_1. The connected in series with switch T_1–D_1 and the other in series with switch T'_1–D'_1. The advantages and disadvantages of such a modification have already been noted.

C.1.4 Full-Bridge Single-Phase Inverters and Three-Phase Inverters

A full-bridge single-phase inverter can be obtained by using two half-bridges, and a three-phase inverter by using three. If all possible flexibility of control has to be maintained at each half-bridge (e.g. in order to operate in Pulse Width Modulation) each half-bridge must have its own turn-off circuit (Fig. C.8).

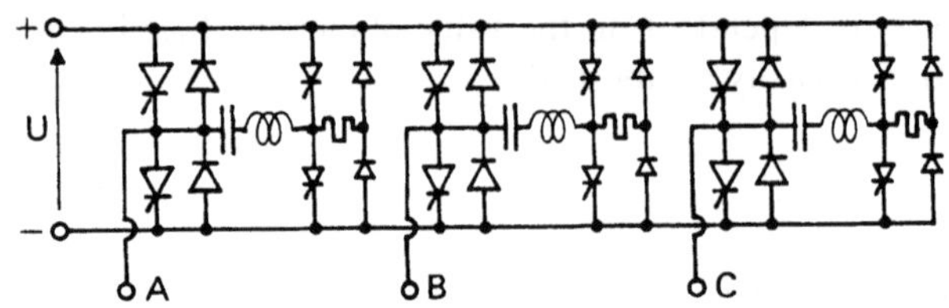

Fig. C.8

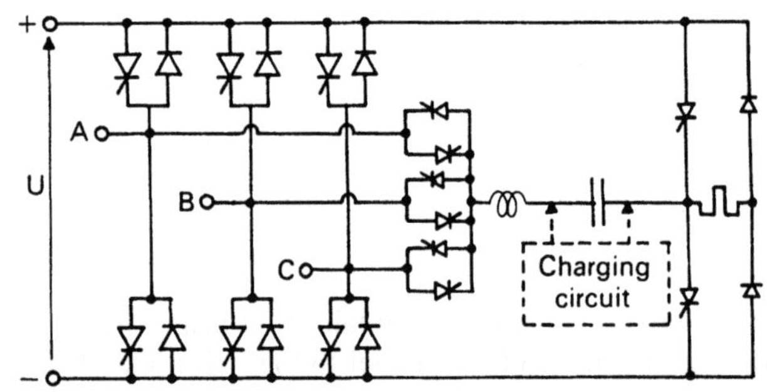

Fig. C.9

On the other hand, in the case of "full-wave" control, the turn-off circuits of the various half-bridges can be grouped together in a single turn-off circuit, by using auxiliary thyristors in order to connect this circuit successively to the various half-bridges (Fig. C.9). This possibility arises from the fact that, with a suitable control, the turn-off of a thyristor connected to the − terminal follows that of a thyristor connected to the + terminal (and vice versa). The commutation circuit capacitor is thus suitably charged after each turn-off to carry out the following turn-off.

When the solution chosen is that of a single turn-off circuit, a separate charging circuit (in broken lines in Fig. C.9) can be used. It enables the turn-off capacitor to be charged at voltage higher than that of the supply and the commutation to be made independent of the supply voltage. This is required when the output voltage amplitude is varied by acting on the input DC voltage value.

C.2 Parallel Commutation by Capacitor

Figure C.10 represents a single-phase half-bridge inverter using the most usual type of parallel turn-off circuit by capacitor.

Capacitor C, two inductors L_1 and L'_1 of a same inductance L, a damping resistance R, two auxiliary thyristors T_{A_1} and T'_{A_1}, and two auxiliary diodes D_{A_1} and D'_{A_1} are added to the semiconductor devices T_1 and D_1, T'_1 and D'_1 which compose the two switches.

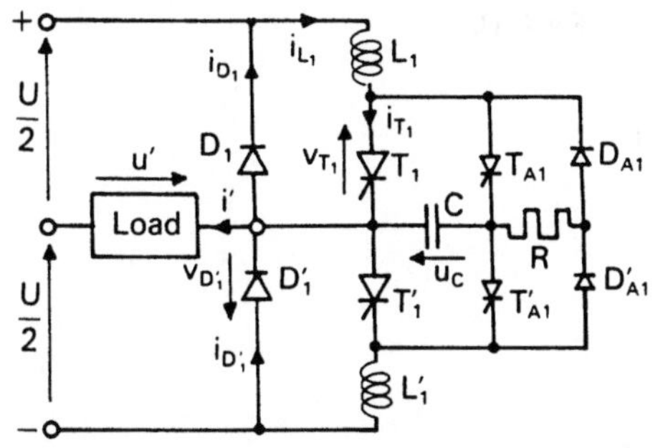

Fig. C.10

C.2.1 Successive Phases of a Commutation

The turn-off to be studied is that of thyristor T_1 when crossed by a current i', assumed to remain constant and equal to I_{OFF} throughout the commutation. Figure C.11 enables the variations of the main variables to be followed.

The initial state is as follows:

- T_1 is the only device in the conducting state

$$i_{T_1} = i_{L_1} = I_{OFF};$$

- the capacitor is charged at voltage $+U$.

- *First phase* ($t_0 < t < t_2$). T_{A_1} *and* D_1 *conducting*

- In order to turn off T_1, T_{A_1} is fired at instant $t = t_0$. This firing applies voltage u_C in reverse across T_1 and turns it off; current I_{OFF} is transferred from T_1 to T_{A_1}.

  If i_{L_1} remained constant, the voltage across D_1 would be equal to u_C and hence to $+U$. At instant $t = t_0$, diode D_1 begins conducting.

- During the first phase, the capacitor is discharged by the joint action of current I_{OFF} and an oscillation arising from the closing of the circuit formed by C, D_1, L_1 and T_{A_1}.

$$u_C = U - \frac{1}{C}\int_{t_0}^{t} i_{L_1}\,dt\,; \quad u_C = L\frac{di_{L_1}}{dt},$$

with

$$u_C(t_0) = U \quad \text{and} \quad i_{L_1}(t_0) = I_{OFF}.$$

It can be deduced that:

$$u_C = U\cos\omega(t - t_0) - L\omega I_{OFF}\sin\omega(t - t_0)$$

$$i_{L_1} = I_{OFF}\cos\omega(t - t_0) + \frac{U}{L\omega}\sin\omega(t - t_0)$$

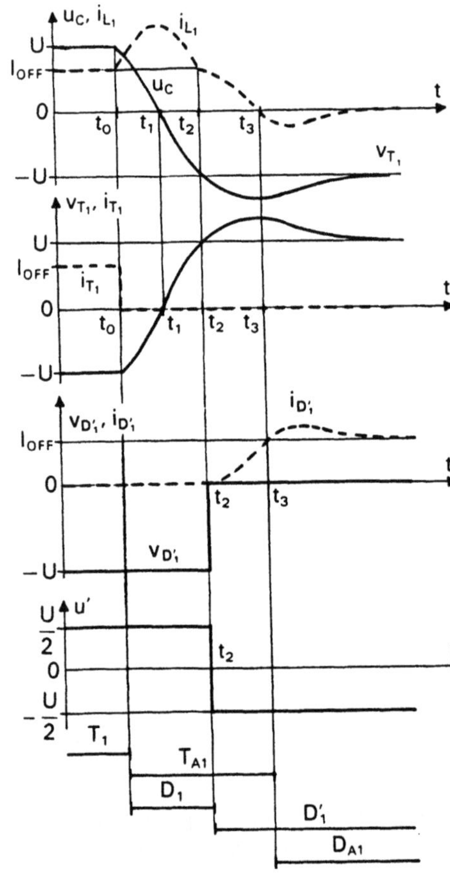

Fig. C.11

with $\omega = \dfrac{1}{\sqrt{LC}}$

u_C and i_{L_1} can be expressed in the following form:

$$u_C = \sqrt{U^2 + (L\omega I_{OFF})^2} \cos[\omega(t - t_0) + \varphi]$$

$$i_{L_1} = \frac{1}{L\omega}\sqrt{U^2 + (L\omega I_{OFF})^2} \sin[\omega(t - t_0) + \varphi]$$

with

$$\varphi = \arctan\left(\frac{L\omega I_{OFF}}{U}\right) \quad \text{and} \quad 0 < \varphi < \frac{\pi}{2}.$$

- Thyristor T_1, which sustains a reverse voltage $-u_C$, is reverse biased up to instant $t = t_1$, when voltage u_C reaches zero. The first expression of u_C gives the reverse bias time t_p.

$$U \cos \omega(t_1 - t_0) - L\omega I_{OFF} \sin \omega(t_1 - t_0) = 0$$

$$t_p = t_1 - t_0 = \frac{1}{\omega} \arctan \frac{U}{L\omega\, I_{OFF}}.$$

- The first phase ends, at instant $t = t_2$, when current i_{L_1} which is now decreasing takes again the value I_{OFF}. Diode D_1, which lets $i_{L_1} - I_{OFF}$ flow through it, then turns off.

  The value of t_2 is such that:

$$\sin[\omega(t_2 - t_0) + \varphi] = \frac{L\omega I_{OFF}}{\sqrt{U^2 + (L\omega I_{OFF})^2}}$$

Giving the value u_C at the end of first phase.

$$u_C(t_2) = -\sqrt{U^2 + (L\omega I_{OFF})^2}\sqrt{1 - \frac{(L\omega I_{OFF})^2}{U^2 + (L\omega I_{OFF})^2}} = -U$$

At instant $t = t_2$, diode D'_1, which is no longer reverse biased, begins conducting and fixes the output voltage u' value to $-U/2$.

- *Phase 2* ($t_2 < t < t_3$). T_{A_1} *and* D'_1 *conducting*

- From $t = t_2$ onwards, the oscillation of circuit L_1C continues through the voltage source U, auxiliary thyristor T_{A_1} and diode D'_1.

$$i_{L_1} = I_{OFF} \cos \omega(t - t_2)$$

$$u_C = -U - L\omega I_{OFF} \sin \omega(t - t_2)$$

$$i_{D'_1} = I_{OFF} - i_{L_1}$$

- For $t = t_3$, such that $\omega(t_3 - t_2) = \pi/2$, i_{L_1} is zero, thyristor T_{A_1} is off and the oscillation stops. u_C then has the following value:

$$u_C(t_3) = -U - L\omega I_{OFF}$$

- From $t = t_3$ onwards, the overvoltage $L\omega I_{OFF}$ across C is resorbed via resistance R, diode D_{A_1}, inductance L_1, voltage source U and diode D'_1.

- *Remarks*

As was the case for the parallel commutation by oscillating circuit configuration, if current i' flows through D_1 when T_{A_1} is fired, this firing leads to a simple reversal of u_C owing to a half-cycle of oscillation in circuit L_1C. The current i' transfer from D_1 towards T'_1 occurs at the instant when the latter is fired.

The damping of circuit L_1C reduces the absolute value of u_C at the end of such a reversal. This reduction can here again be compensated by making the controls of T_{A_1} and T'_1 overlap.

C.2.2 Turn-Off Circuit Rating

The values of L and C can be determined according to the reverse bias time t_p and the ratio between the amplitude of i_{L_1} and that of the current to be commuted I_{OFF}.

By writing

$$t_p = at_q, \text{ with } a > 1 \text{ often taken close to 2,}$$

$$k = \frac{I_{OFF}}{U\sqrt{C/L}}$$

the following is obtained:

$$C = \frac{kI_{OFF}t_p}{\operatorname{arc\,cotan} k} \qquad L = \frac{Ut_p}{kI_{OFF}\operatorname{arc\,cotan} k}$$

C.2.3 Variants

As in the case of the configuration using parallel commutation by oscillating circuit variants of the basic circuit can be obtained by moving the commutation inductances.

The configuration in Fig. C.12 offers the advantage of requiring only a single inductance. However, the latter is no longer in the position which makes it easy to protect the main thyristors against short-circuit. The configuration which was taken as the basic circuit thus offers the most advantageous topological layout.

C.2.4 Three-Phase Inverters

As in the case of parallel commutation by oscillating circuit, it is possible to

- either use a turn-off circuit for each group of two switches in series in an inverter-leg,
- or use a single turn-off circuit with directing thyristors.

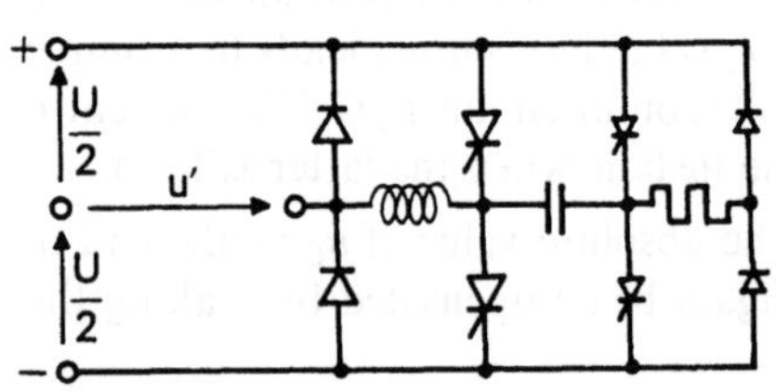

Fig. C.12

C.3 Series Commutation

Series commutations is rarely used in the case of voltage-source inverters.

It should, however, be noted that a single series turn-off circuit ensuring the simultaneous turn-off of all conducting thyristors (Fig. C.13) has occasionally been used to produce small-power light-weight inverters.

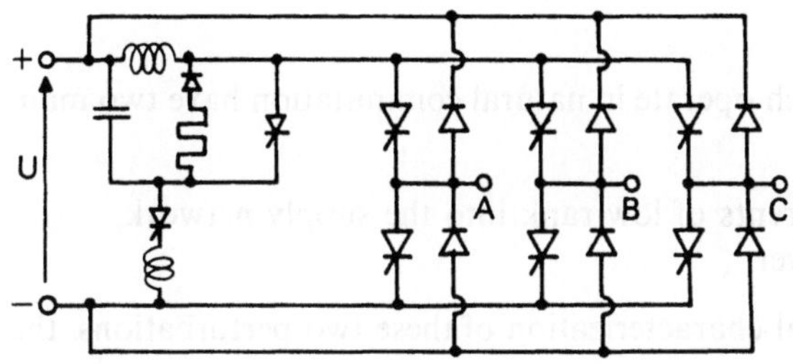

Fig. C.13

Appendix D: PWM Rectifier

- The thyristor rectifiers which operate in natural commutation have two main drawbacks:

  - they inject harmonic currents of low rank into the supply network,
  - they absorb reactive power.

  In order to give a global characterization of these two perturbations, the power factor F, defined by

  $$F = \frac{I_1}{I} \cos \varphi_1 ,$$

  is used, with

  I the rms value of the input current,
  I_1 the rms value of its fundamental,
  φ_1 the lagging shift of the fundamental related to the voltage.

- The problem of the power factor is of crucial importance when operating in heavy traction with an AC-fed catenary. At the locomotive input there is always a transformer followed by a rectifier: the latter supplies either DC motors directly or, through a second converter, polyphase synchronous or asynchronous motors.

  Because of the single-phase supply, harmonics at the input are of ranks $3, 5, 7, \ldots$, while in three-phase operation they are of ranks $5, 7, 11, \ldots$ Furthermore, the ratio of the locomotive power rating to the short-circuit power rating of the mains seen from the input of the locomotive is much higher than in the usual type of equipment.

  As a result the voltage drop on the catenary may become a problem when the locomotive is located between two substations. Moreover, the catenary forms a perfect aerial for disturbing the local residents.

- The improvement in the power factor at the input of a locomotive fed by single-phase AC supply has led to numerous studies, some of which have been already applied.

  The usual solution was to use series-connected rectifiers and to shift their controls (see Vol. 1, Chap. 9).

In certain cases, input filters with high capacitances are used in order to compensate for the reactive power consumed by the rectifiers.

When natural commutation has been replaced by forced commutation, the first step was to use "phase recentring" bridges: the current square-wave taken during each half-cycle is centred on the middle of the voltage half-cycle in order to operate at $\cos\varphi_1$ equal to 1. The result is quite disappointing as any gain made in reactive power is lost in the harmonic content.

Therefore, it was concluded that, for operation in forced commutation, several commutations would have to be carried out per half-cycle, i.e. it would be necessary to work in Pulse Width Modulation[1].

D.1 The Structure

- Figure D.1 gives the basic diagram of the converter and indicates the notations used. Connected to the secondary of the input transformer, there is
  - an input filter made up of an inductor L_1 and a capacitor C,
  - a four diode (D_1, D_2, D_3, D_4) bridge,
  - a chopper made up of a controlled turn-on/turn-off semiconductor device TC and a freewheeling diode D_5.

The load, supplied via the smoothing inductance L_2, may be either a DC motor or another converter.

Inductance L_1 consists of the supply inductance as well as the transformer leakage inductance both referred to the secondary winding of the latter and increased by the filtering inductance.

The diode-bridge-chopper association acts a *series chopper preceded by a current commutator*. It is fed by an AC voltage source owing to capacitor C; it feeds a DC current load owing to inductance L_2. Its main characteristic is that it comprises no intermediary storage element.

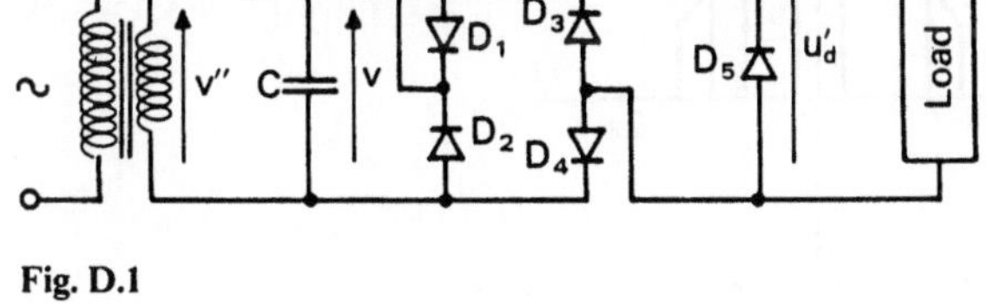

Fig. D.1

[1] The study summarized in this Appendix was carried out by E. Destobbeleer in conjunction with Jeumont-Schneider Company.

- When the controlled device TC is conducting.
  - if voltage v is positive, current i' flows via diodes D_1 and D_4:

$$i = i' = i_d, \qquad u'_d = v$$

  - if voltage v is negative, i' flows via D_2 and D_3:

$$i = -i' = -i_d, \qquad u'_d = -v.$$

When TC is off, diode D_5 conducts:

$$i = i' = 0, \qquad u'_d = 0$$

By centring the positive and negative half-cycles of current i on those of voltage v, the current fundamental can be placed *in phase* with the voltage and the consumption of reactive power can be eliminated.

By suitably chopping up current i', the first *harmonics* of current i can be eliminated and, in particular, the input filter can be made less bulky.

This chopping also enables to *vary the average value* U'_{d_0} of the rectified voltage u'_d.

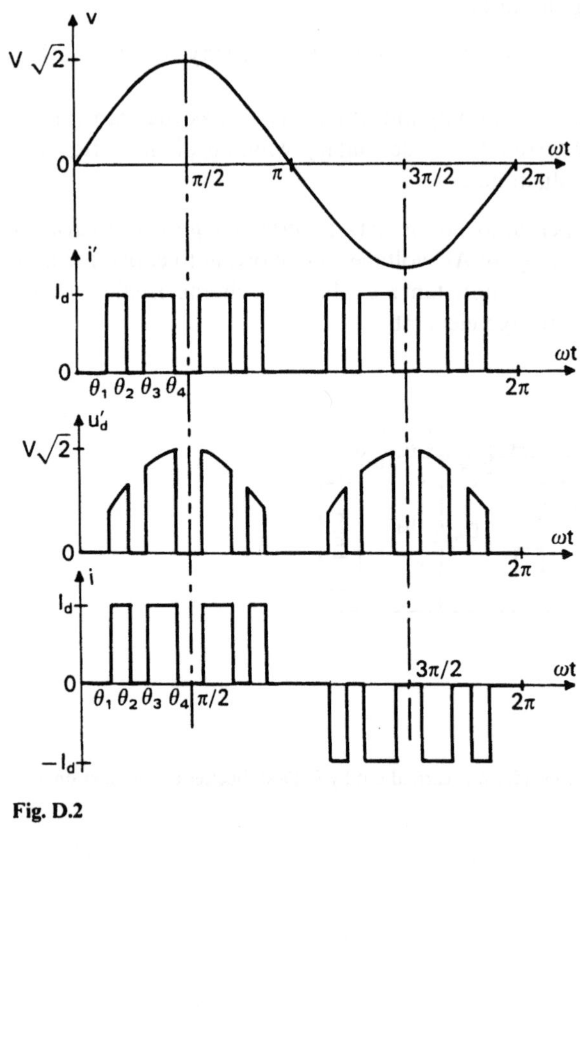

Fig. D.2

Figure D.2 shows the waveforms which correspond to a sinusoidal voltage v, to a perfectly smoothed current i_d (thus equal to its mean value I_d), and to perfect semiconductor devices

The structure is studied in three stages:

- the converter alone,
- the converter with its input filter,
- the converter with its input and output filters.

D.2 The Converter Alone

Initially, the voltage v at the converter input is assumed to be sinusoidal and current i_d at its output to be perfectly smoothed:

$$v = V\sqrt{2}\sin \omega t, \qquad i_d = I_d$$

For various values of the number p of current i pulses per half-cycle, the values to be given to the turn-on and turn-off angles of TC is determined, as a function of the mean value U'_{d_0} of the output voltage.

U'_{d_0} is referred to the mean rectified voltage U_{d_0} which would be produced if TC was permanently on:

$$\frac{U'_{d_0}}{U_{d_0}} = \frac{U'_{d_0}}{\frac{2}{\pi}V\sqrt{2}}.$$

D.2.1 Calculating the Commutation Angles

If p is the number of pulses per half-cycle, since the current i wave must be symmetrical in relation to the midpoint of its half-cycles, only p angles, $\theta_1, \theta_2, \ldots, \theta_p$ need to be determined.

– The rectified voltage has the following mean value:

$$U'_{d_0} = U_{d_0}(\cos\theta_1 - \cos\theta_2 + \ldots \pm \cos\theta_p)$$

The series expansion of current i only contains harmonics of uneven rank, in addition to the fundamental component:

$$i = I_1\sqrt{2}\sin\omega t + I_3\sqrt{2}\sin 3\omega t + \ldots + I_{2k+1}\sqrt{2}\sin(2k+1)\omega t + \ldots$$

with

$$I_{2k+1}\sqrt{2} = \frac{4I_d}{\pi(2k+1)}[\cos(2k+1)\theta_1 - \cos(2k+1)\theta_2 + \ldots \pm \cos(2k+1)\theta_p]$$

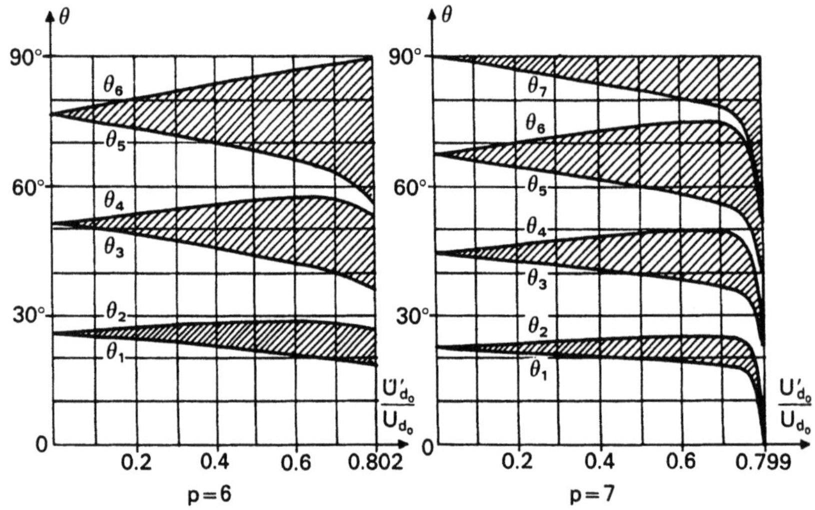

Fig. D.3

- Angles $\theta_1, \theta_2, \ldots, \theta_p$ are chosen in order to eliminate the first harmonics of current *i*. Among the *p* equations giving the *p* angles, the first one is given by the value of U'_{do}/U_{do}, and the $p-1$ others by eliminating the $p-1$ first uneven harmonics:

$$\cos\theta_1 - \cos\theta_2 + \cos\theta_3 - \ldots \pm \cos\theta_p = U'_{do}/U_{do}$$

$$\cos 3\theta_1 - \cos 3\theta_2 + \cos\theta_3 - \ \ldots \pm \cos 3\theta_p = 0$$

$$\cos 5\theta_1 - \cos 5\theta_2 + \cos 5\theta_3 - \ldots \pm \cos 5\theta_p = 0$$

. . . .

$$\cos(2p-1)\theta_1 - \cos(2p-1)\theta_2 + \ldots \pm \cos(2p-1)\theta_p = 0$$

- Figure D.3 shows, for $p = 6$ and $p = 7$ for example, how the *p* angles vary as a function of U'_{do}/U_{do}; the shaded areas correspond to the conduction periods of TC.

D.2.2 "Voltage Loss"

Value of *p* needs to be limited for three reasons:

- the commutation losses are all the higher as *p* increases,
- as *p* increases, so must the precision to be given to the values of the angles,
- the "voltage loss".

The last of these is the decisive factor in this case.

U'_{d_0} cannot be taken up to U_{d_0}. The maximum value of U'_{d_0} is obtained when θ_1 falls to zero for uneven p and when θ_p reaches $\pi/2$ for even p. This gives a "voltage loss" equal to the difference between U_{d_0} and the maximum value of U'_{d_0}. The theoretical maximum value of U'_{d_0} is:

$0.866\, U_{d_0}$ for $p = 2$,

$0.836\, U_{d_0}$ for $p = 3$,

$0.817\, U_{d_0}$ for $p = 4$,

$0.809\, U_{d_0}$ for $p = 5$,

$0.802\, U_{d_0}$ for $p = 6 \ldots$

When p increases, the input voltage must be increased in order to maintain the same maximum value of U'_{d_0} across the load. This means a similar increase in the power rating of all the elements of the converter and of its supplying circuit.

The real voltage loss is much greater. Values which are above given minima must be selected for the on and off periods of switch TC.

This brings about a further slight reduction in the maximum value of U'_{d_0}.

More especially, however, U'_{d_0} cannot fall below a given value which becomes higher as p becomes larger. This minimum value must not exceed the voltage which must be applied when starting the motor supplied by the converter.

D.2.3 The Remaining Harmonics

- The harmonics which remain in input current i are all uneven harmonics of rank equal to or greater than $2p + 1$.

Figure D.4 shows – e.g. for $p = 6$ and $p = 7$ – the variations depending on U'_{d_0}/U_{d_0} of the rms values of the first harmonics,

$I_{13}, I_{15}, I_{17}, I_{19}$ for $p = 6$,

$I_{15}, I_{17}, I_{19}, I_{21}$ for $p = 7$.

All these values are referred to the rms value I_{1_0} which the current fundamental component would have if TC were permanently on.

The relative value of the remaining harmonics, $I_{2p+1}, I_{2p+3}, \ldots$ varies in virtually the same way whatever the value of p. Their frequency simply increases with p.

- The current i fundamental component is in phase with voltage v and its rms value I_1 can thus be deduced directly from the conservation of the active power between the input and output of the converter.

$$VI_1 = U'_{d_0} I_d$$

From

$$I_{1_0} = \frac{2\sqrt{2}}{\pi} I_d \quad \text{and} \quad U_{d_0} = \frac{2\sqrt{2}}{\pi} V$$

it can be deduced:

$$\frac{I_1}{I_{1_0}} = \frac{U'_{d_0}}{U_{d_0}}, \quad \text{whatever the value of } p.$$

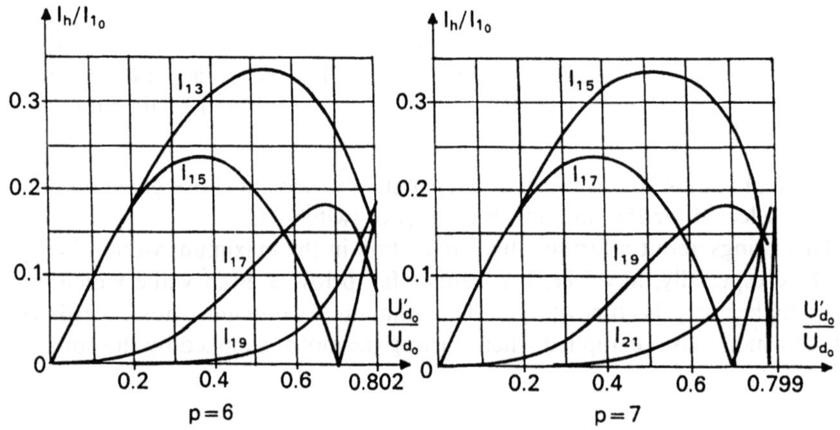

Fig. D.4

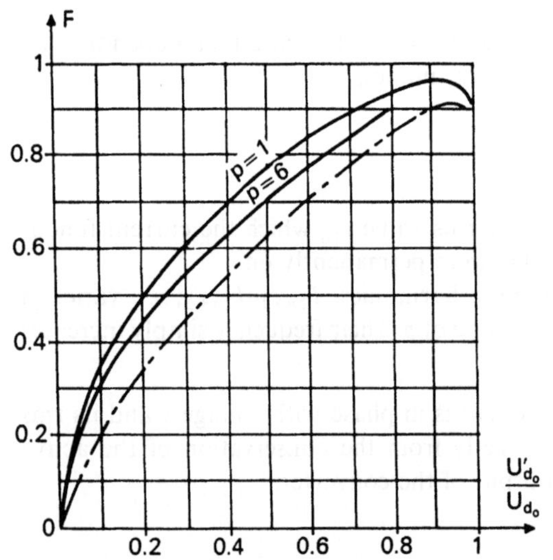

Fig. D.5

Since the displacement factor $\cos \varphi_1$ is equal to 1, the power factor F at the converter input depends only on the harmonics.

$$F = \frac{I_1}{I} = \frac{I_1}{\sqrt{I_1^2 + \sum_{k=1}^{\infty} I_{2k+1}^2}}$$

- Figure D.5 shows the variations of F for $p = 1$ and $p = 6$. For $p = 6$, the quadratic sum of the harmonics is slightly higher than for $p = 1$ and so the power factor is slightly lower. It is only just above that obtained with a single full-bridge thyristor rectifier with shifted control of the two half-bridges (curve in broken line).

This leads to the same conclusion as for the output voltage of a voltage-source inverter, (see Chap. 3): Pulse Width Modulation does not decrease the harmonics but increases their frequency. As it makes filtering easier, it is an interesting solution

D.3 The Converter and its Input Filter

In order to study the converter fed through its input filter (Fig. D.6a), the current at the chopper output is once again assumed to be perfectly smooth. Now, however, voltage v'' at the filter input is considered to be sinusoidal, rather than voltage v at its output.

This study is very similar to that of the voltage-source inverter output filter (Chap. 2, Sect. 2.6). Therefore we will not go into the details of the choice of the value of the capacitor which is the result of a difficult compromise.

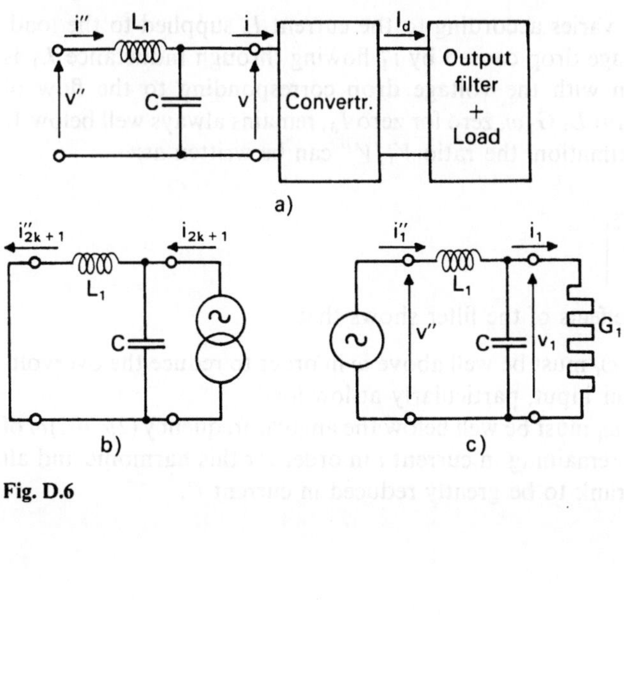

Fig. D.6

D.3.1 Resonant Frequency of the Filter

The filter is firstly characterized by its resonant angular frequency:

$$\omega_f = \frac{1}{\sqrt{L_1 C}}$$

This is determined from the study of the effects of the filter on the harmonic currents and on the fundamental of the voltage.

- For the *harmonic currents*, the converter acts as a current generator feeding a parallel resonant circuit $L_1 C$ (Fig. D.6b).

  The ratio of the rms value I''_{2k+1} of the harmonic current of rank $2k + 1$ in the line current to the rms value I_{2k+1} of this current at the converter input is equal to

$$\frac{I''_{2k+1}}{I_{2k+1}} = \frac{1}{\left|1 - \left[(2k+1)\dfrac{\omega}{\omega_f}\right]^2\right|}$$

- For the *fundamental of the voltage*, the supply feeds capacitor C, via L_1, in parallel with the converter seen from its input (Fig. D.6c). Since fundamental i_1 of current i is in phase with that v_1 of voltage v, as concerns the fundamental components, the converter seen from its input can be assimilated to a conductance G_1.

  This gives the ratio linking the rms value V_1 of the fundamental of v to the rms value V'' of the supply voltage:

$$\frac{V_1}{V''} = \frac{1}{\sqrt{\left[1 - \left(\dfrac{\omega}{\omega_f}\right)^2\right]^2 + L_1^2 G_1^2 \omega^2}}$$

  Conductance G_1 varies according to the current I_d supplied to the load. However, if the voltage drop caused by i''_1 flowing through inductance L_1 is small in comparison with the voltage drop corresponding to the flow of i_1 through G_1, the term $L_1 G_1 \omega$, zero for zero I_d, remains always well below 1.

  As a first approximation, the ratio V_1/V'' can be written as:

$$\frac{V_1}{V''} \simeq \frac{1}{\left|1 - \left(\dfrac{\omega}{\omega_f}\right)^2\right|}.$$

- A study of the two effects of the filter shows that:
  - angular frequency ω_f must be well above ω in order to reduce the overvoltage at the converter input, particularly at low load;
  - angular frequency ω_f must be well below the angular frequency $(2p + 1)\omega$ of the first harmonic remaining in current i in order for this harmonic and all those of a higher rank to be greatly reduced in current i''.

This gives rise to a double condition:

$$1 \ll \frac{\omega_f}{\omega} \ll 2p + 1$$

which is easier to satisfy as the number p of pulses per half-cycle increases.

- In fact, other constraints are added to the above. The value of L_1 is virtually imposed by the minimum value which must be given to the additional inductance placed at the transformer output. It will be seen that the capacitance of capacitor C must be higher than a given value but that this capacitor gives more problems when C is high.

D.3.2 The Need for a Sufficiently High Capacitance

Capacitance C must be sufficiently high in order to ensure that the converter is effectively fed by a voltage source. Voltage v is not sinusoidal, variations caused by the discharge of C when TC is conducting and by the charge of C when TC is off, are added to its fundamental.

- If C is sufficiently high (Fig. D.7), voltage v does not fall to zero during the pulses of current i, and, particularly, during the first pulse of each half-cycle.
- If C is too small, during the period $[\theta_1, \theta_2]$ when TC has to conduct, voltage v falls to a value which provides insufficient compensation for the voltage drop in TC, D_1 and D_4. Current I_d is gradually transferred to diode D_5; this causes a decrease in i which enables the capacitor to be recharged. However, as soon as TC begins to conduct again, v once more decreases . . . the result is a series of short pulses in current i.

  Figure D.8a shows the waveforms of voltage v and current i during one half-cycle. Figure D.8b gives a more precise representation of the variations in v and i during period (θ_1, θ_2). Figure D.8c shows the two types of current path during this period, depending on whether TC in on (full lines) or off (broken lines).

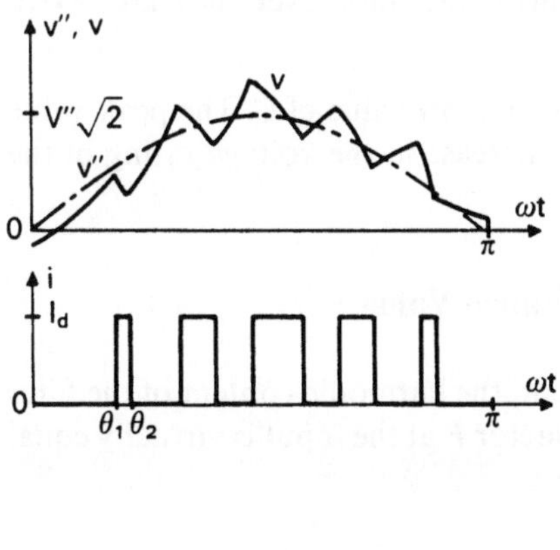

Fig. D.7

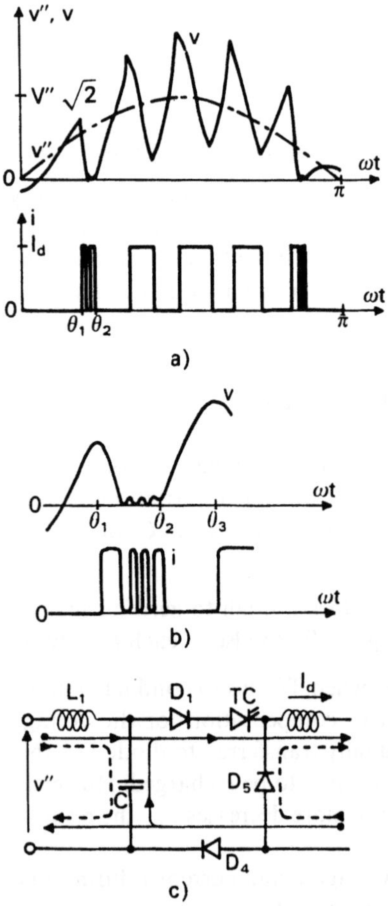

Fig. D.8

These parasitic oscillations which appear on the first pulse of each half-cycle and on the last when the value of C decreases, would affect the others if the value of C continued to decrease. These parasic oscillations affect the quality of current i by reintroducing harmonics; moreover they give extra commutation losses.

- Figure D.8 shows the other disadvantage of a low value of C. The peak value of voltage v increases, thus requiring an increase in the voltage rating of the semiconductor devices.

D.3.3 Disadvantages of a High Capacitance Value

As soon as the ratio $(2p + 1)\omega/\omega_f$ is sufficient, the harmonic content of the filter input current can be neglected. The power factor F at the input is virtually equal to the displacement factor $\cos \varphi_1''$, if φ_1'' is used to denote current i'' fundamental phase shift with respect to voltage v''.

The difference between F and 1 is mainly due to the reactive power Q_1, equal to $V''I_1'' \sin \varphi_1''$, absorbed or created at ω angular frequency.

The reactive power Q_C generated by the capacitor is virtually constant, as is the voltage v fundamental across C. The reactive power Q_L absorbed by inductance L_1 varies depending on the current absorbed by the converter. If the pulse sequences which make up i are centred on the midpoint of the voltage v fundamental half-cycles, the reactive power at the converter input is zero. The reactive power delivered by the supply is then:

$$Q_1 = Q_L - Q_C = L_1 \omega I_1''^2 - C\omega V_1^2.$$

If, for the rated value $I_{d\,nom}$ of output I_d and the maximum value of ratio U'_{d_0}/U_{d_0}, both reactive powers Q_L and Q_C compensate for each other, the capacitor creates too much reactive power when U'_{d_0} decreases. This excess becomes all the more noticeable when I_d decreases.

D.3.4 Pulse Sequence Shift

The deterioration in F due to the excess of reactive power generated by C at low and medium loads can be partly corrected by having reactive power absorbed by the converter.

To achieve this, the pulse sequences are delayed by an angle γ, i.e. the current i fundamental lags the fundamental of v by γ. The reactive power delivered by the supply becomes:

$$Q_1 = L_1 \omega I_1''^2 - C\omega V_1^2 + V_1 I_1 \sin \gamma.$$

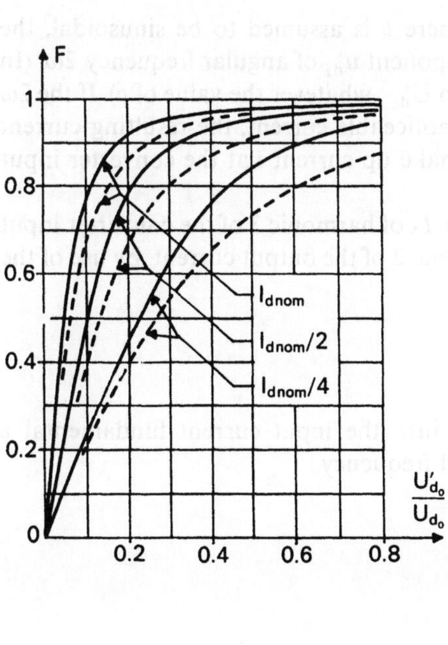

Fig. D.9

The optimal control law $\gamma(I_1)$ can be approximated as follows:

- for low values of I_1, γ is kept constant,
- then γ decreases linearly as a function of I_1,
- for high values of I_1, γ is kept constant.

Figure D.9 shows variations of F as a function of $U'_{\mathrm{d0}}/U_{\mathrm{d0}}$ obtained by using the optimal control, for the rated current, as well as for the half and for the quarter of the latter. The dashed lines show results which would be obtained with γ constantly equal to zero. These curves correspond to $p = 6$, $\omega_{\mathrm{f}}/\omega = 6$, $L_1 \omega I_{\mathrm{d\,nom}} = 0.3 U_{\mathrm{d0}}$.

D.4 The Converter with its Input and Output Filters

D.4.1 Harmonic 2 of the Output Voltage

The converter output voltage u'_{d} comprises, in addition to the mean value term U'_{d0}, a term of angular frequency 2ω and terms with angular frequencies $2p\omega$, $(2p+2)\omega, (2p+4)\omega, \ldots$

Unless inductance L_2 is infinite, the terms of same angular frequencies are found in current i_{d} at the converter output. Harmonic 2 poses the greatest problem since it leads to the presence of a harmonic 3 in currents i and i''. Since the input filter was calculated in order to attenuate the harmonics of a rank equal to or greater than $(2p+1)$, the presence of this harmonic 3 – which angular frequency may be close to the resonance angular frequency of the output filter – could make it useless or even harmful to use Pulse Width Modulation.

Figure D.10 shows, for the case where v is assumed to be sinusoidal, the waveforms of voltage u'_{d} and of its component u'_{d_2} of angular frequency 2ω. (In this case, the amplitude of u'_{d_2} is equal to U'_{d0}, whatever the value of p). If the 2ω angular frequency component gives a noticeable current, the resulting current i_{d} ripple means that the pulses which make up current i at the converter input have a variable amplitude.

It is possible to deduce the rms value I_3 of harmonic 3 of the converter input current from the rms value I_{d_2} of harmonic 2 of the output current, by use of the instantaneous power conservation:

$$I_3 = \frac{U'_{\mathrm{d0}}}{V} \frac{I_{\mathrm{d}_2}}{\sqrt{2}}$$

Moreover, harmonic 2 introduces into the input current fundamental a reactive component at the fundamental frequency.

$$I_1 \sin \varphi_1 = \frac{U'_{\mathrm{d0}}}{V} \frac{I_{\mathrm{d}_2}}{\sqrt{2}}.$$

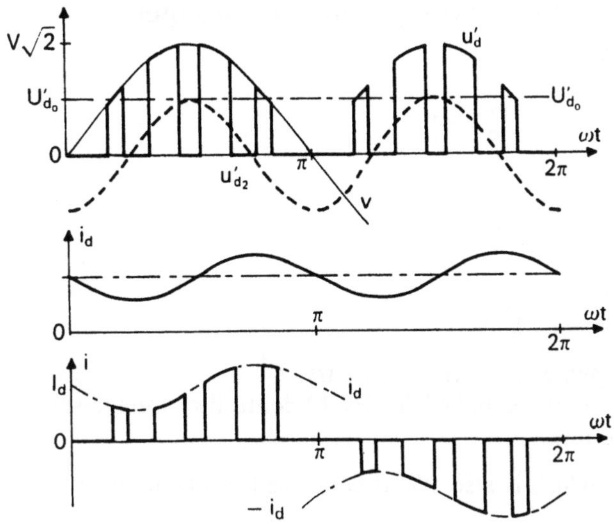

Fig. D.10

D.4.2 Eliminating Harmonic 2 of the Output Current

At the converter output, a simple current smoothing is not sufficient; harmonic 2 of current i_d must be eliminated.

Several selective filtering solutions are possible. The diagram in Fig. D.11 shows one of the best. A parallel resonant circuit $C'_3 L'_3$ is series-connected with the load. An inductance L'_2 is added to reduce the harmonic currents of a high rank which would flow via C'_3[1].

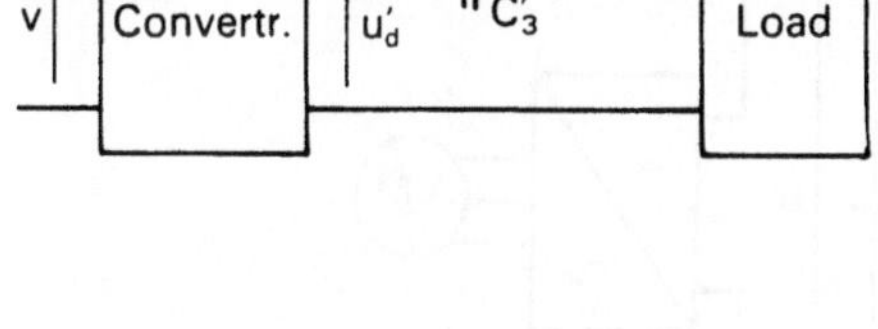

Fig. D.11

[1] The same principle can be used in three-phase operation: the PWM rectifier is then made up of a six-diode bridge and chopper. It is less advantageous to change from a standard form of rectifier to a PWM rectifier than in single-phase operation, as the initial situation is better: harmonic currents of ranks 5, 7, 11, 13, ... instead of 3, 5, 7, 9, ... ; relative power of equipment usually lower.

Moreover, specific conditions are imposed by the need to obtain a balanced system of line currents.

Appendix E: Use of the Four-Quadrant Chopper as a PWM Rectifier

In electric traction equipment which

- is supplied by a single-phase AC-current catenary,
- uses an asynchronous motor supplied by PWM-controlled voltage-source inverters as traction motors,

a four-quadrant chopper which is also PWM-controlled can be used as input rectifier.

Such a solution enables

- the number of filtering elements to be reduced,
- the functional reversibility to be ensured,
- the reactive power absorbed at the catenary to be controlled,
- the harmonic content of the current taken at the catenary to be reduced.

Figure E.1 shows the basic diagram of the converter and the notations which are adopted.

The chopper is made up of four controlled switches, K_1, K'_1, K_2 and K'_2, which are bidirectional in current.

The chopper is directly connected to the input transformer output. L denotes the sum of the mains and the catenary inductances and of the transformer leakage inductance, referred to the secondary winding of the latter. Under these

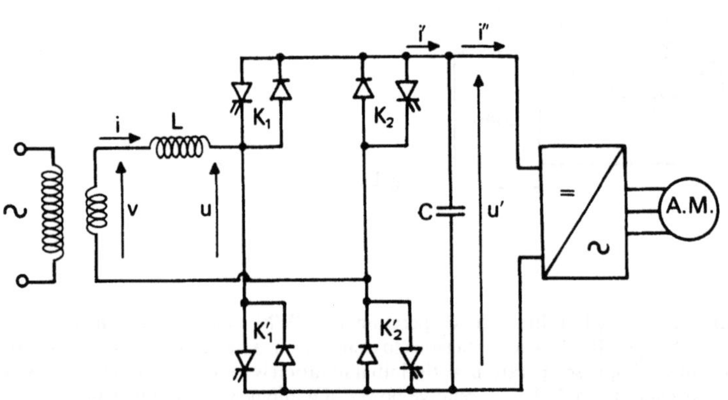

Fig. E.1

conditions, v is the mains voltage – assumed to be sinusoidal – referred to the secondary winding of the transformer.

A capacitor C is connected across the chopper output.

The chopper is fed by a current source, on account of the inductance L, and feeds a voltage source, on account of capacitor C.

E.1 Switch Control

E.1.1 Voltage Ratio Introduced by the Chopper

The switches are fitted with a PWM control synchronised on the frequency of the mains supplying the catenary.

Figure E.2 shows a control sequence in sinusoidal modulation. The turn-on and turn-off instants of the switches are determined by the intersections of a triangular carrier and a sinusoidal reference of the same frequency as voltage v. The ratio between the amplitude of the reference and that of the carrier is denoted by r.

In drawing Fig. E.2, K_2 and K'_2 are taken as directing switches and K_1 and K'_1 as chopping switches. Output voltage u' is assumed to be perfectly smoothed and assimilated to its mean value U'.

During one half-cycle, K'_2 is on:

- if K_1 is on,

$$u = u', \qquad i' = i$$

- if K'_1 is on,

$$u = 0, \qquad i' = 0.$$

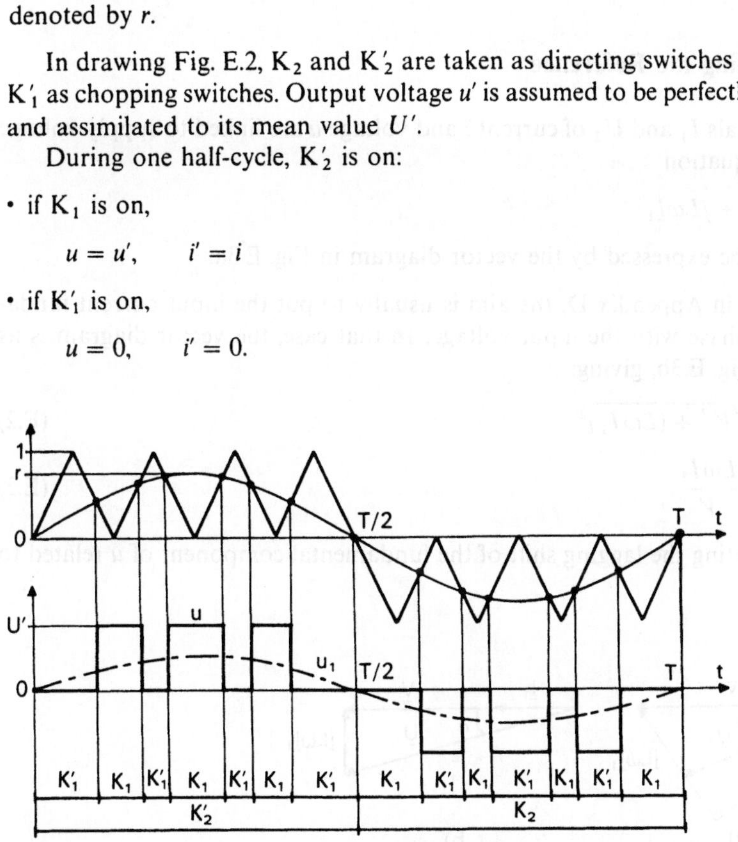

Fig. E.2

During the other half-cycle, K_2 is on:

- if K_1 is on,

$$u = 0, \qquad i' = 0$$

- if K_1' is on,

$$u = -u', \qquad i' = -i.$$

As soon as the frequency-modulation ratio is sufficiently high (see Chap. 3, Sect. 3.3.2.1), fundamental u_1 of the input voltage has an rms value U_1 equal to

$$U_1 = r\frac{U'}{\sqrt{2}}. \tag{E.1}$$

Since r is less than 1, the mean value U' of the output voltage is greater than $U_1\sqrt{2}$. Unlike the standard type of rectifier which operates as a step-down voltage converter, the configuration being studied operates as a step-up voltage converter.

E.1.2 Stalling the Reference

Fundamentals I_1 and U_1 of current i and voltage u are linked to supply voltage V by the equation

$$\underline{V} = \underline{U}_1 + jL\omega\underline{I}_1$$

which can be expressed by the vector diagram in Fig. E.3a.

As seen in Appendix D, the aim is usually to put the input current fundamental in phase with the input voltage. In that case, the vector diagram is as shown in Fig. E.3b, giving:

$$U_1 = \sqrt{V^2 + (L\omega I_1)^2} \tag{E.2}$$

$$\tan\delta = \frac{L\omega I_1}{V}. \tag{E.3}$$

with δ denoting the lagging shift of the fundamental component of u related to voltage v.

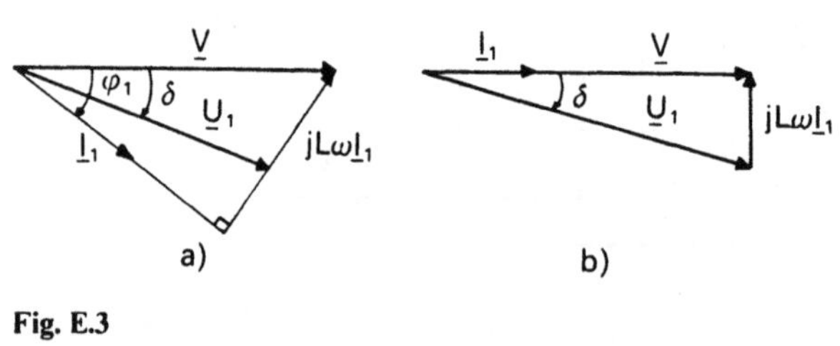

Fig. E.3

- Equation (E.3) shows that the shift δ which corresponds to the reactive power $VI_1 \sin \varphi_1$ falling to zero, is independent of U_1.
  The shift δ of $\underline{U}_1$ related to $\underline{V}$ is equal to the shift, related to $\underline{V}$, of the reference wave which is used to determine the control instants of the switches.
- If the normal values of $L_1 \omega I_1 / V$ are taken into account, shift δ is virtually proportional to I_1:

$$\delta \simeq \tan \delta = \frac{L\omega I_1}{V}.$$

  The direction of the shift depends on the direction of the power flow:
  - it is a lagging shift, as in Fig. E.3b, if the supply delivers power to the motor via the converter;
  - it is a leading shift when the motor sends power back to the supply.
    In Eq. (E.3), I_1 has to be considered as an algebraic quantity, positive in the first case and negative in the second.
- It should be noted that the value of I_1 can be deduced from the mean value I'' of the current i'' absorbed by the inverter supplying the motor. Active power conversion indeed gives:

$$VI_1 = U'I'', \qquad I_1 = \frac{U'}{V} I''$$

  The sign of I_1 is inverted with that of I''.

E.1.3 Rectifier Output Voltage Control

Equation (E.1) gives the mean value U' of the rectifier output voltage as a function of U_1; Eq. (E.2) gives U_1 as a function of V and I_1. The following can be deduced:

$$U' = \frac{V\sqrt{2}}{r} \sqrt{1 + \left(\frac{L\omega I_1}{V}\right)^2}$$

For several values of $U'/V\sqrt{2}$, Fig. E.4 shows how r varies as a function of $L\omega I_1/V$.

E.1.4 Example of Waveforms

Figure E.5 indicates the waveforms of voltage u, as well as of current i and i', assuming voltage u' to be perfectly smoothed.

The reference is lagging v by δ.

The difference between v and u gives i by means of

$$L \frac{\mathrm{d}i}{\mathrm{d}t} = v - u$$

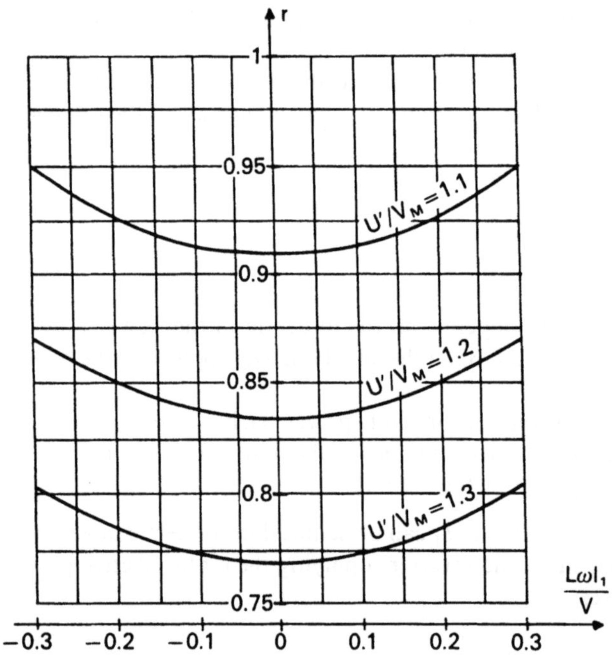

r
1
0.95
0.9
0.85
0.8
0.75
U'/V_M=1.1
U'/V_M=1.2
U'/V_M=1.3
Lωl₁/V
−0.3
−0.2
−0.1
0
0.1
0.2
0.3

Fig. E.4

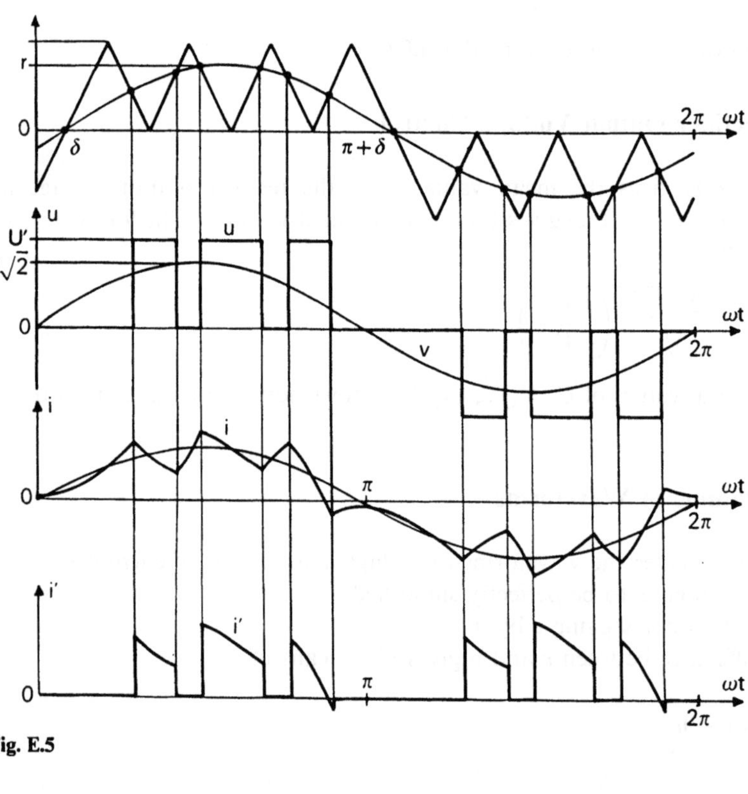

r
0
δ
π+δ
2π
ωt
u
U'
V√2
v
i
π
i'

Fig. E.5

For going from i to i' it should be noted that

i' equals i when K_1 and K'_2 are on and give u equal to U',
i' equals $-i$ when K'_1 and K_2 are on and give u equal to $-U'$,
i' has a value of zero during the other periods.

E.2 Current Harmonics Taken from the Supply

The series expansion of voltage u at the chopper input is the same as was found for the voltage at the full-bridge voltage-source inverter output (see Chap. 3, Sect. 3.3.2.3).

If m is the frequency-modulation ratio and if its value is sufficiently high, the main harmonics of voltage u are those of the following ranks:

$m-1$ and $m+1$, $m-3$ and $m+3$

$2m-1$ and $2m+1$, $2m-3$ and $2m+3$.

The characteristics shown in Fig. 3.21 can be used; they now give $U_h/(2\sqrt{2}U'/\pi)$, if U_h is used to denote the rms value of the $h\omega$ angular frequency harmonic of voltage u.

Since $L\,di/dt = v - u$ and voltage v is assumed to be sinusoidal, the rms value of the h rank harmonic of current i taken from the supply is given by

$$I_h = \frac{U_h}{Lh\omega}$$

If m is sufficiently high, a greatly reduced harmonic rate may be obtained for current i.

Bibliography

PWM rectifiers [1–20] are the subject of an increasing number of studies. In appendices D and E, we have given two examples of single-phase PWM rectifiers supplied, in the first case, by an AC voltage source and, in the second, by an AC current source.

The first studies dealt with PWM rectifiers supplied by an AC voltage source and feeding a DC current receptor. Such assemblies are not reversible. Thus the configurations which are most frequently studied at present are the full-bridge assemblies made up of bidirectional voltage switches. The diagrams are the same as for current-source inverters and the same modulation laws can be used.

Similarly a reversible PWM rectifier powered by an AC-current source and feeding a DC-voltage source uses the same diagram, the same semiconductor devices and the same modulation techniques as the voltage-source inverter.

Active filters [21–28] are also PWM inverters; some publications concerning these are thus also mentioned.

An active filter is placed at the point of the network where the current is to be corrected (suppression of either the harmonics or the reactive component or both). An active filter is a correcting current generator.

The current to be corrected is compared to the wanted current; the gap can be used as a reference for the current generator. The latter comprises an inductive or capacitive element of energy storage and a current or voltage inverter which generates the correcting signals. The inverter is Pulse-Width-Modulated and operates at a frequency well above that of the mains. The gap to be resorbed is used as a reference for the inverter and is determined in real time. A small filter, with a cut-off frequency well above that of the mains but below that of the modulation frequency, stops the harmonics caused by the modulation.

PWM Rectifiers

1. Krishnamurthy KA, Dubey GK, Revankar GN (1978) Convertor control with selective reduction of line harmonics, *Proc. Inst. Electr. Eng.* **125** (2), 141–145.
2. Kataoka T, Mizumachi K, Miyairi S (1979) A pulsewidth controlled AC-to-DC converter to improve power factor and waveform of AC line current, *IEEE Trans Ind. Appl.*, **15** (6), 670–675.
3. Marino P, Picardi C, Russo A (1983) AC characteristics in AC/DC/DC conversion, *Proc. Inst. Electr. Eng., Part B*, **130** (3), 201–206.

4. Ziogas PD, Kang YG, Stefanovic VR (1985) PWM control techniques for rectifier filter minimization, *IEEE Trans. Ind. Appl.*, **21** (5), 1206–1214.
5. Ziogas PD, Kang YG, Stefanovic VR (1985) Optimum system design of a three-phase rectifier-inverter type of frequency changer, *IEEE Trans. Ind. Appl.*, **21** (5), 1215–1225.
6. Doralda SR, Nagamini C, Sanyal S (1985) A sinusoidal pulsewidth modulated three-phase AC to DC converter-fed DC motor drive, *IEEE Trans. Ind. Appl.*, **21** (6), 1394–1408.
7. Biswas SK, Mahesh MS, Iyengar BSH (1986) Simple new PWM patterns for thyristor three-phase AC-DC convertors, *Proc. Inst. Electr. Eng., Part B*, **133** (6), 3546358.
8. Malesani L, Tenti P (1987) Three-phase AC/DC PWM converter with sinusoidal AC currents and minimum filter requirements, *IEEE Trans. Ind. Appl.*, **23** (1), 71–77.
9. Rodriguez J (1987) High performance DC motor drive using a PWM rectifier with power transistors, *Proc. Inst. Electr. Eng., Part B*, **135** (1), 9–13.
10. Destobbeleer E, Seguier G, Castelain A (1987) AC-DC converter minimizing induced harmonics in industrial power systems, *IEEE Trans. Power Electron.*, **2** (4), 320–327.
11. Dixon JW, Kulkarni AB, Nishimoto M, Ooi BT (1987) Characteristics of a controlled-current PWM rectifier-inverter link, *IEEE Trans. Ind. Appl.*, **23** (6), 1022–1028.
12. Stihi O, Ooi BT (1988) A single-phase controlled-current PWM rectifier, *IEEE Trans. Power Electron.*, **3** (4), 453–459.
13. Green AW, Boys JT, Gates GF (1988) 3-phase voltage sourced reversible rectifier, *Proc. Inst. Electr. Eng., Part B*, **135** (6), 362–370.
14. Matsure K, Kubota H (1989) Improved current source GTO inverter-fed induction motor drives with PWM-controlled thyristor converter, *IEEE Trans. Ind.* Appl., **25** (2), 287–291.
15. Habetler TG, Divan DM (1989) Rectifier/inverter reactive component minimization, *IEEE Trans. Ind. Appl.*, **25** (2), 307–315.
16. Green AW, Boys JT (1989) Hysteresis current-forced three-phase voltage-sourced reversible rectifier, *Proc. Inst. Electr. Eng., Part B*, **136** (3), 113–120.
17. Itoh R (1989) Steady-state and transient characteristics of a single-way step-down PWM GTO voltage-source convertor with sinusoidal supply currents, *Proc. Inst. Electr. Eng., Part B*, **136** (4), 168–174.
18. Dixon JW, Ooi BT (1989) Series and parallel operation of hysteresis current-controlled PWM rectifiers, *IEEE Trans. Ind. Appl.*, **25** (4), 644–651.
19. Boys JT, Green AW (1989) Current-forced single-phase reversible rectifier, *Proc. Inst. Electr. Eng., Part B*, **136** (5), 205–211.
20. Itoh R, Ishizaka K (1989) Single-phase sinusoidal convertor using MOSFET's, *Proc. Inst. Electr. Eng., Part B*, **136** (5), 237–242.

Active filters

21. Uceda J, Aldana F, Martinez P. (1983) Active filters for static power convertors, *Proc. Inst. Electr. Eng., Part B*, **130** (5), 347–354.
22. Akagi H, Kanazawa Y, Nabae A (1984) Instantaneous reactive power compensators comprising switching devices without energy storage components, *IEEE Trans. Ind. Appl.*, **20** (3), 625–630.
23. Akagi H, Nabae A, Atoh S (1986) Control strategy of active power filters using multiple voltage-source PWM converters, *IEEE Trans. Ind. Appl.*, **22** (3), 460–465.
24. Cox MD, Mirbod A (1986) A new static VAR compensator for an arc furnace, *IEEE Trans. Power Syst.*, **1** (3), 110–119.
25. Choe GH, Park MH (1988) A new injection method for AC harmonic elimination by active power filter, *IEEE Trans. Ind. Electron.*, **35** (1), 141–147.
26. Enslin JHR, Van Wyck JD (1988) Measurement and compensation of fictitious power under nonsinusoidal voltage and current conditions, *IEEE Trans. Instrum. Meas.*, **37** (3), 403–408.
27. Choe GH, Park MH (1989) Analysis and control of active power filter with optimized injection, *IEEE Trans. Power Electron.*, **4** (4), 427–433.
28. Wong C, Mohan N, Wright SE, Mortensen KN (1989) Feasibility study of AC- and DC- side active filters for HVDC converter terminals, *IEEE Trans. Power Deliv.*, **4** (4), 2067–2073.

Subject Index

MIX
Papier aus verantwortungsvollen Quellen
Paper from responsible sources
FSC® C105338

If you have any concerns about our products,
you can contact us on
ProductSafety@springernature.com

In case Publisher is established outside the EU,
the EU authorized representative is:
Springer Nature Customer Service Center GmbH
Europaplatz 3, 69115 Heidelberg, Germany

Printed by Libri Plureos GmbH
in Hamburg, Germany